continued on back

FORECASTING WITH UNIVARIATE BOX–JENKINS MODELS

Forecasting With Univariate Box–Jenkins Models

CONCEPTS AND CASES

ALAN PANKRATZ

DePauw University

JOHN WILEY & SONS

New York · Chichester · Brisbane · Toronto · Singapore

To my family—
old, new, borrowed, brown,
and otherwise

Library of Congress Cataloging in Publication Data:

Pankratz, Alan, 1944–
    Forecasting with univariate Box–Jenkins models.

    (Wiley series in probability and mathematical
statistics, ISSN 0271-6356. Probability and mathematical
statistics)
    Bibliography: p.
    Includes index.
    1. Time-series analysis. 2. Prediction theory.
I. Title. II. Series.

QA280.P37 1983 519.5′5   83-1404
ISBN 0-471-09023-9

Printed in the United States of America

10 9 8 7 6 5 4 3 2 1

PREFACE

The theory and practice of time-series analysis and forecasting has developed rapidly over the last several years. One of the better known short-term forecasting methods is often referred to as univariate Box–Jenkins analysis, or ARIMA analysis. Several years ago I introduced this method into my undergraduate course in model building and forecasting. I searched in vain for a text that presented the concepts at a level accessible to readers with a modest background in statistical theory and that also showed clearly how the method could be applied to a wide variety of real data sets. That fruitless search became my motivation for writing this text.

The purposes of this text are (1) to present the concepts of univariate Box–Jenkins/ARIMA analysis in a manner that is friendly to the reader lacking a sophisticated background in mathematical statistics, and (2) to help the reader learn the art of ARIMA modeling by means of detailed case studies. Part I (Chapters 1–11) presents the essential concepts underlying the method. Part II (Chapter 12 and Cases 1–15) contains practical rules to guide the analyst, along with case studies showing how the technique is applied.

This book can be used as a basic or supplementary text in graduate courses in time-series analysis and forecasting in MBA programs and in departments of economics, engineering, operations research, or applied statistics. It can serve as a basic text in similar advanced undergraduate courses. Practicing forecasters in business, industry, and government should find the text a helpful guide to the proper construction of ARIMA forecasting models.

The theory is presented at a relatively elementary level. Only a one-semester course in statistical methods is required. The reader should know the fundamentals of probability, estimation, and hypothesis testing, especially the use of the t-distribution and the chi-squared distribution. Some knowledge of regression methods is also helpful. Proofs do not appear in the text, although some results are derived; most technical matters are relegated to appendixes. The reader well-grounded in mathematical statistics will find the discussion of the theory to be quite elementary. The reader with a minimal background in mathematical statistics (at whom the book is aimed) should find the theoretical material to be sufficiently challenging and a helpful stepping-stone to more advanced literature.

The 15 case studies in Part II use real data to show in detail how the univariate Box–Jenkins method is applied. They illustrate the varieties of models that can occur and the problems that arise in practice. The practical rules summarized in Chapter 12 are emphasized and illustrated throughout the case studies. The reader who becomes thoroughly familiar with the case studies should be well-prepared to use the univariate Box–Jenkins method.

The case studies move from easier to more challenging ones. They may be read as a whole following completion of the first 12 chapters, or some of them may be read following the reading of selected chapters. Here is a suggested schedule:

1. After Chapters 1–4, read Cases 1–4.
2. After Chapter 6, review Cases 1–4.
3. After Chapter 7, read Cases 5–8.
4. After Chapter 9, review Cases 1–8.
5. After Chapter 11, read Cases 9–15.

The material in this text is based on the work of many individuals, but especially that of George E. P. Box and Gwilym M. Jenkins. I am deeply indebted to an anonymous reviewer whose painstaking comments on several drafts led to numerous improvements in both substance and style; any remaining errors of fact or judgment are my own. I was also fortunate in having the editorial guidance and encouragement of Beatrice Shube and Christina Mikulak during this project. Rich Lochrie influenced my treatment of the case studies. My colleagues Underwood Dudley and John Morrill were always patient with my questions. Ralph Gray's constant support was invaluable. Rande Holton, Mike Dieckmann, and Debbie Peterman at the DePauw University Computer Center were exceptionally helpful. I have been fortunate in having many students who provided challenging questions, criticism, data, preliminary data analysis, references, and programming assistance, but nine individuals deserve special mention: Carroll Bottum, Jim Coons, Kester Fong, Ed Holub, David Martin, Fred Miller, Barry Nelson, John Tedstrom, and Regina Watson. Lucy Field, Louise Hope, and Vijaya Shetty typed portions of the manuscript; Charity Pankratz deserves an honorary degree, Doctor of Humane Typing. And I thank my children for finally realizing that I will not be able to retire next year just because I have written a book.

All data analysis presented in this text was carried out on a VAX 11-780 at the DePauw University Computer Center using an interactive program. Inquiries about the program should be addressed to me at DePauw University, Greencastle, Indiana 46135.

ALAN PANKRATZ

Greencastle, Indiana
April 1983

CONTENTS

PART I. BASIC CONCEPTS

PART II. THE ART OF ARIMA MODELING

GROUP A   STATIONARY, NONSEASONAL MODELS

GROUP B   NONSTATIONARY, NONSEASONAL MODELS

GROUP C   NONSTATIONARY, SEASONAL MODELS

Part I

BASIC CONCEPTS

1

OVERVIEW

1.1 Planning and forecasting

In December 1981 I made plans to drive to Chicago with my family to visit relatives. The day before we intended to leave, the weather service issued a winter storm warning for that night and the following day. We decided to take a train rather than risk driving in a blizzard. As it turned out, there was a bad storm; but I was able to sleep and read (though not at the same time) on the train instead of developing a tension headache from driving on icy, snow-filled roads.

The weather forecast (highly accurate in this instance) was clearly an important factor in our personal planning and decision making. Forecasting also plays a crucial role in business, industry, government, and institutional planning because many important decisions depend on the anticipated future values of certain variables. Let us consider three more examples of how forecasting can aid in planning.

1. A business firm manufactures computerized television games for retail sale. If the firm does not manufacture and keep in inventory enough units of its product to meet demand, it could lose sales to a competitor and thus have lower profits. On the other hand, keeping an inventory is costly. If the inventory of finished goods is too large, the firm will have higher carrying costs and lower profits than otherwise. This firm can maximize profits (other things equal) by properly balancing the benefits of holding inventory (avoiding lost sales) against the costs (interest charges). Clearly,

3

the inventory level the firm should aim for depends partly on the anticipated amount of future sales. Unfortunately, future sales can rarely be known with certainty so decisions about production and inventory levels must be based on sales forecasts.

2. A nonprofit organization provides temporary room and board for indigent transients in a large city in the northern part of the United States. The number of individuals requesting aid each month follows a complex seasonal pattern. Cold weather drives some potentially needy individuals out of the city to warmer climates, but it also raises the number of requests for aid from those who remain in the city during the winter. Warmer weather reverses this pattern. The directors of the organization could better plan their fund-raising efforts and their ordering of food and clothing if they had reliable forecasts of the seasonal variation in aid requests.

3. A specialty foods wholesaler knows from experience that sales are usually sufficient to warrant delivery runs into a given geographic region if population density exceeds a critical minimum number. Forecasting the exact amount of sales is not necessary for this decision. The wholesaler uses census information about population density to choose which regions to serve.

Forecasts can be formed in many different ways. The method chosen depends on the purpose and importance of the forecasts as well as the costs of the alternative forecasting methods. The food wholesaler in the example above combines his or her experience and judgment with a few minutes looking up census data. But the television game manufacturer might employ a trained statistician or economist to develop sophisticated mathematical and statistical models in an effort to achieve close control over inventory levels.

1.2 What this book is about

As suggested by its title this book is about *forecasting with single-series (univariate) Box–Jenkins (UBJ) models*.* We use the label "Box–Jenkins" because George E. P. Box and Gwilym M. Jenkins are the two people most

---

*Although our focus is on forecasting, univariate Box–Jenkins analysis is often useful for simply explaining the past behavior of a single data series, for whatever reason one may want to do so. For example, if we discover that interest rates have historically shown a certain seasonal pattern, we may better understand the causes and consequences of past policy decisions made by the Open Market Committee of the Federal Reserve System. This information may be valuable to persons having no desire to forecast interest rates.

responsible for formalizing the procedure used in the type of analysis we will study. They have also made important contributions to the underlying theory and practice. The basic theory and modeling procedures presented in this book are drawn largely from their work [1].*

We use the letters UBJ in this text to stand for "univariate Box–Jenkins." UBJ models are also often referred to as ARIMA models. The acronym ARIMA stands for Auto-Regressive Integrated Moving Average. This terminology is explained further in Chapters 3 and 5. We use the labels UBJ, ARIMA, and UBJ–ARIMA more or less interchangeably throughout the book.

"Single-series" means that UBJ–ARIMA forecasts are based *only on past values of the variable being forecast*. They are not based on any other data series. Another word for single-series is "univariate" which means "one variable." We use the terms single-series and univariate interchangeably.

For our purposes a model is an algebraic statement telling how one thing is statistically related to one or more other things. An ARIMA model is an algebraic statement telling how observations on a variable are statistically related to past observations on the same variable. We will see an example of an ARIMA model later in this chapter.

All statistical forecasting methods are *extrapolative* in nature: they involve the projection of past patterns or relationships into the future. In the case of UBJ–ARIMA forecasting we extrapolate past patterns within a single data series into the future.

The purpose of this book is twofold. The first objective is to explain the basic concepts underlying UBJ–ARIMA models. This is done in Part I (Chapters 1–11). These concepts involve the application of some principles of classical probability and statistics to time-sequenced observations in a single data series.

The second objective of this book is to provide enough detailed case studies and practical rules to enable you to build UBJ–ARIMA models properly and quickly. This is done in Part II.

Box and Jenkins propose an entire family of models, called ARIMA models, that seems applicable to a wide variety of situations. They have also developed a practical procedure for choosing an appropriate ARIMA model out of this family of ARIMA models. However, selecting an appropriate ARIMA model may not be easy. Many writers suggest that building a proper ARIMA model is an art that requires good judgment and a lot of experience. The practical rules and case studies in Part II are designed to help you develop that judgment and to make your experiences with UBJ modeling more valuable.

---

*All references are listed by number at the end of the book.

In this chapter we consider some restrictions on the types of data that can be analyzed with the UBJ method. We also summarize the Box–Jenkins three-stage procedure for building good ARIMA models.

In Chapter 2 we present two important tools, the estimated autocorrelation function and the estimated partial autocorrelation function, used in the UBJ method to measure the statistical relationships between observations within a single data series.

In Chapter 3 we go more deeply into the principles underlying UBJ analysis. In Chapter 4 we summarize the characteristics of a good ARIMA model and present two examples of the three-stage UBJ modeling procedure.

The emphasis in Chapter 5 is on special notation used for representing ARIMA models and on the intuitive interpretation of these models.

Chapters 6 through 11 contain more detailed discussion of the basic concepts behind the UBJ method along with some examples.

Chapter 12 contains a list of practical rules for building UBJ–ARIMA forecasting models. This chapter is followed by 15 case studies. The data in the case studies are related largely to economics and business, but modeling procedures are the same regardless of the context from which the data are drawn.

1.3   Time-series data

In this book we are concerned with forecasting *time-series data*. Time-series data refers to observations on a variable that occur in a time sequence. We use the symbol z_t to stand for the numerical value of an observation; the subscript t refers to the time period when the observation occurs. Thus a sequence of n observations could be represented this way: $z_1, z_2, z_3, \ldots, z_n$.

As an example of time-series data consider monthly production of athletic shoes (in thousands) in the United States for the year 1971.* The sequence of observations for that year is as follows:

t	z_t	t	z_t
1	659	7	520
2	740	8	641
3	821	9	769
4	805	10	718
5	687	11	697
6	687	12	696

*Data from various issues of *Business Statistics*, U.S. Commerce Department.

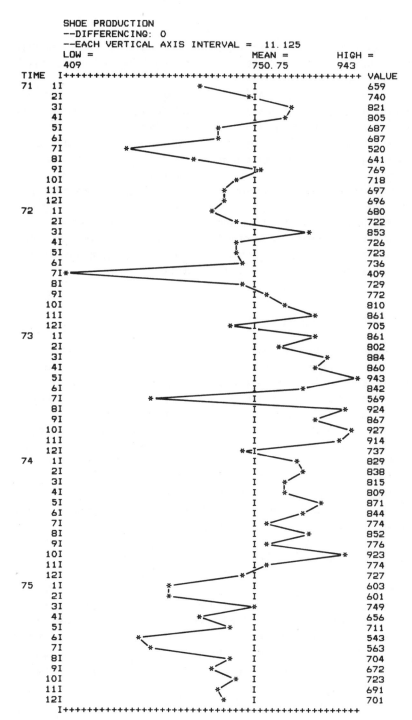

```
           SHOE PRODUCTION
           --DIFFERENCING: 0
           --EACH VERTICAL AXIS INTERVAL =  11.125
           LOW =                        MEAN =       HIGH =
           409                          750.75       943
TIME   I++++++++++++++++++++++++++++++++++++++++++++++++++  VALUE
71  1I                          *———————I                       659
    2I                                *–I                        740
    3I                                I    *                     821
    4I                                I   *                      805
    5I                          *     I                          687
    6I              *           *     I                          687
    7I        *                       I                          520
    8I                    *           I                          641
    9I                                I*                         769
   10I                           *    I                          718
   11I                         *      I                          697
   12I                         *      I                          696
72  1I                        *       I                          680
    2I                           *    I                          722
    3I                                I          *               853
    4I                        *       I                          726
    5I                        *       I                          723
    6I                            *   I                          736
    7I*                                I                          409
    8I                            *   I                          729
    9I                                I *                        772
   10I                                I   *                      810
   11I                                I       *                  861
   12I                         *      I                          705
73  1I                                I         *                861
    2I                                I   *                      802
    3I                                I            *             884
    4I                                I          *               860
    5I                                I              * 943
    6I                                I        *                 842
    7I           *                    I                          569
    8I                                I            *             924
    9I                                I        *                 867
   10I                                I             *            927
   11I                                I            *             914
   12I                       *<–I                                737
74  1I                                I      *                   829
    2I                                I       *                  838
    3I                                I     *                    815
    4I                                I     *                    809
    5I                                I           *              871
    6I                                I        *                 844
    7I                                I *                        774
    8I                                I         *                852
    9I                                I *                        776
   10I                                I              *           923
   11I                                I *                        774
   12I                          *     I                          727
75  1I                       *        I                          603
    2I                       *        I                          601
    3I                                I*                         749
    4I                    *           I                          656
    5I                          *     I                          711
    6I             *                  I                          543
    7I              *                 I                          563
    8I                          *     I                          704
    9I                     *          I                          672
   10I                        *       I                          723
   11I                      *         I                          691
   12I                        *       I                          701
       I++++++++++++++++++++++++++++++++++++++++++++++++++
```

Figure 1.1 Example of time-series data: Production of athletic shoes, in thousands, January 1971 to December 1975.

In this example z_1 is the observation for January 1971, and its numerical value is 659; z_2 is the observation for February 1971, and its value is 740, and so forth.

It is useful to look at time-series data graphically. Figure 1.1 shows 60 monthly observations of production of athletic shoes in the United States covering the period January 1971 to December 1975.

The vertical axis scale in Figure 1.1 measures thousands of pairs of shoes produced per month. The horizontal axis is a time scale. Each asterisk is an observation associated with *both* a time period (directly below the asterisk on the horizontal axis) *and* a number of pairs of shoes in thousands (directly to the left of the asterisk on the vertical axis). As we read the graph from left to right, each asterisk represents an observation which succeeds the previous one in time. These data are recorded at discrete intervals: the lines connecting the asterisks do not represent numerical values, but merely remind us that the asterisks occur in a certain time sequence.

1.4 Single-series (univariate) analysis

The phrase "time-series analysis" is used in several ways. Sometimes it refers to any kind of analysis involving times-series data. At other times it is used more narrowly to describe attempts to explain behavior of time-series data using only past observations on the variable in question. Earlier we referred to this latter activity as *single-series* or *univariate analysis*, and we said that UBJ–ARIMA modeling is a type of univariate analysis.

In some types of statistical analysis the various observations within a single data series are assumed to be statistically independent. Some readers might recall that this is a standard assumption about the error term (and therefore about observations on the dependent variable) in traditional regression analysis. But in UBJ–ARIMA analysis we suppose that the time-sequenced observations in a data series $(\ldots, z_{t-1}, z_t, z_{t+1}, \ldots)$ may be *statistically dependent*. We use the statistical concept of *correlation* to measure the relationships between observations *within* the series. In UBJ analysis we want to examine the correlation between z at time t (z_t) and z at earlier time periods ($z_{t-1}, z_{t-2}, z_{t-3}, \ldots$). In the next chapter we show how to calculate the correlation between observations within a single time series.

We can illustrate the idea of UBJ forecasting in a rough way using Figure 1.2. Suppose we have available 60 time-sequenced observations on a single variable. These are represented on the left-hand side of the graph in Figure 1.2, labeled "Past". By applying correlation analysis to these 60 observations, we build an ARIMA model. This model describes how any given observation (z_t) is related to previous observations ($z_{t-1}, z_{t-2}, \ldots$). We

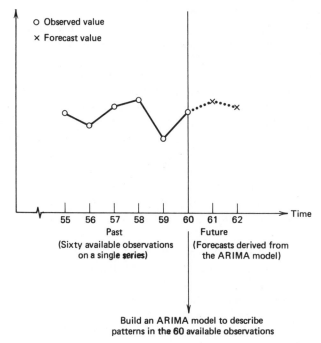

Figure 1.2   The idea of UBJ forecasting.

may use this model to forecast future values (for periods 61, 62, ...) of this
variable.

Thus, if the data series being analyzed is athletic shoe production, our
forecasts of shoe production for period 61 and thereafter are based only on
the information contained in the available (past) data on shoe production.
We make no appeal to additional information contained in other variables
such as Gross National Product, interest rates, average monthly tempera-
tures, and so forth. Instead we start with the idea that *shoe production for
any given time period may be statistically related to production in earlier
periods. We then attempt to find a good way of stating the nature of that
statistical relationship.*

1.5   When may UBJ models be used?

Short-term forecasting.   UBJ–ARIMA models are especially suited to
short-term forecasting. We emphasize short-term forecasting because most
ARIMA models place heavy emphasis on the recent past rather than the

distant past. For example, it is not unusual to see an ARIMA model where z_t is related explicitly to just the two most recent observations (z_{t-1} and z_{t-2}). On the other hand, ARIMA models showing z_t explicitly related to observations very far in the past, such as z_{t-70} or z_{t-115} are rare indeed.

This emphasis on the recent past means that long-term forecasts from ARIMA models are less reliable than short-term forecasts. For example, consider an ARIMA model where z_t is related explicitly to the most recent value z_{t-1}. Let n be the last period for which data are available. To forecast z_{n+1} (one period ahead) we use the most recent value z_n. To forecast z_{n+2} (two periods ahead) we want the most recent observation z_{n+1}, but it is not available; we must use the forecast of z_{n+1} in place of the observed value for that period. Obviously our forecasts for period $n + 2$ and beyond are less reliable than the forecast for period $n + 1$ since they are based on less reliable information (i.e., forecasts rather than observations).

Data types. The UBJ method applies to either *discrete data* or *continuous data*. Discrete data are measured in integers only (e.g., 1, 8, -42), never in decimal amounts. Data that can be measured in decimal amounts (e.g., $4\frac{1}{2}$, $-19.87, 2.4$) are called continuous data.

For example, counting the number of fielding errors committed by each major league baseball team produces discrete data: there is no such thing as part of an error. But measuring the distance in meters from home plate down the third-base line to the left-field wall in each baseball stadium could produce continuous data: it is possible to measure this variable in parts of a meter.

Although the UBJ method can handle either discrete or continuous data, it deals only with data measured at *equally spaced*, *discrete time intervals*. For example, consider an electronic machine that measures the pressure in a tank continuously. A gauge attached to the machine produces a reading at every moment in time, and a mechanical pen continuously records the results on a moving strip of paper. Such data are not appropriate for the UBJ method because they are measured continuously rather than at discrete time intervals. However, if tank pressure were recorded once every hour, the resulting data series could be analyzed with the UBJ technique.

Data measured at discrete time intervals can arise in two ways. First, a variable may be *accumulated* through time and the total recorded periodically. For example, the dollar value of all sales in a tavern may be totaled at the end of each day, while tons of steel output could be accumulated and recorded monthly. Second, data of this type can arise when a variable is *sampled* periodically. Recording tank-pressure readings once every hour (as discussed in the last paragraph) is an example of such sampling. Or suppose an investment analyst records the closing price of a stock at the end of each week. In these last two cases the variable is being sampled at an instant in time rather than being accumulated through time.

UBJ–ARIMA models are particularly useful for forecasting data series that contain *seasonal* (or other periodic) variation, including those with shifting seasonal patterns. Figure 1.3 is an example of seasonal data. It shows monthly cigar consumption (withdrawals from stock) in the United States from 1969 through 1976.* These data repeat a pattern from year to year. For instance, October tends to be associated with a high value and December with a low value during each year. With seasonal data any given observation is similar to other observations in the same season during different years. In the cigar-consumption series, October is similar to other Decembers, February is like other Februarys, and so on. We discuss seasonal data and models in detail in Chapter 11.†

Sample size. Building an ARIMA model requires an adequate sample size. Box and Jenkins [1, p. 33] suggest that about 50 observations is the minimum required number. Some analysts may occasionally use a smaller sample size, interpreting the results with caution. A large sample size is especially desirable when working with seasonal data.

Stationary series. The UBJ–ARIMA method applies only to *stationary* data series. A stationary time series has a mean, variance, and autocorrelation function that are essentially constant through time.‡ (We introduce the idea of an autocorrelation function in Chapter 2. An autocorrelation function is one way of measuring how the observations within a single data series are related to each other.) In this section we illustrate the idea of a constant mean and variance.

The stationarity assumption simplifies the theory underlying UBJ models and helps ensure that we can get useful estimates of parameters from a moderate number of observations. For example, with 50 observations we can get a fairly good estimate of the true mean underlying a data series if there is only one mean. But if the variable in question has a different mean each time period, we could not get useful estimates of each mean since we typically have only one observation per time period.

The mean of a stationary series indicates the overall level of the series. We estimate the true mean (μ) underlying a series with the sample mean

---

*Data from various issues of the *Survey of Current Business*, U.S. Commerce Department.
† The cigar series is analyzed in detail in Case 13.
‡ The formal definition of stationarity is more complicated than this, but the definition given here is adequate for present purposes. We discuss stationarity more formally in Chapter 3.

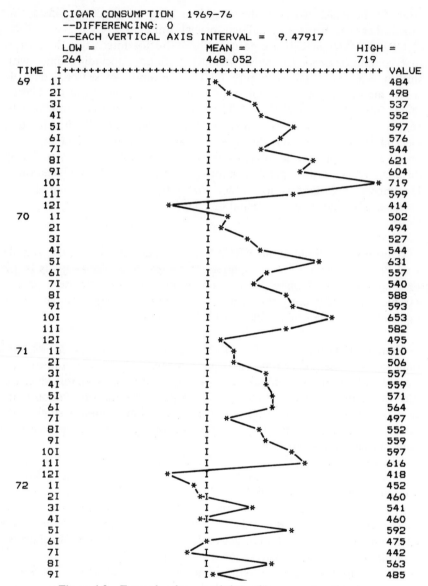

```
CIGAR CONSUMPTION   1969-76
--DIFFERENCING:  0
--EACH VERTICAL AXIS INTERVAL =   9.47917
LOW =                 MEAN =                  HIGH =
264                   468.052                 719
```

Figure 1.3   Example of seasonal data: Cigar consumption.

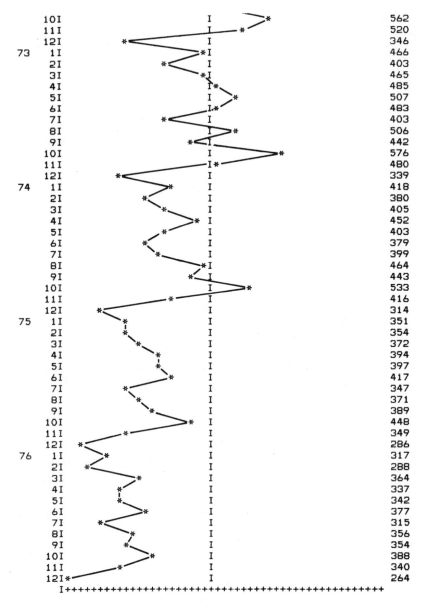

10I	I	562
11I	I	520
12I		346
73 1I		466
2I		403
3I		465
4I		485
5I		507
6I		483
7I		403
8I		506
9I		442
10I		576
11I		480
12I		339
74 1I		418
2I		380
3I		405
4I		452
5I		403
6I		379
7I		399
8I		464
9I		443
10I		533
11I		416
12I		314
75 1I		351
2I		354
3I		372
4I		394
5I		397
6I		417
7I		347
8I		371
9I		389
10I		448
11I		349
12I		286
76 1I		317
2I		288
3I		364
4I		337
5I		342
6I		377
7I		315
8I		356
9I		354
10I		388
11I		340
12I		264

Figure 1.3 (*Continued*).

($\bar{z}$). The sample mean of a time series is calculated just as any ordinary arithmetic mean. That is, sum the observations for each time period (z_t) and divide by the total number of observations (n):

$$\bar{z} = \frac{1}{n} \sum_{t=1}^{n} z_t \tag{1.1}$$

Consider the data in Figure 1.4. By summing the observations and dividing by 60 (the number of observations) we find the mean of this time series to be 100:

$$\bar{z} = \frac{1}{n} \sum z_t = \frac{1}{60} (102 + 99 + 101 + \cdots + 98)$$

$$= \frac{1}{60} (6000)$$

$$= 100$$

If a time series is stationary then the mean of any major subset of the series does not differ significantly from the mean of any other major subset of the series. The series in Figure 1.4 appears to have a mean that is constant through time. For example, the first half of the data set (observations 1 through 30) seems to have about the same mean as the second half of the data set (observations 31 through 60). We should expect the mean to fluctuate somewhat over brief time spans because of sampling variation. In later chapters we consider two methods besides visual inspection for determining if the mean of a series is stationary.

We use the sample variance s_z^2 of a time series to estimate the true underlying variance σ_z^2. As usual the variance measures the dispersion of the observations around the mean. The sample variance of a time series is calculated just as any variance. That is, find the deviation of each observation from the mean, square each deviation, sum the deviations, and divide by the total number of observations (n):

$$s_z^2 = \frac{1}{n} \sum_{t=1}^{n} (z_t - \bar{z})^2 \tag{1.2}$$

Clearly, if the z_t observations gather closely around $\bar{z}$, then s_z^2 will be relatively small since each individual squared deviation $(z_t - \bar{z})^2$ will be small.

Consider again the data in Figure 1.4. If we insert the previously calculated mean (100) into equation (1.2), find the deviation $(z_t - \bar{z})$ of

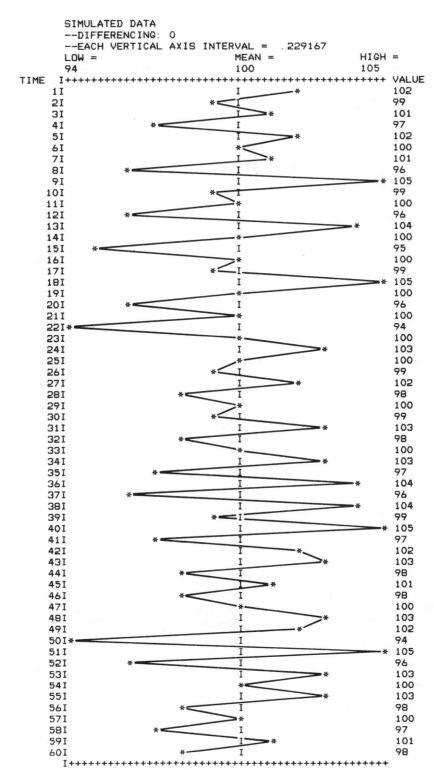

Figure 1.4 Example of a stationary time series (simulated data).

each observation from the mean, square and sum these deviations and divide by 60, we find that the variance of this series (rounded) is 7.97:

$$s_z^2 = \frac{1}{n}\sum(z_t - \bar{z})^2$$

$$= \frac{1}{60}\left[(102 - 100)^2 + (99 - 100)^2 + (101 - 100)^2 + \cdots + (98 - 100)^2\right]$$

$$= \frac{1}{60}(4 + 1 + 1 + \cdots + 4)$$

$$= 7.97$$

If a data series is stationary then the variance of any major subset of the series will differ from the variance of any other major subset only by chance. The variance of the data in Figure 1.4 does not appear to change markedly through time. Of course, we should expect the variance to fluctuate somewhat over short time spans just because of sampling error. In Chapter 7 we refer to a more rigorous method for determining if the variance of a series is stationary, but visual inspection is commonly used in practice.

The stationarity requirement may seem quite restrictive. However, most nonstationary series that arise in practice can be transformed into stationary series through relatively simple operations. We introduce some useful transformations to achieve stationarity in Chapters 2 and 7. These transformations are illustrated in detail in the case studies.

1.6   The Box–Jenkins modeling procedure

The last two sentences at the end of Section 1.4 are important because they summarize the general nature of the UBJ–ARIMA method. Because all aspects of UBJ analysis are related in some way to the ideas contained in those sentences, we repeat the ideas here for emphasis. (i) The observations in a time series may be statistically related to other observations in the same series. (ii) Our goal in UBJ analysis is to find a good way of stating that statistical relationship. That is, we want to find a good model that describes how the observations in a single time series are related to each other.

An ARIMA model is an algebraic statement showing how a time-series variable (z_t) is related to its own past values $(z_{t-1}, z_{t-2}, z_{t-3}, \ldots)$. We discuss the algebraic form of ARIMA models in detail starting in Chapter 3, but it will be helpful to look at one example now. Consider the algebraic expression

$$z_t = C + \phi_1 z_{t-1} + a_t \qquad (1.3)$$

Equation (1.3) is an example of an ARIMA model. It says that z_t is related to its own immediately past value (z_{t-1}). C is a constant term. ϕ_1 is a fixed coefficient whose value determines the relationship between z_t and z_{t-1}. The a_t term is a probabilistic "shock" element.

The terms C, $\phi_1 z_{t-1}$, and a_t are each components of z_t. C is a deterministic (fixed) component, $\phi_1 z_{t-1}$ is a probabilistic component since its value depends in part on the value of z_{t-1}, and a_t is a purely probabilistic component. Together C and $\phi_1 z_{t-1}$ represent the predictable part of z_t while a_t is a residual element that cannot be predicted within the ARIMA model. However, as discussed in Chapter 3, the a_t term is assumed to have certain statistical properties.

We have not yet defined what a "good" model is. In fact a satisfactory model has many characteristics as summarized in Chapter 4 and discussed in detail in later chapters. For now remember that *a good model includes the smallest number of estimated parameters needed to adequately fit the patterns in the available data.*

Box and Jenkins propose a practical three-stage procedure for finding a good model. Our purpose here is to sketch the broad outline of the Box–Jenkins modeling strategy; we consider the details in later chapters. The three-stage UBJ procedure is summarized schematically in Figure 1.5.

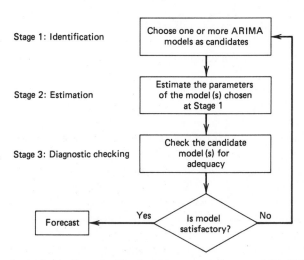

Figure 1.5 Stages in the Box–Jenkins iterative approach to model building. Adapted from Box and Jenkins [1, p. 19] by permission.

Stage 1: identification.   At the identification stage we use two graphical devices to measure the correlation between the observations within a single data series. These devices are called an estimated *autocorrelation function* (abbreviated acf) and an estimated *partial autocorrelation function* (abbreviated pacf). We look at examples of these graphical tools in Chapter 2. The estimated acf and pacf measure the statistical relationships within a data series in a somewhat crude (statistically inefficient) way. Nevertheless, they are helpful in giving us a feel for the patterns in the available data.

The next step at the identification stage is to summarize the statistical relationships within the data series in a more compact way than is done by the estimated acf and pacf. Box and Jenkins suggest a whole family of algebraic statements (ARIMA models) from which we may choose. Equation (1.3) is an example of such a model. We will see more examples of these mathematical statements starting in Chapter 3.

We use the estimated acf and pacf as guides to choosing one or more ARIMA models that seem appropriate. The basic idea is this: every ARIMA model [such as equation (1.3)] has a *theoretical* acf and pacf associated with it. At the identification stage we compare the *estimated* acf and pacf calculated from the available data with various *theoretical* acf's and pacf's. We then tentatively choose the model whose theoretical acf and pacf most closely resemble the estimated acf and pacf of the data series. Note that we do not approach the available data with a rigid, preconceived idea about which model we will use. Instead, we let the available data "talk to us" in the form of an estimated acf and pacf.

Whichever model we choose at the identification stage, we consider it only tentatively: it is only a candidate for the final model. To choose a final model we proceed to the next two stages and perhaps return to the identification stage if the tentatively considered model proves inadequate.

Stage 2: estimation.   At this stage we get precise estimates of the coefficients of the model chosen at the identification stage. For example, if we tentatively choose equation (1.3) as our model, we fit this model to the available data series to get estimates of ϕ_1 and C. This stage provides some warning signals about the adequacy of our model. In particular, if the estimated coefficients do not satisfy certain mathematical inequality conditions, that model is rejected. The method for estimating the coefficients in a model is a technical matter considered in Chapter 8. The inequality conditions the estimated coefficients must satisfy are discussed in Chapter 6.

Stage 3: diagnostic checking.   Box and Jenkins suggest some diagnostic checks to help determine if an estimated model is statistically adequate. A model that fails these diagnostic tests is rejected. Furthermore, the results at this stage may also indicate how a model could be improved. This leads us back to the identification stage. We repeat the cycle of identification, estimation, and diagnostic checking until we find a good final model. As shown in Figure 1.5, once we find a satisfactory model we may use it to forecast.

The iterative nature of the three-stage UBJ modeling procedure is important. The estimation and diagnostic-checking stages provide warning signals telling us when, and how, a model should be reformulated. We continue to reidentify, reestimate, and recheck until we find a model that is satisfactory according to several criteria. This iterative application of the three stages does not guarantee that we will finally arrive at the best possible ARIMA model, but it stacks the cards in our favor.

We return to the three stages of UBJ modeling in Chapter 4 where we consider two examples. The case studies in Part II illustrate the use of the UBJ modeling procedure in detail.

1.7   UBJ models compared with other models

The UBJ approach has three advantages over many other traditional single-series methods. First, the concepts associated with UBJ models are derived from a solid foundation of classical probability theory and mathematical statistics. Many other historically popular univariate methods (though not all) are derived in an *ad hoc* or intuitive way.

Second, ARIMA models are a family of models, not just a single model. Box and Jenkins have developed a strategy that guides the analyst in choosing one or more appropriate models out of this larger family of models.

Third, it can be shown that an appropriate ARIMA model produces *optimal* univariate forecasts. That is, no other standard single-series model can give forecasts with a smaller mean-squared forecast error (i.e., forecast-error variance).*

In sum, there seems to be general agreement among knowledgeable professionals that properly built UBJ models can handle a wider variety of situations and provide more accurate short-term forecasts than any other standard single-series technique. However, the construction of proper UBJ models may require more experience and computer time than some historically popular univariate methods.

---

*The optimal nature of ARIMA forecasts is discussed in Chapter 10.

Single-series (univariate) models differ from multiple-series (multivariate) models. The latter involve a sequence of observations on at least one variable other than the one being forecast.* Multiple-series models should theoretically produce better forecasts than single-series models because multiple-series forecasts are based on more information than just the past values of the series being forecast. But some analysts argue that UBJ models frequently approach or exceed the forecasting accuracy of multiple-series models in practice. This seems especially true for short-term forecasts. Cooper [4], Naylor et al. [5], and Nelson [6] discuss the accuracy of UBJ–ARIMA models compared with multiple-series econometric (regression and correlation) models.[†]

Summary

1. Box and Jenkins propose a family of algebraic models (called ARIMA models) from which we select one that seems appropriate for forecasting a given data series.

2. UBJ–ARIMA models are single-series or univariate forecasting models: forecasts are based only on past patterns in the series being forecast.

3. UBJ–ARIMA models are especially suited to short-term forecasting and to the forecasting of series containing seasonal variation, including shifting seasonal patterns.

4. UBJ–ARIMA models are restricted to data available at discrete, equally spaced time intervals.

5. Construction of an adequate ARIMA model requires a minimum of about 50 observations. A large sample size is especially desirable when seasonal variation is present.

6. The UBJ method applies only to stationary time series.

7. A stationary series has a mean, variance, and autocorrelation function that are essentially constant over time.

8. Although many nonstationary series arise in practice, most can be transformed into stationary series.

9. In UBJ–ARIMA analysis, the observations in a single time series are assumed to be (potentially) statistically dependent—that is, sequentially or serially correlated.

---

*Box and Jenkins [1, Chapters 10 and 11] discuss a certain type of multivariate ARIMA model which they call a transfer function. For more recent developments in multivariate time-series analysis, see Jenkins and Alavi [2] and Tiao and Box [3].
[†]In Appendix 10A of Chapter 10 we discuss how ARIMA models may be used to complement econometric models.

10. The goal of UBJ analysis is to find an ARIMA model with the smallest number of estimated parameters needed to fit adequately the patterns in the available data.

11. The UBJ method for finding a good ARIMA model involves three steps: identification, estimation, and diagnostic checking.

12. At the identification stage we tentatively select one or more ARIMA models by looking at two graphs derived from the available data. These graphs are called an estimated autocorrelation function (acf) and an estimated partial autocorrelation function (pacf). We choose a model whose associated theoretical acf and pacf look like the estimated acf and pacf calculated from the data.

13. At the estimation stage we obtain estimates of the parameters for the ARIMA model tentatively chosen at the identification stage.

14. At the diagnostic-checking stage we perform tests to see if the estimated model is statistically adequate. If it is not satisfactory we return to the identification stage to tentatively select another model.

15. A properly constructed ARIMA model produces optimal univariate forecasts: no other standard single-series technique gives forecasts with a smaller forecast-error variance.

Questions and Problems

1.1 Consider the following sequence of total quarterly sales at a drug store:

Year	Quarter	Sales
1963	1	$12,800
	2	13,400
	3	11,200
	4	14,700
1964	1	13,000
	2	9,400
	3	12,100
	4	15,100
1965	1	11,700
	2	14,000
	3	10,900
	4	14,900

(a)   Do these observations occur at equally spaced, discrete time intervals?

(b)   Did these observations result from daily sales being *accumulated* and recorded periodically, or *sampled* and recorded periodically?

(c)   Could these observations be used to construct an ARIMA model? Explain.

(d)   Plot these observations on a graph with time on the horizontal axis and sales on the vertical axis.

(e)   Does there appear to be a seasonal pattern in this data series? If so, would this fact disqualify the use of an ARIMA model to forecast future sales?

(f)   Let $z_1, z_2, z_3, \ldots, z_n$ stand for the sequence of observations above. What is the numerical value of n? Of z_t? Of z_6? Of z_9?

1.2   What are the restrictions on the type of time-series data to which UBJ–ARIMA analysis may be applied?

1.3   What is meant by a "stationary" time series?

1.4   Summarize the UBJ three-stage modeling procedure.

1.5   What kind of information is contained in an estimated acf?

1.6   What is the difference between a univariate time-series forecasting model and a multivariate time-series forecasting model?

1.7   What advantages do UBJ–ARIMA models have compared with other traditional univariate forecasting models? What disadvantages?

1.8   Consider the following time series:

t	z_t	t	z_t
1	106	13	106
2	107	14	98
3	98	15	99
4	98	16	96
5	101	17	95
6	99	18	99
7	102	19	100
8	104	20	102
9	97	21	108
10	103	22	106
11	107	23	104
12	105	24	98

(a) Does this series appear to be stationary? (It may help if you plot the series on a graph.)

(b) Does this series contain enough observations for you to build a UBJ–ARIMA model from it?

(c) Calculate the mean and variance for this series.

2

INTRODUCTION TO
BOX–JENKINS ANALYSIS
OF A SINGLE DATA SERIES

In the last chapter we referred to the estimated autocorrelation function (acf) and estimated partial autocorrelation function (pacf). They are used in UBJ analysis at the identification stage to summarize the statistical patterns within a single time series. Our chief task in this chapter is to learn how an estimated acf and pacf are constructed. In Chapter 4 we show how an estimated acf and pacf are used in building an ARIMA model.

Before examining the idea of the estimated acf and pacf we briefly consider two other topics. First, we examine a transformation, called *differencing*, that is frequently applied to time-series data to induce a stationary mean. Second, we look at an operation, called calculation of *deviations from the mean*, used to simplify the calculations performed in UBJ–ARIMA analysis.

2.1 Differencing

We pointed out in the last chapter that UBJ–ARIMA analysis is restricted to stationary time series. Fortunately, many *non*stationary series can be transformed into stationary ones. Thus the UBJ method can be used to analyze even nonstationary data. We discuss nonstationary series in detail in Chapter 7. In this section we introduce a common transformation called *differencing.* Differencing is a relatively simple operation that involves calculating *successive changes* in the values of a data series.

Differencing is used when the mean of a series is changing over time. Figure 2.1 shows an example of such a series. (These 52 observations are the weekly closing price of AT & T common stock for 1979.* We analyze them in detail in Part II, Case 6.) Of course, it is possible to calculate a single mean for this series. It is 57.7957, shown in Figure 2.1 as the horizontal line running through the center of the graph. However, this single number is misleading because major subsets of the series appear to have means different from other major subsets. For example, the first half of the data set lies substantially above the second half. The series is nonstationary because its mean is not constant through time.

To difference a data series, define a new variable (w_t) which is the change in z_t, that is,[†]

$$w_t = z_t - z_{t-1}, \qquad t = 2, 3, \dots, n \qquad (2.1)$$

Using the data in Figure 2.1 we get the following results when we difference the observations:

$$w_2 = z_2 - z_1 = 61.625 - 61 = 0.625$$

$$w_3 = z_3 - z_2 = 61 - 61.625 = -0.625$$

$$w_4 = z_4 - z_3 = 64 - 61 = 3$$

$$\vdots \qquad \vdots \qquad \vdots$$

$$w_{52} = z_{52} - z_{51} = 52.25 - 51.875 = 0.375$$

These results are plotted in Figure 2.2. The differencing procedure seems to have been successful: the differenced series in Figure 2.2 appears to have a constant mean. Note that we lost one observation: there is no z_0 available to subtract from z_1 so the differenced series has only 51 observations.

Series w_t is called the *first differences* of z_t. If the first differences do not have a constant mean, we redefine w_t as the first differences of the first differences:

$$w_t = (z_t - z_{t-1}) - (z_{t-1} - z_{t-2}), \qquad t = 3, 4, \dots, n \qquad (2.2)$$

---

*Data from *The Wall Street Journal.*
[†]Differencing for a series with nonstationary seasonal variation is only slightly more complicated. We discuss seasonal differencing in Chapter 11.

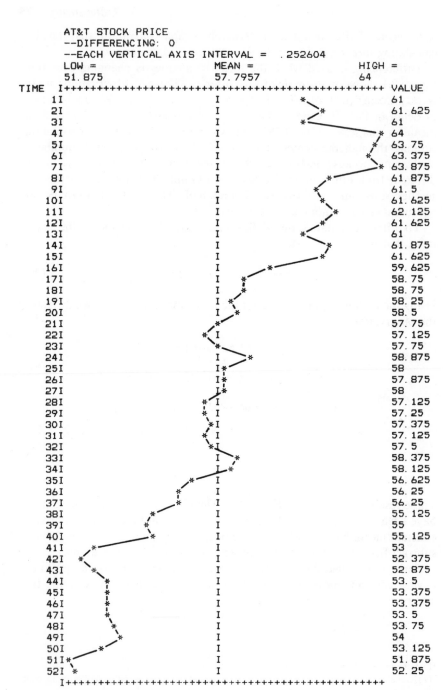

```
          AT&T STOCK PRICE
          --DIFFERENCING: 0
          --EACH VERTICAL AXIS INTERVAL =  .252604
          LOW =                  MEAN =               HIGH =
          51.875                 57.7957              64
 TIME  I++++++++++++++++++++++++++++++++++++++++++++++++++  VALUE
    1I                           I                   *        61
    2I                           I                    *       61.625
    3I                           I                  *  \      61
    4I                           I                      * \   64
    5I                           I                      *     63.75
    6I                           I                     *      63.375
    7I                           I                      *     63.875
    8I                           I                  *         61.875
    9I                           I                *           61.5
   10I                           I                 *          61.625
   11I                           I                  *         62.125
   12I                           I                 *          61.625
   13I                           I              *             61
   14I                           I                *           61.875
   15I                           I               *            61.625
   16I                           I      *                     59.625
   17I                           I    *                       58.75
   18I                           I    *                       58.75
   19I                           I  *                         58.25
   20I                           I  *                         58.5
   21I                           I *                          57.75
   22I                         * I                            57.125
   23I                           *                            57.75
   24I                           I  *                         58.875
   25I                          I*                            58
   26I                          I*                            57.875
   27I                          I*                            58
   28I                        * I                             57.125
   29I                        * I                             57.25
   30I                         *I                             57.375
   31I                        * I                             57.125
   32I                        *I                              57.5
   33I                           I *                          58.375
   34I                          I *                           58.125
   35I                      *  I                              56.625
   36I                    *    I                              56.25
   37I                     *   I                              56.25
   38I                  *      I                              55.125
   39I                 *       I                              55
   40I                   *     I                              55.125
   41I          *              I                              53
   42I        *                I                              52.375
   43I         *               I                              52.875
   44I           *             I                              53.5
   45I           *             I                              53.375
   46I           *             I                              53.375
   47I           *             I                              53.5
   48I           *             I                              53.75
   49I         *               I                              54
   50I       *                 I                              53.125
   51I*                        I                              51.875
   52I *                       I                              52.25
       I+++++++++++++++++++++++++++++++++++++++++++++++++++
```

Figure 2.1 Example of a time series with a nonstationary mean: Weekly closing prices, AT & T common stock, 1979.

26

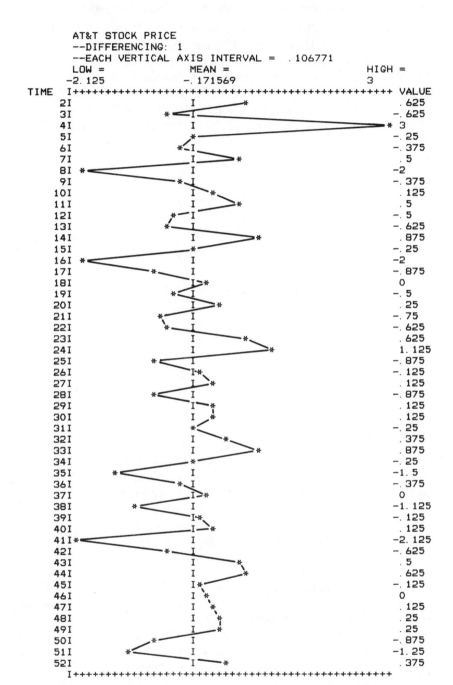

Figure 2.2 First differences of the AT & T stock price series.

Series w_t is now called the *second differences* of z_t because it results from differencing z_t twice. Usually, first differencing is sufficient to bring about a stationary mean. Calculating the second differences for the series in Figure 2.1 does not seem necessary because the first differences in Figure 2.2 appear to have a constant mean. However, we show some calculations for purposes of illustration. Since the second differences are the first differences of the first differences, we merely apply differencing to the data in Figure 2.2:

$$w_3 = (z_3 - z_2) - (z_2 - z_1)$$

$$= (-0.625) - (0.625) = -1.25$$

$$w_4 = (z_4 - z_3) - (z_3 - z_2)$$

$$= (3) - (-0.625) = 3.625$$

$$w_5 = (z_5 - z_4) - (z_4 - z_3)$$

$$= (-0.25) - (3) = -3.25$$

$$\vdots \qquad \qquad \vdots$$

$$w_{52} = (z_{52} - z_{51}) - (z_{51} - z_{50})$$

$$= (0.375) - (-1.25) = 1.625$$

When differencing is needed to induce a stationary mean, we construct a new series w_t that is different from the original series z_t. We then build an ARIMA model for the stationary series w_t. However, usually we are interested in forecasting the original series z_t so we want an ARIMA model for that series. Fortunately, this does not present a serious problem since w_t and z_t are linked by definition (2.1) in the case of first differencing or by definition (2.2) in the case of second differencing. In Chapter 7 and in the case studies (Part II) we will see exactly how an ARIMA model for w_t implies an ARIMA model for z_t.

A final point about differencing: a series which has been made stationary by appropriate differencing frequently has a mean of virtually zero. For example, the nonstationary series in Figure 2.1 has a mean of about 57.8. But the stationary series in Figure 2.2 achieved by differencing the data in Figure 2.1 has a mean of about -0.2, which is obviously much closer to zero than 57.8. This result is especially common for data in the social and behavioral sciences.

2.2   Deviations from the mean

When the mean of a time series is stationary (constant through time) we may treat the mean as a deterministic (meaning fixed, or nonstochastic) component of the series. To focus on the stochastic behavior of the series we express the data in *deviations from the mean*. That is, define a new time series $\tilde{z}_t$ as each z_t minus $\bar{z}$, where the sample mean $\bar{z}$ is an estimate of the parameter μ:

$$\tilde{z}_t = z_t - \bar{z} \tag{2.3}$$

The new series ($\tilde{z}_t$) will behave exactly as the old series (z_t) except that the mean of the $\tilde{z}_t$ series will equal precisely zero rather than $\bar{z}$. Since we know $\bar{z}$ we can always add it back into the $\tilde{z}_t$ series after we have finished our analysis to return to the overall level of the original series.

Consider the stationary simulated series in Figure 1.4. We have already found the mean of that series to be 100. Therefore the $\tilde{z}_t$ values for this series are calculated as follows:

$$\tilde{z}_1 = z_1 - \bar{z} = 102 - 100 = 2$$

$$\tilde{z}_2 = z_2 - \bar{z} = 99 - 100 = -1$$

$$\tilde{z}_3 = z_3 - \bar{z} = 101 - 100 = 1$$

$$\vdots \qquad \vdots \qquad \vdots$$

$$\tilde{z}_{60} = z_{60} - \bar{z} = 98 - 100 = -2$$

Figure 2.3 shows the new series $\tilde{z}_t$. It is indistinguishable from the series z_t in Figure 1.4 except that it has a mean of zero. In fact, the two series z_t and $\tilde{z}_t$ have all the same statistical properties except for their means. For example, the variances of the two series are identical (both are 7.97).

2.3   Two analytical tools: the estimated autocorrelation function (acf) and estimated partial autocorrelation function (pacf)

Several times we have referred to the estimated autocorrelation function (acf) and the estimated partial autocorrelation function (pacf). These tools are very important at the identification stage of the UBJ method. They measure the statistical relationship between observations in a single data series. In this section we discuss how estimated acf's and pacf's are constructed from a sample.

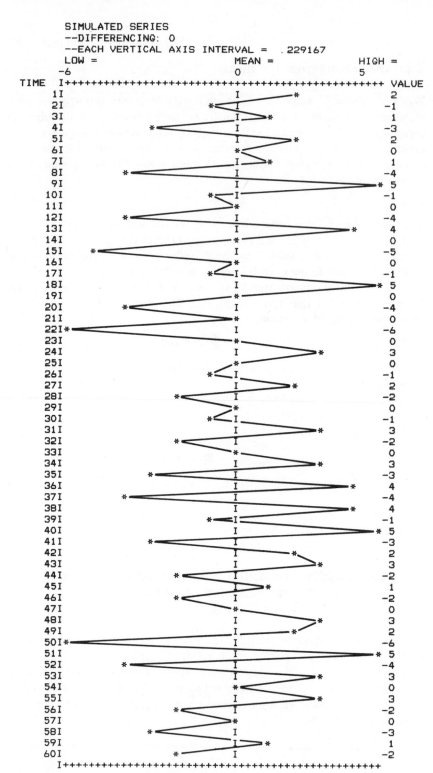

Figure 2.3 The data from Figure 1.4 expressed in deviations from the mean.

30

In this chapter we discuss the estimated acf and pacf primarily as tools for summarizing and describing the patterns within a given data series. But constructing an estimated acf and pacf is not merely an exercise in descriptive statistics. As emphasized in Chapter 3, we use the estimated acf and pacf for statistical inference. That is, we use them to infer the structure of the true, underlying mechanism that has given rise to the available data.

We use the data displayed in Figures 1.4 and 2.3 to illustrate the construction of an estimated acf and pacf. Remember that the data in Figure 2.3 are the data in Figure 1.4 expressed in deviations from the mean; these two series have identical statistical properties (except for their means), including the same acf and pacf.

Graphical analysis. The estimated acf and pacf of a data series are most useful when presented in their graphical forms as well as their numerical forms. To help motivate the ideas behind autocorrelation and partial autocorrelation analysis we first consider some simpler forms of graphical analysis.

One possible type of graphical analysis is merely to look at the observations in Figure 2.3 (or Figure 1.4) in the hope of seeing a pattern. But this is not a very promising approach. Some data series show very clear and obvious patterns to the eye, but many do not. Even if a series does display an obvious visual pattern, estimating its exact nature simply by looking at a plot of the data would be difficult and would give quite subjective results.

A more promising type of graphical analysis is to plot various $\tilde{z}_{t+k}$ values (for $k = 1, 2, \ldots$) against the previous observations $\tilde{z}_t$.* After all, in univariate analysis we are starting with the idea that the observations from different time periods may be related to each other. Perhaps we could see these relationships if we plot each observation $(\tilde{z}_{t+k})$ against the corresponding observation that occurs k periods earlier $(\tilde{z}_t)$.

It will be helpful if we first arrange the data in columns to create ordered pairs: each observation is paired with the corresponding observation k periods earlier. Then we may plot the ordered pairs on a two-space graph.

For example, letting $k = 1$ we can pair $\tilde{z}_{t+1}$ with $\tilde{z}_t$ by first writing all the $\tilde{z}_t$ values in a column. Then create another column, $\tilde{z}_{t+1}$, by shifting every element in column $\tilde{z}_t$ up one space. The results of doing this for a portion of the data in Figure 2.3 are shown in Table 2.1. The arrows indicate the shifting of the data.

---

*The reader should not be confused by the arbitrary use of a positive sign on the subscript k. We may refer to the relationship between $\tilde{z}_t$ and the value that occurs k periods earlier, $\tilde{z}_{t-k}$, or to the relationship between $\tilde{z}_{t+k}$ and the value that occurs k periods earlier, $\tilde{z}_t$. In both cases we are dealing with the relationship between two observations separated by k periods; only the notation is different.

For $t = 1$ we have paired $\tilde{z}_2 = -1$ (in column 3 of Table 2.1) with the observation one period earlier, $\tilde{z}_1 = 2$ (in column 2). For $t = 2$ we have paired $\tilde{z}_3 = 1$ (in column 3) with the observation one period earlier, $\tilde{z}_2 = -1$ (in column 2), and so forth. Note that we have 59 ordered pairs: there is no $\tilde{z}_{61}$ available to pair with $\tilde{z}_{60}$.

Next, we plot each $\tilde{z}_{t+1}$ value in column 3 against its paired $\tilde{z}_t$ value in column 2. This should allow us to see how the observations $\tilde{z}_{t+1}$ are related, on average, to the immediately previous observations $\tilde{z}_t$. The ordered pairs $(\tilde{z}_t, \tilde{z}_{t+1})$ are plotted in Figure 2.4.

There seems to be an inverse relationship between these ordered pairs, that is, as $\tilde{z}_t$ increases (moving to the right along the horizontal axis) there is a tendency for the next observation $(\tilde{z}_{t+1})$ to decrease (moving downward on the vertical axis).

Now suppose we want to see the relationship between observations separated by two time periods. Letting $k = 2$ we want to relate observations $\tilde{z}_{t+2}$ to observations two periods earlier, $\tilde{z}_t$. We do this by again writing down the original observations in a column labeled $\tilde{z}_t$. But now we create a new column $\tilde{z}_{t+2}$ by shifting all the observations in $\tilde{z}_t$ up two spaces. Using a portion of the data in Figure 2.3, the results are shown in Table 2.2. Again the arrows show the shifting procedure.

This time we have 58 ordered pairs: there is no $\tilde{z}_{62}$ to pair with $\tilde{z}_{60}$ and no $\tilde{z}_{61}$ to pair with $\tilde{z}_{59}$. In general, with a sample size of n we will have $n - k$ ordered pairs when we relate observations separated by k time periods. In this instance $n = 60$ and $k = 2$, so we have $60 - 2 = 58$ ordered pairs.

By plotting each $\tilde{z}_{t+2}$ observation from column 3 in Table 2.2 against its paired $\tilde{z}_t$ value in column 2, we can see how observations in this series are

Table 2.1   Ordered pairs $(\tilde{z}_t, \tilde{z}_{t+1})$ for the data in Figure 2.3

t	$\tilde{z}_t$	$\tilde{z}_{t+1}$
1	2	−1
2	−1	1
3	1	−3
4	−3	2
5	2	
⋮	⋮	⋮
59	1	−2
60	−2	n.a.[a]

[a] n.a. = not available.

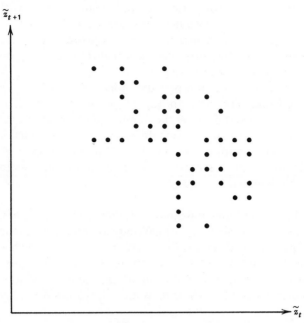

Figure 2.4 A plot of ordered pairs $(\tilde{z}_t, \tilde{z}_{t+1})$ using the data from columns 2 and 3 in Table 2.1.

Table 2.2   Ordered pairs $(\tilde{z}_t, \tilde{z}_{t+2})$ for the data in Figure 2.3

t	$\tilde{z}_t$	$\tilde{z}_{t+2}$
1	2	1
2	-1	-3
3	1	2
4	-3	0
5	2	.
$\vdots$	$\vdots$	$\vdots$
58	-3	-2
59	1	n.a.[a]
60	-2	n.a.

[a] n.a. = not available.

33

related to observations two periods earlier. Figure 2.5 is a plot of the ordered pairs $(\tilde{z}_t, \tilde{z}_{t+2})$. On average, there appears to be a positive relationship between them. That is, higher $\tilde{z}_t$ values (moving to the right on the horizontal axis) seem to be associated with higher values two periods later, $\tilde{z}_{t+2}$ (moving up on the vertical axis).

We could now let $k = 3$ and plot the ordered pairs $(\tilde{z}_t, \tilde{z}_{t+3})$. Then we could let $k = 4$ and plot the pairs $(\tilde{z}_t, \tilde{z}_{t+4})$, and so forth. The practical upper limit on k is determined by the number of observations in the series. Remember that as k increases by one the number of ordered pairs decreases by one. For example, with $n = 60$ and $k = 40$ we have only $n - k = 60 - 40 = 20$ pairs to plot. This number is too small to provide a useful guide to the relationship between $\tilde{z}_t$ and $\tilde{z}_{t+40}$.

Estimated autocorrelation functions.   Rather than plotting more ordered pairs, return to the two diagrams we have already produced, Figures 2.4 and 2.5. Visual analysis of diagrams like these might give a rough idea about how the observations in a time series are related to each other. However, there are two other graphical tools that *summarize many* relationships like those in Figures 2.4 and 2.5. These tools are called an estimated *autocorrela-*

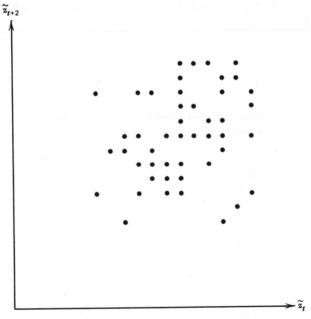

Figure 2.5   A plot of ordered pairs $(\tilde{z}_t, \tilde{z}_{t+2})$ using the data from columns 2 and 3 in Table 2.2.

tion function (acf) and an estimated *partial autocorrelation function* (pacf). In this section we examine the estimated acf.

The idea in autocorrelation analysis is to calculate a correlation coefficient for each set of ordered pairs $(\tilde{z}_t, \tilde{z}_{t+k})$. Because we are finding the correlation between sets of numbers that are part of the *same* series, the resulting statistic is called an *auto*correlation coefficient (*auto* means *self*). We use the symbol r_k for the estimated autocorrelation coefficient of observations separated by k time periods within a time series. (Keep in mind that the r_k's are statistics; they are calculated from a sample and they provide estimates of the true, underlying autocorrelation coefficients designated ρ_k.) After calculating estimated autocorrelation coefficients, we plot them graphically in an estimated autocorrelation function (acf), a diagram that looks something like a histogram. We will see an example of an estimated acf shortly.

An estimated autocorrelation coefficient (r_k) is not fundamentally different from any other sample correlation coefficient. It measures the direction and strength of the statistical relationship between ordered pairs of observations on two random variables. It is a dimensionless number that can take on values only between -1 and $+1$. A value of -1 means perfect negative correlation and a value of $+1$ means perfect positive correlation. If $r_k = 0$ then z_{t+k} and z_t are not correlated at all in the available data. (Of course, sampling error could cause an r_k value to be nonzero even though the corresponding parameter ρ_k is zero. We deal with this matter further in Chapter 3.) Figure 2.6 illustrates various degrees of autocorrelation that might arise in a given sample.

The standard formula for calculating autocorrelation coefficients is*

$$r_k = \frac{\sum_{t=1}^{n-k} (z_t - \bar{z})(z_{t+k} - \bar{z})}{\sum_{t=1}^{n} (z_t - \bar{z})^2} \qquad (2.4)$$

Equation (2.4) can be written more compactly since $\tilde{z}_t$ is defined as $z_t - \bar{z}$. Substitute accordingly and (2.4) becomes

$$r_k = \frac{\sum_{t=1}^{n-k} \tilde{z}_t \tilde{z}_{t+k}}{\sum_{t=1}^{n} (\tilde{z}_t)^2} \qquad (2.5)$$

*There are other ways of calculating r_k. Jenkins and Watts [7] discuss and evaluate some of the alternatives. Equation (2.4) seems to be most satisfactory and is commonly used.

Apply (2.5) to the ordered pairs in Table 2.1 to find r_1 as follows:

$$r_1 = \frac{\tilde{z}_1\tilde{z}_2 + \tilde{z}_2\tilde{z}_3 + \cdots + \tilde{z}_{59}\tilde{z}_{60}}{(\tilde{z}_1)^2 + (\tilde{z}_2)^2 + \cdots + (\tilde{z}_{60})^2}$$

$$= \frac{(2)(-1) + (-1)(1) + \cdots + (1)(-2)}{(2)^2 + (-1)^2 + \cdots + (-2)^2}$$

$$= -0.51$$

Likewise, r_2 for the same data set is calculated by applying (2.5) to the

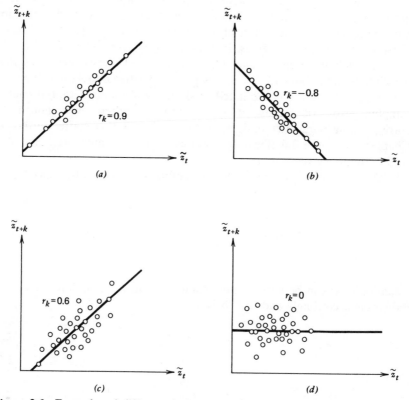

(a)

(b)

(c)

(d)

Figure 2.6 Examples of different degrees of autocorrelation that could arise in a sample: (*a*) strong positive autocorrelation; (*b*) strong negative autocorrelation; (*c*) moderately strong positive autocorrelation; (*d*) zero autocorrelation.

ordered pairs in Table 2.2 in this manner:

$$r_2 = \frac{\tilde{z}_1\tilde{z}_3 + \tilde{z}_2\tilde{z}_4 + \cdots + \tilde{z}_{58}\tilde{z}_{60}}{(\tilde{z}_1)^2 + (\tilde{z}_2)^2 + \cdots + (\tilde{z}_{60})^2}$$

$$= \frac{(2)(1) + (-1)(-3) + \cdots + (-3)(-2)}{(2)^2 + (-1)^2 + \cdots + (-2)^2}$$

$$= 0.22$$

Other r_k are calculated in a similar fashion. It is more convenient to use a computer to calculate autocorrelation coefficients than to find them by hand. Below we apply a computer program to the data in Figure 1.4. The program first finds the mean of the series ($\bar{z}$). It then finds the deviations ($\tilde{z}_t$) of each observation from the mean as shown in Figure 2.3. The r_k are then calculated by applying equation (2.5) to the $\tilde{z}_t$ values in Figure 2.3.

Box and Jenkins [1, p. 33] suggest that the maximum number of useful estimated autocorrelations is roughly $n/4$, where n is the number of observations. In our example $n = 60$, so $n/4 = 15$. Using a computer to calculate r_k (for $k = 1, 2, \ldots, 15$) for the data in Figure 1.4 gives the estimated acf shown in Figure 2.7.

The third column in Figure 2.7 (LAG) is k, the number of time periods separating the ordered pairs used to calculate each r_k. The first column

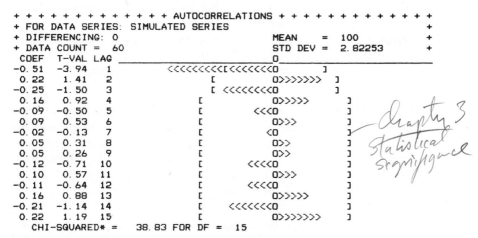

Figure 2.7 Estimated autocorrelation function (acf) calculated from the data in Figure 1.4.

(COEF) is the set of r_k, the estimated autocorrelation coefficients calculated by applying (2.5) to the data in Figure 1.4. Note that the first two autocorrelation coefficients ($-0.51, 0.22$) are identical to the ones we found by hand. The second column (T-VAL) measures the statistical significance of each r_k. (We discuss the topic of statistical significance in Chapter 3. For now note that a large absolute t-value indicates that the corresponding r_k is significantly different from zero, suggesting that the parameter ρ_k is non-zero.) The diagram next to the three columns of numbers in Figure 2.7 is a plot of the various r_k values. All positive r_k values are represented to the right of the zero line and all negative r_k values are shown to the left of the zero line. The length of each spike ($\lll$ or $\ggg$) is proportional to the value of the corresponding r_k. The square brackets [ ] show how large each r_k would have to be to have an absolute t-value of approximately 2.0. Any r_k whose spike extends past the square brackets has an absolute t-value larger than 2.0.

Looking at the pattern in an estimated acf is a key element at the identification stage of the UBJ method. The analyst must make a *judgment* about what ARIMA model(s) might fit the data by examining the patterns in the estimated acf. Thus, there is an element of subjectivity in the UBJ method. However, the acf offers great advantages over other types of graphical analysis. It summarizes in a single graph information about *many* different sets of ordered pairs ($\tilde{z}_t, \tilde{z}_{t+k}$), whereas a graph like Figure 2.4 gives information about only one set of ordered pairs. Furthermore, the acf provides mathematically objective measures of the relationship between each set of ordered pairs. In addition, we can perform statistical tests of significance using the acf and pacf that we cannot perform using visual analysis of plots of ordered pairs.

Before you read further be sure you understand (i) what estimated autocorrelations (r_k) measure, (ii) how the r_k are calculated, and (iii) how the r_k are represented graphically in an estimated acf diagram.

Estimated partial autocorrelation functions.* An estimated partial auto-correlation function (pacf) is broadly similar to an estimated acf. An estimated pacf is also a graphical representation of the statistical relation-ship between sets of ordered pairs ($\tilde{z}_t, \tilde{z}_{t+k}$) drawn from a single time series. The estimated pacf is used as a guide, along with the estimated acf, in choosing one or more ARIMA models that might fit the available data.

---

*In this section we assume the reader is familiar with the rudiments of multiple regression analysis. Those wanting to review the fundamentals may consult an introductory statistics text such as Mansfield [8, Chapters 11 and 12] or an intermediate text such as Wonnacott and Wonnacott [9, Chapter 3].

The idea of partial autocorrelation analysis is that we want to measure how $\tilde{z}_t$ and $\tilde{z}_{t+k}$ are related, but *with the effects of the intervening $\tilde{z}$'s accounted for*. For example, we want to show the relationship between the ordered pairs $(\tilde{z}_t, \tilde{z}_{t+2})$ taking into account the effect of $\tilde{z}_{t+1}$ on $\tilde{z}_{t+2}$. Next, we want the relationship between the pairs $(\tilde{z}_t, \tilde{z}_{t+3})$, but with the effects of both $\tilde{z}_{t+1}$ and $\tilde{z}_{t+2}$ on $\tilde{z}_{t+3}$ accounted for, and so forth, each time adjusting for the impact of any z's that fall between the ordered pairs in question. The estimated partial autocorrelation coefficient measuring this relationship between $\tilde{z}_t$ and $\tilde{z}_{t+k}$ is designated $\hat{\phi}_{kk}$. (Recall that $\hat{\phi}_{kk}$ is a statistic. It is calculated from sample information and provides an estimate of the true partial autocorrelation coefficient ϕ_{kk}.)

In constructing an estimated acf we examine ordered pairs of $\tilde{z}$'s, but we do not account for the effects of the intervening $\tilde{z}$'s. Thus in calculating estimated autocorrelation coefficients, we deal with only two sets of variables at a time, so autocorrelation analysis is easy to picture graphically. To do so we just look at a two-dimensional scatter diagram (e.g., Figure 2.4 or 2.5) and think of an autocorrelation coefficient as measuring how closely the matched pairs are related to each other.

By comparison, in partial autocorrelation analysis we must deal with more than two variables at once. That is, we have to contend not only with the ordered pairs $(\tilde{z}_t, \tilde{z}_{t+k})$, but also with all the $\tilde{z}$'s that fall between these matched pairs $(\tilde{z}_{t+1}, \tilde{z}_{t+2}, \ldots, \tilde{z}_{t+k-1})$. Thus visualizing partial autocorrelations on a two-dimensional graph is not possible. [The only exception is the first partial $\hat{\phi}_{11}$. Since there are no $\tilde{z}$'s falling between $\tilde{z}_t$ and $\tilde{z}_{t+1}$ we can use a scatter diagram of the ordered pairs $(\tilde{z}_t, \tilde{z}_{t+1})$ to picture the idea of the first partial autocorrelation coefficient. In fact, $\hat{\phi}_{11} = r_1$.]

The most accurate way of calculating partial autocorrelation coefficients is to estimate a series of least-squares regression coefficients. An estimated regression coefficient is interpreted as a measure of the relationship between the "dependent" variable and the "independent" variable in question, with *effects of other variables in the equation taken into account*. That is exactly the definition of a partial autocorrelation coefficient: $\hat{\phi}_{kk}$ measures the relationship between $\tilde{z}_t$ and $\tilde{z}_{t+k}$ while the effects of the $\tilde{z}$'s falling between these ordered pairs are accounted for.

Let us show how estimated partial autocorrelation coefficients are found by applying regression techniques to the data in Figure 2.3. First consider the true regression relationship between $\tilde{z}_{t+1}$ and the preceding value $\tilde{z}_t$:

$$\tilde{z}_{t+1} = \phi_{11}\tilde{z}_t + u_{t+1} \tag{2.6}$$

where ϕ_{11} is the partial autocorrelation coefficient to be estimated for $k = 1$.

In equation (2.6), $\tilde{z}_{t+1}$ and $\tilde{z}_t$ are all the possible ordered pairs of observations whose statistical relationship we want to measure. ϕ_{11} is the true partial autocorrelation coefficient to be estimated by the regression. u_{t+1} is the error term representing all the things affecting $\tilde{z}_{t+1}$ that do not appear elsewhere in the regression equation. Since there are no other $\tilde{z}$'s between $\tilde{z}_t$ and $\tilde{z}_{t+1}$, we can visualize an estimate of equation (2.6) as an estimated regression line running through the scatter of data in Figure 2.4. Using a least-squares regression computer program to estimate ϕ_{11}, we find $\hat{\phi}_{11} = -0.513$.

Now we want to find $\hat{\phi}_{22}$. This entails estimating the multiple regression

$$\tilde{z}_{t+2} = \phi_{21}\tilde{z}_{t+1} + \phi_{22}\tilde{z}_t + u_{t+2} \qquad (2.7)$$

where ϕ_{22} is the partial autocorrelation coefficient to be estimated for $k = 2$. Note that ϕ_{22} is estimated with $\tilde{z}_{t+1}$ included in the equation. Therefore, $\hat{\phi}_{22}$ estimates the relationship between $\tilde{z}_t$ and $\tilde{z}_{t+2}$ with $\tilde{z}_{t+1}$ accounted for. Estimating regression (2.7) with the data in Figure 2.3, we find $\hat{\phi}_{22} = -0.047$.

Next, we estimate the following regression to find $\hat{\phi}_{33}$:

$$\tilde{z}_{t+3} = \phi_{31}\tilde{z}_{t+2} + \phi_{32}\tilde{z}_{t+1} + \phi_{33}\tilde{z}_t + u_{t+3} \qquad (2.8)$$

where ϕ_{33} is the partial autocorrelation coefficient to be estimated for $k = 3$. By including $\tilde{z}_{t+1}$ and $\tilde{z}_{t+2}$ in this regression we are accounting for their effects on $\tilde{z}_{t+3}$ while estimating ϕ_{33}. Therefore, $\hat{\phi}_{33}$ is the estimated partial autocorrelation for $k = 3$. Using the data in Figure 2.3 gives this estimate of ϕ_{33}: $\hat{\phi}_{33} = -0.221$.

There is a slightly less accurate though computationally easier way to estimate the ϕ_{kk} coefficients. It involves using the previously calculated autocorrelation coefficients (r_k). As long as a data series is stationary the following set of recursive equations gives fairly good estimates of the partial autocorrelations:*

$$\hat{\phi}_{11} = r_1$$

$$\hat{\phi}_{kk} = \frac{r_k - \sum\limits_{j=1}^{k-1} \hat{\phi}_{k-1, j} r_{k-j}}{1 - \sum\limits_{j=1}^{k-1} \hat{\phi}_{k-1, j} r_j} \qquad (k = 2, 3, \ldots) \qquad (2.9)$$

where

$$\hat{\phi}_{kj} = \hat{\phi}_{k-1, j} - \hat{\phi}_{kk}\hat{\phi}_{k-1, k-j} \qquad (k = 3, 4, \ldots; j = 1, 2, \ldots, k - 1)$$

---

*This method of estimating partial autocorrelations is based on a set of equations known as the Yule–Walker equations. The method for solving the Yule–Walker equations for the $\hat{\phi}_{kk}$ values embodied in (2.9) is due to Durbin [10].

Illustrating how equations (2.9) can be used to calculate partial autocorrelations by hand is cumbersome and we will not do that here. But using a computer program to apply these equations to the estimated autocorrelations in Figure 2.7 (derived from the data in Figure 2.3) gives the estimated partial autocorrelation function (pacf) shown in Figure 2.8.

The column labeled LAG is the sequence of values for $k = 1, 2, 3, \ldots$ indicating which set of ordered pairs $(\tilde{z}_t, \tilde{z}_{t+k})$ we are examining. Column COEF is the set of estimated partial autocorrelation coefficients $(\hat{\phi}_{kk})$ for each set of ordered pairs $(\tilde{z}_t, \tilde{z}_{t+k})$ calculated using equation (2.9). The column labeled T-VAL shows the t-statistic associated with each $\hat{\phi}_{kk}$. We discuss these t-statistics in Chapter 3. For now, remember that any $\hat{\phi}_{kk}$ with an absolute t-value larger than 2.0 is considered to be significantly different from zero, suggesting that the parameter ϕ_{kk} is nonzero.

The graph toward the right-hand side of the pacf is a visual representation of the $\hat{\phi}_{kk}$ coefficients. Positive $\hat{\phi}_{kk}$ coefficients are represented to the right of the zero line and negative $\hat{\phi}_{kk}$ coefficients are shown to the left of the zero line. Each graphical spike ($\lll$ or $\ggg$) is proportional to the value of the corresponding $\hat{\phi}_{kk}$ coefficient. Any $\hat{\phi}_{kk}$ with a spike extending past the square brackets [ ] has an absolute t-value greater than 2.0. Note that the first three estimated partial autocorrelations in Figure 2.8 (-0.51, -0.05, -0.20) are very close to the estimates we obtained earlier for the same data using regression analysis (-0.513, -0.047, -0.221).

We make extensive use of estimated pacf's in applying the UBJ method. For now be sure you understand (i) how to interpret estimated partial autocorrelation coefficients, (ii) how estimated partial autocorrelation coefficients can be calculated, (iii) how partial autocorrelation coefficients differ

```
+ + + + + + + + + + PARTIAL AUTOCORRELATIONS + + + + + + + + + + +
  COEF   T-VAL  LAG _____0_____
 -0. 51  -3. 94   1         <<<<<<<<<<[<<<<<<<<0              ]
 -0. 05  -0. 36   2                   [       <<0             ]
 -0. 20  -1. 59   3                   [<<<<<<<<0              ]
 -0. 06  -0. 50   4                   [      <<0              ]
 -0. 03  -0. 21   5                   [       <0              ]
  0. 02   0. 17   6                   [        0>             ]
  0. 07   0. 51   7                   [        0>>            ]
  0. 09   0. 73   8                   [        0>>>           ]
  0. 18   1. 41   9                   [        0>>>>>>        ]
 -0. 03  -0. 19  10                   [       <0              ]
  0. 05   0. 37  11                   [        0>>            ]
 -0. 04  -0. 31  12                   [       <0              ]
  0. 05   0. 41  13                   [        0>>            ]
 -0. 14  -1. 07  14                   [   <<<<<0              ]
  0. 03   0. 24  15                   [        0>             ]
```

Figure 2.8 Estimated partial autocorrelation function (pacf) calculated from the data in Figure 1.4.

from autocorrelation coefficients, and (iv) how estimated partial autocorrelations are represented graphically in an estimated pacf.

Stationarity and estimated autocorrelation functions. In Chapter 1 we pointed out that a data series had to be rendered stationary (have a mean, variance, and acf that are essentially constant through time) before the UBJ method could be applied. It happens that the estimated acf is useful in deciding whether the mean of a series is stationary. If the mean is stationary the estimated acf *drops off rapidly to zero*. If the mean is not stationary the estimated acf drops off slowly toward zero.

Consider the estimated acf in Figure 2.7. It was calculated using the data series in Figure 1.4 which appears to have a stationary (constant) mean. This conclusion is reinforced by the appearance of the estimated acf in Figure 2.7 since it drops off to zero quite rapidly. That is, the estimated autocorrelations quickly become insignificantly different from zero. Only one acf spike (at lag 1) extends past the square brackets and only the first three spikes have absolute t-values greater than 1.2.

In contrast, consider the estimated acf in Figure 2.9. It was calculated using the AT&T stock price data in Figure 2.1. The mean of that series appears to be shifting through time. We therefore expect the estimated acf for this series to drop slowly toward zero. This is what we find. The first four autocorrelations in Figure 2.9 have absolute t-values greater than 2.0, and the first six have absolute t-values exceeding 1.6. This is fairly typical for a data series with a nonstationary mean. If estimated autocorrelations

```
+ + + + + + + + + + + + AUTOCORRELATIONS + + + + + + + + + + + + +
+ FOR DATA SERIES: AT&T STOCK PRICE                                 +
+ DIFFERENCING: 0                         MEAN    =  57. 7957       +
+ DATA COUNT =  52                        STD DEV =  3. 4136        +
  COEF   T-VAL LAG _____0_____
  0. 93   6. 74   1  |                           [   0>>>>>]>>>>>>>>>>>>>>>>>>>
  0. 86   3. 75   2  |                      [         0>>>>>>>>>>>>>]>>>>>>>>>>
  0. 81   2. 85   3  |                   [            0>>>>>>>>>>>>]>>>>>>>
  0. 75   2. 29   4  |                [               0>>>>>>>>>>>>>]>>>
  0. 68   1. 91   5  |            [                   0>>>>>>>>>>>>>>>]
  0. 62   1. 62   6  [                                0>>>>>>>>>>>>>>>    ]
  0. 55   1. 38   7  [                                0>>>>>>>>>>>>    ]
  0. 49   1. 19   8  [                                0>>>>>>>>>>>    ]
  0. 44   1. 03   9  [                                0>>>>>>>>>>        ]
  0. 38   0. 87  10  [                                0>>>>>>>>>         ]
  0. 29   0. 65  11  [                                0>>>>>>>          ]
  0. 22   0. 49  12  [                                0>>>>>           ]
  0. 18   0. 39  13  [                                0>>>>           ]
      CHI-SQUARED* =   280. 32 FOR DF =   13
```

Figure 2.9 Estimated acf calculated from the AT&T stock price data in Figure 2.1.

have absolute t-values greater than roughly 1.6 for the first five to seven lags, this is a warning that the series may have a nonstationary mean and may need to be differenced. The estimated autocorrelations need not start from a high level to indicate nonstationarity. See Part II, Case 8 for an estimated acf that starts from relatively small autocorrelations but whose slow decay indicates the data are nonstationary.

Summary

1. A data series with a nonstationary mean can often be transformed into a stationary series through a differencing operation.

2. To difference a series once, calculate the period-to-period changes: $w_t = z_t - z_{t-1}$. To difference a series twice, calculate the changes in the first differences: $w_t = (z_t - z_{t-1}) - (z_{t-1} - z_{t-2})$.

3. In practice first differencing is required fairly often; second differencing is called for only occasionally; third differencing (or more) is virtually never needed.

4. To focus on the stochastic (nondeterministic) components in a stationary time series, we subtract out the sample mean $\bar{z}$, which is an estimate of the parameter μ. We then analyze these data expressed in deviations from the mean: $\tilde{z}_t = z_t - \bar{z}$.

5. A series expressed in deviations from the mean has the same statistical properties as the original series (e.g., it has the same variance and estimated acf) except the mean of the differenced series is identically zero.

6. An estimated autocorrelation function (acf) shows the correlation between ordered pairs $(\tilde{z}_t, \tilde{z}_{t+k})$ separated by various time spans ($k = 1, 2, 3, \ldots$), where the ordered pairs are drawn from a single time series. Each estimated autocorrelation coefficient r_k is an estimate of the corresponding parameter ρ_k.

7. An estimated partial autocorrelation function (pacf) shows the correlation between ordered pairs $(\tilde{z}_t, \tilde{z}_{t+k})$ separated by various time spans ($k = 1, 2, 3, \ldots$) with the effects of intervening observations $(\tilde{z}_{t+1}, \tilde{z}_{t+2}, \ldots, \tilde{z}_{t+k-1})$ accounted for. The ordered pairs are drawn from a single time series. Each estimated partial autocorrelation coefficient $\hat{\phi}_{kk}$ is an estimate of the corresponding parameter ϕ_{kk}.

8. The estimated acf for a series whose mean is stationary drops off rapidly to zero. If the mean is nonstationary the estimated acf falls slowly toward zero.

Questions and Problems

2.1   In Section 2.2 we assert that expressing a data series in deviations from the mean shifts the series so its mean is identical to zero. Prove this assertion. That is, prove $\Sigma(z_t - \bar{z}) = 0$. *Hint*: Use the following two rules about summation: $\Sigma(x + y) = \Sigma x + \Sigma y$ and $\Sigma K = nK$ when K is a constant.

2.2   Does a series expressed in deviations from the mean always have a mean of zero? Does a differenced series always have a mean of zero? Discuss.

2.3   Consider the time series in Problem 1.8.
(a)   Express those data in deviations from the mean.
(b)   Calculate r_1, r_2, and r_3 for this series. Plot these values on an acf diagram.
(c)   Calculate $\hat{\phi}_{11}$, $\hat{\phi}_{22}$, and $\hat{\phi}_{33}$ for this series. Plot these values on a pacf diagram.

2.4   How can you tell if the mean of a time series is stationary?

2.5   Calculate the first differences of the following time series. Does the estimated acf of the original series confirm that differencing is required? If so, is first differencing sufficient to induce a stationary mean?

t	z_t	t	z_t	t	z_t	t	z_t	t	z_t
1	23	13	29	25	39	37	48	49	41
2	21	14	31	26	38	38	50	50	39
3	23	15	30	27	40	39	49	51	39
4	25	16	35	28	40	40	52	52	35
5	22	17	36	29	39	41	46	53	38
6	27	18	34	30	42	42	48	54	35
7	26	19	32	31	40	43	50	55	37
8	29	20	36	32	45	44	47	56	32
9	28	21	35	33	46	45	45	57	33
10	27	22	35	34	47	46	46	58	34
11	30	23	38	35	45	47	42	59	32
12	31	24	40	36	44	48	40	60	33

2.6   How many useful estimated autocorrelation coefficients can one obtain from a given sample?

3

UNDERLYING STATISTICAL
PRINCIPLES

In Chapter 1 we discussed the nature of time-series data and introduced the three-stage modeling framework proposed by Box and Jenkins (identification, estimation, diagnostic checking). Then in Chapter 2 we constructed two useful graphs from a set of time-series observations—an estimated autocorrelation function (acf) and an estimated partial autocorrelation function (pacf). An estimated acf and pacf show how the observations in a single time series are correlated.

In Chapter 4 we show how ARIMA models are constructed by applying the three-stage UBJ procedure to two data sets. But first we must establish some terminology and introduce some principles that underlie the UBJ method. These principles are similar to those in an introductory statistics course, although the terminology may be different.

3.1 Process, realization, and model

An important question is: From where do observations (like those shown in Figures 1.1 and 1.4) come? A quick answer is that the shoe production data in Figure 1.1 came from a U.S. Commerce Department publication called *Business Statistics*, and the simulated series in Figure 1.4 came from a computer. Another similar answer is to say that the Commerce Department data came from a survey of shoe manufacturers.

45

But the question about "where the observations come from" is meant to get at a different, more abstract matter. A better way to ask it is: What kind of underlying mechanism produced these observations?

In UBJ–ARIMA analysis observations are assumed to have been produced by an ARIMA *process*. The corresponding concept in classical statistics is the *population*. The population is the set of all possible observations on a variable; correspondingly, an ARIMA process consists of all possible observations on a time-sequenced variable.

Now we must add something to the preceding definition of a process to clarify it. An ARIMA process consists not only of all possible observations on a time-sequenced variable, but it also includes an algebraic statement, sometimes called a *generating mechanism*, specifying how these possible observations are related to each other. We examine two such algebraic statements later in this chapter.

In classical statistics we distinguish between the population (all possible observations) and a *sample* (a set of actual observations). A sample is a particular subset of the population. In UBJ–ARIMA analysis we usually refer to a sample as a *realization*. A realization is one subset of observations coming from the underlying process. For example, the shoe production data

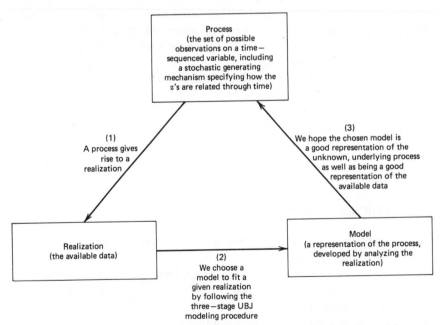

Figure 3.1   The relationship among a process, realization, and model.

in Figure 1.1 are a realization; these data are only 60 observations out of many possible observations.

If we could discover the underlying process that has generated a realization, then maybe we could forecast future values of each series with some accuracy, assuming the same mechanism continues to produce future observations. Unfortunately, in practice we never know the underlying process.

Our goal in UBJ analysis is to find a good representation of the process generating mechanism that has produced a given realization. This representation is called a *model*. An ARIMA model is an algebraic statement chosen in light of the available realization. Our hope is that a model which fits the available data (the realization) will also be a good representation of the unknown underlying generating mechanism. The three-stage UBJ procedure is designed to guide us to an appropriate model. We return to this important topic in Chapter 4, where we introduce the characteristics of a good model and present two examples of UBJ–ARIMA modeling. Figure 3.1 summarizes the relationship among a process, a realization, and a model.

3.2 Two common processes

An ARIMA process refers to the set of possible observations on a time-sequenced variable, along with an algebraic statement (a generating mechanism) describing how these observations are related. In this section we introduce two common ARIMA processes. We examine their algebraic form and discuss their stochastic (probabilistic) nature.

In Chapter 2 we learned how to construct an *estimated* acf and pacf from a realization. Every ARIMA process has an associated *theoretical* acf and pacf. We will examine the theoretical acf's and pacf's of two common processes in this section. Then in Chapter 4 we see how to identify an ARIMA model by comparing the estimated acf and pacf calculated from a realization with various theoretical acf's and pacf's.

Their algebraic form. The generating mechanisms for two common ARIMA processes are written as follows:

$$z_t = C + \phi_1 z_{t-1} + a_t \tag{3.1}$$

$$z_t = C - \theta_1 a_{t-1} + a_t \tag{3.2}$$

Consider process (3.1). Processes with past (time-lagged) z terms are called *autoregressive* (abbreviated AR) processes. The longest time lag associated with a z term on the right-hand-side (RHS) is called the AR order of the process. Equation (3.1) is an AR process of order 1, abbreviated AR(1), because the longest time lag attached to a past z value is one period. That is, the subscript of the RHS z is $t - 1$. On the left-hand-side (LHS), z_t represents the set of possible observations on a time-sequenced random variable z_t.*

Process (3.1) tells us how observed values of z_t are likely to behave through time. It states that z_t is related to the immediately past value of the same variable (z_{t-1}). The coefficient ϕ_1 has a fixed numerical value (not specified here) which tells how z_t is related to z_{t-1}.† C is a constant term related to the mean of the process. The variable a_t stands for a random-shock element. Although z_t is related to z_{t-1}, the relationship is not exact: it is probabilistic rather than deterministic. The random shock represents this probabilistic factor. We discuss the random-shock term a_t in more detail later.

Now consider process (3.2). Processes with past (time-lagged) random shocks are called *moving-average* (abbreviated MA) processes.‡ The longest time lag is called the *MA order* of the process. Equation (3.2) is an MA process of order 1, abbreviated MA(1), since the longest time lag attached to a past random shock is $t - 1$. Once again z_t on the LHS is the set of possible observations on the time-sequenced random variable z_t, C is a constant related to the mean of the process, and a_t is the random-shock term.

The negative sign attached to θ_1 is merely a convention. It makes no difference whether we use a negative or a positive sign, as long as we are consistent. We follow the convention used by Box and Jenkins by prefixing all θ coefficients with negative signs.

---

*It is common to denote a random variable with an upper-case letter (Z_t) and a particular value of that random variable with a lower-case letter (z_t). However, the common practice in Box–Jenkins literature is to use lower-case letters for both random variables and specific observations, letting the context determine the interpretation. We follow this practice in the text; the symbol z_t refers to a random variable when we speak of a process, and the same symbol refers to a specific observation when we speak of a realization.

†The ϕ and θ coefficients in ARIMA processes are assumed to be fixed parameters. It is possible to postulate variable-parameter models, with coefficients changing through time in some specified manner. However, such models are beyond the scope of this book. Standard UBJ–ARIMA models are fixed-parameter models, and we restrict our inquiry to these standard types.

‡The label "moving average" is technically incorrect since the MA coefficients may be negative and may not sum to unity. This label is used by convention.

One aspect of process (3.2) is sometimes confusing for students at first glance. We have emphasized that ARIMA models are univariate; they deal with the relationship between observations within a single data series. Process (3.1) is consistent with this since the set of possible observations at time t (z_t) is related to the set of possible observations on the same variable at time $t - 1$ (z_{t-1}).

However, in process (3.2) z_t is related to a past random shock. How can we think of (3.2) as a univariate process if it does not describe how z_t is related to other past z elements in the same series? The answer is that any MA process, including equation (3.2), is a univariate process because past random-shock terms can be replaced by past z terms through algebraic manipulation. In Chapter 5 we show how this is done. Alternatively, consider that a_t is simply part of z_t. Thus, an MA term represents a relationship between z_t and a component of a past z term, where the component is the appropriately lagged random shock.

Their stochastic nature: the random shock a_t.   Because of the random shock a_t, an ARIMA process generates realizations in a *stochastic* (meaning chance or probabilistic) manner. The a_t terms in an ARIMA process are usually assumed to be Normally, identically, and independently distributed random variables with a mean of zero and a constant variance. Such variables are often called "white noise." Figure 3.2 illustrates this idea with a Normal distribution centered on zero. The horizontal axis shows the values of a_t that could occur. The area under the curve between any two a_t values (such as the shaded area between a_{t1} and a_{t2}) equals the probability

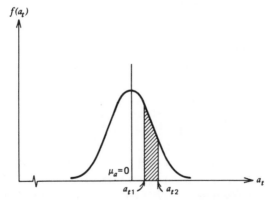

Figure 3.2   A normal distribution for the random shock a_t.

of a_t falling within that range. Since the a_t are identically distributed for all t, this distribution characterizes a_t for all t. Since the a_t are independently distributed they are not autocorrelated, that is, knowing the set of past random shocks $(a_{t-1}, a_{t-2}, a_{t-3}, \ldots)$ would not help us predict the current shock a_t.

To see the importance of the chance element a_t, consider two extreme situations. First, suppose the random shocks were absent from equations (3.1) and (3.2). Then they would be deterministic rather than stochastic relationships. In (3.1) z_t would be exactly known from C, ϕ_1, and z_{t-1}. In (3.2) z_t would be exactly known from C. We will not consider such deterministic relationships but deal only with stochastic processes whose random shocks meet the above assumptions.

Alternatively, suppose we would have $C = \phi_1 = \theta_1 = 0$ in equations (3.1) and (3.2). Then, in both cases, $z_t = a_t$ so that z_t would have no identifiable univariate time structure. Realizations generated by this process would be white noise, a sequence of uncorrelated values.

Consider again the shoe production data in Figure 1.1. In UBJ–ARIMA analysis we assume these data were generated by an unknown stochastic process. Thus, we think of any specific observed value (e.g., $z_1 = 659$ for January 1971) as composed of three parts: a deterministic part represented by the constant term C, another part reflecting past observations represented by AR and MA terms, and a pure chance component represented by a_t. Since the value of each observation z_t is determined at least partly by chance, we think of this particular realization as *only one which might have occurred*. Mere chance could have produced many realizations other than the one actually observed.

In practice we cannot return to January 1971 to observe how athletic shoe production might have been different due to chance. But we can imagine conducting such an experiment and recording the results. As we conduct this experiment suppose the random shock a_t for each month is drawn from a probability distribution like the one in Figure 3.2.

We can imagine conducting this experiment for the year 1971 over and over again, hundreds or thousands of times, each time drawing a different value of a_t for each month. Since a_t is a component of z_t, z_t for each month would also have a different value for each experiment and thus each experiment would generate a different realization.

This idea is illustrated in Figure 3.3. We have reproduced the first 12 months of the realization shown in Figure 1.1 along with two other (imaginary) realizations that might have occurred instead because of the stochastic nature of the underlying process. The heavy line with the asterisks is the original realization; the other lines are the two imaginary realizations.

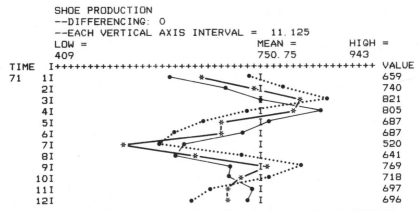

```
       SHOE PRODUCTION
       --DIFFERENCING:  0
       --EACH VERTICAL AXIS INTERVAL =   11.125
       LOW =                      MEAN =        HIGH =
       409                        750.75        943
TIME   I++++++++++++++++++++++++++++++++++++++++++++++++++ VALUE
71    1I                                                   659
      2I                                                   740
      3I                                                   821
      4I                                                   805
      5I                                                   687
      6I                                                   687
      7I                                                   520
      8I                                                   641
      9I                                                   769
     10I                                                   718
     11I                                                   697
     12I                                                   696
```

Figure 3.3 Actual realization for athletic shoe production, January 1971–December 1971, and two imaginary realizations.

Their stochastic nature: joint probability functions. Another way of discussing the stochastic nature of an ARIMA process such as (3.1) or (3.2) is to describe it in terms of a *stationary, Normal, joint probability distribution function.*

Consider the realization $z_1, \ldots, z_n$. Let us suppose these observations are drawn from a joint probability distribution

$$P(z_1, \ldots, z_n) \qquad (3.3)$$

where $P(\ )$ is a joint probability density function that assigns a probability to each possible combination of values for the random variables $z_1, \ldots, z_n$.

Our goal in forecasting is to make statements about the likely values of future z's. Now, if we know the joint density function $P(z_1, \ldots, z_{n+1})$, including the relevant marginal probabilities, we could form the conditional distribution

$$P(z_{n+1} | z_1, \ldots, z_n) \qquad (3.4)$$

Then from knowledge of the past values $(z_1, \ldots, z_n)$ we could use (3.4) to make a probability statement about the future value z_{n+1}.

Recall that UBJ–ARIMA analysis is restricted to stationary processes and realizations. (Keep in mind that many nonstationary realizations can be rendered stationary with suitable transformations.) For a process to be stationary, the joint distribution function describing that process must be *invariant with respect to time*. That is, if we displace each random variable $(z_t, \ldots, z_{t+k})$ by m time periods, we have the stationarity condition

$$P(z_{t+m}, \ldots, z_{t+k+m}) = P(z_t, \ldots, z_{t+k}) \qquad (3.5)$$

Condition (3.5) is sometimes referred to as *strong* or *strict* stationarity. It shows the entire probability structure of our joint function constant through time. *Weak* stationarity requires only that certain characteristics of our joint function be time-invariant. But now we encounter a pleasing simplification: if our joint function is a joint *Normal* distribution, then it is strongly stationary if its mean (first moment) and variance and covariances (second moment) are constant over time. In fact our assumption that the random shocks a_t are Normally distributed is equivalent to the assumption that the joint distribution for the z's is a joint Normal distribution.

If we have a stationary joint Normal distribution for the z's, then we have a constant mean, $\mu = E(z_t)$ for all z's,*

$$\mu = E(z_t) = E(z_{t+m}) \qquad (3.6)$$

a constant variance, $\sigma_z^2 = \gamma_0 = E(z_t - \mu)^2$, for all z's,

$$\sigma_z^2 = \gamma_0 = E(z_t - \mu)^2 = E(z_{t+m} - \mu)^2 \qquad (3.7)$$

and constant covariances, $\gamma_k = E[(z_t - \mu)(z_{t+k} - \mu)]$, for any two z's separated by k time periods,

$$\gamma_k = E\left[(z_t - \mu)(z_{t+k} - \mu)\right] = E\left[(z_{t+m} - \mu)(z_{t+k+m} - \mu)\right] \quad (3.8)$$

(Since we are talking about the covariances between random variables occurring within the *same* time series, these covariances are called *autocovariances*.)

We can conveniently summarize variances and covariances in matrix form. A matrix is simply an array of numbers. In this case we want to present the variances and covariances of the random variables $(z_1, \ldots, z_n)$ in an organized array. The variance–covariance matrix for a stationary joint

---

*E is the expected value operator.

distribution function for $(z_1, \ldots, z_n)$ is a square array:

$$
\begin{bmatrix}
\gamma_0 & \gamma_1 & \gamma_2 & \cdot & \cdots & \gamma_{n-1} \\
\gamma_1 & \gamma_0 & \gamma_1 & \gamma_2 & \cdots & \gamma_{n-2} \\
\gamma_2 & \textcircled{γ_1} & \gamma_0 & \gamma_1 & \cdots & \gamma_{n-3} \\
\cdot & \gamma_2 & \gamma_1 & \gamma_0 & \cdots & \cdot \\
\vdots & \vdots & \vdots & \vdots & \cdots & \vdots \\
\gamma_{n-1} & \gamma_{n-2} & \gamma_{n-3} & \cdot & \cdots & \gamma_0
\end{bmatrix}
\tag{3.9}
$$

Row 1 (the top row) and column 1 (the left-most column) refer to random variable z_1, row 2 (second from the top) and column 2 (second from the left) refer to random variable z_2, and so forth. The covariance between any two z variables is the γ value corresponding to the appropriate row and column. For example, the covariance between z_3 and z_2 is circled; it is found where row 3 (for z_3) intersects column 2 (for z_2). The subscript k attached to γ refers to the number of time periods separating the two variables whose covariance is being considered. Since z_3 and z_2 are separated by one time period, their γ_k has the subscript $k = 1$.

Note that the covariance between z_t and itself is the variance of z_t. When $k = 0$, $(z_t - \mu)$ times $(z_{t+k} - \mu)$ is simply $(z_t - \mu)^2$, in which case $\gamma_k = \gamma_0 = \sigma_z^2$.

Matrix (3.9) is a useful vehicle for discussing the idea of stationarity. For example, since the variance σ_z^2 (equal to γ_0) does not vary with time for a stationary process, we find the same element along the entire main diagonal. That is, the variance of z_1 is γ_0, the variance of z_2 is also γ_0, the variance of z_3 is also γ_0, and so on.

Similarly, stationary covariances depend only on the number of time periods separating the variables in question, not on the particular time subscripts attached to them. For example, the covariance between z_1 and z_2 is γ_1; they are separated by one time period. The covariance between z_2 and z_3 is also γ_1 because they are also separated by one time period. The same is true for z_3 and z_4, z_4 and z_5, and so forth. Therefore, the diagonals immediately above and below the main diagonal contain the constant γ_1. The stationarity assumption likewise explains why every other diagonal is made up of a constant.

Autocovariances are awkward to use because their sizes depend on the units in which the variables are measured. It is convenient to standardize autocovariances so their values fall in the range between -1 and $+1$ regardless of the units in which the variables are measured. This is accomplished by dividing each autocovariance (γ_k) by the variance of the process ($\gamma_0 = \sigma_z^2$). Such standardized autocovariances are autocorrelation coefficients and, for a process, are denoted by the symbol ρ. By definition,

$$\rho_k = \frac{\gamma_k}{\gamma_0} \tag{3.10}$$

As with autocovariances, autocorrelations can be conveniently represented in matrix form. Start with matrix (3.9) and divide each element by γ_0. All elements on the main diagonal become one, indicating that each z_t is perfectly correlated with itself. All other γ_k values become ρ_k values as indicated by equation (3.10):

$$
\begin{bmatrix}
1 & \rho_1 & \rho_2 & \cdot & \cdots & \rho_{n-1} \\
\rho_1 & 1 & \rho_1 & \rho_2 & \cdots & \rho_{n-2} \\
\rho_2 & \rho_1 & 1 & \rho_1 & \cdots & \rho_{n-3} \\
\cdot & \rho_2 & \rho_1 & 1 & \cdots & \cdot \\
\vdots & \vdots & \vdots & \vdots & \cdots & \vdots \\
\rho_{n-1} & \rho_{n-2} & \rho_{n-3} & \cdot & \cdots & 1
\end{bmatrix}
\tag{3.11}
$$

Once again, stationarity dictates that each diagonal be composed of a constant.*

Although we may discuss the stochastic nature of ARIMA processes in terms of joint probability distribution functions like (3.3) and (3.4), in practice we do not specify such distribution functions in detail. Instead we summarize the behavior of a process with a stochastic generating mechanism, like the AR(1) or MA(1) in equation (3.1) or (3.2). We may then use these generating mechanisms to derive the mean, variance, and autocovariances (and corresponding autocorrelation coefficients) and the conditional distribution of future z's for that process. Such derivations are presented in Chapters 6 and 10.

Theoretical acf's and pacf's. Each time-dependent process has a *theoretical* acf and pacf associated with it. These are derived by applying certain

---

*It can be shown that for stationary processes, the autocorrelation matrix (3.11) and the autocovariance matrix (3.9) are positive definite. It follows that the determinant of the autocorrelation matrix and all principal minors are positive, so the autocorrelation coefficients for a stationary process must satisfy numerous conditions. For linear processes these stationarity conditions can be stated conveniently in the form of restrictions on the AR coefficients. The restrictions are discussed in Chapter 6.

definitions and rules to the process in question. In Chapter 6 we derive the theoretical acf's for processes (3.1) and (3.2). For now we simply present the theoretical acf's and pacf's associated with these two processes.

Remember that theoretical acf's and pacf's are different from *estimated* ones. Estimated acf's and pacf's (like the ones we constructed in Chapter 2) are found by applying equation (2.5) to the n observations in a realization. On the other hand, theoretical acf's and pacf's are found by applying definitions and rules about mathematical expectation to specific processes.

As we shall see in Chapter 4, a critical part of the identification stage in UBJ–ARIMA modeling involves the comparison of estimated acf's and pacf's with theoretical acf's and pacf's. Thus the UBJ analyst must become thoroughly familiar with the most common theoretical acf's and pacf's.

Following are the most important general characteristics of theoretical AR and MA acf's and pacf's, summarized in Table 3.1. (We discuss *mixed* ARMA processes, which contain both AR and MA terms, starting in Chapter 6.)

1. Stationary autoregressive (AR) processes have theoretical *acf's that decay or "damp out" toward zero*. But they have theoretical *pacf's that cut off sharply to zero* after a few spikes. The lag length of the last pacf spike equals the AR order (p) of the process.

2. Moving-average (MA) processes have theoretical *acf's that cut off to zero* after a certain number of spikes. The lag length of the last acf spike equals the MA order (q) of the process. Their theoretical *pacf's decay or "die out" toward zero*.

Figure 3.4 shows the kinds of theoretical acf's and pacf's associated with an AR(1) process like (3.1). Note that the acf decays toward zero whether ϕ_1

Table 3.1 General characteristics of theoretical acf's and pacf's for AR and MA processes

Process	acf	pacf
AR	Decays toward zero	Cuts off to zero (lag length of last spike equals AR order of process)
MA	Cuts off to zero (lag length of last spike equals MA order of process)	Decays toward zero

is positive (Example I) or negative (Example II). When ϕ_1 is negative the autocorrelations alternate in sign, starting on the negative side. But, as with all stationary AR processes, the absolute values of the autocorrelations in Example II die out toward zero rather than cut off to zero.

In Examples I and II the theoretical pacf has a spike at lag 1 followed by a cutoff to zero. This cutoff in the pacf is typical for AR processes. There is only one spike in the pacf's in Figure 3.4 because they are associated with AR(1) processes. That is, the lag length of the last spike in the theoretical pacf of an AR process is equal to the AR order (the maximum lag length of the z terms) of the process.

Figure 3.5 shows examples of theoretical acf's and pacf's associated with MA(1) processes. The lag length of the last spike in the theoretical acf of an MA process equals the order of the MA process. Thus, an MA(1) theoretical acf has a spike at lag 1 followed by a cutoff to zero. This is an example of the general rule for MA theoretical acf's: they always cut off to zero rather than decay toward zero. For the MA(1) the sign and size of θ_1 determine the

Example I: ϕ_1 is positive

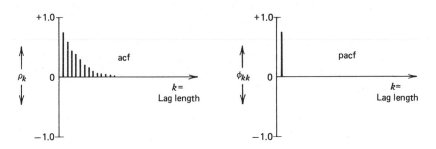

Example II: ϕ_1 is negative

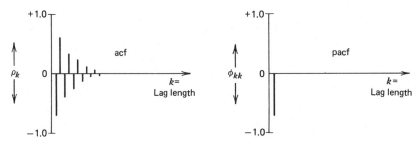

Figure 3.4   Theoretical acf's and pacf's for stationary AR(1) processes.

sign and size of the acf spike. The MA(1) acf spike is positive if θ_1 is negative, but negative if θ_1 is positive.

All theoretical pacf's for MA processes decay toward zero unlike the pacf's for AR processes, which cut off to zero. The pacf for an MA process may or may not alternate in sign. As shown in Figure 3.5, when θ_1 is negative the theoretical pacf for an MA(1) starts on the positive side and alternates in sign. But when θ_1 is positive the pacf decays entirely on the negative side.

In Chapter 5 we consider three additional common processes and their associated acf's and pacf's. Familiarity with the common theoretical acf's and pacf's is essential for the analyst who wants to use the UBJ method effectively. The various theoretical acf's and pacf's may differ substantially from each other in detail, and this may seem confusing initially. But keep in mind these two points: (i) Thorough knowledge of just a few common theoretical acf's and pacf's is sufficient for building proper ARIMA models

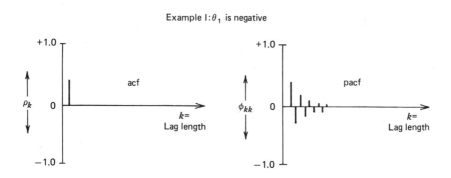

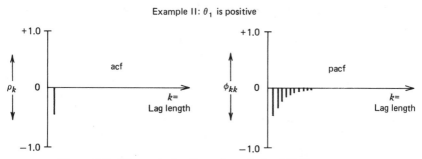

Figure 3.5 Theoretical acf's and pacf's for MA(1) processes.

for the vast majority of data sets. (ii) Unusual ARIMA processes share certain general characteristics with the more common ones. Thus, knowledge of the common theoretical acf's and pacf's gives good guidance even when the appropriate model is not common.

Estimated acf's and pacf's.   At the identification stage of UBJ–ARIMA analysis we first calculate an estimated acf and pacf by applying (2.5) and (2.9) to the available data. Then we compare the estimated acf and pacf with some common theoretical acf's and pacf's to find a reasonably good "match." We then select as a tentative model for our data the process associated with the matching theoretical acf and pacf.

Suppose, for example, that an estimated acf (calculated from a given realization) has a single spike at lag 1. We know that a theoretical acf with a spike at lag 1 characterizes an MA(1) process, represented algebraically by equation (3.2). Therefore, we would tentatively choose equation (3.2) as a model to represent the realization in question. We then go to the estimation and diagnostic-checking stages to estimate the parameters and to test the adequacy of the chosen model. Clearly, an important step in this procedure is the tentative matching of estimated acf's and pacf's with theoretical acf's and pacf's. In this section we use simulation methods to develop a feel for how closely estimated acf's and pacf's might, or might not, match theoretical acf's and pacf's.

We specify two processes—an AR(1) and an MA(1)—assuming we know their parameters. (In practice we never know what process has generated a given realization. But as a learning exercise we can pretend that we know these processes exactly.) Then we use a computer to produce a series of Normally and independently distributed random shocks with zero mean and a constant variance. Using the known processes and the random shocks produced by the computer, we generate five simulated realizations for each process. Finally, we compute estimated acf's and pacf's from these simulated realizations to see how closely they match the known theoretical acf's and pacf's associated with the known AR(1) and MA(1) processes by which the realizations were generated.

As a numerical illustration of how these simulated realizations are generated, consider the following AR(1) process:

$$z_t = 0.5z_{t-1} + a_t \qquad (3.12)$$

In this example, $\phi_1 = 0.5$ and $C = 0$. Let the starting value for z_{t-1} be 0. Now suppose we draw at random a sequence of a_t values, for $t = 1, 2, 3,$ and 4, from a collection of Normally and independently distributed numbers having a mean of zero and a constant variance [designated $a_t \sim$

NID(0, σ_a^2)]. Let these a_t values be $(3, -2, -1, 2)$. Then we can calculate the values of z_t for $t = 1, 2, 3,$ and 4 recursively as follows:

$$z_1 = 0.5z_0 + a_1$$

$$= 0.5(0) + 3$$

$$= 3$$

$$z_2 = 0.5z_1 + a_2$$

$$= 0.5(3) - 2$$

$$= -0.5 \qquad (3.13)$$

$$z_3 = 0.5z_2 + a_3$$

$$= 0.5(-0.5) - 1$$

$$= -1.25$$

$$z_4 = 0.5z_3 + a_4$$

$$= 0.5(-1.25) + 2$$

$$= 1.375$$

Keep in mind that we generated this realization $(3, -0.5, -1.25, 1.375)$ artificially. In practice the random shocks are not observable, and C and ϕ_1 are unknown. Here we are merely trying to illustrate how process (3.12) could generate one particular series of observations (3.13). All the simulated realizations considered in this section were generated in a similar fashion, though using a computer for convenience.

The following AR(1) process was used to simulate five different realizations:

$$z_t = 0.7z_{t-1} + a_t \qquad (3.14)$$

In all cases $\phi_1 = 0.7$, $C = 0$, the starting value for z_{t-1} is zero, the variance of the random shocks (σ_a^2) is one, and $n = 100$.

The theoretical acf and pacf for process (3.14) are shown in Figure 3.6. As with all stationary AR(1) processes, the theoretical acf decays toward zero and the theoretical pacf cuts off to zero after a spike at lag 1. The estimated acf and pacf for each of the five realizations are shown in Figure 3.7.

The important thing to note is this: although the estimated acf's and pacf's are similar to the theoretical acf and pacf in Figure 3.6, in some cases the similarity is vague. This is because the estimated acf's and pacf's are based on a realization and therefore contain sampling error. Thus we cannot expect an estimated acf and pacf to match the corresponding theoretical acf and pacf exactly.

This suggests ambiguities at the identification stage as we try to match the estimated acf and pacf with a theoretical acf and pacf. While estimated acf's and pacf's are extremely helpful, they are only rough guides to model selection. That is why model selection at the identification stage is only tentative. We need the more precise parameter estimates obtained at the

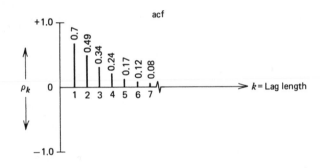

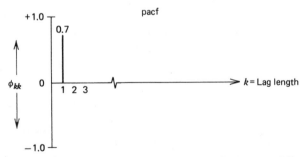

Figure 3.6 Theoretical acf and pacf for process (3.14): $z_t = 0.7z_{t-1} + a_t$, with $a_t \sim \text{NID}(0, 1)$.

```
+ + + + + + + + + + + + AUTOCORRELATIONS + + + + + + + + + + + + +
+ FOR DATA SERIES: SIMULATED DATA                                +
+ DIFFERENCING: 0                      MEAN      =   .1947       +
+ DATA COUNT =  100                    STD DEV =  1.35519         +
  COEF  T-VAL  LAG _____0_____
  0.69   6.92   1                    [       0>>>>>]>>>>>>>>>>>>>
  0.53   3.78   2                    [       0>>>>>]>>>>>>>
  0.37   2.34   3                [       0>>>>>>>]>
  0.31   1.88   4                [       0>>>>>>>]
  0.28   1.61   5                [       0>>>>>>>]
  0.30   1.69   6                [       0>>>>>>>]
  0.28   1.56   7            [       0>>>>>>>    ]
  0.16   0.84   8            [       0>>>>>      ]
  0.10   0.55   9            [       0>>>        ]
  0.09   0.46  10            [       0>>         ]
  0.05   0.26  11            [       0>          ]
  0.06   0.33  12            [       0>>         ]
    CHI-SQUARED* =  135.89 FOR DF =   12

+ + + + + + + + + + + PARTIAL AUTOCORRELATIONS + + + + + + + + + + +
  COEF  T-VAL  LAG _____0_____
  0.69   6.92   1                    [       0>>>>>]>>>>>>>>>>>>>>>>
  0.10   0.97   2                    [       0>>   ]
 -0.05  -0.52   3                    [      <0     ]
  0.09   0.92   4                    [       0>>   ]
  0.06   0.57   5                    [       0>    ]
  0.12   1.19   6                    [       0>>>  ]
  0.02   0.18   7                    [       0     ]
 -0.20  -2.03   8                    [<<<<<0       ]
  0.03   0.25   9                    [       0>    ]
  0.07   0.67  10                    [       0>>   ]
 -0.08  -0.80  11                    [     <<0     ]
  0.05   0.46  12                    [       0>    ]


+ + + + + + + + + + + + AUTOCORRELATIONS + + + + + + + + + + + + +
+ FOR DATA SERIES: SIMULATED DATA                                +
+ DIFFERENCING: 0                      MEAN      =   .0639       +
+ DATA COUNT =  100                    STD DEV =  1.52114         +
  COEF  T-VAL  LAG _____0_____
  0.74   7.41   1                    [       0>>>>>>]>>>>>>>>>>>>>>>>>
  0.50   3.45   2                [       0>>>>>>>]>>>>>
  0.29   1.78   3                [       0>>>>>>>]
  0.17   1.02   4                [       0>>>>   ]
  0.16   0.97   5                [       0>>>>   ]
  0.14   0.80   6                [       0>>>    ]
  0.15   0.89   7                [       0>>>>   ]
  0.09   0.50   8                [       0>>     ]
  0.01   0.08   9                [       0       ]
 -0.04  -0.24  10                [      <0       ]
 -0.06  -0.37  11                [     <<0       ]
 -0.01  -0.04  12                [       0       ]
    CHI-SQUARED* =  103.32 FOR DF =   12

+ + + + + + + + + + + PARTIAL AUTOCORRELATIONS + + + + + + + + + + +
  COEF  T-VAL  LAG _____0_____
  0.74   7.41   1                    [       0>>>>>]>>>>>>>>>>>>>>>>
 -0.11  -1.10   2                    [     <<<0    ]
 -0.10  -0.98   3                    [     <<<0    ]
  0.05   0.54   4                    [       0>    ]
  0.14   1.45   5                    [       0>>>> ]
 -0.07  -0.70   6                    [     <<0     ]
  0.09   0.88   7                    [       0>>   ]
 -0.13  -1.31   8                    [    <<<0     ]
 -0.04  -0.39   9                    [      <0     ]
 -0.01  -0.12  10                    [       0     ]
  0.02   0.17  11                    [       0     ]
  0.09   0.89  12                    [       0>>   ]
```

Figure 3.7 Five estimated acf's and pacf's for realizations generated by process (3.14).

```
+ + + + + + + + + + + + AUTOCORRELATIONS + + + + + + + + + + + + + +
+ FOR DATA SERIES: SIMULATED DATA                                         +
+ DIFFERENCING: 0                        MEAN     =   .3676               +
+ DATA COUNT =  100                      STD DEV =  1.60827               +
  COEF   T-VAL  LAG _____0_____
  0.76   7.58   1                        [        0>>>>>]>>>>>>>>>>>>>>>
  0.54   3.68   2                         [       0>>>>>>>]>>>>>
  0.36   2.17   3                         [       0>>>>>>>]>
  0.27   1.57   4                         [       0>>>>>>>]
  0.16   0.90   5                         [       0>>>    ]
  0.08   0.47   6                         [       0>>     ]
  0.02   0.09   7                         [       0       ]
 -0.03  -0.15   8                         [      <0       ]
 -0.06  -0.34   9                         [      <0       ]
 -0.10  -0.56  10                         [     <<<0      ]
 -0.09  -0.52  11                         [      <<0      ]
 -0.11  -0.60  12                       [       <<<0          ]
      CHI-SQUARED* =   118.13 FOR DF =   12

+ + + + + + + + + + + PARTIAL AUTOCORRELATIONS + + + + + + + + + + +
  COEF   T-VAL  LAG _____0_____
  0.76   7.58   1                        [        0>>>>>]>>>>>>>>>>>>>>>
 -0.08  -0.83   2                         [     <<0       ]
 -0.05  -0.49   3                         [      <0       ]
  0.09   0.91   4                         [       0>>     ]
 -0.12  -1.24   5                         [     <<<0      ]
  0.01   0.07   6                         [       0       ]
 -0.04  -0.37   7                         [      <0       ]
 -0.03  -0.29   8                         [      <0       ]
 -0.01  -0.10   9                         [       0       ]
 -0.08  -0.76  10                         [     <<0       ]
  0.07   0.70  11                         [       0>>     ]
 -0.08  -0.81  12                         [     <<0       ]


+ + + + + + + + + + + + AUTOCORRELATIONS + + + + + + + + + + + + + +
+ FOR DATA SERIES: SIMULATED DATA                                         +
+ DIFFERENCING: 0                        MEAN     =   .3993               +
+ DATA COUNT =  100                      STD DEV =  1.38105               +
  COEF   T-VAL  LAG _____0_____
  0.68   6.79   1                        [        0>>>>>]>>>>>>>>>>>>
  0.47   3.38   2                        [        0>>>>>]>>>>>>
  0.41   2.68   3                         [       0>>>>>>>]>>
  0.35   2.13   4                         [       0>>>>>>>]>
  0.23   1.34   5                         [       0>>>>>> ]
  0.06   0.34   6                         [       0>      ]
 -0.01  -0.04   7                         [       0       ]
 -0.05  -0.27   8                         [      <0       ]
 -0.09  -0.51   9                         [     <<0       ]
 -0.08  -0.48  10                         [     <<0       ]
 -0.07  -0.39  11                         [     <<0       ]
 -0.09  -0.49  12                         [     <<0       ]
      CHI-SQUARED* =   110.75 FOR DF =   12

+ + + + + + + + + + + PARTIAL AUTOCORRELATIONS + + + + + + + + + + +
  COEF   T-VAL  LAG _____0_____
  0.68   6.79   1                        [        0>>>>>]>>>>>>>>>>>>
  0.01   0.13   2                         [       0     ]
  0.17   1.66   3                         [       0>>>> ]
  0.02   0.15   4                         [       0     ]
 -0.09  -0.90   5                         [     <<0     ]
 -0.19  -1.93   6                       [<<<<<0       ]
  0.00  -0.04   7                         [       0     ]
 -0.05  -0.50   8                         [      <0     ]
  0.00   0.01   9                         [       0     ]
  0.07   0.73  10                         [       0>>   ]
  0.03   0.33  11                         [       0>    ]
 -0.06  -0.62  12                         [      <<0    ]
```

Figure 3.7 (*Continued*).

```
+ + + + + + + + + + + + AUTOCORRELATIONS + + + + + + + + + + + +
+ FOR DATA SERIES: SIMULATED DATA                                    +
+ DIFFERENCING: 0                       MEAN     = -.0378            +
+ DATA COUNT = 100                      STD DEV =  1.45072           +
  COEF   T-VAL LAG _____.___0_____
  0.69   6.88   1                              [    0>>>>>]>>>>>>>>>>>>
  0.40   2.83   2                              [    0>>>>>]>>>>>
  0.21   1.41   3                         [    0>>>>>  ]
  0.13   0.82   4                         [    0>>>   ]
  0.06   0.41   5                         [    0>>    ]
  0.00   0.02   6                         [    0      ]
 -0.01  -0.09   7                         [    0      ]
  0.07   0.47   8                         [    0>>    ]
  0.12   0.80   9                         [    0>>>   ]
  0.09   0.60  10                         [    0>>    ]
  0.04   0.25  11                         [    0>     ]
  0.01   0.08  12                         [    0      ]
      CHI-SQUARED* =    75.47 FOR DF =  12

+ + + + + + + + + + PARTIAL AUTOCORRELATIONS + + + + + + + + + + +
  COEF   T-VAL LAG _____0_____
  0.69   6.88   1                              [    0>>>>>]>>>>>>>>>>>>
 -0.15  -1.50   2                         [ <<<<0     ]
  0.00   0.05   3                         [    0      ]
  0.03   0.29   4                         [    0>     ]
 -0.03  -0.35   5                         [    <0     ]
 -0.04  -0.41   6                         [    <0     ]
  0.03   0.26   7                         [    0>     ]
  0.17   1.67   8                         [    0>>>>  ]
 -0.01  -0.08   9                         [    0      ]
 -0.06  -0.61  10                         [ <<0       ]
 -0.02  -0.19  11                         [    0      ]
  0.01   0.10  12                         [    0      ]
```

Figure 3.7 (*Continued*).

estimation stage and the results of the diagnostic-checking stage to help us choose a final model for forecasting.

The following MA(1) process was used to simulate five realizations:

$$z_t = -0.8a_{t-1} + a_t \qquad (3.15)$$

For all five realizations $\theta_1 = 0.8$, $C = 0$, $\sigma_a^2 = 1$, and $n = 100$. Note that θ_1 is positive so in process (3.15) it has a negative sign; we are following the convention of writing MA coefficients with negative signs.

The theoretical acf and pacf derived from process (3.15) are shown in Figure 3.8. The estimated acf and pacf for each realization are shown in Figure 3.9. Again we find that while the estimated acf's and pacf's are similar to the theoretical ones (in Figure 3.8), they do not match perfectly because the estimated ones contain sampling error.

Box and Jenkins comment on the imperfect relationship between estimated and theoretical acf's in this way:

... *detailed* adherence to the theoretical autocorrelation function cannot be expected in the estimated function. In particular, moderately

large estimated autocorrelations can occur after the theoretical autocorrelation function has damped out, and apparent ripples and trends
can occur in the estimated function which have no basis in the
theoretical function. In employing the estimated autocorrelation function as a tool for identification, it is usually possible to be fairly sure
about broad characteristics, but more subtle indications may or may
not represent real effects, and two or more related models may need to
be entertained and investigated further at the estimation and diagnostic checking stages of model building. [1, p. 177; emphasis in original.
Quoted by permission.]

This point is emphasized by a result due to Bartlett [11] showing that
estimated autocorrelation coefficients at different lags $(r_k, r_{k+i}, i \neq 0)$ may
be correlated with each other. Thus, a large estimated autocorrelation
coefficient might also induce neighboring coefficients to be rather large. (See
Part II, Cases 2 and 4 for examples of this phenomenon.)

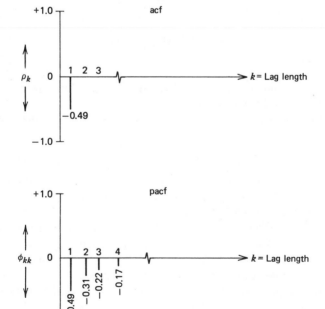

Figure 3.8   Theoretical acf and pacf for process (3.15): $z_t = -0.8a_{t-1} + a_t$, with
$a_t \sim NID(0, 1)$.

```
+ + + + + + + + + + + + AUTOCORRELATIONS + + + + + + + + + + + + +
+ FOR DATA SERIES: SIMULATED DATA                                  +
+ DIFFERENCING: 0                        MEAN    = .0212           +
+ DATA COUNT = 100                       STD DEV = 1.14373          +
   COEF  T-VAL LAG _____O_____
 -0.43  -4.31  1    <<<<<<<<<<<<<[<<<<<<<<<<O               ]
 -0.04  -0.37  2              [        <<O               ]
  0.12   1.01  3              [        O>>>>>>            ]
 -0.13  -1.12  4              [    <<<<<<<O               ]
  0.04   0.37  5              [        O>>                ]
 -0.17  -1.42  6              [   <<<<<<<<O               ]
  0.12   0.96  7              [        O>>>>>>            ]
 -0.04  -0.35  8              [        <<O               ]
  0.09   0.77  9              [        O>>>>>            ]
 -0.03  -0.25  10             [        <<O               ]
  0.02   0.15  11             [        O>                 ]
  0.07   0.58  12             [        O>>>               ]
      CHI-SQUARED* =   29.51 FOR DF =  12

+ + + + + + + + + + + PARTIAL AUTOCORRELATIONS + + + + + + + + + + +
   COEF  T-VAL LAG _____O_____
 -0.43  -4.31  1    <<<<<<<<<<<<<[<<<<<<<<<<O               ]
 -0.28  -2.80  2        <<<<[<<<<<<<<<O               ]
 -0.03  -0.34  3              [        <<O               ]
 -0.13  -1.25  4              [   <<<<<<O               ]
 -0.07  -0.67  5              [        <<<O               ]
 -0.29  -2.88  6        <<<<[<<<<<<<<<<O               ]
 -0.13  -1.35  7              [   <<<<<<O               ]
 -0.17  -1.66  8              [  <<<<<<<<O               ]
  0.04   0.37  9              [        O>>                ]
 -0.04  -0.39  10             [        <<O               ]
  0.02   0.16  11             [        O>                 ]
  0.05   0.50  12             [        O>>>               ]


+ + + + + + + + + + + + AUTOCORRELATIONS + + + + + + + + + + + + +
+ FOR DATA SERIES: SIMULATED DATA                                  +
+ DIFFERENCING: 0                        MEAN    = .0003           +
+ DATA COUNT = 100                       STD DEV = 1.3403           +
   COEF  T-VAL LAG _____O_____
 -0.60  -6.02  1    <<<<<<<<<<<<<<<<<[<<<<<O        ]
  0.29   2.21  2              [        O>>>>>>>]>>
 -0.25  -1.82  3       [ <<<<<<<<O               ]
  0.07   0.53  4              [        O>>                ]
 -0.08  -0.54  5              [        <<<O               ]
  0.19   1.30  6              [        O>>>>>>            ]
 -0.13  -0.88  7              [        <<<<O               ]
 -0.01  -0.07  8              [        O                 ]
  0.02   0.10  9              [        O>                 ]
 -0.04  -0.26  10             [        <O                 ]
  0.09   0.61  11             [        O>>>               ]
  0.04   0.27  12             [        O>                 ]
      CHI-SQUARED* =   60.84 FOR DF =  12

+ + + + + + + + + + + PARTIAL AUTOCORRELATIONS + + + + + + + + + + +
   COEF  T-VAL LAG _____O_____
 -0.60  -6.02  1    <<<<<<<<<<<<<<<<<[<<<<<O        ]
 -0.11  -1.13  2              [  <<<<O        ]
 -0.20  -1.97  3            <[<<<<<<O        ]
 -0.24  -2.41  4           <<[<<<<<<O        ]
 -0.24  -2.40  5           <<[<<<<<<O        ]
  0.04   0.39  6              [     O>        ]
  0.01   0.07  7              [     O        ]
 -0.19  -1.95  8            [<<<<<<O        ]
 -0.10  -1.00  9              [  <<<O        ]
 -0.08  -0.81  10             [  <<<O        ]
 -0.02  -0.16  11             [     <O        ]
  0.11   1.10  12             [     O>>>> ]
```

Figure 3.9 Five estimated acf's and pacf's for realizations generated by process (3.15).

```
+ + + + + + + + + + + + AUTOCORRELATIONS + + + + + + + + + + + + +
+ FOR DATA SERIES: SIMULATED DATA                                    +
+ DIFFERENCING: 0                      MEAN     =  .0049             +
+ DATA COUNT =  100                    STD DEV =  1.33952            +
  COEF   T-VAL  LAG                          0
 -0.45  -4.48    1   <<<<<<<<<<<<[<<<<<<<<<<0              ]
 -0.12  -1.05    2              [    <<<<<<0              ]
  0.12   1.02    3              [    0>>>>>>              ]
 -0.04  -0.31    4              [      <<0              ]
 -0.08  -0.62    5              [    <<<<0              ]
  0.10   0.80    6              [    0>>>>>              ]
 -0.01  -0.06    7              [      0              ]
  0.07   0.57    8              [    0>>>              ]
 -0.09  -0.70    9              [   <<<<0              ]
 -0.14  -1.16   10              [  <<<<<<0              ]
  0.18   1.42   11              [    0>>>>>>>>>         ]
  0.01   0.08   12              [    0>              ]
    CHI-SQUARED* =    32.87 FOR DF =   12

+ + + + + + + + + + + PARTIAL AUTOCORRELATIONS + + + + + + + + + + +
  COEF   T-VAL  LAG                          0
 -0.45  -4.48    1   <<<<<<<<<<<<[<<<<<<<<<<0              ]
 -0.41  -4.06    2   <<<<<<<<<<<[<<<<<<<<<<0               ]
 -0.21  -2.07    3              [<<<<<<<<<<0               ]
 -0.17  -1.72    4              [<<<<<<<<<0                ]
 -0.23  -2.34    5           <<[<<<<<<<<<<0                ]
 -0.13  -1.34    6           [  <<<<<<<0                   ]
 -0.09  -0.91    7           [   <<<<<0                    ]
  0.11   1.07    8           [      0>>>>>                 ]
  0.06   0.63    9           [      0>>>                   ]
 -0.19  -1.94   10           [<<<<<<<<<0                   ]
 -0.09  -0.88   11           [   <<<<0                     ]
 -0.01  -0.10   12           [     <0                      ]

+ + + + + + + + + + + + AUTOCORRELATIONS + + + + + + + + + + + + +
+ FOR DATA SERIES: SIMULATED DATA                                    +
+ DIFFERENCING: 0                      MEAN     = -.0092             +
+ DATA COUNT =  100                    STD DEV =  1.23945            +
  COEF   T-VAL  LAG                          0
 -0.40  -4.01    1   <<<<<<<<<<<[<<<<<<<<<<0              ]
 -0.14  -1.25    2              [   <<<<<<0              ]
  0.11   0.90    3              [    0>>>>>              ]
 -0.06  -0.50    4              [     <<<0              ]
  0.05   0.41    5              [     0>>              ]
 -0.12  -0.98    6              [   <<<<<0              ]
  0.06   0.51    7              [     0>>>              ]
  0.05   0.43    8              [     0>>>              ]
 -0.06  -0.52    9              [     <<<0              ]
  0.03   0.25   10              [     0>              ]
  0.07   0.57   11              [     0>>>              ]
 -0.04  -0.35   12              [     <<0              ]
    CHI-SQUARED* =    23.95 FOR DF =   12

+ + + + + + + + + + + PARTIAL AUTOCORRELATIONS + + + + + + + + + + +
  COEF   T-VAL  LAG                          0
 -0.40  -4.01    1   <<<<<<<<<<<[<<<<<<<<<<0              ]
 -0.36  -3.64    2    <<<<<<<<<[<<<<<<<<<<0               ]
 -0.16  -1.64    3              [ <<<<<<<<0               ]
 -0.18  -1.78    4              [<<<<<<<<<0               ]
 -0.06  -0.63    5              [    <<<0                 ]
 -0.21  -2.06    6              [<<<<<<<<<0               ]
 -0.13  -1.30    7              [   <<<<<0                ]
 -0.07  -0.72    8              [    <<<<0                ]
 -0.08  -0.84    9              [    <<<0                 ]
 -0.06  -0.56   10              [    <<<0                 ]
  0.06   0.63   11              [      0>>>               ]
  0.05   0.48   12              [      0>>                ]
```

Figure 3.9   (*Continued*).

66

```
+ + + + + + + + + + + + AUTOCORRELATIONS + + + + + + + + + + + +
+ FOR DATA SERIES: SIMULATED DATA                                    +
+ DIFFERENCING: 0                         MEAN     = .0099           +
+ DATA COUNT =  100                       STD DEV = 1.08161          +
  COEF   T-VAL LAG _____0_____
 -0.55  -5.50   1            <<<<<<<<<<<<<<[<<<<<0       ]
  0.12   0.93   2                         [     0>>>>    ]
 -0.02  -0.15   3                         [    <0        ]
 -0.13  -1.03   4                         [  <<<<0       ]
  0.25   1.95   5                         [     0>>>>>>>]
 -0.18  -1.31   6                        [ <<<<<<0       ]
  0.04   0.31   7                  [          0>        ]
 -0.04  -0.31   8                  [         <0         ]
  0.08   0.59   9                  [          0>>>       ]
 -0.06  -0.43  10                  [         <<0         ]
 -0.02  -0.14  11                  [         <0          ]
  0.11   0.80  12                  [          0>>>>       ]
         CHI-SQUARED* =   47.63 FOR DF =   12

+ + + + + + + + + + PARTIAL AUTOCORRELATIONS + + + + + + + + + +
  COEF   T-VAL LAG _____0_____
 -0.55  -5.50   1            <<<<<<<<<<<<<<<[<<<<<0       ]
 -0.27  -2.66   2                    <<<[<<<<<0           ]
 -0.13  -1.29   3                       [ <<<<0           ]
 -0.27  -2.75   4                    <<<[<<<<<0           ]
  0.05   0.49   5                       [    0>>          ]
  0.01   0.08   6                       [    0            ]
 -0.05  -0.47   7                       [   <<0           ]
 -0.11  -1.06   8                       [ <<<<0           ]
  0.06   0.63   9                       [    0>>          ]
 -0.04  -0.42  10                       [   <0            ]
 -0.09  -0.93  11                       [  <<<0           ]
  0.10   0.96  12                       [    0>>>         ]
```

Figure 3.9 (*Continued*).

3.3 Statistical inference at the identification stage

Recall that in classical statistics we may want to know something about the population, but getting all relevant information about the population is frequently impossible or too costly. Therefore, we infer something about the population by using a sample, along with some probability concepts and formulas and theories from mathematical statistics. The sample may be used to *estimate* a characteristic of the population or to *test a hypothesis* about the population. This type of inductive reasoning is known as *statistical inference*.

In UBJ–ARIMA analysis we engage in statistical inference at all three stages of the method. That is, we infer something about the unknown process by using the realization, along with some probability principles and statistical concepts. At the identification stage we use the estimated acf and pacf, calculated from the realization, to help us tentatively select one or more models to represent the unknown process that generated the realization. The purpose of this section is to discuss and illustrate statistical estimation and hypothesis testing, as they occur at the identification stage.*

Testing autocorrelation coefficients.   In Chapter 2 we saw how autocorrelation coefficients are calculated from a realization. An estimated autocorrelation coefficient (r_k) is an estimate of the corresponding (unknown) theoretical autocorrelation coefficient (ρ_k). We do not expect each r_k to be exactly equal to its corresponding ρ_k because of sampling error. Thus, with $k = 1$ and $n = 100$, we would get a certain value for r_1 using (2.5). But if we could then obtain another realization with $n = 100$ and recalculate r_1, we would probably get a different value. In turn, another realization would give us yet another value for r_1. These differences among the various r's are due to sampling error. This was illustrated in the last section where we saw a series of estimated acf's and pacf's that were similar to, but not identical to, the corresponding theoretical acf's and pacf's.

If we could calculate r_1 for *all possible* realizations with $n = 100$ we would have a collection of all possible r_1 values. The distribution of these possible values is called a *sampling distribution.*† As with other sample statistics, these different possible sample values (r_1) will be distributed around the parameter (ρ_1) in some fashion. R. L. Anderson [13] has shown that the r_k values are approximately Normally distributed when $\rho_k = 0$ if n is not too small.

Bartlett [11] has derived an approximate expression for the standard error of the sampling distribution of r_k values. (The standard error of a sampling distribution is the square root of its estimated variance.) This estimated standard error, designated $s(r_k)$, is calculated as follows:

$$s(r_k) = \left(1 + 2\sum_{j=1}^{k-1} r_j^2\right)^{1/2} n^{-1/2} \tag{3.16}$$

This approximation is appropriate for stationary processes with Normally distributed random shocks where the true MA order of the process is $k - 1$.

This expression may be applied to the first three autocorrelations in Figure 3.10 in the following way. First, let $k = 1$. Then sum inside the parentheses of expression (3.16) from $j = 1$ to $j = 0$; since j must increase

---

*Calling the second stage the estimation stage is misleading because statistical estimation also takes place at the other two stages.

†You may find it helpful at this point to review the concept of a sampling distribution in an introductory statistics textbook such as Wonnacott and Wonnacott [12].

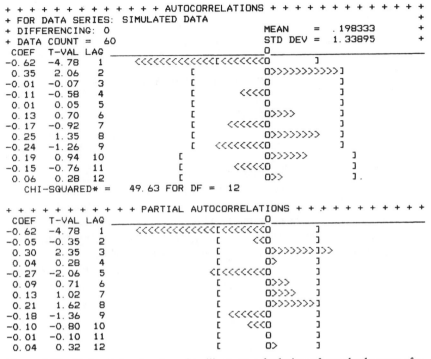

Figure 3.10 Estimated acf and pacf to illustrate calculation of standard errors of r_k and $\hat{\phi}_{kk}$.

by 1 to perform summation, there is no summation in this case, and we replace the summation term with a zero. Thus with $n = 60$ and $k = 1$,

$$s(r_1) = (1 + 0)^{1/2} n^{-1/2}$$

$$= (60)^{-1/2}$$

$$= 0.13$$

Next, let $k = 2$. Then (3.16) gives this result:

$$s(r_2) = (1 + 2r_1^2)^{1/2} n^{-1/2}$$

$$= \left[1 + 2(-0.62)^2\right]^{1/2} 60^{-1/2}$$

$$= (1.33)(0.13)$$

$$= 0.17$$

Then letting $k = 3$ we obtain

$$s(r_3) = \left(1 + 2r_1^2 + 2r_2^2\right)^{1/2} n^{-1/2}$$

$$= \left[1 + 2(-0.62)^2 + 2(0.35)^2\right]^{1/2} 60^{-1/2}$$

$$= (1.42)(0.13)$$

$$= 0.18$$

Other $s(r_k)$ values are calculated in a similar fashion.

Now use these estimated standard errors to test the null hypothesis H_0: $\rho_k = 0$ for $k = 1, 2, 3, \ldots$. It is common when using the estimate $s(r_k)$ in place of the true standard error $\sigma(r_k)$ to refer to the t-distribution rather than the Normal distribution. We test the null hypothesis by finding out how far away the sample statistic r_k is from the hypothesized value $\rho_k = 0$, where "how far" is a t-statistic equal to a certain number of estimated standard errors. Thus we find an approximate t-statistic in this way:

$$t_{r_k} = \frac{r_k - \rho_k}{s(r_k)} \qquad (3.17)$$

Let ρ_k in (3.17) equal its hypothesized value of zero and insert each calculated r_k along with its corresponding estimated standard error $s(r_k)$. Using the r_k values in the acf in Figure 3.10, for $k = 1$ we find

$$t_{r_1} = \frac{r_1 - \rho_1}{s(r_1)}$$

$$= \frac{-0.62 - 0}{0.13}$$

$$= -4.78.$$

This result says that r_1 falls 4.78 estimated standard errors below zero.* Using a rule of thumb that only about 5% of the possible r_k would fall two or more estimated standard errors away from zero if $\rho_k = 0$, we reject the null hypothesis $\rho_1 = 0$ since r_1 is significantly different from zero at about the 5% level.

---

*Hand calculations may give results slightly different from those printed in Figure 3.10 because of rounding.

Figure 3.11 illustrates these ideas. The label on the horizontal axis shows that this is a distribution of all possible values of r_1 for a certain sample size n. That is, Figure 3.11 is a sampling distribution for r_1. This distribution is centered on the parameter ρ_1, which is unknown. Since this is approximately a Normal (or t) distribution with an estimated standard error $s(r_1)$ given by (3.16), the interval $\rho_1 \pm 2s(r_1)$ contains about 95% of all possible r_1 values. This is represented by the shaded area under the curve. If $\rho_1 = 0$, then r_1 in our example (-0.62) is 4.78 estimated standard errors below zero. Instead of calculating the t-value, we might look at the square brackets [ ] at lag 1 in Figure 3.10. These brackets are about two standard errors above and below zero. Since the acf spike at lag 1 extends beyond the bracket on the negative side, the autocorrelation at that lag is more than two standard errors below zero. Thus whether we use the two standard error limits (square brackets) printed on the acf, or calculate a t-value as we did above, we conclude that r_1 is significantly different from zero at better than the 5% level. Similar calculations for $k = 2$ and $k = 3$ give these results:*

$$t_{r_2} = \frac{r_2 - \rho_2}{s(r_2)}$$

$$= \frac{0.35 - 0}{0.17}$$

$$= 2.06$$

$$t_{r_3} = \frac{r_3 - \rho_3}{s(r_3)}$$

$$= \frac{-0.01 - 0}{0.18}$$

$$= -0.17$$

It must be emphasized that these calculations are only approximate since they are based on Bartlett's approximation (3.16) for the standard error of the sampling distribution of r_k. We are taking a practical approach to a difficult mathematical problem, giving up some precision to achieve a useful procedure.

In the preceding example, we implicitly were supposing that the true MA order of the underlying process was first zero, then one, then two, and so forth. That is, equation (3.16) applies when the true MA order of the underlying process is $k - 1$. When calculating $s(r_2)$ above, we let $k = 1$, implying that the MA order of the process was $k - 1 = 1 - 1 = 0$; when calculating $s(r_1)$ we let $k = 2$, implying a true MA order of $k - 1 = 2 -$

---

*Hand calculations may give a slightly different answer due to rounding.

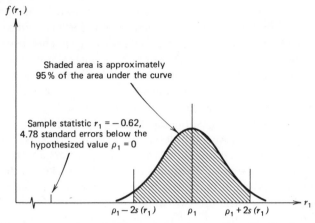

$f(r_1)$

Shaded area is approximately
95 % of the area under the curve

Sample statistic $r_1 = -0.62$,
4.78 standard errors below the
hypothesized value $\rho_1 = 0$

$\rho_1 - 2s(r_1)$ ρ_1 $\rho_1 + 2s(r_1)$ r_1

Figure 3.11 An approximately Normal (or t) sampling distribution for r_1, with estimated standard error $s(r_1)$ given by equation (3.16).

$1 = 1$; when calculating $s(r_3)$ we let $k = 3$, implying a true MA order of two. All the acf t-values printed in this text are based on standard errors calculated in this manner.

If we let the assumed true MA order increase by one each time we test an additional r_k coefficient, we see from (3.16) that $s(r_k)$ tends to increase as k increases. This is illustrated in Figure 3.10, where the square brackets trace out a gradually widening band. This occurs because we add an additional r_k^2 term in (3.16) each time k increases. These additional terms may be quite small so that $s(r_k)$ might not increase noticeably until we cumulate several r_k^2 terms. This gradual increase in $s(r_k)$ is illustrated in Figure 3.10 where the two standard error limits are virtually constant from lags 2 through 9.

At times we may want to assume that the true MA order of an underlying process is a single value. For example, we might maintain the hypothesis that the underlying process is white noise. There would be no autocorrelation within the process and the true MA order would be zero. Then we must replace $k - 1$ at the top of the summation sign in (3.16) with the fixed value of zero. In that case (3.16) tells us that $s(r_k) = n^{-1/2} = 60^{-1/2} = 0.13$ for *all* k. Then we would use the square brackets at lag 1 in Figure 3.10 as the two standard error limits for other lags as well. For all acf's in this text, the assumed true MA order can be fixed at any desired value by extending the printed two standard error limits appropriately. For example, if we suppose that the true MA order $(k - 1)$ is fixed at 2, then $k = 3$. We would then extend the square brackets printed at lag 3 to other lags.

Testing partial autocorrelation coefficients. We can also test the statistical significance of estimated partial autocorrelation coefficients. The required estimated standard error* is

$$s(\hat{\phi}_{kk}) = n^{-1/2} \qquad (3.18)$$

Let us apply (3.18) to test the significance of the first three partial autocorrelation coefficients in Figure 3.10. The $\hat{\phi}_{kk}$ shown there were calculated from a sample of 60 observations. Inserting $n = 60$ into (3.18) we find that $s(\hat{\phi}_{kk}) = 0.13$ for all k. Testing the null hypothesis H_0: $\phi_{11} = 0$, we get this t-statistic:

$$t_{\hat{\phi}_{11}} = \frac{\hat{\phi}_{11} - \phi_{11}}{s(\hat{\phi}_{11})}$$

$$= \frac{-0.62 - 0}{0.13}$$

$$= -4.78$$

Since the absolute value of this t-statistic is greater than 2.0 we conclude that $\hat{\phi}_{11}$ is different from zero at about the 5% significance level and we reject the null hypothesis $\phi_{11} = 0$.

Now letting $k = 2$ and testing the null hypothesis H_0: $\phi_{22} = 0$, we get this t-statistic:

$$t_{\hat{\phi}_{22}} = \frac{\hat{\phi}_{22} - \phi_{22}}{s(\hat{\phi}_{22})}$$

$$= \frac{-0.05 - 0}{0.13}$$

$$= -0.35$$

For $k = 3$, testing the null hypothesis H_0: $\phi_{33} = 0$,

$$t_{\hat{\phi}_{33}} = \frac{\hat{\phi}_{33} - \phi_{33}}{s(\hat{\phi}_{33})}$$

$$= \frac{0.30 - 0}{0.13}$$

$$= 2.35$$

Again we find that our calculations agree with the results printed by

*For discussion of this result see Quenouille [14], Jenkins [15], and Daniels [16].

computer (Figure 3.10) though calculations by hand may give slightly different results due to rounding. As with estimated acf's, the square brackets printed on estimated pacf's throughout this book are approximately two standard errors above and below zero. These brackets provide a fast way to find estimated partial autocorrelations that are significantly different from zero at about the 5% level. In Figure 3.10, we see immediately that the estimated partial autocorrelations at lags 1, 3, and 5 are different from zero at about the 5% significance level because their printed spikes extend beyond the square brackets.

Summary

1. In UBJ–ARIMA analysis a set of time-series observations is called a realization.

2. A realization is assumed to have been produced by an underlying mechanism called a process.

3. A process includes all possible observations on a time-sequenced variable along with an algebraic statement (a generating mechanism) describing how these possible observations are related. In practice, generating mechanisms are not known.

4. We use the UBJ three-stage procedure (identification, estimation, and diagnostic checking) to find a model that fits the available realization. Our hope is that such a model is also a good representation of the unknown underlying generating mechanism.

5. AR means *autogressive*. Each AR term in an ARIMA process has a fixed coefficient (ϕ) multiplied by a past z term.

6. MA means *moving average*. Each MA term in an ARIMA process has a fixed coefficient (θ) multiplied by a past random shock.

7. Two common processes are

$$\text{AR(1)}: z_t = C + \phi_1 z_{t-1} + a_t$$

$$\text{MA(1)}: z_t = C - \theta_1 a_{t-1} + a_t$$

where z_t is the variable whose time structure is described by the process; C is a constant term related to the mean μ; $\phi_1 z_{t-1}$ is an AR term; $\theta_1 a_{t-1}$ is an MA term; and a_t is a current random shock. The label AR(1) means that the longest time lag attached to an AR term in that process is one time period; the label MA(1) means that the longest time lag attached to an MA term in that process is one time period.

8. The longest time lag associated with an AR term is called the AR order of a process. The longest time lag attached to an MA term is called the MA order of a process.

9. a_t is a random-shock term that follows a probability distribution. The usual assumption is that the a_t's are identically (for all t), independently, and Normally distributed random variables with a mean of zero and a constant variance.

10. MA terms (past random-shock terms) in an ARIMA process can be replaced by AR terms through algebraic manipulation. Thus, all ARIMA processes are, directly or indirectly, univariate processes; z_t is a function of its own past values. We illustrate this point in Chapter 5.

11. We may think of a realization as observations drawn from a stationary, joint, Normal probability distribution function. Such a function is fully characterized by its mean, variance, and covariances. Because it is stationary, it has a constant mean, a constant variance, and constant covariances (covariances that depend only on the time span separating the variables in question, not on their particular time subscripts.)

12. Rather than specifying joint distribution functions in detail, we summarize a process in the form of a generating mechanism. From this generating mechanism we may derive the mean, variance, autocovariances (and autocorrelation coefficients), and the conditional distribution of future z's for that process.

13. A theoretical autocorrelation coefficient (ρ_k) is an autocovariance (γ_k) divided by the variance of the process ($\gamma_0 = \sigma_z^2$):

$$\rho_k = \frac{\gamma_k}{\gamma_0}$$

14. The diagonals of the variance–covariance matrix and the autocorrelation coefficient matrix for a stationary process are each composed of a constant.

15. Every ARIMA process has an associated theoretical acf and pacf. Stationary AR processes have theoretical acf's that decay toward zero and theoretical pacf's that cut off to zero. MA processes have theoretical acf's that cut off to zero and theoretical pacf's that decay toward zero.

16. Estimated acf's and pacf's do not match theoretical acf's and pacf's in every detail because the estimated ones are contaminated with sampling error.

17. Because z_t is stochastic, any given realization is only one which might have occurred. Likewise, any estimated autocorrelation coefficient r_k is only one which might have occurred; that is, a different realization would produce a different value for each r_k. The distribution of possible values for r_k is a sampling distribution.

18. For large n, when $\rho_k = 0$, the sampling distribution for r_k is approximately Normal with a standard error estimated by equation (3.16). The estimated standard error for partial autocorrelation coefficients is given by equation (3.18).

19. Estimated autocorrelation or partial autocorrelation coefficients with absolute t-values larger than 2.0 are statistically different from zero at roughly the 5% significance level.

Appendix 3A: expected value rules and definitions*

Rule I-E: expected value of a discrete random variable

$$E(x) = \sum_{x} xf(x) = \mu_x$$

where x is a discrete random variable; E is the expected value operator; $f(x)$ is the probability density function of x; μ_x is the mean of x.

Rule II-E: expected value of a constant

$$E(C) = C$$

where C is a constant.

Rule III-E: expected value of a finite linear combination of random variables. If m is a finite integer,

$$E(C_1 x_1 + C_2 x_2 + \cdots + C_m x_m)$$
$$= C_1 E(x_1) + C_2 E(x_2) + \cdots + C_m E(x_m)$$

where $C_1, C_2, \ldots, C_m$ are constants; $x_1, x_2, \ldots, x_m$ are random variables.

Rule IV-E: expected value of an infinite linear combination of random variables. If $m = \infty$, Rule III-E holds only if $\sum_{i=0}^{\infty} C_i$ (where $C_0 = 1$) converges (is equal to some finite number).

---

*For simplicity, these rules are stated for discrete random variables. For continuous random variables, summation signs are replaced by integral signs.

Rule V-E: covariance

$$\gamma_{xv} = \text{cov}(x, v) = E\big[(x - \mu_x)(v - \mu_v)\big]$$
$$= \sum_v \sum_x (x - \mu_x)(v - \mu_v) f(x, v)$$

where γ_{xv} is the covariance of x and v; x, v are discrete random variables; μ_x, μ_v are means of x and v, respectively. If $x = z_t$ and $v = z_{t-i}$, then γ_{xv} is an autocovariance denoted as γ_k, where $k = |i|$.

Rule VI-E: variance

$$\gamma_{xx} = \sigma_x^2 = \text{var}_x = \text{cov}(x, x)$$
$$= E\big[(x - \mu_x)(x - \mu_x)\big]$$
$$= E(x - \mu_x)^2$$
$$= \sum_x (x - \mu_x)^2 f(x)$$

Following the notation for autocovariances noted under Rule V-E, if $x = z_t$, $\gamma_{xx} = \gamma_0 = \sigma_z^2$.

Questions and Problems

3.1   Explain the relationship among an ARIMA process, a realization, and an ARIMA model.

3.2   Suppose you want to forecast a variable whose ARIMA process is known. Would you first have to build an ARIMA model?

3.3   Consider this ARIMA process:

$$z_t = C - \theta_1 a_{t-1} - \theta_2 a_{t-2} + a_t$$

(a)   How can you tell that this is a process rather than a model?
(b)   What is the AR order of this process? Explain.
(c)   What is the MA order of this process? Explain.
(d)   Is this a mixed process? Explain.
(e)   Why are the θ coefficients written with negative signs?
(f)   Is this a univariate process? Explain.
(g)   Contrast the statistical characteristics or attributes of θ_1 and a_t.

(h)   What are the usual assumptions about a_t in ARIMA analysis? Illustrate graphically. Are these assumptions always satisfied in practice? Why are these assumptions made?

3.4   Explain the difference between a deterministic relationship and a stochastic process.

3.5   How are estimated acf's and pacf's found? How are theoretical acf's and pacf's found?

3.6   Consider the following pairs of theoretical acf's and pacf's. In each case indicate whether the pair of diagrams is associated with an AR or an

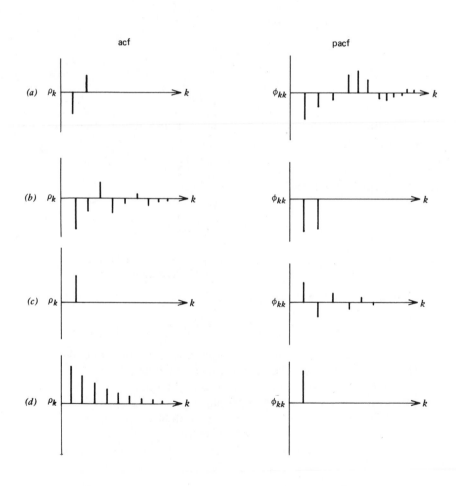

MA process, state the AR or MA order of the process, and write out the process generating mechanism. Explain your reasoning.

3.7 Explain why estimated acf's and pacf's do not match theoretical acf's and pacf's in every detail.

3.8 Consider the following estimated autocorrelation coefficients calculated from a realization with $n = 100$:

k	r_k
1	0.50
2	0.28
3	0.10
4	0.05
5	−0.01

(a) Calculate an approximate t-statistic for each r_k on the assumption that the true MA order increases by one with each additional calculation.

(b) Plot the r_k values on an acf diagram. Indicate on the acf how large each r_k would have to be if it were to be significantly different from zero at roughly the 5% level.

(c) Repeat parts (a) and (b) above on the assumption that the true MA order is fixed at zero.

3.9 Consider these estimated partial autocorrelation coefficients calculated from a realization with $n = 100$:

k	$\hat{\phi}_{kk}$
1	−0.60
2	−0.32
3	−0.21
4	0.11
5	0.03

(a) Calculate an approximate t-statistic for each $\hat{\phi}_{kk}$.

(b) Plot the $\hat{\phi}_{kk}$ values on a pacf diagram. Indicate on the pacf how large each $\hat{\phi}_{kk}$ would have to be if it were to be significantly different from zero at about the 5% level.

4

AN INTRODUCTION
TO THE PRACTICE
OF ARIMA MODELING

In this chapter we first discuss the general characteristics of a good ARIMA model. Then we apply the strategy of identification, estimation, and diagnostic checking to two realizations.

After reading this chapter you should be ready to start on the case studies in Part II. (See the Preface or the Introduction to the cases for a suggested reading schedule.) You must read Chapters 5–12 to, have a full grasp of the notation and procedures employed in the case studies. But the first few cases are not too complicated; starting them after this chapter will help you develop a better understanding of the practice of ARIMA modeling.

4.1 What is a good model?

Several times in the first three chapters we have referred to the goal of building a good model. Before looking at two examples of UBJ model building in the next section, we summarize the qualities of a good ARIMA model.

It is important to remember the difference between a model and a process. In practice we never know which ARIMA process has generated a given realization, so we must follow a trial-and-error procedure. In the "trial" part of the procedure (the identification stage) we are guided by the estimated acf and pacf calculated from the realization. We select some hypothetical ARIMA generating mechanisms, like the AR(1) or MA(1) mechanisms shown in equations (3.1) and (3.2), in the hope that they will fit the available data adequately. These possible or "trial" generating mechanisms are models. A model is different from a process: a process is the true but unknown mechanism that has generated a realization, while a model is only an *imitation or representation* of the process. Because the process is unknown, we never know if we have selected a model that is essentially the same as the true generating process. All we can do is select a model that seems adequate in light of the available data.

How do we decide if a model is a good one? Following are some important points to remember. They are summarized in Table 4.1.

(1) A good model is *parsimonious*.   Box and Jenkins emphasize a key principle of model building called the *principle of parsimony*, meaning "thrift." A parsimonious model fits the available data adequately without using any unnecessary coefficients. For example, if an AR(1) model and an AR(2) model are essentially the same in all other respects, we would select the AR(1) model because it has one less coefficient to estimate.

The principle of parsimony is important because, in practice, parsimonious models generally produce better forecasts. The idea of parsimony gives our modeling procedure a strong practical orientation. In particular, we are not necessarily trying to find the true process responsible for generating a given realization. Rather, we are happy to find a model which only

Table 4.1   Characteristics of a good ARIMA model

1. It is parsimonious (uses the smallest number of coefficients needed to explain the available data).

2. It is stationary (has AR coefficients which satisfy some mathematical inequalities; see Chapter 6).

3. It is invertible (has MA coefficients which satisfy some mathematical inequalities; see Chapter 6).

4. It has estimated coefficients ($\hat{\phi}$'s and $\hat{\theta}$'s) of high quality (see Chapter 8):
   (a)   absolute *t*-values about 2.0 or larger,
   (b) $\hat{\phi}$'s and $\hat{\theta}$'s not too highly correlated.

5. It has uncorrelated residuals (see Chapter 9).

6. It fits the available data (the past) well enough to satisfy the analyst:
   (a)   root-mean-squared error (RMSE) is acceptable,
   (b)   mean absolute percent error (MAPE) is acceptable.

7. It forecasts the future satisfactorily.

approximates the true process as long as the model explains the behavior of the available realization in a parsimonious and statistically adequate manner. The importance of the principle of parsimony cannot be overemphasized.

Matrix (3.11) is a convenient vehicle for discussing the principle of parsimony. For example, consider the theoretical acf for an AR(1) process with $\phi_1 > 0$ in Figure 3.4, Example I. This is simply a plot of the ρ_k values ($k = 1, \ldots, n - 1$) found in the upper-right triangle of matrix (3.11) for that particular process. (We need only the ρ_k's in the upper-right triangle since (3.11) is symmetric.) For the n random variables ($z_1, \ldots, z_n$) we can show that an AR(1) process has n nonzero ρ_k values ($\rho_0, \ldots, \rho_{n-1}$). However, we can represent all this information in a highly parsimonious manner with the AR(1) generating mechanism (3.1), a process containing only two parameters (C and ϕ_1). In fact, as we show in Chapter 6, all the autocorrelation coefficients for an AR(1) process are a function of ϕ_1.

In later chapters and in the case studies in Part II we see that many realizations characterized by a large number of statistically significant autocorrelation coefficients can be represented parsimoniously by generating mechanisms having just a few parameters.

(2) A good AR model is *stationary*.   As noted in Chapter 1 the UBJ–ARIMA method applies only to a realization that is (or can be made) stationary, meaning it has a constant mean, variance, and acf. Any model we choose must also be stationary. In Chapter 6 we learn that we can check a model for stationarity by seeing if the estimated AR coefficients satisfy some mathematical inequalities.

(3) A good MA model is *invertible*.   Invertibility is algebraically similar to stationarity. We check a model for invertibility by seeing if the estimated MA coefficients satisfy some mathematical inequalities. These inequalities and the idea behind invertibility are discussed in Chapter 6.

(4) A good model has *high-quality estimated coefficients* at the estimation stage.   (This refers to the estimated ϕ's and θ's, designated $\hat{\phi}$ and $\hat{\theta}$, not the autocorrelation coefficients r_k and partial autocorrelation coefficients $\hat{\phi}_{kk}$ found at the identification stage.) We want to avoid a forecasting model which represents only a chance relationship, so we want each $\hat{\phi}$ or $\hat{\theta}$ coefficient to have an absolute t-statistic of about 2.0 or larger. This means each estimated $\hat{\phi}$ or $\hat{\theta}$ coefficient should be about two or more standard errors away from zero. If this condition is met, each $\hat{\phi}$ or $\hat{\theta}$ is statistically different from zero at about the 5% level.

In addition, estimated ϕ and θ coefficients should not be too highly correlated with each other. If they are they tend to be somewhat unstable even if they are statistically significant. This topic is discussed in Chapter 8.

(5) A good model has *statistically independent residuals*. An important assumption stated in Section 3.2 is that the random shocks (a_t) are independent in a process. We cannot observe the random shocks, but we can get estimates of them (designated $\hat{a}_t$) at the estimation stage. The $\hat{a}_t$ are called *residuals* of a model. We test the shocks for independence by constructing an acf using the residuals as input data. Then we apply t-tests to each estimated residual autocorrelation coefficient and a chi-squared test to all of them as a set. These t-tests and chi-squared test are primary tools at the diagnostic-checking stage, discussed in detail in Chapter 9. If the residuals are statistically independent, this is important evidence that we cannot improve the model further by adding more AR or MA terms.

(6) A good model *fits the available data sufficiently well* at the estimation stage. Of course no model can fit the data perfectly because there is a random-shock element present in the data. We use two measures of closeness of fit: the root-mean-squared error (RMSE) and the mean absolute percent error (MAPE). These two ideas are discussed in Chapter 8.

How well is "sufficiently well?" This is a matter of judgment. Some decisions require very accurate forecasts while others require only rough estimates. The analyst must decide in each case if an ARIMA model fits the available data well enough to be used for forecasting.

(7) Above all, a good model has *sufficiently small forecast errors*. Although a good forecasting model will usually fit the past well, it is even more important that it forecast the future satisfactorily. To evaluate a model by this criterion we must monitor its forecast performance.

4.2 Two examples of UBJ–ARIMA modeling

In this section we present two examples of the complete UBJ modeling cycle of identification, estimation, and diagnostic checking. In both examples the data are simulated with a computer: first, a generating mechanism is chosen; then, a set of random shocks are generated to represent the purely stochastic part of the mechanism.

Example 1. Consider the realization in Figure 4.1. Inspection suggests that the variance of the series is approximately constant through time. But the mean could be fluctuating through time, so the series may not be

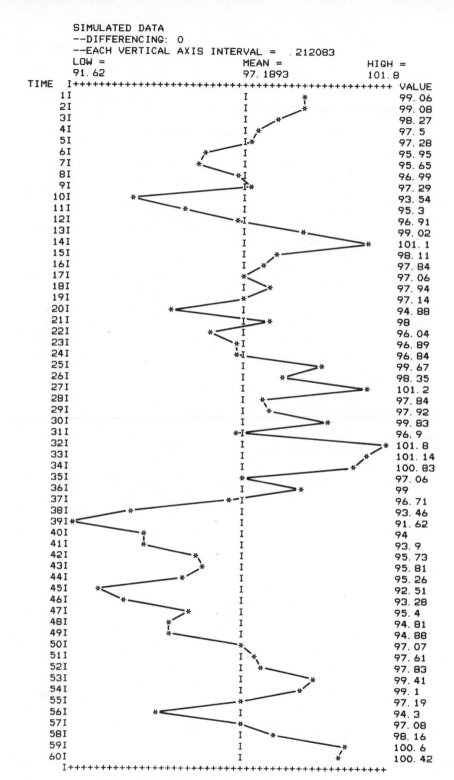

```
          SIMULATED DATA
          --DIFFERENCING: 0
          --EACH VERTICAL AXIS INTERVAL =   .212083
          LOW =                      MEAN =              HIGH =
          91.62                      97.1893             101.8
    TIME  I+++++++++++++++++++++++++++++++++++++++++++++++++++ VALUE
      1I                                  I         *              99.06
      2I                                  I        *               99.08
      3I                                  I     *                  98.27
      4I                                  I *                      97.5
      5I                                  I.*                      97.28
      6I                             *     I                       95.95
      7I                          *        I                       95.65
      8I                             *.I                           96.99
      9I                               .I.*                        97.29
     10I              *                   I                        93.54
     11I                     *            I                        95.3
     12I                              *.I                          96.91
     13I                                  I      *                 99.02
     14I                                  I                  *     101.1
     15I                                  I    *                   98.11
     16I                                  I  *                     97.84
     17I                               *   I                       97.06
     18I                                  I *                      97.94
     19I                               *   I                       97.14
     20I                  *               I                        94.88
     21I                                 .I *                      98
     22I                        *        I                         96.04
     23I                             *.I                           96.89
     24I                             *.I                           96.84
     25I                                  I       *                99.67
     26I                                  I    *                   98.35
     27I                                  I             *          101.2
     28I                                  I *                      97.84
     29I                                  I *                      97.92
     30I                                  I        *               99.83
     31I                             *.I                           96.9
     32I                                  I                   *    101.8
     33I                                  I                 *      101.14
     34I                                  I              *         100.83
     35I                              *   I                        97.06
     36I                                  I      *                 99
     37I                             *.I                           96.71
     38I              *                   I                        93.46
     39I*                                 I                        91.62
     40I               *                  I                        94
     41I               *                  I                        93.9
     42I                    *             I                        95.73
     43I                    *             I                        95.81
     44I                  *               I                        95.26
     45I          *                       I                        92.51
     46I            *                     I                        93.28
     47I                  *               I                        95.4
     48I                *                 I                        94.81
     49I                *                 I                        94.88
     50I                                 *I                        97.07
     51I                                  I *                      97.61
     52I                                  I *                      97.83
     53I                                  I       *                99.41
     54I                                  I      *                 99.1
     55I                                 *I                        97.19
     56I           *                      I                        94.3
     57I                                 *I                        97.08
     58I                                  I *                      98.16
     59I                                  I               *        100.6
     60I                                  I              *         100.42
    I+++++++++++++++++++++++++++++++++++++++++++++++++++
```

Figure 4.1 A simulated realization: example 1.

84

stationary. The estimated acf will offer additional clues about the stationarity of the mean of this realization. If the estimated acf drops quickly to zero, this is evidence that the mean of the data is stationary; if the estimated acf falls slowly to zero, the mean of the data is probably not stationary.

Even if the mean is not stationary it is still possible to calculate a single sample mean ($\bar{z}$) for a given realization. The mean of our realization is $\bar{z} = 97.1893$, shown in Figure 4.1 as the line running through the center of the data. If the mean is stationary it is a fixed, nonstochastic element in the data. Our next step is to remove this element temporarily by expressing the data in deviations from the mean $\tilde{z}_t = z_t - \bar{z}$. This allows us to focus on the stochastic components of the data. We then employ equations (2.5) and (2.9) to find the estimated autocorrelation and partial autocorrelation coefficients r_k and $\hat{\phi}_{kk}$.

The estimated acf and pacf are shown in Figure 4.2. The estimated autocorrelations drop to zero fairly quickly; absolute t-values fall below 1.6 by lag 3. We conclude that the mean of the realization is stationary and we do not difference the data.

At the identification stage our task is to compare the estimated acf and pacf with some common theoretical acf's and pacf's. If we find a match between the estimated and theoretical functions, we then select the process

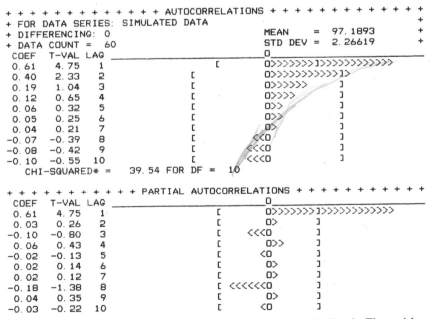

```
+ + + + + + + + + + + + AUTOCORRELATIONS + + + + + + + + + + + +
+ FOR DATA SERIES:  SIMULATED DATA                                         +
+ DIFFERENCING:  0                          MEAN     =   97. 1893          +
+ DATA COUNT =   60                         STD DEV  =   2. 26619           +
  COEF   T-VAL  LAG _____0_____
  0. 61   4. 75   1                    [           0>>>>>>> 1>>>>>>>>>>>>>>>
  0. 40   2. 33   2              [           0>>>>>>>>>>>1>
  0. 19   1. 04   3              [           0>>>>>>>            ]
  0. 12   0. 65   4              [           0>>>>              ]
  0. 06   0. 32   5              [           0>>               ]
  0. 05   0. 25   6              [           0>>               ]
  0. 04   0. 21   7              [            0>               ]
 -0. 07  -0. 39   8              [           <<0               ]
 -0. 08  -0. 42   9              [           <<0               ]
 -0. 10  -0. 55  10              [           <<<0              ]
        CHI-SQUARED* =     39. 54 FOR DF =   10

+ + + + + + + + + + + PARTIAL AUTOCORRELATIONS + + + + + + + + + + +
  COEF   T-VAL  LAG _____0_____
  0. 61   4. 75   1                    [           0>>>>>>> 1>>>>>>>>>>>>>>>
  0. 03   0. 26   2              [           0>                ]
 -0. 10  -0. 80   3              [           <<<0              ]
  0. 06   0. 43   4              [           0>>               ]
 -0. 02  -0. 13   5              [           <0                ]
  0. 02   0. 14   6              [           0>                ]
  0. 02   0. 12   7              [           0>                ]
 -0. 18  -1. 38   8              [  <<<<<<0                    ]
  0. 04   0. 35   9              [           0>                ]
 -0. 03  -0. 22  10              [           <0                ]
```

Figure 4.2   Estimated acf and pacf calculated from the realization in Figure 4.1.

associated with the matching theoretical functions as a tentative model for the available data. Keep in mind that we want a parsimonious as well as a statistically adequate model. That is, we want a model that fits the data adequately and requires the smallest possible number of estimated parameters.

The only theoretical acf's and pacf's we have seen so far are those for the AR(1) and MA(1) processes shown in Figures 3.4 and 3.5. Can you find a theoretical acf and pacf that match the estimated functions in Figure 4.2?

The closest match is an AR(1) with $\phi_1 > 0$. As shown in Figure 3.4, the theoretical acf for this process decays toward zero on the positive side, while the theoretical pacf has a single positive spike at lag 1. This is similar to the estimated acf and pacf in Figure 4.2. Therefore, we tentatively select this AR(1) model for our data:

$$z_t = C + \phi_1 z_{t-1} + a_t \qquad (4.1)$$

We now move to the estimation stage. Model (4.1) has two parameters, C and ϕ_1, requiring estimation. At the estimation stage we obtain accurate estimates of these parameters. We now make better use of the available data than we did at the identification stage since we estimate only two parameters. At the identification stage, by contrast, we estimated 21 values (the mean plus 10 autocorrelation coefficients plus 10 partial autocorrelation coefficients).

Figure 4.3 shows the results of fitting model (4.1) to the realization in Figure 4.1. We get these estimates:

$$\hat{\phi}_1 = 0.635$$

$$\hat{C} = 35.444$$

```
+ + + + + + + + +ECOSTAT UNIVARIATE B-J RESULTS+ + + + + + + + + +
+ FOR DATA SERIES:   SIMULATED DATA                                    +
+ DIFFERENCING:      0                              DF    = 57         +
+ AVAILABLE:         DATA = 60    BACKCASTS = 0     TOTAL = 60         +
+ USED TO FIND SSR:  DATA = 59    BACKCASTS = 0     TOTAL = 59         +
+ (LOST DUE TO PRESENCE OF AUTOREGRESSIVE TERMS:              1)       +

COEFFICIENT    ESTIMATE       STD ERROR       T-VALUE
PHI    1       0. 635         0. 104          6. 09
CONSTANT       35. 444        10. 1338        3. 49762

MEAN           97. 1978       . 642506        151. 279

ADJUSTED RMSE =  1. 79903   MEAN ABS % ERR =    1. 45
     CORRELATIONS
        1      2
1     1. 00
2     0. 03   1. 00
```

Figure 4.3   Estimation results for model (4.1).

It happens that $\hat{C}$ is equal to $\hat{\mu}(1 - \hat{\phi}_1) = 97.1978(1 - 0.635)$, where $\hat{\mu}$ is found simultaneously along with $\hat{\phi}_1$ by the computer estimation routine.* The relationship between the constant term and the mean is discussed in more detail in Chapter 5.

Because the absolute value of $\hat{\phi}_1$ is less than 1.0, we conclude the model is stationary. (This topic is discussed further in Chapter 6.) The absolute t-values (3.49 and 6.09) attached to $\hat{C}$ and $\hat{\phi}_1$ are greater than 2.0, so we conclude that these estimates are different from zero at better than the 5% significance level.

Next, we subject our tentative model to some diagnostic checks to see if it fits the data adequately. Diagnostic checking is related primarily to the assumption that the random shocks (a_t) are independent. If the shocks in a given model are correlated, the model must be reformulated because it does not fully capture the statistical relationship among the z's. That is, the shocks are part of the z's; if the shocks of a model are significantly correlated, then there is an important correlation among the z's that is not adequately explained by the model.

In practice we cannot observe the random shocks. But the residuals ($\hat{a}_t$) of an estimated model are estimates of the random shocks. To see how the residuals are calculated, solve (4.1) for a_t, that is, $a_t = z_t - C - \phi_1 z_{t-1}$. Although the z values are known, C and ϕ_1 are unknown; substitute the estimates $\hat{C} = 35.444$ and $\hat{\phi}_1 = 0.635$. The resulting equation gives estimates of the random shocks based on the known z's and the estimated parameters: $\hat{a}_t = z_t - \hat{C} - \hat{\phi}_1 z_{t-1}$. For $t = 1$, we cannot find $\hat{a}_1$ since there is no $z_{t-1} = z_0$ available. But for $t = 2$ and $t = 3$, we get the following results:

$$\hat{a}_2 = z_2 - \hat{C} - \hat{\phi}_1 z_1$$

$$= 99.08 - 35.444 - 0.635(99.06)$$

$$= 0.733$$

$$\hat{a}_3 = z_3 - \hat{C} - \hat{\phi}_1 z_2$$

$$= 98.27 - 35.444 - 0.635(99.08)$$

$$= -0.090$$

Other residuals are calculated in a similar manner.

In diagnostic checking we construct an acf, called a residual acf, using the residuals ($\hat{a}_t$) of the model as observations. This estimated acf is used to test the hypothesis that the random shocks (a_t) are independent. Figure 4.4

---

*Some programs first estimate μ with the realization mean $\bar{z}$ and then estimate the ϕ and θ coefficients.

shows the residual acf for model (4.1). Because the absolute t-values and the chi-squared statistic are all relatively small (none are significant at the 5% level), we conclude that the random shocks are independent and that model (4.1) is statistically adequate. The diagnostic-checking stage is discussed more fully in Chapter 9.

The preceding three-stage procedure is potentially iterative because the diagnostic checks might suggest a return to the identification stage and the tentative selection of a different model. In this introductory example our first try produced an adequate model. Examples of the repetitive application of these stages are presented in the case studies in Part II. Case 2 illustrates very well how diagnostic-checking results can send us back to the identification stage to choose an alternative model. When all diagnostic checks are satisfied the model is used for forecasting. In Chapter 10 and in several of the case studies we show how forecasts are produced.

Example 2.   Consider the data in Figure 4.5. As with the previous example, these data were simulated using a computer. Inspection of the data indicates that this series has a constant mean and variance.

Figure 4.6 shows the estimated acf and pacf for this realization. We conclude that the mean of the realization is stationary because the estimated acf falls off quickly to zero. At the identification stage we tentatively choose a model whose theoretical acf and pacf look like the estimated acf and pacf calculated from the data. The estimated acf and pacf (in Figure 4.6) are similar to the theoretical acf and pacf (in Figure 3.5) associated with an MA(1) process where $\theta_1 > 0$. The estimated acf cuts off to virtually zero after lag 1 while the estimated pacf decays toward zero. Therefore, we tentatively choose an MA(1) model to represent this realization. This model is written as

$$z_t = C - \theta_1 a_{t-1} + a_t \tag{4.2}$$

```
++RESIDUAL ACF++
  COEF   T-VAL LAG _____0_____
 -0. 04  -0. 31   1                            <<<<0
  0. 11   0. 84   2                                0>>>>>>>>>>>
 -0. 09  -0. 66   3                     <<<<<<<<<0
  0. 04   0. 31   4                                0>>>>
 -0. 06  -0. 44   5                       <<<<<<0
  0. 00   0. 00   6                                0
  0. 12   0. 88   7                                0>>>>>>>>>>>>
 -0. 11  -0. 84   8                     <<<<<<<<<<<0
  0. 05   0. 33   9                                0>>>>>
 -0. 01  -0. 07  10                               <0
       CHI-SQUARED* =     3. 71 FOR DF =   8
```

Figure 4.4   Residual acf for model (4.1).

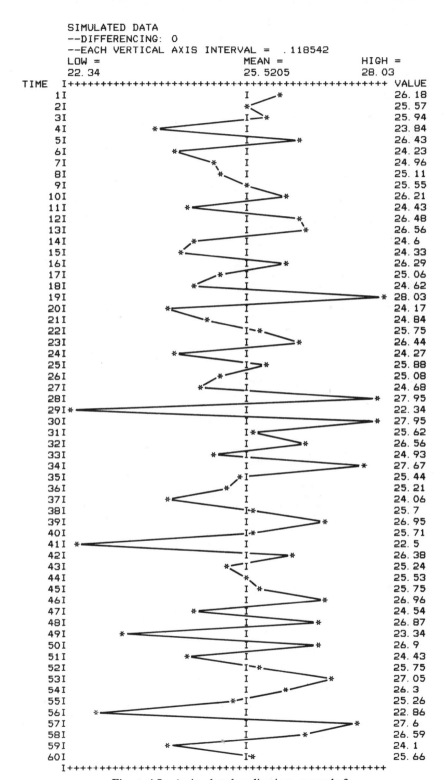

```
SIMULATED DATA
--DIFFERENCING: 0
--EACH VERTICAL AXIS INTERVAL =  .118542
   LOW =                        MEAN =              HIGH =
   22.34                        25.5205             28.03
TIME  I++++++++++++++++++++++++++++++++++++++++++++++++++  VALUE
   1I                                  I   ·*                26.18
   2I                                  *·*                   25.57
   3I                                  I·*                   25.94
   4I                 *                I                     23.84
   5I                                  I          *          26.43
   6I                   *              I                     24.23
   7I                      *·          I                     24.96
   8I                       *·         I                     25.11
   9I                                  *·                    25.55
  10I                                  I  ·*                 26.21
  11I                    *             I                     24.43
  12I                                  I      ·*             26.48
  13I                                  I        ·*           26.56
  14I                   *·             I                     24.6
  15I                  *·              I                     24.33
  16I                                  I     *               26.29
  17I                  *·*             I                     25.06
  18I                 *                I                     24.62
  19I                                  I                 *   28.03
  20I                *                 I                     24.17
  21I                   *·             I                     24.84
  22I                                  I·*                   25.75
  23I                                  I         *           26.44
  24I                 *                I                     24.27
  25I                                  I·*                   25.88
  26I                    *·            I                     25.08
  27I                  *·              I                     24.68
  28I                                  I              *      27.95
  29I*                                 I                     22.34
  30I                                  I               *     27.95
  31I                              I*                        25.62
  32I                              I      *                  26.56
  33I                  *·I                                   24.93
  34I                                  I            *        27.67
  35I                            *·I                         25.44
  36I                          *·      I                     25.21
  37I                 *·               I                     24.06
  38I                              I·*                        25.7
  39I                                  I       *             26.95
  40I                              I·*                        25.71
  41I  *                               I                      22.5
  42I                                  I     *               26.38
  43I                            *·I                         25.24
  44I                              *·                        25.53
  45I                                  I ·*                  25.75
  46I                                  I           *         26.96
  47I                   *              I                     24.54
  48I                                  I          *          26.87
  49I            *                     I                     23.34
  50I                                  I          *           26.9
  51I                   *              I                     24.43
  52I                              I·*·                      25.75
  53I                                  I              *      27.05
  54I                                  I      ·*              26.3
  55I                            *·I                         25.26
  56I        *                         I                     22.86
  57I                                  I              *       27.6
  58I                                  I         *           26.59
  59I                 *·               I                     24.1
  60I                              I·*                        25.66
      I++++++++++++++++++++++++++++++++++++++++++++++++++
```

Figure 4.5 A simulated realization: example 2.

```
+ + + + + + + + + + + + AUTOCORRELATIONS + + + + + + + + + + + + +
+ FOR DATA SERIES:  SIMULATED DATA                                    +
+ DIFFERENCING:  O                         MEAN    =  25.5205         +
+ DATA COUNT =  60                         STD DEV =   1.28263        +
  COEF   T-VAL LAG _____0_____
 -0.46   -3.53   1   <<<<<<<<<<<<[<<<<<<<<<<<<O                ]
  0.01    0.06   2          [              O                   ]
 -0.05   -0.34   3          [          <<<O                    ]
  0.16    1.01   4          [              O>>>>>>>>>           ]
 -0.13   -0.85   5          [         <<<<<<<O                  ]
  0.01    0.06   6          [              O                    ]
  0.00    0.01   7          [              O                    ]
  0.00    0.00   8          [              O                    ]
  0.06    0.36   9          [              O>>>                 ]
 -0.16   -1.03  10          [         <<<<<<<O                  ]
    CHI-SQUARED* =    18.30 FOR DF =   10

+ + + + + + + + + + + PARTIAL AUTOCORRELATIONS + + + + + + + + + + +
  COEF   T-VAL LAG _____0_____
 -0.46   -3.53   1   <<<<<<<<<<<<[<<<<<<<<<<<<O                ]
 -0.25   -1.94   2       <[<<<<<<<<<<<<O                       ]
 -0.22   -1.67   3        [<<<<<<<<<<<<O                       ]
  0.04    0.32   4          [              O>>                 ]
 -0.05   -0.39   5          [          <<<O                    ]
 -0.07   -0.53   6          [          <<<O                    ]
 -0.05   -0.41   7          [          <<<O                    ]
 -0.07   -0.56   8          [         <<<<O                    ]
  0.05    0.41   9          [              O>>>                 ]
 -0.15   -1.18  10          [         <<<<<<<O                  ]
```

Figure 4.6 Estimated acf and pacf calculated from the realization in Figure 4.5.

```
+ + + + + + + + + +ECOSTAT UNIVARIATE B-J RESULTS+ + + + + + + + + +
+ FOR DATA SERIES:   SIMULATED DATA                                   +
+ DIFFERENCING:      O                          DF    = 58            +
+ AVAILABLE:         DATA = 60   BACKCASTS = 0   TOTAL = 60            +
+ USED TO FIND SSR:  DATA = 60   BACKCASTS = 0   TOTAL = 60            +
+ (LOST DUE TO PRESENCE OF AUTOREGRESSIVE TERMS:            0)        +

COEFFICIENT     ESTIMATE        STD ERROR       T-VALUE
  THETA   1      0.631           0.101           6.27
  CONSTANT      25.5194          .537192E-01    475.053

  MEAN          25.5194          .537192E-01    475.053

ADJUSTED RMSE =  1.10313    MEAN ABS % ERR =    3.59
      CORRELATIONS
         1      2
  1    1.00
  2    0.00   1.00
```

Figure 4.7 Estimation results for model (4.2).

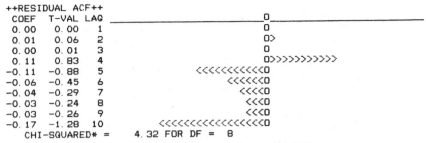

```
++RESIDUAL ACF++
 COEF   T-VAL  LAG _____O_____
 0.00   0.00    1                                        O
 0.01   0.06    2                                        O>
 0.00   0.01    3                                        O
 0.11   0.83    4                                        O>>>>>>>>>>>
-0.11  -0.88    5                            <<<<<<<<<<<O
-0.06  -0.45    6                                  <<<<<<O
-0.04  -0.29    7                                    <<<<O
-0.03  -0.24    8                                     <<<O
-0.03  -0.26    9                                     <<<O
-0.17  -1.28   10                      <<<<<<<<<<<<<<<<<O
         CHI-SQUARED* =   4.32 FOR DF =  8
```

Figure 4.8   Residual acf for model (4.2).

Now we are ready for the estimation stage, where we reuse the realization to estimate C and θ_1. Estimation results appear in Figure 4.7. The estimated parameters are

$$\hat{\theta}_1 = 0.631$$

$$\hat{C} = 25.5194$$

In this case, $\hat{C}$ is equal to $\hat{\mu}$, the estimated mean of the realization.

Because the absolute value of $\hat{\theta}_1$ is less than one, we conclude that this model is invertible. (The concept of invertibility is explained in Chapter 6.) The t-values of 475.05 for $\hat{C}$ and 6.27 for $\hat{\theta}_1$ indicate that these estimates are significantly different from zero at better than the 5% level. Thus far the model is acceptable.

Next, we do some diagnostic checking. As in the previous example we test the hypothesis that the shocks (a_t) are statistically independent by constructing an acf using the model's residuals ($\hat{a}_t$). The residual acf is shown in Figure 4.8. The t-values and the chi-squared statistic are all relatively small, allowing us to conclude that the random shocks of model (4.2) are independent. We have found a satisfactory model—one that is both parsimonious and statistically adequate.

You should now be able to read the first several case studies in Part II, although some of the points will not be clear until you have read Chapters 5–12.

Summary

1. A model is an imitation of the underlying process derived from analysis of the available data.

2. A good ARIMA model
    (a)   is parsimonious;
    (b)   is stationary;
    (c)   is invertible;
    (d)   has high-quality estimated coefficients;
    (e)   has statistically independent residuals;
    (f)   fits the available data satisfactorily; and
    (g)   forecasts the future satisfactorily.

Questions and Problems

4.1   State the principle of parsimony.

4.2   Why is the principle of parsimony considered important?

4.3   How do we test whether the random shocks of a tentatively selected model are statistically independent?

4.4   Construct an ARIMA model for the following three realizations by subjecting them to the UBJ procedure of identification, estimation, and diagnostic checking. Defend your final model in terms of the characteristics of a good model (see Table 4.1). You may find this to be a difficult exercise since we have not yet discussed all the relevant details. Nevertheless, analyzing these realizations should help you develop a better understanding of the three stages in the UBJ method.

Realization I

t	z_t	t	z_t	t	z_t	t	z_t
1	0.61	16	−1.45	31	−0.42	46	0.10
2	0.85	17	−2.77	32	−0.64	47	0.01
3	0.41	18	−1.19	33	−1.58	48	0.24
4	1.24	19	0.17	34	−0.74	49	−1.46
5	0.95	20	−0.43	35	0.77	50	−0.85
6	0.94	21	−1.82	36	0.17	51	−1.28
7	0.65	22	−0.47	37	0.75	52	−0.92
8	−0.04	23	0.5	38	−0.18	53	0.96
9	0.76	24	−0.33	39	−0.34	54	1.24
10	1.47	25	−1.33	40	1.29	55	0.17
11	1.39	26	0.46	41	1.01	56	0.20
12	0.66	27	0.43	42	−0.23	57	0.92
13	0.17	28	−0.36	43	−2.12	58	0.86
14	−0.73	29	−1.42	44	0.48	59	−0.56
15	−0.52	30	−1.65	45	0.59	60	0.12

Realization II

t	z_t	t	z_t	t	z_t	t	z_t
1	1.36	16	−0.54	31	1.10	46	−2.62
2	−0.19	17	0.49	32	−0.16	47	2.61
3	0.52	18	−0.31	33	−0.49	48	0
4	−0.07	19	0.30	34	−0.33	49	1.67
5	−0.17	20	−2.02	35	−0.62	50	−1.68
6	−0.58	21	1.58	36	1.68	51	1.78
7	1.58	22	−0.16	37	−1.54	52	−0.53
8	−0.46	23	−0.68	38	0.75	53	−0.21
9	0.33	24	−0.12	39	−0.10	54	0.76
10	−0.76	25	−0.41	40	−0.99	55	−2.25
11	0.71	26	0.37	41	1.38	56	2.47
12	−0.03	27	−1.75	42	−1.49	57	−0.93
13	1.23	28	0.04	43	0.50	58	0.50
14	−1.19	29	−1.31	44	−0.65	59	1.26
15	1.37	30	1.03	45	0.79	60	−0.16

Realization III

t	z_t	t	z_t	t	z_t	t	z_t
1	−0.36	16	−2.49	31	0.04	46	1.78
2	0.55	17	−2.57	32	1.08	47	3.24
3	0.13	18	−1.63	33	1.23	48	2.29
4	−1.27	19	−2.03	34	0.13	49	0.52
5	1.36	20	−1.10	35	−0.50	50	2.22
6	−0.18	21	−1.09	36	2.19	51	1.34
7	0.11	22	−0.42	37	0.99	52	1.35
8	−0.79	23	−2.13	38	0.72	53	0.17
9	−0.11	24	−0.25	39	1.35	54	1.59
10	0.53	25	1.58	40	−0.54	55	−1.04
11	0.39	26	0.29	41	1.24	56	1.21
12	0.47	27	1.20	42	0.57	57	0.06
13	−1.30	28	0.78	43	1.35	58	−0.26
14	−2.38	29	0.29	44	0.30	59	−0.66
15	−2.31	30	−1.09	45	1.13	60	−0.24

5

NOTATION AND
THE INTERPRETATION
OF ARIMA MODELS

In Chapters 1–4 we introduced the basic definitions, statistical concepts, and modeling procedures of the UBJ–ARIMA forecasting method.

In the present chapter we first discuss some of the terminology and notation associated with ARIMA models. You must become familiar with this notation in order to read UBJ–ARIMA literature with ease. If you are familiar with basic high school algebra you should not have serious difficulty understanding and using ARIMA notation. The only requirement is that you *perform the indicated algebraic manipulations yourself*. This takes time, but it is necessary and not difficult.

Next, we consider how ARIMA models can be interpreted in common-sense ways. This is especially important for the practicing forecaster who must give managers an intuitive explanation of ARIMA models.

In discussing the interpretation of ARIMA models, we demonstrate a very important result—AR terms and MA terms are algebraically interchangeable (though not on a simple one-for-one basis). In Section 5.3 we show that an MA(1) process is equivalent to an AR process of infinitely high order; in Section 5.4 we show that an AR(1) process is equivalent to an MA process of infinitely high order. This interchangeability of AR and MA terms is important in practice since our objective is to find the most parsimonious model that adequately represents a data series. If we can substitute a few AR terms for many MA terms, or a few MA terms for many AR terms, we satisfy the principle of parsimony.

In Chapter 6 we return to a discussion of the concepts underlying the identification stage of the UBJ method.

5.1   Three processes and ARIMA(p, d, q) notation

In Chapter 3 we presented the ordinary algebraic form of two common ARIMA processes, the AR(1) and the MA(1):

$$z_t = C + \phi_1 z_{t-1} + a_t \tag{5.1}$$

$$z_t = C - \theta_1 a_{t-1} + a_t \tag{5.2}$$

Here are three additional processes to consider:

$$z_t = C + \phi_1 z_{t-1} + \phi_2 z_{t-2} + a_t \tag{5.3}$$

$$z_t = C - \theta_1 a_{t-1} - \theta_2 a_{t-2} + a_t \tag{5.4}$$

$$z_t = C + \phi_1 z_{t-1} - \theta_1 a_{t-1} + a_t \tag{5.5}$$

Equation (5.3) is called an AR(2) process because it contains only AR terms (in addition to the constant term and the current random shock), and the maximum time lag on the AR terms is two. Process (5.4) is called an MA(2) since it has only MA terms, with a maximum time lag on the MA terms of two. Equation (5.5) is an example of a *mixed* process—it contains both AR and MA terms. It is an ARMA(1, 1) process because the AR order is one and the MA order is also one.

We may generalize from these examples. Let the AR order of a process be designated p, where p is some non-negative integer. Let q, also a non-negative integer, be the MA order of a process. Let d, another non-negative integer, stand for the number of times a realization must be differenced to achieve a stationary mean.* After a differenced series has been modeled, it is *integrated d* times to return the data to the appropriate overall level. (Integration is discussed in Chapter 7.) The letter "I" in the acronym ARIMA refers to this integration step, and it corresponds to the number of times (d) the original series has been differenced; if a series has been differenced d times, it must subsequently be integrated d times to return it to its original overall level.

---

*Stationarity and differencing were discussed in Chapters 1–3.

ARIMA processes are characterized by the values of p, d, and q in this manner: ARIMA(p, d, q). For example, equation (5.3) is an ARIMA(2, 0, 0) process, or simply an AR(2). Equation (5.4) is an ARIMA(0, 0, 2) process, or an MA(2). And (5.5) is an ARIMA(1, 0, 1) or an ARMA(1, 1). This notation becomes more complicated when we deal with a certain type of seasonal process; but as we will see in Chapter 11 the basic idea remains the same for seasonal models also.

Some coefficients with lags less than the order of a process could be zero. For example, if θ_1 in process (5.4) is zero, that process is written more simply as

$$z_t = C - \theta_2 a_{t-2} + a_t. \tag{5.6}$$

Equation (5.6) is still an MA(2) process because the maximum lag on past random shock terms is two.

5.2 Backshift notation

ARIMA models are often written in *backshift notation*. You must become familiar with this notation if you want to thoroughly understand time-series literature.

Some students find backshift notation difficult at first. However, with a little patience and some practice you should find backshift notation relatively easy to understand and convenient to use. The important thing is to practice translating ARIMA models written in backshift form into both ARIMA(p, d, q) form and common algebraic form. (Take out a pencil and some scratch paper. You will need them as you read this chapter.)

Keep in mind that backshift notation involves no new statistical concepts. It is merely a convenient way of writing ARIMA processes and models.

We utilize the backshift operator B, which operates in this way: if we multiply z_t by B, we get z_{t-1}. That is,

$$Bz_t = z_{t-1} \tag{5.7}$$

The operator may be unlike any other you have seen in algebra. You will avoid confusion if you do *not* think of B as a number. Although we treat B like other algebraic terms (e.g., we may raise it to a power), it does not stand for a number.

To make common sense of the B symbol, recall equation (5.7). It states B is meaningful because it shifts time subscripts. When you see B in an algebraic expression, remember that B must be multiplied by some other variable, such as z_t or a_t. B is meaningful, then, because it alters the time subscript on the variable by which it is multiplied, as stated in (5.7).*

In equation (5.7) the exponent of B is one. Since any number raised to the power one is that same number, we need not explicitly write the exponent when it is one. But the exponent of B might be two, for example. Multiplying z_t by B^2 gives

$$B^2 z_t = z_{t-2} \tag{5.8}$$

The same pattern holds for other exponents of B. In general, multiplying z_t by B^k gives z_{t-k}. Thus, by definition,

$$B^k z_t = z_{t-k} \tag{5.9}$$

Multiplying a constant by B^k does not affect the constant, regardless of the value of k, because constants lack time subscripts. For example, let C be a constant. Then

$$BC = C$$

$$B^2 C = C$$

$$\vdots$$

$$B^k C = C \tag{5.10}$$

We can extend the above definitions of how B operates to write the *differencing operator* $(1 - B)$. (Recall that some data series must be differenced to induce a stationary mean before being modeled with the UBJ–ARIMA method.) Multiplying z_t by $(1 - B)$ produces the first differences of z_t:

$$(1 - B)z_t = z_t - z_{t-1} \tag{5.11}$$

There is really nothing new in equation (5.11). It is merely an extension of equation (5.7). If we expand the LHS of (5.11) and recall from (5.7) that $Bz_t = z_{t-1}$, we get

$$(1 - B)z_t = z_t - Bz_t$$

$$= z_t - z_{t-1} \tag{5.12}$$

thus showing that (5.11) is indeed correct.

---

*Note that B may not operate on a function of z_t or a_t, such as z_t^2.

Again, you should not think of B as a number. Thus $(1 - B)$ is not a numerical value; it is an operator. $(1 - B)$ has a common-sense meaning only when it is multiplied by a variable. When you see the operator $(1 - B)$, recall equation (5.11). It shows that $(1 - B)$ multiplied by a time-sequenced variable is just another way of writing the first differences of that variable.

In (5.11) the differencing operator $(1 - B)$ is raised to the power one. Multiplying z_t by $(1 - B)^2$ would produce the second differences of z_t. In general, multiplying z_t by $(1 - B)^d$ gives the dth differences of z_t. Of course, if $d = 0$, then $(1 - B)^d$ is equal to one. In that case we need not explicitly multiply z_t by the differencing operator.

Let us demonstrate that $(1 - B)^2 z_t$ is the same as the second differences of z_t (designated w_t). The first differences of z_t are $z_t - z_{t-1}$. The second differences of z_t are the first differences of the first differences; that is, subtract from any given first difference $(z_t - z_{t-1})$ the previous first difference $(z_{t-1} - z_{t-2})$:

$$w_t = (z_t - z_{t-1}) - (z_{t-1} - z_{t-2})$$

$$= z_t - 2z_{t-1} + z_{t-2} \tag{5.13}$$

To show that $(1 - B)^2 z_t$ is identical to (5.13) expand the operator $(1 - B)^2$ and apply definition (5.9):

$$(1 - B)^2 z_t = (1 - 2B + B^2)z_t$$

$$= z_t - 2Bz_t + B^2 z_t$$

$$= z_t - 2z_{t-1} + z_{t-2} \tag{5.14}$$

We see that $(1 - B)^2 z_t$ is a compact and convenient way of writing the second differences of z_t.

In Chapter 2 we introduced the idea of expressing data in deviations from the realization mean $(\bar{z})$, defining $\tilde{z}_t = z_t - \bar{z}$. When writing a process in backshift form we write the random variable z_t in deviations from the process mean (μ), defining $\tilde{z}_t = z_t - \mu$. Thus the symbol $\tilde{z}_t$ does double duty. When referring to a realization or a model based on a realization, $\tilde{z}_t$ stands for deviations from the realization mean $\bar{z}$. When referring to a process, $\tilde{z}_t$ stands for deviations from the process mean μ.

We are now ready to write some nonseasonal processes in backshift form. The procedure is given by these six steps:

1. Start with a variable z_t that has been transformed (if necessary) so it has a constant variance.

2. Write z_t in deviations from its mean: $\tilde{z}_t = z_t - \mu$.

3. Multiply $\tilde{z}_t$ by the differencing operator $(1 - B)^d$ to ensure that we have a variable whose mean is stationary.

4. Multiply the result from step 3 by the *AR operator* whose general form is $(1 - \phi_1 B - \phi_2 B^2 - \cdots - \phi_p B^p)$. For a specific process, assign the appropriate numerical value to p, the order of the AR part of the process. If any ϕ coefficients with subscripts less than p are zero, exclude those terms from the AR operator.

5. Multiply the random shock a_t by the *MA operator* whose general form is $(1 - \theta_1 B - \theta_2 B^2 - \cdots - \theta_q B^q)$. For a specific process, assign the appropriate numerical value to q, the order of the MA portion of the process. If any θ coefficients at lags less than q are zero, exclude them from the MA operator.

6. Equate the results from steps 4 and 5.

Combining the above six steps, a nonseasonal process in backshift notation has this general form (numbers in parentheses represent steps):

$$\underbrace{\left(1 - \phi_1 B - \phi_2 B^2 - \cdots - \phi_p B^p\right)}_{(4)} \underbrace{\left(1 - B\right)^d}_{(3)} \underbrace{\tilde{z}_t}_{(1) \text{ and } (2)}$$

$$\underbrace{=}_{(6)} \underbrace{\left(1 - \theta_1 B - \theta_2 B^2 - \cdots - \theta_q B^q\right) a_t}_{(5)}$$

(5.15)

Equation (5.15) can be written in a compact form that often appears in time-series literature. Define the following symbols:

$$\nabla^d = \left(1 - B\right)^d$$

$$\phi(B) = \left(1 - \phi_1 B - \phi_2 B^2 - \cdots - \phi_p B^p\right)$$

$$\theta(B) = \left(1 - \theta_1 B - \theta_2 B^2 - \cdots - \theta_q B^q\right)$$

Substituting each of these definitions into (5.15) we get

$$\phi(B)\nabla^d \tilde{z}_t = \theta(B) a_t \tag{5.16}$$

Although we do not use this compact notation very often in this book (it is useful in Chapter 11), you may see similar notation in other texts or in professional journal articles dealing with ARIMA models. Remember that (5.16) is merely a compact way of saying that the random variable z_t evolves according to an ARIMA(p, d, q) process.

We now consider some examples of models in backshift form. We first show the common algebraic form of a process, and then we follow the six steps to write the process in backshift notation. We then apply rules (5.9) and (5.10) to demonstrate that the backshift form and the common algebraic form are identical.

Example 1.   Consider an AR(2) process. Let z_t have a constant mean and variance so no transformations are necessary. The common algebraic form of a stationary AR(2) process, seen earlier in (5.3), is

$$z_t = C + \phi_1 z_{t-1} + \phi_2 z_{t-2} + a_t \tag{5.17}$$

To write this in backshift notation, follow the six steps. Step 1 is satisfied by assumption. At step 2, express z_t in deviations from the mean $\tilde{z}_t = z_t - \mu$. At step 3, multiply this result by the differencing operator $(1 - B)^d$. Because z_t already has a constant mean, no differencing is required, so $d = 0$. Therefore, $(1 - B)^d = 1$ and we need not write out this term explicitly. Now multiply $\tilde{z}_t$ by the appropriate AR operator (step 4). For an AR(2) process, $p = 2$. Therefore, the required AR operator is $(1 - \phi_1 B - \phi_2 B^2)$. Next, multiply a_t by the appropriate MA operator (step 5). Since $q = 0$ for an AR(2), the MA operator collapses to 1. Finally (step 6), equate the results from steps 4 and 5 to obtain

$$\left(1 - \phi_1 B - \phi_2 B^2\right)\tilde{z}_t = a_t \tag{5.18}$$

We want to show that (5.17), the common algebraic form, and (5.18), the backshift form, are identical. First, expand the LHS of (5.18) and move all terms except $\tilde{z}_t$ to the RHS to obtain

$$\tilde{z}_t = \phi_1 B \tilde{z}_t + \phi_2 B^2 \tilde{z}_t + a_t \tag{5.19}$$

Next, apply rule (5.9) to get

$$\tilde{z}_t = \phi_1 \tilde{z}_{t-1} + \phi_2 \tilde{z}_{t-2} + a_t \tag{5.20}$$

The only difference between (5.20) and (5.17) is that in (5.20) z_t is expressed in deviations from the mean. Substituting $(z_t - \mu)$ for $\tilde{z}_t$ in (5.20) and rearranging terms we get

$$z_t = \mu(1 - \phi_1 - \phi_2) + \phi_1 z_{t-1} + \phi_2 z_{t-2} + a_t$$

Now let $C = \mu(1 - \phi_1 - \phi_2)$ and we have (5.17).

In this example the constant term is not the same as the mean, but it is related to the mean. This is true for all processes containing AR terms. The constant term of an ARIMA process is equal to the mean times the quantity one minus the sum of the AR coefficients:*

$$C = \mu(1 - \phi_1 - \phi_2 - \cdots - \phi_p)$$

$$= \mu\left(1 - \sum_{i=1}^{p} \phi_i\right) \tag{5.21}$$

According to (5.21) if no AR terms are present the constant term C is equal to the mean μ. This is true for all pure MA processes.

Example 2. Consider a process for a variable (z_t) which must be differenced once because its mean is not constant. Suppose the first differences of z_t are a series of independent random shocks. That is,

$$z_t - z_{t-1} = a_t \tag{5.22}$$

To write this in backshift form, begin by expressing z_t in deviations from the mean. Then, to account for the differencing, multiply $\tilde{z}_t$ by the backshift operator with $d = 1$. Both the AR and MA operators collapse to one in this example because $p = q = 0$, so we may ignore them. Equating terms (step 6)

$$(1 - B)\tilde{z}_t = a_t \tag{5.23}$$

To show that (5.22) and (5.23) are identical, substitute $z_t - \mu$ for $\tilde{z}_t$ and expand the LHS of (5.23):

$$z_t - Bz_t - \mu + B\mu = a_t \tag{5.24}$$

Apply (5.9) and (5.10) to (5.24). The μ terms add to zero so

$$z_t - z_{t-1} = a_t \tag{5.25}$$

which is identical to (5.22).

---

*As noted in Chapter 11, for certain kinds of seasonal models the constant term is somewhat different from (5.21), but the basic idea remains the same.

Note that μ dropped out when the differencing operator was applied to $(z_t - \mu)$. This happens with any process when $d > 0$. This result for processes parallels an earlier statement in Chapter 2 that differencing a realization usually produces a new series with a mean that is not statistically different from zero.

If (5.25) were intended to represent a data series whose first differences had a mean significantly different from zero, we could insert a constant term on the RHS of (5.25). This point is discussed more fully in Chapter 7. For now, we emphasize this point: in practice, differencing a realization often induces a mean of zero so that insertion of a constant term in the model after differencing is not needed. This result is especially common for data in business, economics, and other social science disciplines. The corresponding algebraic result for processes, as shown in equations (5.23)–(5.25), is that the μ terms add to zero (and the constant term is therefore zero) when $d > 0$.

Example 3.   Consider an ARIMA(1, 1, 1) process. Let the variable z_t have a constant variance. Because $d = 1$, the AR terms apply to the first differences of z_t rather than to z_t itself. Therefore, this process is written in common algebra as

$$z_t - z_{t-1} = C + \phi_1(z_{t-1} - z_{t-2}) - \theta_1 a_{t-1} + a_t \qquad (5.26)$$

Because $d = 1$, the LHS variable is not z_t but the first differences of z_t. Likewise, the AR coefficient is attached to the first differences of z_{t-1} rather than to z_{t-1}.

To write the ARIMA(1, 1, 1) in backshift notation, follow the same six steps defined earlier. Work through those steps yourself to see if you arrive at the following result:

$$(1 - \phi_1 B)(1 - B)\tilde{z}_t = (1 - \theta_1 B)a_t \qquad (5.27)$$

Start with the deviations of z_t from μ (step 2). Multiplying $\tilde{z}_t$ by $(1 - B)$ gives the first differences of z_t (step 3). Apply the AR operator to this result (step 4) and multiply a_t by the MA operator (step 5). Step 6 gives (5.27).

To show that (5.26) and (5.27) are identical, substitute $z_t - \mu$ for $\tilde{z}_t$ and expand both sides of (5.27) to get

$$z_t - Bz_t + B\mu - \mu - \phi_1 Bz_t + \phi_1 B^2 z_t + \phi_1 B\mu - \phi_1 B^2 \mu = a_t - \theta_1 a_{t-1}$$

$$(5.28)$$

Apply rules (5.9) and (5.10) to (5.28). The μ terms add to zero. Rearrange and collect terms to get (5.26).

Note that $C = 0$, implying from equation (5.21) that the first differences have a mean of zero. This is the same result we obtained in Example 2. If this result were not true for a data series whose behavior is otherwise well-represented by (5.27), we could insert a nonzero constant term on the RHS of that model to reflect the nonzero mean of the differenced data.

You should check your understanding of ARIMA(p, d, q) and backshift notation by doing the exercises at the end of this chapter.

5.3   Interpreting ARIMA models I: optimal extrapolation of past values of a single series

In this section and the next two sections we discuss the interpretation of ARIMA models. This is important since many analysts and managers who use statistical forecasts prefer techniques that can be interpreted in a common-sense or intuitive way.

There is no general, definitive interpretation of ARIMA models. However, we will discuss some ideas to help you see that many ARIMA models can be rationalized.

We discuss how ARIMA forecasts can be interpreted as optimal *extrapolations of past values of the given series.* "Optimal" refers to the fact that a properly constructed ARIMA model has a smaller forecast-error variance than any other linear univariate model. This optimal quality of ARIMA forecasts is discussed further in Chapter 10.

In this section we assume that we have a properly constructed ARIMA model (one that is parsimonious and statistically adequate). We want to show that all three major components of such a model—the constant term, the AR terms, and the MA terms—represent past z values with certain weights attached. Therefore, we may interpret ARIMA forecasts as optimal extrapolations of past values of the series.

This point is most easily demonstrated for the AR terms. We have already seen that the AR portion of an ARIMA model is simply the sum of selected past z values, each with a weight (a ϕ coefficient) assigned to it. In practice we must use estimated ϕ's (designated $\hat{\phi}$) found at the estimation stage, so we have the AR part of a model equal to

$$\hat{\phi}_1 z_{t-1} + \hat{\phi}_2 z_{t-2} + \cdots + \hat{\phi}_p z_{t-p}$$

Clearly, a forecast of z_t based on this portion of an ARIMA model involves the extrapolation of past z's ($z_{t-1}, z_{t-2}, \ldots, z_{t-p}$) into the future.

It is also easy to show that the constant term reflects only past z's. From (5.21) we know that (for a nonseasonal process) the constant term is

$$C = \mu\left(1 - \sum_{i=1}^{p} \phi_i\right)$$

In practice we have only an estimated mean $\hat{\mu}$ and we replace each ϕ with its estimated value $(\hat{\phi})$. Therefore, the estimated constant term is

$$\hat{C} = \hat{\mu}\left(1 - \sum_{i=1}^{p} \hat{\phi}_i\right)$$

An estimated mean $\hat{\mu}$ is clearly a combination of past z's. The quantity $(1 - \sum\hat{\phi}_i)$ is simply a weight assigned to $\hat{\mu}$. Therefore, a forecast of z_t based on the constant term in an ARIMA model also represents an extrapolation of past z's into the future.

It is more difficult to show that MA terms represent past z's. Rather than prove it rigorously for the general case, we demonstrate it for the MA(1). We will find that the MA(1) process can be interpreted as an AR process of infinitely high order. (We show this result using a theorem about geometric series, but it can also be shown using ordinary algebraic substitution. See problem 5.3 at the end of the chapter.)

Start with an MA(1) process. In backshift form this is

$$\tilde{z}_t = (1 - \theta_1 B)a_t \tag{5.29}$$

Dividing both sides of (5.29) by $(1 - \theta_1 B)$ gives

$$(1 - \theta_1 B)^{-1}\tilde{z}_t = a_t \tag{5.30}$$

Now, a theorem about geometric series states that if $|\theta_1| < 1$, then $(1 - \theta_1 B)^{-1}$ is the sum of a convergent infinite series:

$$(1 - \theta_1 B)^{-1} = \left(1 + \theta_1 B + \theta_1^2 B^2 + \theta_1^3 B^3 + \cdots\right), \quad \text{if } |\theta_1| < 1 \tag{5.31}$$

Substituting (5.31) into (5.30), we get

$$\left(1 + \theta_1 B + \theta_1^2 B^2 + \theta_1^3 B^3 + \cdots\right)\tilde{z}_t = a_t \tag{5.32}$$

You should be able to see that (5.32) is an AR process of infinitely high order with the ϕ coefficients following this pattern:

$$\phi_1 = -\theta_1$$

$$\phi_2 = -\theta_1^2$$

$$\phi_3 = -\theta_1^3$$

$$\vdots$$

In practice we have only an estimate of θ_1 (designated $\hat{\theta}_1$), and estimates of the a_t series (designated $\hat{a}_t$). Furthermore, we would use the more compact MA form (5.29) rather than the expanded AR form (5.32) to produce forecasts because the principle of parsimony dictates that we use a few MA terms in place of many AR terms whenever possible. Nevertheless, we see from (5.32) that an MA(1) model can be *interpreted* as an AR model of infinitely high order. The same is true for any pure MA model, and for the MA portion of any mixed ARIMA model. Therefore, we can interpret the MA portion of a properly constructed ARIMA model as representing a large number of past z's with certain weights attached. Thus all three parts of a properly constructed ARIMA model—the AR terms, the constant term, and the MA terms—taken together, can be interpreted as providing an optimal extrapolation of past values of the given series.

5.4   Interpreting ARIMA models II: rationalizing them from their context

The emphasis in ARIMA forecasting tends to be on finding statistical patterns regardless of the reason for those patterns. But it is also true that ARIMA models can often provide a *reasonable* representation of the behavior of a data series; that is, they can be interpreted in a common-sense way based on insight into the nature of the data. In this section we present some examples to show how ARIMA models can sometimes be rationalized from their context.

Example 1.   Suppose there are hundreds of financial analysts studying a corporation. They decide to buy or sell shares of this company by comparing it with the alternatives. These analysts attempt to use all available information (as long as the expected benefits exceed the expected costs of acquiring the information) as they monitor the price of the company's stock and try to estimate the size and stability of future earnings. News about events affecting this firm is disseminated quickly and at low cost to all interested parties. The shares are traded continuously in a market with low transaction costs. Under these conditions, it is reasonable to suppose that new information will be reflected in the price of the shares quite rapidly.

These circumstances imply that past prices contain virtually no information that would allow an analyst to forecast future price changes so as to *regularly* make above-normal trading profits. After all, if past price patterns could be exploited in this way, people watching this stock would learn about the patterns and try to take advantage of them. This collective action would quickly raise or lower the stock price to a level where the chance for unusual gain would disappear. Therefore, past prices would cease to show patterns that could be exploited with any consistency.

From the preceding argument, a stock-price forecasting model *based only on past prices* would state that the change in price ($z_t - z_{t-1}$) is independent of past prices; the model would consist of a series of independent random shocks. Price changes would reflect only things other than past prices, plus the current irregular errors of judgment that the market participants cannot avoid entirely. We are not saying that all information about this firm is useless; rather, we are saying that knowledge of past prices would not help in forecasting future price changes. In backshift form this model is

$$(1 - B)\tilde{z}_t = a_t \qquad (5.33)$$

Expanding the LHS and applying the rules for the backshift operator gives the following common algebraic form for this model:

$$z_t - z_{t-1} = a_t$$

or

$$z_t = z_{t-1} + a_t \qquad (5.34)$$

(What happened to the μ term implicit in $\tilde{z}_t$?) Equation (5.34) is the famous *random-walk* model implied by the efficient-markets hypothesis. It has been found to be a good model for many stock-price series (see Part II, Case 6).

The concept of a random walk plays an important role in the analysis of nonstationary data series, discussed further in Chapter 7. According to (5.33), the series z_t is differenced once: the exponent d of the differencing operator $(1 - B)^d$ is one. In Chapter 2 we noted that many series without a fixed mean can be transformed into stationary series by differencing. Equation (5.34) describes a series z_t that requires differencing because it does not have a fixed mean. It says that z_t moves at random starting from the immediately prior value (z_{t-1}) rather than starting from a fixed central value.

Example 2.* A national computer-dating service has a pool of clients. The list is updated each week. The number in each week's pool (z_t) is composed of several parts. We begin with a constant fraction (ϕ_1) of last week's pool (z_{t-1}) that remains in the pool this week. The fraction of last week's pool no longer in the pool this week is therefore $(1 - \phi_1)$. Thus $\phi_1 z_{t-1}$ represents the number of people from last week remaining in the pool, and $(1 - \phi_1)z_{t-1}$ represents the sum of those who cancel their registration plus those who are successfully matched with someone else by the computer.

Next, we add a number (C') of new clients registering each week. The number of new clients fluctuates randomly about a fixed central value. That is, C' has a fixed component C and an additive white-noise (Normal random-shock) element a_t, and is defined as $C' = C + a_t$.

Let the fixed number C added each week be equal to the overall mean of the weekly pool (μ) times the fraction $(1 - \phi_1)$ lost each week due to cancellation or a successful match, $C = \mu(1 - \phi_1)$. In other words, we let the series mean be stationary; the fixed number added each week (C) is just enough to keep the mean level of the pool (μ) constant through time. But because a_t is part of C', z_t will fluctuate randomly around μ.

Combining the above elements, we get the following AR(1) model:

$$z_t = \mu(1 - \phi_1) + \phi_1 z_{t-1} + a_t \tag{5.35}$$

which in backshift form is

$$(1 - \phi_1 B)\tilde{z}_t = a_t \tag{5.36}$$

Example 3. A chemical process generates an hourly yield (z_t) of an output. The yield is fixed at a certain level (μ) if the two input chemicals are combined in a 3:1 ratio. But z_t varies around μ because the input ratio varies randomly around 3:1 due to measurement error. Furthermore, when the input ratio is not 3:1, there are several trace by-products left in the processing tank that are dispersed gradually over many hours. The exact combination of by-products depends on the input ratio. Some of the by-products raise future hourly yields, while other by-products lower future yields.

Any trace by-products are effectively dispersed after q hours. Then the hourly yield follows an MA(q) process:

$$z_t = \mu - \theta_1 a_{t-1} - \theta_2 a_{t-2} - \cdots - \theta_q a_{t-q} + a_t$$

*The next two examples are adapted from Granger and Newbold [17, pp. 15 and 23–24]. Adapted by permission of the authors and publisher.

or

$$\tilde{z}_t = \left(1 - \theta_1 B - \theta_2 B^2 - \cdots - \theta_q B^q\right) a_t \tag{5.37}$$

If the input ratio is always exactly 3:1, all the random shocks are zero and $z_t = \mu$. When the input ratio varies around 3:1, it causes the yield to deviate from μ by amount a_t. The resulting by-products cause further deviations from μ for up to q hours.

We have rationalized the use of an MA(q) process to represent the above situation. But in practice it might be possible to represent realizations generated by (5.37) more parsimoniously with an AR model. In order to see how this is possible, write the AR(1) in backshift form:

$$(1 - \phi_1 B)\tilde{z}_t = a_t \tag{5.38}$$

Divide both sides of (5.38) by $(1 - \phi_1 B)$:

$$\tilde{z}_t = (1 - \phi_1 B)^{-1} a_t \tag{5.39}$$

Now apply a mathematical theorem about geometric series, which states that if $|\phi_1| < 1$, then $(1 - \phi_1 B)^{-1}$ is equivalent to a convergent infinite series, that is,

$$(1 - \phi_1 B)^{-1} = \left(1 + \phi_1 B + \phi_1^2 B^2 + \phi_1^3 B^3 + \cdots\right), \qquad \text{if } |\phi_1| < 1 \tag{5.40}$$

Substitute (5.40) into (5.39) to get an MA process of infinitely high order:

$$\tilde{z}_t = \left(1 + \phi_1 B + \phi_1^2 B^2 + \phi_1^3 B^3 + \cdots\right) a_t \tag{5.41}$$

where

$$\theta_1 = -\phi_1$$
$$\theta_2 = -\phi_1^2$$
$$\theta_3 = -\phi_1^3$$
$$\vdots \tag{5.42}$$

Therefore, if q in process (5.37) is large, and if (5.42) is a reasonable representation of the pattern of the θ coefficients in (5.37), an AR(1) model will fit the data generated by (5.37) about as well as an MA(q) model even though (5.37) is, strictly speaking, an MA(q) process. In fact, the AR(1) model would likely produce more accurate forecasts because it is more parsimonious.

Example 3 illustrates two important points. First, any pure MA process can be written as an AR process of infinitely high order. Second, our objective in UBJ modeling is not necessarily to find the true process that has generated the data, but rather to find a good model (a parsimonious and statistically adequate imitation of the process) as discussed in Chapter 4.

5.5   Interpreting ARIMA Models III: ARIMA(0, d, q) models as exponentially weighted moving averages

Many practicing forecasters are familiar with a univariate method called the *exponentially weighted moving average*, abbreviated EWMA. This method is often used in business planning, especially when forecasts are needed for hundreds or thousands of inventory items. It is relatively easy to use and is intuitively appealing. But the method is sometimes used largely out of habit, with little consideration given to whether an EWMA model is appropriate for the data. In this section we explain the idea behind the simplest kind of EWMA and show that the ARIMA(0, 1, 1) model can be interpreted as an EWMA.

The EWMA model involves a certain kind of averaging of past observations. It may be helpful if we start with a simpler, more familiar averaging procedure that could be used for forecasting—the ordinary arithmetic mean $\bar{z}$. As usual, $\bar{z}$ is formed by summing n available observations on z_t and dividing by the number of observations:

$$\bar{z} = \frac{1}{n} \left(\sum_{t=1}^{n} z_t \right) \tag{5.43}$$

Table 5.1   Calculation of the arithmetic mean for a short realization

t	$z_t{}^a$
1	10
2	9
3	9
4	12
5	9
6	11
7	9
8	8
9	11
10	12

$^a\Sigma z_t = 100$; $\bar{z} = 100/10 = 10$.

Table 5.1 illustrates these calculations with a short data series. Figure 5.1 shows a graph of the data. The ten available observations sum to 100. Dividing this sum by the number of observations (ten) gives an arithmetic mean of 10. It appears that the data are stationary (see Figure 5.1); in particular, they seem to move about a constant mean, estimated from the available data to be 10.

Using this method, the forecast for period 11 is the previously calculated mean (10). The intuitive idea behind this type of forecast is that the future values of z_t may be something like the past values. If the data tend to fluctuate around a fixed central value, perhaps the arithmetic mean will give fairly good forecasts. (Recall that the mean appears in an ARIMA model as part of the constant term.)

The arithmetic mean is a specific case of a more general idea, the weighted mean. In a weighted mean ($\bar{z}_c$), each observation is multiplied by a weight (c_t). Then the weighted observations are summed, and this sum is divided by the sum of the weights:

$$\bar{z}_c = \frac{\sum c_t z_t}{\sum c_t} \qquad (5.44)$$

In the special case of the ordinary arithmetic mean, each weight is equal to one and there are n weights; the weights sum to n.

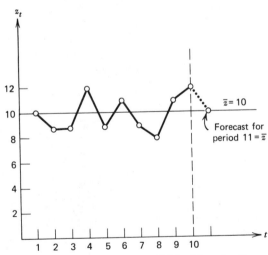

Figure 5.1   Plot of the realization shown in Table 5.1 and a forecast based on the arithmetic mean $\bar{z}$.

Now consider the EWMA. Forecasting with an EWMA model also involves averaging past observations, but the weights are not all equal to one. Instead, the weights applied to recent observations are larger than the weights applied to earlier observations. This weighting structure has a common-sense appeal; it seems reasonable that the recent past would be a better guide to the immediate future than would the distant past. Forecasting with an EWMA model allows for this possibility. Such an emphasis on the recent past is especially appealing if a data series does not fluctuate around a fixed central value. For this type of data, a forecast emphasizing the last few observations seems more sensible than a forecast emphasizing all past observations equally. That is, if the data show little tendency to return to the level of earlier observations, then we should not use a forecast (such as the arithmetic mean) that gives much weight to early values.

A common form of the EWMA expresses the forecast for time t (designated $\hat{z}_t$) as a weighted mean of the latest observation (z_{t-1}) and the last forecast ($\hat{z}_{t-1}$):

$$\hat{z}_t = (1 - \theta_1)z_{t-1} + \theta_1\hat{z}_{t-1} \qquad (5.45)$$

where

$$0 < \theta_1 < 1$$

θ_1 is a positive fraction. It is the weight applied to the last forecast $\hat{z}_{t-1}$. $(1 - \theta_1)$, also a positive fraction, is the weight applied to the last observation z_{t-1}. These weights sum to one, so we need not divide explicitly by the sum of the weights on the RHS of (5.45) to find our weighted average, since dividing by one would not alter the result.

Equation (5.45) is a computationally convenient form of the EWMA because it requires knowledge of only three items: the weight θ_1, the last observation z_{t-1}, and the last forecast $\hat{z}_{t-1}$. By making some algebraic substitutions, we can express z_t in a form less computationally convenient, but which shows that the EWMA is, in fact, a weighted average of all past observations.* The result is the following infinite series:

$$\hat{z}_t = (1 - \theta_1)z_{t-1} + \theta_1(1 - \theta_1)z_{t-2} + \theta_1^2(1 - \theta_1)z_{t-3}$$

$$+ \theta_1^3(1 - \theta_1)z_{t-4} + \cdots \qquad (5.46)$$

---

*Consider that (5.45) implies

$$\hat{z}_{t-1} = (1 - \theta_1)z_{t-2} + \theta_1\hat{z}_{t-2}$$

As long as $|\theta_1| < 1$, it can be shown that the weights in (5.46) decline geometrically and sum to one. As an example, suppose $\theta_1 = 0.6$.[†] Then the weights, rounded to two decimals, are

$$(1 - \theta_1) = (1 - 0.6) = 0.40$$

$$\theta_1(1 - \theta_1) = 0.6(0.4) = 0.24$$

$$\theta_1^2(1 - \theta_1) = (0.6)^2(0.4) = 0.14$$

$$\theta_1^3(1 - \theta_1) = (0.6)^3(0.4) = 0.09$$

$$\theta_1^4(1 - \theta_1) = (0.6)^4(0.4) = 0.05$$

$$\theta_1^5(1 - \theta_1) = (0.6)^5(0.4) = 0.03$$

$$\theta_1^6(1 - \theta_1) = (0.6)^6(0.4) = 0.02$$

$$\theta_1^7(1 - \theta_1) = (0.6)^7(0.4) = 0.01$$

$$\theta_1^8(1 - \theta_1) = (0.6)^8(0.4) = 0.01$$

$$\vdots \qquad \vdots$$

All subsequent weights round to zero.

Substitute this into (5.45) to obtain

$$\hat{z}_t = (1 - \theta_1)z_{t-1} + \theta_1(1 - \theta_1)z_{t-2} + \theta_1^2\hat{z}_{t-2}$$

(5.45) also implies

$$\hat{z}_{t-2} = (1 - \theta_1)z_{t-3} + \theta_1\hat{z}_{t-3}$$

Substituting this into the previous result, we get

$$\hat{z}_t = (1 - \theta_1)z_{t-1} + \theta_1(1 - \theta_1)z_{t-2} + \theta_1^2(1 - \theta_1)z_{t-3} + \theta_1^3\hat{z}_{t-3}$$

Continue altering the time subscripts on (5.45) and substituting as above to get the infinite series (5.46).

[†] We have selected this value arbitrarily. In practice various values of θ_1 are tried and the one which best fits the available data, according to some criterion, is selected.

Figure 5.2 is a graph of these weights. The weights give the appearance of an exponential decay as the time lag increases, which is why this model is said to be "exponentially weighted." The "moving-average" part of the name reflects the idea that a new weighted average is calculated as each new observation becomes available.

Table 5.2 shows how these weights could be used to forecast the data in Table 5.1 using an EWMA. These are the calculations needed to find $\hat{z}_t$ from equation (5.46). Columns 1 and 2 are a reproduction of Table 5.1. Column 3 is the set of EWMA weights calculated previously. (These weights sum to 0.99 instead of 1.0 because weights after lag 9 are rounded to zero.) Column 4 is each weight times each z_{t-i} observation, for $i = 1, 2, \ldots, 10$. At the bottom of the table is the sum of the weighted past z's, $\hat{z}_t = \hat{z}_{11}$, which is the EWMA forecast for z_{11} as stated in equation (5.46).

We now show that the EWMA in equation (5.45) is an ARIMA(0, 1, 1) model. To do this, first consider an ARIMA(0, 1, 1) in common algebraic form:

$$z_t = z_{t-1} - \theta_1 a_{t-1} + a_t \tag{5.47}$$

For simplicity, let θ_1 be known. Since a_t is not known when a forecast of z_t is formed at time $t - 1$, assign a_t its expected value of zero. Then the forecast of z_t based on (5.47) is

$$\hat{z}_t = z_{t-1} - \theta_1 a_{t-1} \tag{5.48}$$

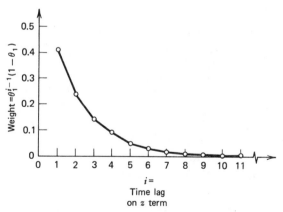

Figure 5.2   The weights on past z's for an EWMA with $\theta_1 = 0.6$.

Table 5.2   An EWMA forecast for the data in Table 5.1,
with $\theta_1 = 0.6^a$

t	z_t	Weight	Weight Multiplied by z_t
1	10	0.00	0.00
2	9	0.01	0.09
3	9	0.01	0.09
4	12	0.02	0.24
5	9	0.03	0.27
6	11	0.05	0.55
7	9	0.09	0.81
8	8	0.14	1.12
9	11	0.24	2.64
10	12	0.40	4.80

^{a}The EWMA forecast is equal to the summation of the weighted
z's which is equal to 10.61.

Subtracting (5.48) from (5.47), we see that an observed z value minus a
forecast z value is simply the random shock for that period:

$$z_t - \hat{z}_t = a_t \qquad (5.49)$$

Now return to the EWMA in (5.45). Expand the RHS and rearrange
terms to get

$$\hat{z}_t = z_{t-1} - \theta_1(z_{t-1} - \hat{z}_{t-1}) \qquad (5.50)$$

Using (5.49), substitute a_{t-1} into (5.50) for $(z_{t-1} - \hat{z}_{t-1})$. The result shows
that a forecast from the EWMA (5.50) is identical to a forecast from the
ARIMA(0, 1, 1) in equation (5.48).

Thus the EWMA in (5.45) may be interpreted as an ARIMA(0, 1, 1) and
vice versa.* This fact may be useful to practicing forecasters who must
interpret ARIMA models for managers. (See Part II, Case 9 for an example
of an ARIMA model that is a combination of two EWMA's, one explaining
the seasonal part of the data and the other explaining the nonseasonal part.)

EWMA models are sometimes used simply by habit. When using the
UBJ method, we attempt to find one or more ARIMA models that are
appropriate in light of the data. This procedure may or may not lead us to

*There are other types of EWMA forecasting models which are ARIMA(0, d, q) models of
various orders, as discussed by Cogger [18].

an EWMA. A clear strength of the UBJ method is that models are not chosen arbitrarily. Instead, the UBJ method guides the forecaster to a proper model based on some classical statistical estimation procedures applied to the available data. The present drawback of the UBJ–ARIMA method is that its proper use requires more experience and computer time than the habitual application of the EWMA to a forecasting problem. However, advances in computer technology tend to reduce that particular cost element associated with UBJ–ARIMA modeling. Furthermore, a thorough reading of this book and practice with numerous data sets should make one skilled at building proper UBJ–ARIMA models.

Summary

1. Three common ARIMA processes, in addition to the AR(1) and MA(1) are

$$\text{AR(2)}: z_t = C + \phi_1 z_{t-1} + \phi_2 z_{t-2} + a_t$$

$$\text{MA(2)}: z_t = C - \theta_1 a_{t-1} - \theta_2 a_{t-2} + a_t$$

$$\text{ARMA(1, 1)}: z_t = C + \phi_1 z_{t-1} - \theta_1 a_{t-1} + a_t$$

2. ARIMA models may be characterized this way: ARIMA(p, d, q), where p is the AR order; d is the number of times the data series must be differenced to induce a stationary mean; and q is the MA order.

3. It is convenient to write ARIMA models in backshift notation using the multiplicative backshift operator B. B is defined such that any variable which it multiplies has its time subscript shifted back by the power of B:

$$B^k z_t = z_{t-k}$$

A constant is unaffected when multiplied by B since a constant has no time subscript:

$$B^k C = C$$

4. It can be shown that $(1 - B)z_t$ represents the first differences of z_t; $(1 - B)^2 z_t$ represents the second differences of z_t; generally, $(1 - B)^d z_t$ represents the dth differences of z_t.

5. A nonseasonal ARIMA process in backshift notation has this general form:

$$(1 - \phi_1 B - \phi_2 B^2 - \cdots - \phi_p B^p)(1 - B)^d \tilde{z}_t$$
$$= (1 - \theta_1 B - \theta_2 B^2 - \cdots - \theta_q B^q)a_t$$

6. To write an ARIMA model in backshift notation,
   (a) transform z_t so it has a constant variance;
   (b) write z_t in deviations from the mean, $\tilde{z}_t = z_t - \mu$;
   (c) multiply z_t by the differencing operator $(1 - B)^d$, with d assigned the appropriate value;
   (d) multiply the last result by the AR operator $(1 - \phi_1 B - \phi_2 B^2 - \cdots - \phi_p B^p)$, with p assigned the proper value;
   (e) multiply a_t by the MA operator $(1 - \theta_1 B - \theta_2 B^2 - \cdots - \theta_q B^q)$, with q assigned an appropriate value;
   (f) equate the results of the last two steps.

7. The constant term in a nonseasonal ARIMA process is related to the mean μ of the process and the AR coefficients in this way:

$$C = \mu\left(1 - \sum_{i=1}^{p} \phi_i\right)$$

For a pure MA model, $p = 0$ and $C = \mu$.

8. Differencing ($d > 0$) causes the mean μ to drop out of an ARIMA process. The process will therefore have a constant term of zero unless the differenced variable is assumed to have a nonzero mean.

9. The constant term, AR terms, and MA terms all represent weighted past z values. Thus ARIMA processes are univariate, and a forecast from an ARIMA model may be interpreted as an extrapolation of past observations into the future.

10. Any MA process is algebraically equivalent to an AR process of infinitely high order. Any AR process is algebraically equivalent to an MA process of infinitely high order.

11. ARIMA models can sometimes be rationalized (interpreted in a common-sense way) through insight into the nature of the situation that has produced the data.

12.   A commonly used univariate forecasting technique, the exponentially weighted moving average (EWMA), is algebraically equivalent to an ARIMA(0, 1, 1) model. An advantage of the UBJ method is that we are guided to a proper model through analysis of the available data. The appropriate model may, or may not, be an ARIMA(0, 1, 1).

Questions and Problems

5.1   Write the following in both ARIMA(p, d, q) notation and common algebraic form:

(a) $(1 - \phi_1 B)(1 - B)\tilde{z}_t = a_t$
(b) $(1 - B)^2 \tilde{z}_t = (1 - \theta_1 B)a_t$
(c) $(1 - \phi_1 B)\tilde{z}_t = (1 - \theta_1 B)a_t$
(d) $(1 - B)\tilde{z}_t = (1 - \theta_1 B - \theta_2 B^2)a_t$
(e) $\tilde{z}_t = (1 - \theta_2 B^2)a_t$

5.2   Write the following in backshift notation:

(a)   ARIMA(1, 1, 1)
(b)   ARIMA(0, 2, 1)
(c)   ARIMA(2, 0, 2)
(d) $z_t = \mu(1 - \phi_1 - \phi_2) + \phi_1 z_{t-1} + \phi_2 z_{t-2} + a_t$
(e) $z_t = \mu(1 - \phi_1) + \phi_1 z_{t-1} - \theta_1 a_{t-1} + a_t$
(f) $z_t = z_{t-1} + \phi_1(z_{t-1} - z_{t-2}) + \phi_2(z_{t-2} - z_{t-3}) + a_t$
(g) $z_t = z_{t-1} - \theta_1 a_{t-1} - \theta_2 a_{t-2} + a_t$
(h) $z_t = 2z_{t-1} - z_{t-2} - \theta_1 a_{t-1} + a_t$

5.3   Show that an MA(1) process is equivalent to an AR process of infinitely high order in the following way: (i) write the MA(1) in common algebraic form; (ii) solve this form for a_t; (iii) use the expression for a_t to write expressions for $a_{t-1}, a_{t-2}, a_{t-3}, \ldots$; (iv) substitute the expressions for $a_{t-1}, a_{t-2}, \ldots$ into the original MA(1) model one at a time.

5.4   Consider the AR form of the MA(1). How is ϕ_4 related to θ_1?

5.5   Is an EWMA an AR model or an MA model? How does the principle of parsimony influence your answer?

5.6   Suppose an analyst presents this AR(4) model for a given realization:

$$(1 + 0.53B + 0.22B^2 + 0.14B^3 + 0.05B^4)\tilde{z}_t = a_t$$

In this model,

$$\hat{\phi}_1 = -0.53$$

$$\hat{\phi}_2 = -0.22$$

$$\hat{\phi}_3 = -0.14$$

$$\hat{\phi}_4 = -0.05$$

Can you suggest an alternative, more parsimonious model? Explain.

6

IDENTIFICATION: STATIONARY MODELS

In Chapters 1–4 we introduced the fundamental statistical concepts and modeling procedures of UBJ–ARIMA forecasting. In Chapter 5 we examined the special notation used for ARIMA models and considered how ARIMA models can be interpreted.

In this chapter we return to a discussion of the iterative, three–stage UBJ modeling procedure (identification, estimation, and diagnostic checking). Our emphasis in this chapter is on the identification of models for stationary realizations. Until Chapter 11 we will focus on models that do not have a seasonal component.

Before getting into the detail of this chapter, it may help to review some of the basic ideas presented in Chapters 1–4.

1. We begin with a set of n time-sequenced observations on a single variable $(z_1, z_2, z_3, \dots, z_n)$. Ideally, we have at least 50 observations. The realization is assumed to have been generated by an unknown ARIMA process.

2. We suppose the observations might be autocorrelated. We measure the statistical relationship between pairs of observations separated by various time spans (z_t, z_{t+k}), $k = 1, 2, 3, \dots$ by calculating estimated autocorrelation and partial autocorrelation coefficients. These coefficients are displayed graphically in an estimated autocorrelation function (acf) and partial autocorrelation function (pacf).

3. The UBJ–ARIMA method is appropriate only for a data series that is stationary. A stationary series has a mean, variance, and autocorrelation coefficients that are essentially constant through time. Often, a nonstationary series can be made stationary with appropriate transformations. The most common type of nonstationarity occurs when the mean of a realization changes over time. A nonstationary series of this type can frequently be rendered stationary by differencing.

4. Our goal is to find a good model. That is, we want a statistically adequate and parsimonious representation of the given realization. (The major characteristics of a good model are introduced in Chapter 4.)

5. At the identification stage we compare the estimated acf and pacf with various theoretical acf's and pacf's to find a match. We choose, as a tentative model, the ARIMA process whose theoretical acf and pacf best match the estimated acf and pacf. In choosing a tentative model, we keep in mind the principle of parsimony: we want a model that fits the given realization with the smallest number of estimated parameters.

6. At the estimation stage we fit the model to the data to get precise estimates of its parameters. We examine these coefficients for stationarity, invertibility, statistical significance, and other indicators of their quality.

7. At the diagnostic-checking stage we examine the residuals of the estimated model to see if they are independent. If they are not, we return to the identification stage to tentatively select another model.

Identification is clearly a critical stage in UBJ–ARIMA modeling, and a thorough knowledge of the most common theoretical acf's and pacf's is required for effective identification. Knowing the association between the common theoretical acf's and pacf's and their corresponding processes does not guarantee that we will identify the best model for any given realization, especially not at the first try. But familiarity with the common theoretical acf's and pacf's greatly improves our chances of finding a good model quickly.

There is an infinite number of possible processes within the family of ARIMA models proposed by Box and Jenkins. Fortunately, however, there also seems to be a relatively small number of models that occur commonly in practice. Furthermore, studying the common processes carries a substantial spillover benefit: uncommon ARIMA processes display certain characteristics broadly similar to those of the more ordinary ones. Thus we need to examine the properties of only a few common processes to be able to intelligently identify even unusual models.

In this chapter we first present and discuss the theoretical acf's and pacf's for these five common models: AR(1), AR(2), MA(1), MA(2), and ARMA(1, 1). Next, we discuss the ideas of stationarity and invertibility. We then derive the theoretical acf's for the MA(1) and AR(1).

6.1 Theoretical acf's and pacf's for five common processes

We have already encountered five common ARIMA models for stationary, nonseasonal data. In Chapter 3 we introduced the AR(1) and MA(1) models. Then in Chapter 5 we introduced the AR(2), MA(2), and ARMA(1, 1) models. In backshift form these five models are written as follows:*

$$AR(1): \ (1 - \phi_1 B)\tilde{z}_t = a_t \tag{6.1}$$

$$AR(2): \ (1 - \phi_1 B - \phi_2 B^2)\tilde{z}_t = a_t \tag{6.2}$$

$$MA(1): \ \tilde{z}_t = (1 - \theta_1 B)a_t \tag{6.3}$$

$$MA(2): \ \tilde{z}_t = (1 - \theta_1 B - \theta_2 B^2)a_t \tag{6.4}$$

$$ARMA(1, 1): \ (1 - \phi_1 B)\tilde{z}_t = (1 - \theta_1 B)a_t \tag{6.5}$$

In this section we examine the theoretical acf's and pacf's associated with each of these processes. We discussed the acf's and pacf's associated with the AR(1) and MA(1) in Chapter 3, but we present them again for convenience.

Keep in mind that in this section we are looking at *theoretical* acf's and pacf's derived from processes. Estimated acf's and pacf's calculated from realizations never match theoretical acf's and pacf's in every detail because of sampling error.

Table 6.1 states the major characteristics of theoretical acf's and pacf's for stationary AR, MA, and mixed (ARMA) processes. As we proceed we will discuss the acf's and pacf's of the above five processes in greater detail. In practice, however, a UBJ analyst must sometimes temporarily ignore the details and focus on the broader characteristics of an estimated acf and

*$\tilde{z}_t$ is z_t expressed in deviations from the mean: $\tilde{z}_t = z_t - \mu$.

Table 6.1   Primary distinguishing characteristics of theoretical acf's and pacf's for stationary processes

Process	acf	pacf
AR	Tails off toward zero (exponential decay or damped sine wave)	Cuts off to zero (after lag p)
MA	Cuts off to zero (after lag q)	Tails off toward zero (exponential decay or damped sine wave)
ARMA	Tails off toward zero	Tails off toward zero

pacf. As Table 6.1 shows, the three major types of ARIMA models have some primary distinguishing characteristics:

1. Stationary AR processes have theoretical acf's that decay toward zero rather than cut off to zero. (The words "decay", "die out", "damp out", and "tail off" are used interchangeably.) The autocorrelation coefficients may alternate in sign frequently, or show a wavelike pattern, but in all cases they tail off toward zero. By contrast, AR processes have theoretical pacf's that cut off to zero after lag p, the AR order of the process.

2. The theoretical acf's of MA processes cut off to zero after lag q, the MA order of the process. However, their theoretical pacf's tail off toward zero.

3. Stationary mixed (ARMA) processes show a mixture of AR and MA characteristics. Both the theoretical acf and the pacf of a mixed process tail off toward zero.

Now we consider each of the three major process types in greater detail. Table 6.2 summarizes the detailed characteristics of the five common processes we are considering in this chapter.

AR processes. All AR processes have theoretical acf's which tail off toward zero. This tailing off might follow a simple exponential decay pattern, a damped sine wave, or more complicated decay or wave patterns. But in all cases, there is a damping out toward zero.

An AR theoretical pacf has spikes up to lag p followed by a cutoff to zero. (Recall that p is the maximum lag length for the AR terms in a

Table 6.2 Detailed characteristics of five common stationary processes

Process	acf	pacf
AR(1)	Exponential decay: (i) on the positive side if $\phi_1 > 0$; (ii) alternating in sign starting on the negative side if $\phi_1 < 0$.	Spike at lag 1, then cuts off to zero; (i) spike is positive if $\phi_1 > 0$; (ii) spike is negative if $\phi_1 < 0$.
AR(2)	A mixture of exponential decays or a damped sine wave. The exact pattern depends on the signs and sizes of ϕ_1 and ϕ_2.	Spikes at lags 1 and 2, then cuts off to zero.
MA(1)	Spike at lag 1, then cuts off to zero: (i) spike is positive if $\theta_1 < 0$; (ii) spike is negative if $\theta_1 > 0$.	Damps out exponentially: (i) alternating in sign, starting on the positive side, if $\theta_1 < 0$; (ii) on the negative side, if $\theta_1 > 0$.
MA(2)	Spikes at lags 1 and 2, then cuts off to zero.	A mixture of exponential decays or a damped sine wave. The exact pattern depends on the signs and sizes of θ_1 and θ_2.
ARMA(1, 1)	Exponential decay from lag 1: (i) sign of ρ_1 = sign of $(\phi_1 - \theta_1)$; (ii) all one sign if $\phi_1 > 0$; (iii) alternating in sign if $\phi_1 < 0$.	Exponential decay from lag 1: (i) $\phi_{11} = \rho_1$; (ii) all one sign if $\theta_1 > 0$; (iii) alternating in sign if $\theta_1 < 0$.

process; it is also called the AR order of a process.) In practice, p is usually not larger than two or three for nonseasonal models.

Figure 6.1 shows the theoretical acf's and pacf's for two types of stationary AR(1) processes. The key point to remember is that any stationary AR(1) process has a theoretical acf showing exponential decay and a pacf with a spike at lag 1. If ϕ_1 is positive, the acf decays on the positive side and the pacf spike is positive. This is illustrated by Example I at the top of Figure 6.1. If ϕ_1 is negative, the AR(1) acf decays with alternating signs, starting from the negative side, while the pacf spike is negative. This is illustrated at the bottom of Figure 6.1 by Example II. (See Part II, Case 1, for an example of an estimated acf and pacf that resemble the theoretical ones in Figure 6.1 with $\phi_1 > 0$.)

The exact numerical values of the coefficients in both the theoretical acf and pacf of an AR(1) are determined by the value of ϕ_1. At lag 1, both the autocorrelation coefficient (ρ_1) and the partial autocorrelation coefficient (ϕ_{11}) are equal to ϕ_1. All other theoretical partial autocorrelations are zero. The theoretical autocorrelations at subsequent lags are equal to ϕ_1 raised to

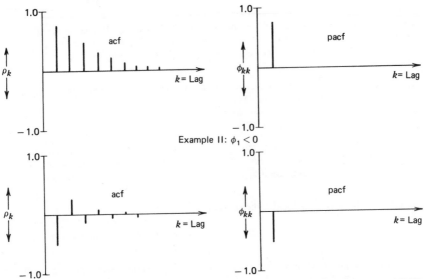

Figure 6.1   Examples of theoretical acf's and pacf's for two stationary AR(1) processes.

the power k, where k is the lag length. For example, if $\phi_1 = 0.8$, then $\rho_1 = \phi_{11} = 0.8$, $\rho_2 = (0.8)^2 = 0.64$, $\rho_3 = (0.8)^3 = 0.51$, $\rho_4 = (0.8)^4 = 0.41$, and so on. In general, $\rho_k = \phi_1^k$. This particular process is written in backshift form as

$$(1 - 0.8B)\tilde{z}_t = a_t \qquad (6.6)$$

A greater variety of patterns is possible with AR(2) processes than with AR(1) processes. Figure 6.2 shows the theoretical acf's and pacf's for four types of AR(2) processes. In general, a stationary AR(2) process has an acf with either a mixture of exponential decays or a damped sine wave, and a pacf with spikes at lags 1 and 2. The exact pattern depends on the signs and sizes of ϕ_1 and ϕ_2.* (See Part II, Case 3, for an example of an estimated acf and pacf resembling the theoretical ones in Figure 6.2.)

---

*For the mathematically inclined: to determine the general nature of these patterns, use the AR(2) operator to create the characteristic equation $(1 - \phi_1 B - \phi_2 B^2) = 0$, where B is now treated as an ordinary variable. Then the following can be shown for the acf of the AR(2):

Note that some AR(2) acf's are roughly similar in appearance to AR(1) acf's. In particular, the first two AR(2) acf's at the top of Figure 6.2 look much like the two AR(1) acf's in Figure 6.1. These broad similarities between AR(1) and AR(2) acf's can cause difficulties at the identification stage: we may not be able to tell from an estimated acf whether to consider an AR(1) or an AR(2) model. This is where the estimated pacf is especially useful: an AR(1) process is associated with only one spike in the pacf, while an AR(2) has two pacf spikes. In general, the lag length of the last pacf spike is equal to the order (p) of an AR process. In practice, p is usually not larger than two for nonseasonal data.

MA processes. An MA process has a theoretical acf with spikes up to lag q followed by a cutoff to zero. (Recall that q is the maximum MA lag, also called the MA order of the process.) Furthermore, an MA process has a theoretical pacf which tails off to zero after lag q. This tailing off may be either some kind of exponential decay or some type of damped wave pattern. In practice, q is usually not larger than two for nonseasonal data.

Figure 6.3 shows two MA(1) theoretical acf's and pacf's. They illustrate the rule that any MA(1) process has a theoretical acf with a spike at lag 1 followed by a cutoff to zero, and a theoretical pacf which tails off toward zero. If θ_1 is negative, the spike in the acf is positive, whereas the pacf decays exponentially, with alternating sign, starting on the positive side. This is illustrated by Example I at the top of Figure 6.3. Alternatively, if θ_1 is positive, the acf spike is negative, while the pacf decays exponentially on the negative side. This is illustrated at the bottom of Figure 6.3. [Cases 7–9 in Part II show estimated acf's and pacf's similar to the MA(1) theoretical acf's and pacf's in Figure 6.3.]

The exact numerical values of the coefficients in the theoretical acf and pacf of the MA(1) depend on the value of θ_1. Unlike the AR(1) process, which has $\rho_1 = \phi_1$, the absolute value of ρ_1 for the MA(1) is not equal to θ_1.

In Figure 6.4 we have examples of theoretical acf's and pacf's for MA(2) processes. All illustrate the rule that an MA(q) acf has spikes up to lag q ($q = 2$ in these examples) followed by a cutoff to zero, while the pacf tails

---

(i) If the roots of $(1 - \phi_1 B - \phi_2 B^2) = 0$ are real, so that $\phi_1^2 + 4\phi_2 \geq 0$, and the dominant root is positive, then the acf decays toward zero from the positive side.

(ii) If the roots are real, but the dominant root is negative, the acf decays toward zero while alternating in sign.

(iii) If the roots are complex, so that $\phi_1^2 + 4\phi_2 < 0$, and ϕ_1 is positive, the acf has the appearance of a damped sine wave starting from the positive side.

(iv) If the roots are complex, but ϕ_1 is negative, the acf has the appearance of a damped sine wave starting from the negative side.

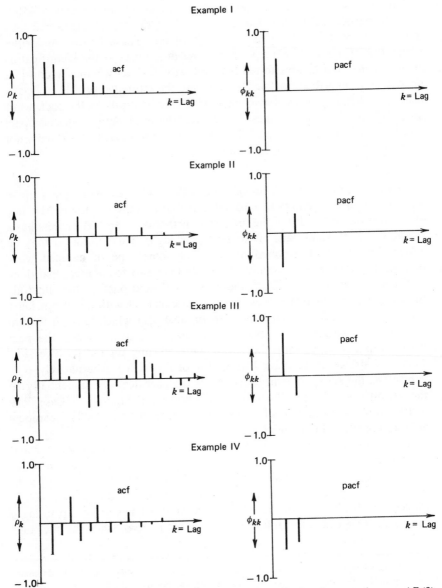

Figure 6.2 Examples of theoretical acf's and pacf's for four stationary AR(2) processes.

off toward zero. [Case 11 in Part II shows an estimated acf and pacf suggestive of an MA(2) process.]

ARMA processes. Mixed processes have theoretical acf's with both AR and MA characteristics. The acf tails off toward zero after the first $q - p$ lags with either exponential decay or a damped sine wave. The theoretical pacf tails off to zero after the first $p - q$ lags. In practice, p and q are usually not larger than two in a mixed model for nonseasonal data.

Figure 6.5 shows theoretical acf's and pacf's for six types of ARMA(1, 1) processes. The important thing to note is that *both* the acf and pacf tail off toward zero (rather than cut off to zero) in all cases. The acf and pacf may alternate in sign.

Because $q = 1$ and $p = 1$ for these examples, $q - p = 0$, and each acf in Figure 6.5 tails off toward zero starting from lag 1. Likewise, $p - q = 0$ in these examples, so each pacf in Figure 6.5 also tails off toward zero starting from lag 1.

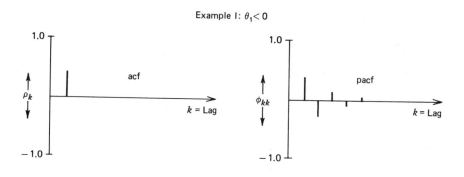

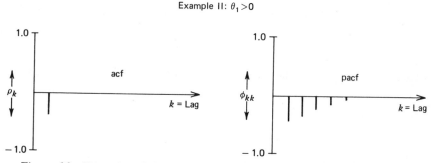

Figure 6.3 Examples of theoretical acf's and pacf's for two MA(1) processes.

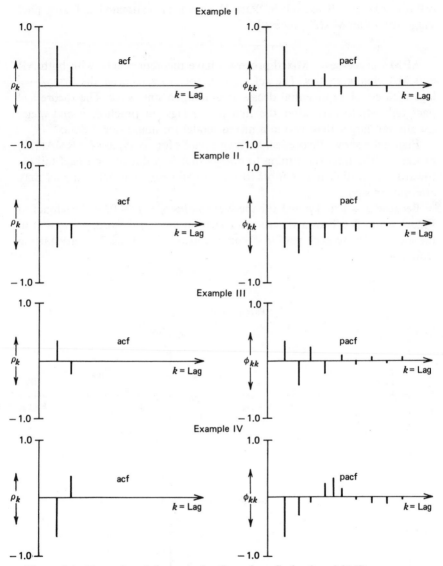

Figure 6.4 Examples of theoretical acf's and pacf's for four MA(2) processes.

Example I

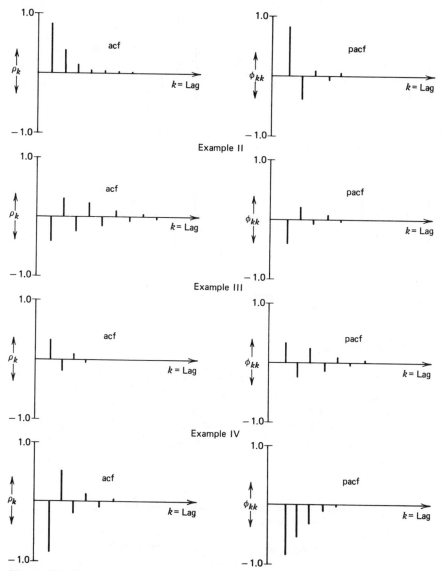

Figure 6.5 Examples of theoretical acf's and pacf's for six stationary ARMA(1, 1) processes.

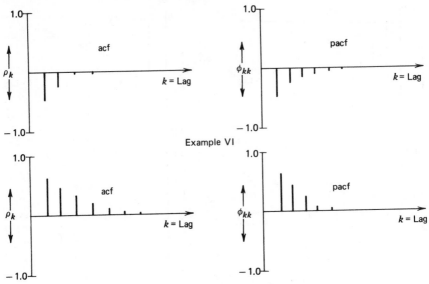

Figure 6.5 (*Continued*).

6.2 Stationarity

In Chapter 2 we stated that the UBJ method applies only to stationary realizations, or to those which can be made stationary by suitable transformation. In this section we discuss the conditions that AR coefficients must satisfy for an ARIMA model to be stationary, reasons for the stationarity requirement, and how to determine if a realization or model is stationary in practice.

Table 6.3   Summary of stationarity conditions for AR coefficients

Model Type	Stationarity Conditions
ARMA(0, q)	Always stationary
AR(1) or ARMA(1, q)	$|\phi_1| < 1$
AR(2) or ARMA(2, q)	$|\phi_2| < 1$
	$\phi_2 + \phi_1 < 1$
	$\phi_2 - \phi_1 < 1$

Conditions on the AR coefficients. Stationarity implies that the AR coefficients must satisfy certain conditions. These conditions, summarized in Table 6.3, are of great practical importance in UBJ modeling. You should regularly check the estimated AR coefficients (at the estimation stage) to see if they satisfy the appropriate stationarity conditions.

If $p = 0$, we have either a pure MA model or a white-noise series. All pure MA models and white noise are stationary, so there are no stationarity conditions to check.

For an AR(1) or ARMA(1, q) process, the stationarity requirement is that the absolute value of ϕ_1 must be less than one:

$$|\phi_1| < 1 \tag{6.7}$$

In practice we do not know ϕ_1. Instead, we find an estimate of it, designated $\hat{\phi}_1$, at the estimation stage. Therefore, in practice we apply condition (6.7) to $\hat{\phi}_1$ rather than to ϕ_1. [Case 1 in Part II is an example of a model where $\hat{\phi}_1$ satisfies condition (6.7). Case 5 shows a model where $\hat{\phi}_1$ meets condition (6.7), but it is not significantly different from 1.0, so the data are differenced.]

For an AR(2) or ARMA(2, q) process, the stationarity requirement is a set of three conditions:

$$|\phi_2| < 1$$

$$\phi_2 + \phi_1 < 1 \tag{6.8}$$

$$\phi_2 - \phi_1 < 1$$

All three conditions must be satisfied for an AR(2) or ARMA(2, q) model to be stationary. Again, in practice we apply conditions (6.8) to the estimates of ϕ_1 and ϕ_2 ($\hat{\phi}_1$ and $\hat{\phi}_2$) obtained at the estimation stage. [Cases 3 and 13 in Part II contain models satisfying the AR(2) stationarity conditions. Case 15 shows an AR(2) model that fails to meet these conditions.]

The stationarity conditions become complicated when $p > 2$. Fortunately, ARIMA models with $p > 2$ do not occur often in practice. When p exceeds 2 we can at least check this necessary (but not sufficient) stationarity condition:

$$\phi_1 + \phi_2 + \cdots + \phi_p < 1 \tag{6.9}$$

(See Appendix 6A for a discussion of the formal mathematical requirements for stationarity for any value of p.)

Now consider an ARMA(1, 1) model: $(1 - \phi_1 B)\tilde{z}_t = (1 - \theta_1 B)a_t$. Suppose we fit this model to a realization and get these estimation results: $\hat{\phi}_1 = -0.6$ and $\hat{\theta}_1 = 0.5$. Then the model can be written this way: $(1 + 0.6B)\tilde{z}_t = (1 - 0.5B)\hat{a}_t$. Is this model stationary? The answer is yes, because $|\hat{\phi}_1| = 0.6 < 1$, thus satisfying condition (6.7). We need not check any conditions on $\hat{\theta}_1$ to ensure stationarity; stationarity conditions apply only to AR coefficients. (However, we must check $\hat{\theta}_1$ to see that it satisfies the invertibility requirement. This is discussed in the next section.)

As another example, consider an AR(2) model: $(1 - \phi_1 B - \phi_2 B^2)\tilde{z}_t = a_t$. Fitting this model to a realization gives these estimation results: $\hat{\phi}_1 = 1.5$ and $\hat{\phi}_2 = -0.4$. Thus, our fitted model is $(1 - 1.5B + 0.4B^2)\tilde{z}_t = \hat{a}_t$. Inserting the estimated values of ϕ_1 and ϕ_2 into (6.8) gives

$$|\hat{\phi}_2| = 0.4 < 1$$

$$\hat{\phi}_2 + \hat{\phi}_1 = -0.4 + 1.5 = 1.1 > 1$$

$$\hat{\phi}_2 - \hat{\phi}_1 = -0.4 - 1.5 = -1.9 < 1$$

This model is not stationary. Although the first and third conditions in (6.8) are satisfied, the second condition is not met since the sum of ϕ_2 and ϕ_1 is greater than 1.

Reasons for the stationarity requirement.   There is a common-sense reason for requiring stationarity: we could not get useful estimates of the parameters of a process otherwise. For example, suppose a process has a mean that is different each time period. How could we estimate these means? As usual, we must use sample information. But typically we have only one observation per time period for time-series data. Therefore, we have only one observation at time t to estimate the mean at time t, one observation at time $t + 1$ to estimate the mean at time $t + 1$, and so forth. An estimate of a mean based on only one observation is not useful.

The situation becomes even worse if the variance also is not constant through time. In this case we would have to estimate up to $2n$ parameters (n means and n variances) with only n observations.*

It can also be shown that a model which violates the stationarity

---

*If the mean and variance are changing according to a known pattern, then it might be possible to get useful estimates of the n means and n variances from only n observations. This leads to the idea of variable-parameter ARIMA models, an area of research beyond the scope of this text.

restrictions will produce forecasts whose variance increases without limit, an undesirable result.

Checking for stationarity in practice.   Suppose we have a realization in hand and we want to develop an ARIMA model to forecast future values of this variable. We have three ways to determine if the stationarity requirement is met:

(i) Examine the realization visually to see if either the mean or the variance appears to be changing over time.

(ii) Examine the estimated acf to see if the autocorrelations move rapidly toward zero. In practice, "rapidly" means that the absolute *t*-values of the estimated autocorrelations should fall below roughly 1.6 by about lag 5 or 6. These numbers are only guidelines, not absolute rules. If the acf does not fall rapidly to zero, we should suspect a nonstationary mean and consider differencing the data.

(iii) Examine any estimated AR coefficients to see if they satisfy the stationarity conditions (6.7), (6.8), and (6.9).

You should rely most heavily on the appearance of the estimated acf and on the values of any estimated AR coefficients in deciding if the mean of a series is stationary. The only exception is when $p > 2$, so that the set of stationarity conditions on the AR coefficients becomes complicated. In that case, rely more on visual inspection of the data and the estimated acf, while also checking the necessary condition on the AR coefficients (6.9). Visual inspection of the data is perhaps the most practical way of gauging stationarity of the variance. The identification of models for nonstationary realizations is discussed more fully in Chapter 7.

6.3   Invertibility

Conditions on the MA coefficients.   There is another condition that ARIMA models must satisfy called *invertibility*. This requirement implies that the MA coefficients must satisfy certain conditions. These conditions, summarized in Table 6.4, are algebraically identical to the stationarity requirements on AR coefficients.

If $q = 0$, we have a pure AR process or a white-noise series. All pure AR processes (or white noise) are invertible, and no further checks are required.

For an MA(1) or ARMA(p, 1) process, invertibility requires that the absolute value of θ_1 be less than one:

$$|\theta_1| < 1 \qquad (6.10)$$

Table 6.4 Summary of invertibility conditions for MA coefficients

Model Type	Invertibility Conditions
ARMA($p,0$)	Always invertible
MA(1) or ARMA($p,1$)	$\|\theta_1\| < 1$
MA(2) or ARMA($p,2$)	$\|\theta_2\| < 1$
	$\theta_2 + \theta_1 < 1$
	$\theta_2 - \theta_1 < 1$

For an MA(2) or ARMA($p,2$) process the invertibility requirement is a set of conditions on θ_1 and θ_2:

$$|\theta_2| < 1$$

$$\theta_2 + \theta_1 < 1 \qquad\qquad (6.11)$$

$$\theta_2 - \theta_1 < 1$$

All three of the conditions in (6.11) must be met for an MA(2) or ARMA($p,2$) process to be invertible.

In practice the invertibility conditions are applied to the estimates of θ_1 and θ_2 ($\hat{\theta}_1$ and $\hat{\theta}_2$) obtained at the estimation stage because θ_1 and θ_2 are unknown. (See Cases 5, 7–9, and 11 in Part II for examples.) The invertibility conditions become complicated when $q > 2$, but ARIMA models with $q > 2$ do not occur frequently in practice. If $q > 2$, we can at least check this necessary (but not sufficient) condition for invertibility:

$$\theta_1 + \theta_2 + \cdots + \theta_q < 1 \qquad\qquad (6.12)$$

(See Appendix 6A for a discussion of the formal mathematical conditions for invertibility for any value of q.)

Suppose we estimate an ARMA(1, 2) model, $(1 - \phi_1 B)\tilde{z}_t = (1 - \theta_1 B - \theta_2 B^2)a_t$. Fitting this model to a realization produces these results at the estimation stage: $\hat{\phi}_1 = 0.4$, $\hat{\theta}_1 = 0.8$, and $\hat{\theta}_2 = -0.5$. It is easy to show that this model is stationary since $|\hat{\phi}_1| = 0.4 < 1$; thus condition (6.7) is satisfied. As with all ARIMA models, the stationarity conditions apply only to the AR coefficients.

To check this model for invertibility, we must apply the three conditions in (6.11) to the estimated MA coefficients; invertibility conditions apply only to MA coefficients, not AR coefficients. We find that all three conditions are met, and hence the model is invertible:

$$|\hat{\theta}_2| = 0.5 < 1$$

$$\hat{\theta}_2 + \hat{\theta}_1 = -0.5 + 0.8 = 0.3 < 1$$

$$\hat{\theta}_2 - \hat{\theta}_1 = -0.5 - 0.8 = -1.3 < 1$$

A reason for invertibility. There is a common-sense reason for the invertibility condition: a noninvertible ARIMA model implies that the weights placed on past z observations do not decline as we move further into the past; but common sense says that larger weights should be attached to more recent observations. Invertibility ensures that this result holds.*

It is easy to see the common sense of the invertibility condition as it applies to the MA(1). In Chapter 5 we showed how the MA(1) could be written as an AR process of infinitely high order:

$$\left(1 + \theta_1 B + \theta_1^2 B^2 + \theta_1^3 B^3 + \cdots \right)\tilde{z}_t = a_t$$

or

$$z_t = C - \theta_1 z_{t-1} - \theta_1^2 z_{t-2} - \theta_1^3 z_{t-3} - \cdots \qquad (6.13)$$

The θ coefficients in (6.13) are weights attached to the lagged z terms. If condition (6.10) for the MA(1) is not met, then the weights implicitly assigned to the z's in (6.13) get larger as the lag length increases. For example, for $\theta_1 = 2$, the weights (θ_1^k) have the following values (k is lag length):

k	θ_1^k
1	$\theta_1 = 2$
2	$\theta_1^2 = (2)^2 = 4$
3	$\theta_1^3 = (2)^3 = 8$
4	$\theta_1^4 = (2)^4 = 16$
$\vdots$	$\vdots \qquad \vdots$

On the other hand, suppose condition (6.10) is satisfied. For example, let $\theta_1 = 0.8$. Then the weights on the time-lagged z's in (6.13) decline as we

---

*Invertibility also ensures a unique association between processes and theoretical acf's. See Appendix 6B for a discussion of this point.

move further into the past:

k	θ_1^k
1	$\theta_1 = 0.8$
2	$\theta_1^2 = (0.8)^2 = 0.64$
3	$\theta_1^3 = (0.8)^3 = 0.51$
4	$\theta_1^4 = (0.8)^4 = 0.41$
$\vdots$	$\vdots$ $\vdots$

We could show the same result for any MA process that we have shown here for the MA(1). First, we could write it as an AR process of infinitely high order. Then, we could show that the coefficients on the past z's will not decline as we move further into the past unless the invertibility conditions are met.

6.4   Deriving theoretical acf's for the MA(1) process

We saw in Chapter 2 how *estimated* acf's are calculated from realizations using equation (2.5). In the next two sections we derive the *theoretical* acf's for the MA(1) and AR(1) processes. These derivations require numerous algebraic manipulations. We will explain the derivations in detail, but you should write out each step to make sure you understand it.

Throughout these two sections we apply certain rules about mathematical expectations stated earlier, in Appendix 3A. Since we make extensive use of three of them, we repeat them here for convenience:

Rule II-E: expected value of a constant

$$E(C) = C$$

where C is a constant.

Rule III-E: expected value of a finite linear combination of random variables.   If m is a finite integer,

$$E(C_1x_1 + C_2x_2 + \cdots + C_mx_m) = C_1E(x_1)$$

$$+ C_2E(x_2) + \cdots + C_mE(x_m),$$

where $C_1, C_2, \ldots, C_m$ are constants; $x_1, x_2, \ldots, x_m$ are random variables.

Rule IV-E: expected value of an infinite linear combination of random variables. If $m = \infty$, Rule III-E holds only if $\sum_{i=0}^{\infty} C_i$ (where $C_0 = 1$) converges (is equal to a finite number).

We also make extensive use of the assumptions about the random shocks stated in Chapter 3. We repeat them more formally for convenience:

Ia: The a_t are Normally distributed.
IIa: $E(a_t) = 0$.
IIIa: $\text{cov}(a_t, a_{t-k}) = 0$; that is $E(a_t a_{t-k}) = 0$.
IVa: $E(a_t)^2 = \sigma_a^2$ (a finite constant for all t).

First, consider the MA(1) process with a constant term C and a θ_1 coefficient that are each a finite constant and with random shocks satisfying Assumptions Ia–IVa above. In backshift notation this process is

$$\tilde{z}_t = (1 - \theta_1 B) a_t \qquad (6.14)$$

Using the rules for backshift notation stated in Chapter 5 and replacing $\tilde{z}_t$ with $z_t - \mu$, we can write this process in common algebraic form as

$$z_t = C - \theta_1 a_{t-1} + a_t \qquad (6.15)$$

where the constant term (C) is equal to μ.

Recall that (6.15) is a population function: it is a *process* which is the true mechanism generating observations of z_t. Therefore, (6.15) is the source from which we can derive the theoretical acf for the MA(1).

We begin by finding the mean and the variance of the MA(1). Then we find its autocorrelation function. We expect to find that the MA(1) has a theoretical acf with a spike at lag 1 followed by a cutoff to zero. All pure MA processes have theoretical acf's described by the following equations:

$$\rho_k = \frac{-\theta_k + \theta_1 \theta_{k+1} + \cdots + \theta_{q-k} \theta_q}{\left(1 + \theta_1^2 + \theta_2^2 + \cdots + \theta_q^2\right)}, \qquad k = 1, 2, \ldots, q$$

$$\rho_k = 0, \qquad k > q \qquad (6.16)$$

As we will see, the MA(1) acf is described by these equations when $q = 1$. We also show that the MA(1) process is stationary if its mean, variance, and MA coefficient (θ_1) are finite constants. This result holds for all MA processes.

Mean.   We have assumed that process (6.15) has a finite constant term C. Therefore, it seems we already know the mean of the process, $C = \mu$, and we seem to have guaranteed that this mean will be stationary since C is a finite constant. However, it will be an instructive exercise to show that C is, indeed, the mean of the process by finding the mathematical expectation (μ) of (6.15). We can do this in a straightforward way, without imposing any special conditions on θ_1 beyond the assumption that it is a finite constant.

We find the mathematical expectation (μ) of (6.15) by applying the expected value operator to both sides of the equation. Because the RHS terms are additive, the operator is applied separately to each term according to Rule III-E:

$$E(z_t) = \mu = E(C) - \theta_1 E(a_{t-1}) + E(a_t) \tag{6.17}$$

Because C is fixed, $E(C) = C$ from Rule II-E. Applying Assumption IIa, that $E(a_t) = 0$ for all t, the last two RHS terms in (6.17) are zero and we are left with

$$E(z_t) = \mu = C \tag{6.18}$$

Since C is by assumption a finite constant, (6.18) states that the mathematical expectation (μ) of process (6.14), which is the mean, exists and does not change over time. This is a necessary condition for stationarity.

Variance and autocovariances. Next, we find the variance and autocovariances of process (6.14). We need them to derive the theoretical acf, and we want to determine the conditions under which the variance and autocovariances of the MA(1) are stationary.

For convenience we work with z_t expressed in deviations from μ: $\tilde{z}_t = z_t - \mu$. The process generating $\tilde{z}_t$ is identical to the process generating z_t, except that the mean of the $\tilde{z}_t$ process is zero. To see this, write (6.14) in common algebraic form:

$$\tilde{z}_t = a_t - \theta_1 a_{t-1} \tag{6.19}$$

Inspection shows that (6.19) is identical to (6.15) except that the mean (i.e., the constant term) of (6.19) is zero. Therefore, (6.19) and (6.15) have identical variances, autocovariances, and autocorrelations since these measures depend on the size of deviations from the mean and not on the size of the mean itself.

To find the variance and autocovariances of (6.19), use Rules V-E and VI-E (stated in Appendix 3A), to write the variance–covariance function for $\tilde{z}_t$:

$$\gamma_k = E\{[\tilde{z}_t - E(\tilde{z}_t)][\tilde{z}_{t-k} - E(\tilde{z}_{t-k})]\} \tag{6.20}$$

We know that $E(\tilde{z}_t) = 0$, and since we have just shown that (6.19) has a stationary mean, this condition holds for all t. Thus (6.20) simplifies to

$$\gamma_k = E(\tilde{z}_t \tilde{z}_{t-k}) \tag{6.21}$$

Now use (6.19) to substitute $a_t - \theta_1 a_{t-1}$ for $\tilde{z}_t$ and $a_{t-k} - \theta_1 a_{t-1-k}$ for $\tilde{z}_{t-k}$, and apply Rule III-E:

$$\gamma_k = E\left[(a_t - \theta_1 a_{t-1})(a_{t-k} - \theta_1 a_{t-1-k})\right]$$

$$= E\left(a_t a_{t-k} - \theta_1 a_{t-1} a_{t-k} - \theta_1 a_t a_{t-1-k} + \theta_1^2 a_{t-1} a_{t-1-k}\right)$$

$$= E(a_t a_{t-k}) - \theta_1 E(a_{t-1} a_{t-k}) - \theta_1 E(a_t a_{t-1-k}) + \theta_1^2 E(a_{t-1} a_{t-1-k}) \tag{6.22}$$

To find the variance of (6.19), let $k = 0$. Applying Assumptions IIIa and IVa to (6.22) gives

$$\gamma_0 = \sigma_z^2 = \sigma_a^2 + \theta_1^2 \sigma_a^2$$

$$= \sigma_a^2 (1 + \theta_1^2) \tag{6.23}$$

Note from (6.23) that the variance $(\gamma_0 = \sigma_z^2)$ of $\tilde{z}_t$ depends on the variance (σ_a^2) of a_t. This is not surprising since it is the presence of the random-shock component that makes z_t stochastic in the first place. Note also that σ_z^2 exists and is constant through time because both σ_a^2 and θ_1 are finite constants by assumption. So (6.23) says that an MA(1) process satisfies a necessary condition for stationarity—the variance is a finite constant.

Next, we find the autocovariances of (6.19). Let $k = 1, 2, \ldots$ and apply Assumptions IIIa and IVa to (6.22) to find that all the autocovariances except γ_1 are zero:

$$\gamma_1 = -\theta_1 \sigma_a^2 \tag{6.24a}$$

$$\gamma_k = 0, \quad k > 1 \tag{6.24b}$$

Note that all the autocovariances are finite constants (since θ_1 and σ_a^2 are finite constants). Thus the MA(1) satisfies a necessary condition for stationarity.*

---

*Equations (6.23) and (6.24) give the elements in the variance–covariance matrix of the MA(1). We have the variance (σ_z^2) which is a constant on the main diagonal. There is one autocovariance (γ_1) which is a constant on the first diagonal, both above and below the main diagonal. All other elements in the MA(1) variance–covariance matrix are zero.

Autocorrelations. Dividing the variance (6.23) and the autocovariances (6.24) by the variance (6.23) translates the γ_k's into autocorrelation coefficients:

$$\rho_0 = \frac{\gamma_0}{\gamma_0} = 1$$

$$\rho_1 = \frac{\gamma_1}{\gamma_0} = \frac{-\theta_1}{1 + \theta_1^2} \tag{6.25}$$

$$\rho_k = \frac{\gamma_k}{\gamma_0} = 0, \qquad k > 1$$

We see that the theoretical acf for an MA(1) process has a distinct pattern: the autocorrelation at lag zero (ρ_0) is always one; ρ_1 is nonzero because γ_1 is nonzero; all other autocorrelations are zero because the relevant γ_k are zero.

Consider an MA(1) process with $\theta_1 = -0.8$. In backshift form this process is

$$\tilde{z}_t = (1 + 0.8B)a_t \tag{6.26}$$

Although θ_1 is negative, it appears in (6.26) with a positive sign since we follow the convention of writing MA coefficients with negative signs. Thus the negative of our negative coefficient is positive.

Use (6.25) to calculate the theoretical autocorrelations for process (6.26):

$$\rho_0 = 1$$

$$\rho_1 = \frac{-\theta_1}{1 + \theta_1^2} = \frac{0.8}{1.64} = 0.49$$

$$\rho_k = 0, \qquad k > 1$$

These values are graphed in an acf in Example I at the top of Figure 6.6. Note that this theoretical acf looks like the MA(1) theoretical acf's presented in Chapter 3 and earlier in this chapter. They all have a spike at lag 1 followed by a cutoff to zero.

As another example consider an MA(1) process with $\theta_1 = 0.5$:

$$\tilde{z}_t = (1 - 0.5B)a_t \tag{6.27}$$

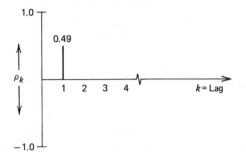

Example I: $\tilde{z}_t = (1 + 0.8B)a_t$

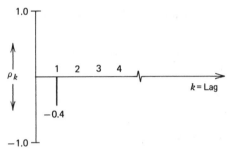

Example II: $\tilde{z}_t = (1 - 0.5B)a_t$

Figure 6.6   Theoretical acf's for two MA(1) processes.

From (6.25) we find these theoretical autocorrelation coefficients for process (6.27):

$$\rho_0 = 1$$

$$\rho_1 = \frac{-\theta_1}{1 + \theta_1^2} = \frac{-0.5}{1.25} = -0.4$$

$$\rho_k = 0, \qquad k > 1$$

These values are plotted in Example II at the bottom of Figure 6.6. Once again, we see that the theoretical acf for an MA(1) process has a single spike at lag 1.

[*Note*: Estimated acf's never match theoretical acf's exactly because of sampling error. We saw examples of this in Chapter 3 where we simulated some realizations. However, if we see an estimated acf with a spike at lag 1 followed by statistically insignificant autocorrelations at the remaining lags, we should consider representing the available data with an MA(1) model. See Part II, Cases 5 and 7–9 for examples.]

6.5   Deriving theoretical acf's for the AR(1) process

In this section we consider a stationary AR(1) process with a mean μ and ϕ_1 coefficient that are each a finite constant, and with a random-shock term satisfying Assumptions Ia–IVa. This process is sometimes referred to as a *Markov process*. In backshift notation the process is

$$(1 - \phi_1 B)\tilde{z}_t = a_t \qquad (6.28)$$

Applying the rules for backshift notation, we can also write the process as

$$z_t = C + \phi_1 z_{t-1} + a_t \qquad (6.29)$$

where

$$C = \mu(1 - \phi_1)$$

As we did with the MA(1) generating mechanism in the last section, we derive the mean, variance, autocovariances, and acf of the AR(1).

Mean.   In the last section we demonstrated rather easily that the expected value (μ) of z_t for the MA(1) process was the finite constant (C). In particular, we did not have to impose any special conditions on θ_1 except that it be a finite constant. By contrast, the expected value of the AR(1) is equal to $C/(1 - \phi_1)$ rather than C. Furthermore, the AR coefficient ϕ_1 must meet an additional condition or the assumption that μ is a finite constant (i.e., stationary) becomes untenable. In fact, we have already seen in Section 6.2 that ϕ_1 must satisfy this condition for stationarity:

$$|\phi_1| < 1 \qquad (6.30)$$

We can justify condition (6.30) intuitively with an example. Consider process (6.29). Suppose $\phi_1 = 2$ and $\mu = 0$, and suppose the initial value for z is $z_0 = 1$. With $|\phi_1| > 1$, the subsequent values of z_t tend to "explode" away from the initial value; realizations generated by this process will not return to a fixed central value, and the assumption that the mean μ is fixed at zero is contradicted.

To show this, suppose the first six random shocks have these values: $(4, -1, 1, -3, -1, -5)$. Process (6.29) along with the above conditions would produce this realization:

$$z_0 = 1$$

$$z_1 = 2(1) + 4 = 6$$

$$z_2 = 2(6) - 1 = 11$$

$$z_3 = 2(11) + 1 = 23$$

$$z_4 = 2(23) - 3 = 43$$

$$z_5 = 2(43) - 1 = 85$$

$$z_6 = 2(85) - 5 = 165$$

This realization is moving further and further away from zero, suggesting that it was generated by a nonstationary process. In this section we show formally that the mean of an AR(1) process is not stationary unless condition (6.30) is met.

To find the mean μ of the AR(1), find the mathematical expectation of (6.29). In doing this we encounter a problem we did not meet when finding the mean of the MA(1): one of the terms to be evaluated on the RHS of (6.29), $E(z_{t-1})$, is unknown. To solve the problem, use (6.29) to write expressions for $z_{t-1}, z_{t-2}, \dots$ and substitute these back recursively into (6.29). Rearrange terms to arrive at this infinite series:

$$z_t = C\left(1 + \phi_1 + \phi_1^2 + \phi_1^3 + \cdots\right) + a_t + \phi_1 a_{t-1} + \phi_1^2 a_{t-2} + \cdots$$

$$(6.31)$$

We can find the expected value of an infinite series by taking the expected value of each term separately only if the sum of the coefficients in that series converges. From Rule IV-E, we require that

$$\sum_{i=0}^{\infty} \phi_1^i = K \qquad (6.32)$$

where K is a finite constant.

If condition (6.32) does not hold, then (6.31) is an explosive (divergent) infinite series; its sum does not exist and we cannot find its mathematical expectation. It can be shown that if the stationarity condition (6.30) holds, then condition (6.32) also holds. Then the first term in (6.31) converges to $C/(1 - \phi_1)$, and we may apply Assumption IIa separately to each remaining term. By doing so we find*

$$E(z_t) = \mu = \frac{C}{1 - \phi_1}, \qquad |\phi_1| < 1 \tag{6.33}$$

From (6.33) we find $C = \mu(1 - \phi_1)$. This is a specific case of the more general result, shown in Chapter 5, that $C = \mu(1 - \Sigma\phi_i)$.

We began by writing a supposedly stationary AR(1) process. A stationary process has, among other things, a finite, constant mean. We have shown that this supposition is contradicted unless $|\phi_1| < 1$, so (6.30) is a condition for stationarity of an AR(1). Similar restrictions on θ_1 are not necessary to ensure that the mean of the MA(1) is stationary.

Variance and autocovariances.   Next, we derive the variance and autocovariances of the AR(1) to see if they are stationary. We also need them to find the theoretical acf of the AR(1). We find the variance and autocovariances simultaneously. The variance $\gamma_0 = \sigma_z^2$ and first autocovariance γ_1 are found by solving two simultaneous equations; the remaining autocovariances are then found recursively.

These derivations are easier if we work with the deviations of z_t from μ, that is, $\tilde{z}_t = z_t - \mu$. The process generating $\tilde{z}_t$ is identical to the process generating z_t, except the mean and constant term of the $\tilde{z}_t$ process are zero:

$$\tilde{z}_t = \phi_1\tilde{z}_{t-1} + a_t \tag{6.34}$$

Since the two processes are identical except for the means, the variances and autocovariances for the two processes are identical because the variances and the autocovariances depend on deviations from the mean rather than the value of the mean.

We now use (6.34) to find the variance and autocovariances of the AR(1). As we did in the last section, use Rules V-E and VI-E (from Appendix 3A) and the fact that the expected value of $\tilde{z}_t$ is zero to write the variance–covariance function for $\tilde{z}_t$:

$$\gamma_k = E(\tilde{z}_t\tilde{z}_{t-k}) \tag{6.35}$$

---

*A faster way to arrive at (6.33) is to find the mathematical expectation of (6.29) and substitute $E(z_t)$ for $E(z_{t-1})$ on the assumption that these must be equal if the process is stationary. The result is easily solved to arrive at (6.33). Our purpose above, however, was not only to find (6.33) but also to demonstrate that (6.30) is required for stationarity.

To evaluate (6.35), multiply both sides of (6.34) by $\tilde{z}_{t-k}$ and find the expected value of the result using Rule III-E:

$$\begin{aligned}
\gamma_k &= E(\tilde{z}_t \tilde{z}_{t-k}) \\
&= E(\phi_1 \tilde{z}_{t-1} \tilde{z}_{t-k} + a_t \tilde{z}_{t-k}) \\
&= \phi_1 E(\tilde{z}_{t-1} \tilde{z}_{t-k}) + E(a_t \tilde{z}_{t-k})
\end{aligned} \tag{6.36}$$

Letting $k = 0, 1, 2, \ldots$, (6.36) yields the following sequence:

$$\gamma_0 = \sigma_z^2 = \phi_1 \gamma_1 + \sigma_a^2 \tag{6.37a}$$

$$\gamma_1 = \phi_1 \gamma_0 \tag{6.37b}$$

$$\gamma_2 = \phi_1 \gamma_1 \tag{6.37c}$$

$$\gamma_3 = \phi_1 \gamma_2 \tag{6.37d}$$

$$\vdots$$

To see how we arrive at equations (6.37), recall from Chapter 5 that the AR(1) can be written in MA form. Thus, each of the following is an acceptable way of writing (6.34):

$$\tilde{z}_t = a_t + \phi_1 a_{t-1} + \phi_1^2 a_{t-2} + \cdots \tag{6.38a}$$

$$\tilde{z}_{t-1} = a_{t-1} + \phi_1 a_{t-2} + \phi_1^2 a_{t-3} + \cdots \tag{6.38b}$$

$$\tilde{z}_{t-2} = a_{t-2} + \phi_1 a_{t-3} + \phi_1^2 a_{t-4} + \cdots \tag{6.38c}$$

These equations differ only in their time subscripts.

To find the variance of (6.34) let $k = 0$. Then the first term in (6.36) is $\phi_1 E(\tilde{z}_t \tilde{z}_{t-1}) = \phi_1 \gamma_1$. Making this substitution, and substituting the MA form of $\tilde{z}_t$ using (6.38a) in the second RHS term in (6.36) gives

$$\begin{aligned}
\gamma_0 &= \phi_1 \gamma_1 + E\big[a_t(a_t + \phi_1 a_{t-1} + \phi_1^2 a_{t-2} + \cdots)\big] \\
&= \phi_1 \gamma_1 + E(a_t a_t + \phi_1 a_t a_{t-1} + \phi_1^2 a_t a_{t-2} + \cdots)
\end{aligned} \tag{6.39}$$

By Assumption IIIa, all expectations except $E(a_t a_t)$ in the second RHS term in (6.39) are zero; $E(a_t a_t)$ is σ_a^2 by Assumption IVa. Thus we have arrived at (6.37a).

To find γ_1 let $k = 1$. Then the first term in (6.36) is $\phi_1 E(\tilde{z}_{t-1}\tilde{z}_{t-1}) = \phi_1\gamma_0$ $= \phi_1\sigma_z^2$. Making this substitution, and substituting the MA form of $\tilde{z}_{t-1}$ using (6.38b) in the second RHS term in (6.36) gives

$$\gamma_1 = \phi_1\gamma_0 + E\left[a_t\left(a_{t-1} + \phi_1 a_{t-2} + \phi_1^2 a_{t-3} + \cdots\right)\right]$$

$$= \phi_1\gamma_0 + E\left(a_t a_{t-1} + \phi_1 a_t a_{t-2} + \phi_1^2 a_t a_{t-3} + \cdots\right) \qquad (6.40)$$

By Assumption IIIa, all expectations in the second RHS term are zero and we have arrived at (6.37b).

Letting $k = 2, 3, \ldots$ and following similar reasoning leads to the expressions for the remaining autocovariances in (6.37).

Now solve (6.37a) and (6.37b) simultaneously for the variance γ_0:

$$\gamma_0 = \sigma_z^2 = \frac{\sigma_a^2}{1 - \phi_1^2} \qquad (6.41)$$

From (6.41), ϕ_1 and σ_a^2 must both be finite constants, as we assume they are, if the variance of the AR(1) process is to exist and be stationary. In addition, the stationarity condition $|\phi_1| < 1$ must be satisfied if γ_0 is to exist. That is, if $|\phi_1| = 1$, the denominator of (6.41) is zero and the variance is undefined. If $|\phi_1| > 1$, the denominator of (6.41) is negative and therefore γ_0 is negative—an impossible result for a variance.

All autocovariances $\gamma_1, \gamma_2, \ldots$ are now found recursively. Having found γ_0 we substitute it into (6.37b) to find γ_1. This result is substituted into (6.37c) to find γ_2, which is then substituted into (6.37d) to find γ_3, and so forth. Thus the variance and all autocovariances for the AR(1) can be written compactly as a function of ϕ_1 and γ_0:

$$\gamma_k = \phi_1^k\gamma_0, \qquad k = 0, 1, 2, \ldots \qquad (6.42)$$

Since γ_0 does not exist if $|\phi_1| \geq 1$, then from (6.42) *none* of the autocovariances exist if the stationarity condition $|\phi_1| < 1$ is violated because they all depend on γ_0. And just as with γ_0, all autocovariances are stationary only if σ_a^2 and ϕ_1 are finite constants. Again we see the importance of our assumptions about the properties of the random shocks and the constancy of parameters in UBJ–ARIMA models.

Autocorrelations.   Dividing (6.42) by $\gamma_0 = \sigma_z^2$ yields a compact autocorrelation function for the AR(1):

$$\rho_k = \phi_1^k, \qquad k = 0, 1, 2, \ldots \qquad (6.43)$$

If $|\phi_1| < 1$, the autocorrelation coefficients for an AR(1) decline exponentially as k increases. This important result states that the theoretical acf for a *stationary* AR(1) process has a pattern of exponential decay as the autocorrelation lag length (k) grows. Figure 6.7 shows two examples of this. In Example I at the top of the figure, $\phi_1 = 0.7$. From (6.43), the autocorrelations for this AR(1) process have these values:

$$\rho_0 = \phi_1^0 = (0.7)^0 = 1$$

$$\rho_1 = \phi_1^1 = (0.7)^1 = 0.7$$

$$\rho_2 = \phi_1^2 = (0.7)^2 = 0.49$$

$$\rho_3 = \phi_1^3 = (0.7)^3 = 0.34$$

$$\vdots \qquad \vdots \qquad \vdots$$

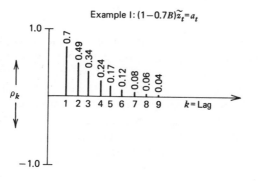

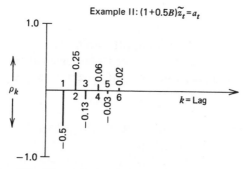

Figure 6.7 Theoretical acf's for two AR(1) processes.

In Example II at the bottom of Figure 6.7, $\phi_1 = -0.5$. Using (6.43) we find

$$\rho_0 = \phi_1^0 = (-0.5)^0 = 1$$

$$\rho_1 = \phi_1^1 = (-0.5)^1 = -0.5$$

$$\rho_2 = \phi_1^2 = (-0.5)^2 = 0.25$$

$$\rho_3 = \phi_1^3 = (-0.5)^3 = -0.125$$

$$\vdots \qquad\qquad \vdots$$

Note that if ϕ_1 is positive, all decay is on the positive side of the acf. But if ϕ_1 is negative the autocorrelation coefficients alternate in sign, while their absolute values decay exponentially. Thus if we see an estimated acf that decays exponentially, either from the positive side or with alternating signs starting from the negative side, we may make an educated guess that a stationary AR(1) is a good model to represent the data. (See Part II, Case 1 for an example.)

Equation (6.43) also suggests that a *nonstationary* AR(1) process will produce a theoretical acf which does not damp out. For example, if $\phi_1 = 1$, the ρ_k follow the pattern 1, 1, 1, ... Thus if we see an estimated acf whose autocorrelations *die out slowly* at higher lags, this is a clue that the underlying process may be nonstationary. Note that the estimated autocorrelations need not all be near 1.0 to suggest nonstationarity; they need merely damp out slowly. (Case 8 in Part II shows an example of an estimated acf which decays slowly from relatively small values.)

Summary

1. Stationary AR processes have
   (a) theoretical acf's that tail off toward zero with either some type of exponential decay or a damped sine wave pattern; and
   (b) theoretical pacf's that cut off to zero after lag p (the AR order of the process).

2. MA processes have
   (a) theoretical acf's that cut off to zero after lag q (the MA order of the process); and
   (b) theoretical pacf's that tail off toward zero with either some type of exponential decay or a damped sine wave pattern.

3. Stationary mixed (ARMA) processes have
    (a)   theoretical acf's that tail off toward zero after the first q-p lags; and
    (b)   theoretical pacf's that tail off toward zero after the first p-q lags.

4. An AR(1) or ARMA(1, q) process must meet the following condition to be stationary: $|\phi_1| < 1$.

5. An AR(2) or ARMA(2, q) process must meet the following three conditions to be stationary:

$$|\phi_2| < 1$$

$$\phi_2 + \phi_1 < 1$$

$$\phi_2 - \phi_1 < 1$$

6. The stationarity requirement ensures that we can obtain useful estimates of the mean, variance, and acf from a sample. If a process mean were different each time period, we could not obtain useful estimates since we typically have only one observation available per time period.

7. To check for stationarity in practice:
    (a)   examine the realization visually to see if the mean and variance appear to be constant;
    (b)   examine the estimated acf to see if it drops to zero rapidly; if it does not, the mean may not be stationary and differencing may be needed; and
    (c)   check any estimated AR coefficients to see that they meet the relevant stationarity conditions.

8. The stationarity conditions on the ϕ coefficients are complicated when $p > 2$. We can at least use this necessary (but not sufficient) condition to check for stationarity when $p > 2$:

$$\phi_1 + \phi_2 + \cdots + \phi_p < 1$$

9. If $p > 2$, we rely primarily on visual inspection of the data and the behavior of the estimated acf to check for stationarity. If the estimated acf does not fall rapidly to zero at longer lags, we suspect nonstationarity.

10. An MA(1) or ARMA(p, 1) process must meet the following condition to be invertible: $|\theta_1| < 1$.

11. An MA(2) or ARMA(p, 2) process must meet the following three conditions to be invertible:

$$|\theta_2| < 1$$

$$\theta_2 + \theta_1 < 1$$

$$\theta_2 - \theta_1 < 1$$

12. The invertibility requirement produces the common-sense implication that smaller weights are attached to observations further in the past.

13. Theoretical acf's and pacf's are derived from processes by applying expected value rules and the assumptions about the random shocks.

14. The autocorrelation function for a pure MA process is

$$\rho_k = \frac{-\theta_k + \theta_1\theta_{k+1} + \cdots + \theta_{q-k}\theta_q}{\left(1 + \theta_1^2 + \theta_2^2 + \cdots + \theta_q^2\right)}, \qquad k = 1, 2, \ldots, q$$

$$\rho_k = 0, \qquad k > q$$

15. The autocorrelation function for an AR(1) process is

$$\rho_k = \phi_1^k, \qquad k = 0, 1, 2, \ldots$$

Appendix 6A: The formal conditions for stationarity and invertibility

In this appendix we discuss the formal mathematical conditions for stationarity and invertibility of any ARIMA process.

Stationarity.   Use the AR operator to form the *characteristic equation*

$$\left(1 - \phi_1 B - \phi_2 B^2 - \cdots - \phi_p B^p\right) = 0 \qquad (6A.1)$$

where B is now treated as an ordinary algebraic variable. Stationarity requires that all roots of (6A.1) lie outside the unit circle (in the complex plane).

Although this formal condition for stationarity is conceptually clear, the implied restrictions on the AR coefficients may not be easy to find in practice. For the AR(1), it is easy to show that $|\phi_1| < 1$ must hold if B is to lie outside the unit circle. For the AR(2), we may apply the standard quadratic formula to (6A.1) to derive the conditions on ϕ_1 and ϕ_2 shown in equation (6.8) and Table 6.2.

For $p = 3$, 4, or 5 there are general solutions for the roots of (6A.1), but they are cumbersome. There are no general solutions for polynomials of degree six or higher. In these cases, the range of root values of (6A.1) satisfying stationarity may be found numerically. Then the implied range of acceptable values for the ϕ coefficients may be found. This procedure is relatively difficult and time-consuming and is often not done.

Occasionally, some analysts express ARIMA models in a multiplicative rather than additive form to ease examination of stationarity and invertibility conditions. Writing the AR operator in multiplicative form gives this characteristic equation:

$$(1 - \phi_1 B)(1 - \phi_2 B^2)(1 - \phi_3 B^3) \cdots (1 - \phi_p B^p) = 0 \quad (6A.2)$$

In this case, the set of stationarity conditions on the coefficients reduces to

$$|\phi_i| < 1, \quad \text{for all } i \quad (6A.3)$$

In Chapter 11 we discuss a common type of multiplicative ARIMA model containing both seasonal and nonseasonal elements.

Invertibility. The formal mathematical requirements for invertibility are identical to those for stationarity except we begin with the MA operator to form this characteristic equation:

$$(1 - \theta_1 B - \theta_2 B^2 - \cdots - \theta_q B^q) = 0 \quad (6A.4)$$

Invertibility requires that all roots of (6A.4) lie outside the unit circle (in the complex plane). All earlier comments about the ease or difficulty of finding the restrictions on ϕ coefficients apply here, but in this case they pertain to θ coefficients. Likewise, the MA operator may be expressed in multiplicative form:

$$(1 - \theta_1 B)(1 - \theta_2 B^2) \cdots (1 - \theta_q B^q) = 0 \quad (6A.5)$$

where the set of invertibility conditions becomes

$$|\theta_i| < 1, \quad \text{for all } i \quad (6A.6)$$

Appendix 6B: Invertibility, uniqueness, and forecast performance

In Section 6.3 we emphasized the common-sense appeal of the invertibility requirement. Some further comments about this requirement are in order, though they involve mathematical proofs beyond the scope of this book.

Invertibility guarantees that, for stationary processes, any given theoretical acf will correspond *uniquely* to some ARIMA generating mechanism. (This result holds only up to a multiplicative factor—a point discussed further under the topic of parameter redundancy in Chapter 8.) This unique correspondence increases the attractiveness of the Box–Jenkins identification procedure: there is only one stationary ARIMA process consistent with any particular theoretical acf. Of course, uniqueness does not ensure that correctly choosing the theoretical acf corresponding to an estimated acf will be easy in practice.

An example may help to clarify the idea of uniqueness. Suppose we have an MA(1) theoretical acf with $\rho_1 = 0.4$ and $\rho_k = 0$ for $k > 1$. We see from (6.25) that these autocorrelations are consistent with either $\theta_1 = 0.5$ or $\theta_1 = 2$. Restricting θ_1 to satisfy the invertibility condition $|\theta_1| < 1$ provides a unique correspondence between the autocorrelations and the value of θ_1. The same conclusion holds for any MA process: the ρ_k may give multiple solutions for MA coefficients, but a unique correspondence between the theoretical acf structure and the process is assured if the invertibility conditions are satisfied. The problem of multiple solutions does not arise with AR models. It can be proven that the coefficients of a pure AR process are uniquely determined by the corresponding theoretical acf.

Achieving uniqueness by restricting our analysis to invertible processes may seem arbitrary. In particular, perhaps we should consider dropping the invertibility requirement whenever a noninvertible form produces better forecasts than the invertible one. It turns out, however, that this cannot happen. There is a theorem, proven elsewhere, which states the following: in practice, the noninvertible form of a model cannot produce better forecasts than the invertible form based on a minimum mean-squared forecast-error criterion.

Questions and Problems

6.1 Consider the following pairs of theoretical acf's and pacf's. Indicate whether each pair is associated with an AR, MA, or ARMA process, and state the orders of each process. Explain your reasoning in each case.

acf   pacf

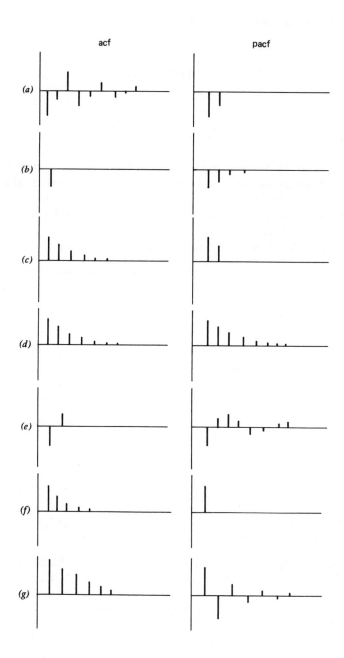

6.2   Consider this AR(1) process with $\phi_1 = 0.8$:

$$(1 - 0.8B)\tilde{z}_t = a_t$$

where

$$\sigma_a^2 = 1$$

(a)   What are the numerical values of the first five autocovariances?
(b)   What are the numerical values of the first five theoretical autocorrelation coefficients?
(c)   What are the numerical values of the first five theoretical partial autocorrelation coefficients?
(d)   Sketch the theoretical acf and pacf of this process.

6.3   Which of the following processes are stationary and invertible? Explain.

(a) $(1 - 1.05B + 0.4B^2)\tilde{z}_t = a_t$
(b) $(1 - 1.05B)\tilde{z}_t = a_t$
(c) $(1 + 0.8B)\tilde{z}_t = (1 - 0.25B)a_t$
(d) $\tilde{z}_t = (1 + 0.7B - 0.5B^2)a_t$
(e) $\tilde{z}_t = (1 - 0.8B)a_t$
(f) $(1 - 0.4B^2)\tilde{z}_t = a_t$
(g) $(1 + 0.6B)\tilde{z}_t = (1 + 0.9B^2)a_t$

6.4   Calculate and plot in an acf the first five theoretical autocorrelation coefficients for this MA(1) process:

$$z_t = (1 + 0.6B)a_t$$

6.5   Show that $C = \mu$ for the MA(2) process.

6.6   Show that $C = \mu(1 - \phi_1 - \phi_2)$ for a stationary AR(2) process. Use the faster method referred to in the footnote in Section 6.5.

6.7   Derive the theoretical acf for the MA(2) process. Calculate and plot in an acf the first five theoretical autocorrelations for this MA(2) process:

$$\tilde{z}_t = (1 + 0.8B - 0.4B^2)a_t$$

7

IDENTIFICATION:
NONSTATIONARY MODELS

In Chapter 1 we said that a data series had to be stationary before we could properly apply the UBJ modeling strategy, and till now we have focused on stationary models. Stationary realizations are generated by stationary processes. If the random shocks (a_t) in a process are Normally distributed, then the process will be stationary if its mean, variance, and (theoretical) autocorrelation function are constant through time. Thus, if we consider segments of a realization generated by a stationary process (the first and second halves, for example), the different segments will typically have means, variances, and autocorrelation coefficients that do not differ significantly.

In practice, however, many realizations are nonstationary. In this chapter we consider how we can (often, but not always) transform a nonstationary realization into a stationary one. If such transformations can be found, we may then apply the three-stage UBJ strategy of identification, estimation, and diagnostic checking to the transformed, stationary series. After modeling the transformed series, we may reverse the transformation procedure to obtain forecasts of the original, nonstationary series.

7.1 Nonstationary mean

The most common type of nonstationarity is when the mean of a series is not constant. Figures 7.1, 7.2, and 7.3 are examples of three such realiza-

155

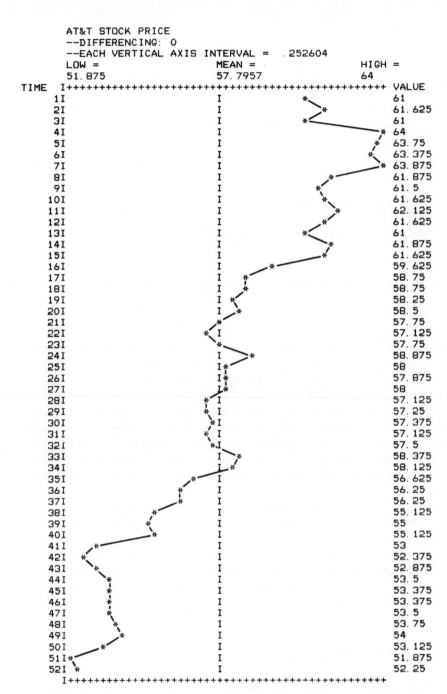

Figure 7.1 Example of a realization with a nonstationary mean: AT & T closing stock prices.

tions. Figure 7.1 shows weekly observations of the closing price of a stock whose overall level (mean) is trending downward through time. Figure 7.2 shows the weekly availability of an industrial part. The level of this series appears to rise and fall episodically rather than trending in one direction. Figure 7.3 is a series of commercial bank real-estate loans. The level of this series shows a trend, much like the stock-price series in Figure 7.1, but the loan series is trending up rather than down. Note also that the loan series changes both slope and level, whereas the stock-price series has a roughly constant slope.

Homogeneous nonstationarity. Each of the above three realizations has an important characteristic called *homogeneous nonstationarity*. That is, different segments of each series behave much like the rest of the series *after* we allow for changes in level and/or slope. This characteristic is important because a homogeneously nonstationary realization can be transformed into a stationary series simply by differencing.

We can visualize the idea of homogeneous nonstationarity by considering the rectangular frames superimposed on the three series, shown in Figures 7.4, 7.5, and 7.6. In Figure 7.4 the observations in the left-hand frame trace out a path very similar to the data path in the right-hand frame. The only difference is that the two frames are drawn at different levels. The same is true of the three frames superimposed on the data in Figure 7.5.

Likewise, the two frames in Figure 7.6 are drawn at different levels. But we must also draw these frames at different angles to make the data paths within each frame look similar. When these two frames are placed next to each other, as shown in Figure 7.7, both segments of this realization appear to have the same level and slope. The similar behavior of the data within the two frames in Figure 7.7 suggests that the nonstationarity in the loans series is of the homogeneous variety; different segments of this series are similar *after* we remove the differences in level and slope.

Differencing. Realizations that are homogeneously nonstationary can be rendered stationary by *differencing*. (Remember that differencing is a procedure for dealing with a nonstationary mean, not a nonstationary variance.) We introduced the mechanics of differencing in Chapter 2 and the associated notation in Chapter 5. For convenience we review the fundamental ideas here.

To difference a series once ($d = 1$), calculate the period-to-period changes. To difference a series twice ($d = 2$), calculate the period-to-period changes in the first-differenced series. For example, consider the short realization (z_t) shown in Table 7.1, column 2. The first differences of z_t (designated

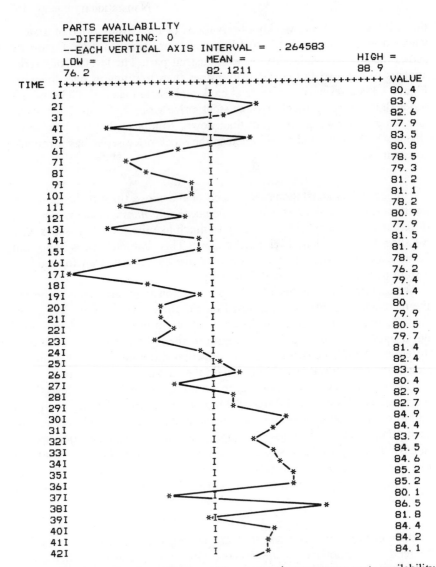

Figure 7.2 Example of a realization with a nonstationary mean: parts availability.

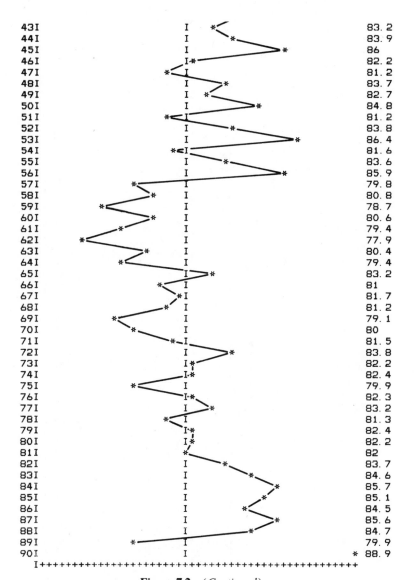

43I		83. 2
44I		83. 9
45I		86
46I		82. 2
47I		81. 2
48I		83. 7
49I		82. 7
50I		84. 8
51I		81. 2
52I		83. 8
53I		86. 4
54I		81. 6
55I		83. 6
56I		85. 9
57I		79. 8
58I		80. 8
59I		78. 7
60I		80. 6
61I		79. 4
62I		77. 9
63I		80. 4
64I		79. 4
65I		83. 2
66I		81
67I		81. 7
68I		81. 2
69I		79. 1
70I		80
71I		81. 5
72I		83. 8
73I		82. 2
74I		82. 4
75I		79. 9
76I		82. 3
77I		83. 2
78I		81. 3
79I		82. 4
80I		82. 2
81I		82
82I		83. 7
83I		84. 6
84I		85. 7
85I		85. 1
86I		84. 5
87I		85. 6
88I		84. 7
89I		79. 9
90I		88. 9

Figure 7.2 (*Continued*)

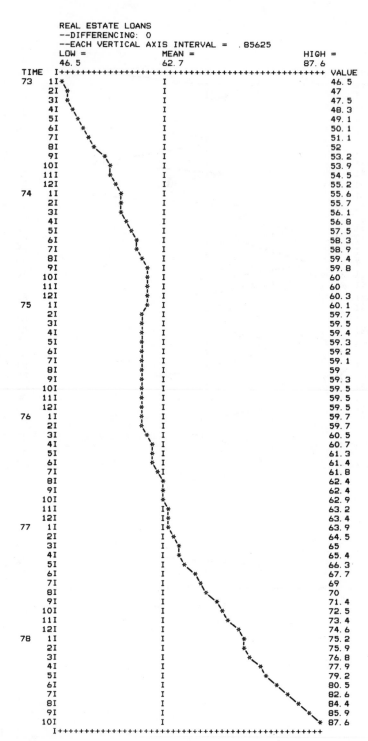

```
REAL ESTATE LOANS
--DIFFERENCING: 0
--EACH VERTICAL AXIS INTERVAL =  .85625
     LOW =              MEAN =              HIGH =
     46.5               62.7                87.6
TIME I++++++++++++++++++++++++++++++++++++++++++++++++++ VALUE
73  1I*                 I                                 46.5
    2I *                I                                 47
    3I *                I                                 47.5
    4I  *               I                                 48.3
    5I   *              I                                 49.1
    6I    *             I                                 50.1
    7I    *             I                                 51.1
    8I     *            I                                 52
    9I      *           I                                 53.2
   10I      *           I                                 53.9
   11I       *          I                                 54.5
   12I       *          I                                 55.2
74  1I        *         I                                 55.6
    2I        *         I                                 55.7
    3I        *         I                                 56.1
    4I         *        I                                 56.8
    5I         *        I                                 57.5
    6I          *       I                                 58.3
    7I          *       I                                 58.9
    8I           *      I                                 59.4
    9I           *      I                                 59.8
   10I           *      I                                 60
   11I           *      I                                 60
   12I            *     I                                 60.3
75  1I            *     I                                 60.1
    2I           *      I                                 59.7
    3I           *      I                                 59.5
    4I           *      I                                 59.4
    5I           *      I                                 59.3
    6I           *      I                                 59.2
    7I           *      I                                 59.1
    8I          *       I                                 59
    9I           *      I                                 59.3
   10I           *      I                                 59.5
   11I           *      I                                 59.5
   12I           *      I                                 59.5
76  1I            *     I                                 59.7
    2I            *     I                                 59.7
    3I            *     I                                 60.5
    4I             *    I                                 60.7
    5I             * *  I                                 61.3
    6I              *   I                                 61.4
    7I              *   I                                 61.8
    8I               *  I                                 62.4
    9I               *  I                                 62.4
   10I                * I                                 62.9
   11I                 I*                                 63.2
   12I                 I*                                 63.4
77  1I                 I*                                 63.9
    2I                 I *                                64.5
    3I                 I  *                               65
    4I                 I  *                               65.4
    5I                 I   *                              66.3
    6I                 I    *                             67.7
    7I                 I      *                           69
    8I                 I       *                          70
    9I                 I        *                         71.4
   10I                 I         *                        72.5
   11I                 I          *                       73.4
   12I                 I           *                      74.6
78  1I                 I            *                     75.2
    2I                 I             *                    75.9
    3I                 I              *                   76.8
    4I                 I               *                  77.9
    5I                 I                *                 79.2
    6I                 I                 *                80.5
    7I                 I                   *              82.6
    8I                 I                     *            84.4
    9I                 I                      *           85.9
   10I                 I                       *          87.6
    I++++++++++++++++++++++++++++++++++++++++++++++++++
```

Figure 7.3 Example of a realization with a nonstationary mean: real-estate loans.

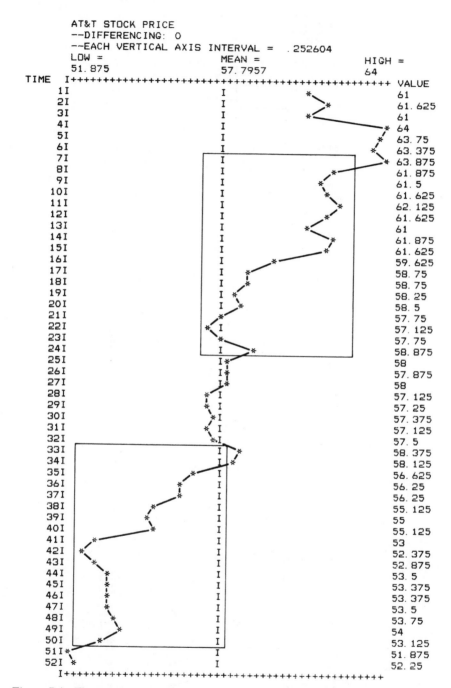

Figure 7.4 Figure 7.1 with superimposed rectangular frames to illustrate homogeneous nonstationarity.

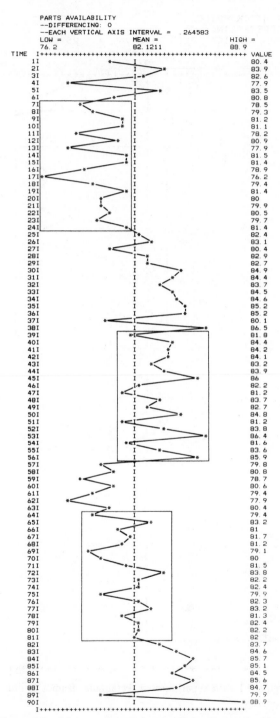

Figure 7.5 Figure 7.2 with superimposed rectangular frames to illustrate homogeneous nonstationarity.

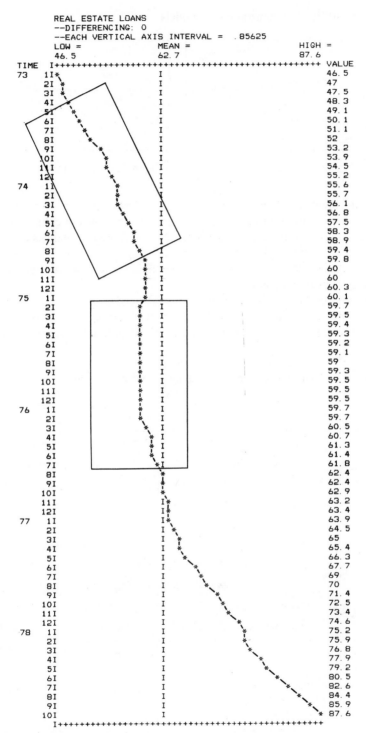

```
REAL ESTATE LOANS
--DIFFERENCING: 0
--EACH VERTICAL AXIS INTERVAL =  .85625
    LOW =              MEAN =              HIGH =
    46.5               62.7                87.6
TIME  I++++++++++++++++++++++++++++++++++++++++++++++++  VALUE
 73   1I*                    I                           46.5
      2I  *                  I                           47
      3I  *                  I                           47.5
      4I   *.                I                           48.3
      5I     *               I                           49.1
      6I      *              I                           50.1
      7I       **            I                           51.1
      8I        **           I                           52
      9I         **          I                           53.2
     10I          **         I                           53.9
      1I           *         I                           54.5
     12I           **        I                           55.2
 74   1I            *        I                           55.6
      2I            **       I                           55.7
      3I             *       I                           56.1
      4I             **      I                           56.8
      5I              *      I                           57.5
      6I               *     I                           58.3
      7I               **    I                           58.9
      8I                *    I                           59.4
      9I                *    I                           59.8
     10I                *    I                           60
     11I                *    I                           60
     12I                *    I                           60.3
 75   1I                *    I                           60.1
      2I               *     I                           59.7
      3I               *     I                           59.5
      4I               *     I                           59.4
      5I               *     I                           59.3
      6I              *      I                           59.2
      7I              *      I                           59.1
      8I              *      I                           59
      9I              *      I                           59.3
     10I              *      I                           59.5
     11I              *.     I                           59.5
     12I               *     I                           59.5
 76   1I                *    I                           59.7
      2I                *    I                           59.7
      3I                 *   I                           60.5
      4I                 *   I                           60.7
      5I                  *  I                           61.3
      6I                  *  I                           61.4
      7I                   *.I                           61.8
      8I                    *                            62.4
      9I                    *                            62.4
     10I                     *                           62.9
     11I                     I*                          63.2
     12I                     I*                          63.4
 77   1I                     I*                          63.9
      2I                     I  *                        64.5
      3I                     I   *                        65
      4I                     I    *                       65.4
      5I                     I     *                      66.3
      6I                     I      **                    67.7
      7I                     I        **                   69
      8I                     I          *                  70
      9I                     I           **               71.4
     10I                     I            **              72.5
     11I                     I             **             73.4
     12I                     I              *             74.6
 78   1I                     I               **           75.2
      2I                     I                *           75.9
      3I                     I                 **         76.8
      4I                     I                  **        77.9
      5I                     I                   *        79.2
      6I                     I                    **      80.5
      7I                     I                      **    82.6
      8I                     I                       **   84.4
      9I                     I                        **  85.9
     10I                     I                         *  87.6
      I++++++++++++++++++++++++++++++++++++++++++++++++
```

Figure 7.6 Figure 7.3 with superimposed rectangular frames to illustrate homogeneous nonstationarity.

163

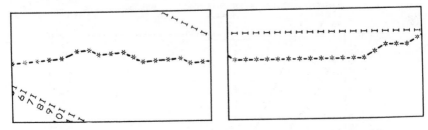

Figure 7.7 Rectangular frames from Figure 7.6 placed side-by-side.

∇z_t) are the changes in z_t: $\nabla z_t = z_t - z_{t-1}$. Thus $\nabla z_2 = z_2 - z_1 = 11 - 8 = 3$. Other calculations of the first differences are shown in column 3.

The second differences of z_t (designated $\nabla^2 z_t$) are the changes in the first differences: $\nabla^2 z_t = \nabla z_t - \nabla z_{t-1} = (z_t - z_{t-1}) - (z_{t-1} - z_{t-2})$. Thus $\nabla^2 z_3 = \nabla z_3 - \nabla z_2 = 4 - 3 = 1$. Further calculations of second differences appear in column 4.

We lose one observation each time we difference a series. For example, there is no observation z_0 to subtract from z_1, so we have only nine first differences in Table 7.1 although there are 10 original observations on z_t.

Table 7.1   Numerical examples of first and second differencing

t	$z_t{}^a$	First Differences of z_t: $\nabla z_t = z_t - z_{t-1}{}^b$	Second Differences of z_t: $\nabla^2 z_t = \nabla z_t - \nabla z_{t-1}{}^d$
1	$z_1 = 8$	$\nabla z_1 = z_1 - z_0 = $ n.a.^{c}	$\nabla^2 z_1 = \nabla z_1 - \nabla z_0 = $ n.a.
2	$z_2 = 11$	$\nabla z_2 = z_2 - z_1 = 11 - 8 = 3$	$\nabla^2 z_2 = \nabla z_2 - \nabla z_1 = $ n.a.
3	$z_3 = 15$	$\nabla z_3 = z_3 - z_2 = 15 - 11 = 4$	$\nabla^2 z_3 = \nabla z_3 - \nabla z_2 = 4 - 3 = 1$
4	$z_4 = 16$	$\nabla z_4 = z_4 - z_3 = 16 - 15 = 1$	$\nabla^2 z_4 = \nabla z_4 - \nabla z_3 = 1 - 4 = -3$
5	$z_5 = 17$	$\nabla z_5 = z_5 - z_4 = 17 - 16 = 1$	$\nabla^2 z_5 = \nabla z_5 - \nabla z_4 = 1 - 1 = 0$
6	$z_6 = 19$	$\nabla z_6 = z_6 - z_5 = 19 - 17 = 2$	$\nabla^2 z_6 = \nabla z_6 - \nabla z_5 = 2 - 1 = 1$
7	$z_7 = 23$	$\nabla z_7 = z_7 - z_6 = 23 - 19 = 4$	$\nabla^2 z_7 = \nabla z_7 - \nabla z_6 = 4 - 2 = 2$
8	$z_8 = 28$	$\nabla z_8 = z_8 - z_7 = 28 - 23 = 5$	$\nabla^2 z_8 = \nabla z_8 - \nabla z_7 = 5 - 4 = 1$
9	$z_9 = 27$	$\nabla z_9 = z_9 - z_8 = 27 - 28 = -1$	$\nabla^2 z_9 = \nabla z_9 - \nabla z_8 = -1 - 5 = -6$
10	$z_{10} = 29$	$\nabla z_{10} = z_{10} - z_9 = 29 - 27 = 2$	$\nabla^2 z_{10} = \nabla z_{10} - \nabla z_9 = 2 - (-1) = 3$

^{a}Mean = 18.3.
^{b}Mean = 2.3.
^{c}n.a. = not available.
^{d}Mean = -0.13.

Since there are only nine first differences, there are only eight second differences. That is, there is no first difference $\nabla z_1 = (z_1 - z_0)$ available at $t = 1$ to subtract from the first difference $\nabla z_2 = (z_2 - z_1)$ at $t = 2$, so we cannot calculate a second difference for $t = 2$.

Note that the means of the three series $(z_t, \nabla z_t, \nabla^2 z_t)$ get closer to zero the more we difference. (The means are shown at the bottom of Table 7.1.) This is a common result, especially for data in business, economics, and other social sciences. We discuss this point further in Section 7.3.

Let w_t stand for a differenced series. Although we may build an ARIMA model for the differenced series (w_t) when the original series (z_t) is not stationary, we are often interested in forecasting the undifferenced (z_t) series. While the w's are differences of the z's, the z's are *sums* of w's. We can obtain the z_t series by *integrating* (summing) successive w_t values. Integration is discussed further in Appendix 7A.

Backshift notation for differencing. Backshift notation for differenced variables is as follows: $(1 - B)z_t$ or $(1 - B)\tilde{z}_t$ represents the first differences of z_t. $(1 - B)^2 z_t$ or $(1 - B)^2 \tilde{z}_t$ represents the second differences of z_t. In general, $(1 - B)^d z_t$ or $(1 - B)^d \tilde{z}_t$ represents the dth differences of z_t.

It is easy to demonstrate these conclusions by applying the rules for the backshift operator B stated in Chapter 5. For example, we can show that $(1 - B)\tilde{z}_t$ is the same as $z_t - z_{t-1}$ by expanding the expression $(1 - B)\tilde{z}_t$ and applying rules (5.9) and (5.10) from Chapter 5:

$$(1 - B)\tilde{z}_t = (1 - B)(z_t - \mu) = z_t - z_t B - \mu + B\mu$$

$$= z_t - z_{t-1} - \mu + \mu$$

$$= z_t - z_{t-1} \tag{7.1}$$

The μ terms add to zero when z_t is differenced, so we could write the differenced series as either $(1 - B)^d z_t$ or $(1 - B)^d \tilde{z}_t$. This result is the algebraic analog of our statement above that a differenced realization often has a mean that is statistically zero. This topic is discussed further in Section 7.3.

Identification procedures. Let w_t represent a differenced series:

$$w_t = \nabla^d z_t$$

$$= (1 - B)^d \tilde{z}_t \tag{7.2}$$

After a nonstationary series z_t has been transformed into a differenced, stationary series w_t, then w_t is modeled with the same UBJ–ARIMA procedures that apply to any stationary series. For example, suppose the estimated acf of the differenced series w_t decays exponentially while the estimated pacf has a spike at lag 1. According to our discussion of theoretical acf's and pacf's in Chapters 3 and 6, we should then entertain an AR(1) model for w_t:

$$(1 - \phi_1 B)w_t = a_t \tag{7.3}$$

Since w_t and z_t are linked deterministically by definition (7.2), (7.3) also implies a model for z_t. Use (7.2) to substitute $(1 - B)^d \tilde{z}_t$ for w_t in (7.3) to see that (7.3) implies an ARIMA $(1, d, 0)$ model for z_t:

$$(1 - \phi_1 B)(1 - B)^d \tilde{z}_t = a_t \tag{7.4}$$

In general, any ARMA(p, q) model for a differenced series w_t is also an *integrated* ARIMA(p, d, q) model for the undifferenced or integrated series z_t, with p and q having the same values for both models. In fact, the AR and MA coefficients are the same for the two models. The link between the two models is definition (7.2), which states that the w's are obtained from differencing the z's d times, and the z's are obtained by integrating the w's d times.

For any realization we must select an appropriate degree of differencing (the value of d) before choosing the AR and MA terms to include in the model. If the original series z_t is stationary, we do not difference, so $d = 0$. When segments of a series differ only in level, as with the stock-price series in Figure 7.1 or the parts availability series in Figure 7.2, differencing once is sufficient to induce a stationary mean, so $d = 1$. When a series has a time-varying level *and* slope, as with the loans series in Figure 7.3, differencing twice will induce a stationary mean, so $d = 2$.

In practice, first differencing is required frequently while second differencing is needed only occasionally. Differencing more than twice is virtually never needed. We must be careful not to difference a series more than is needed to achieve stationarity. Unnecessary differencing creates artificial patterns in a series and tends to reduce forecast accuracy. On the other hand, Box and Jenkins suggest that, in a forecasting situation, a series should be differenced if there is serious doubt as to whether the stationary or nonstationary formulation is appropriate:

In doubtful cases there may be advantage in employing the nonstationary model rather than the stationary alternative (for example, in treating a ϕ_1, whose estimate is close to unity, as being *equal* to unity). This is particularly true in forecasting and control problems. Where ϕ_1 is close to unity, we do not really know whether the mean of the series has meaning or not. Therefore, it may be advantageous to employ the nonstationary model which does not include a mean μ. If we use such a model, forecasts of future behavior will not in any way depend on an estimated mean, calculated from a previous period, which may have no relevance to the future level of the series. [1, p. 192, emphasis in original. Quoted by permission.]

How do we choose the value of d? As noted in Chapter 6, there are three complementary procedures:

1. Examine the data visually. This often gives a clue to the appropriate degree of differencing. For example, it is difficult to look at the stock-price series in Figure 7.1 without seeing that the level of the data is trending down. However, the slope of the series does not appear to be changing through time. Therefore, setting $d = 1$ (differencing once) would seem appropriate. While such visual analysis can be helpful, we should not rely on it exclusively to determine the degree of differencing.

2. Examine the estimated acf's of the original series and the differenced series. The estimated acf of a nonstationary series will *decay only slowly*. While the estimated acf for a series with a nonstationary mean might decay slowly from a very high level, with r_1 close to 1.0, this is not a necessary characteristic of such series. The estimated acf could start out with rather small values of r_1 (less than 0.5, for example). The critical point is that the estimated acf decays toward zero very slowly.*

3. Check any estimated AR coefficients at the estimation stage to see if they satisfy the stationarity conditions discussed in Chapter 6.

Example 1. The above modeling steps may be illustrated using the three realizations in Figures 7.1, 7.2, and 7.3. Visual analysis of the stock-price series suggests that its mean is nonstationary because the series trends down. The estimated acf for this realization, shown in Figure 7.8, declines very slowly. This behavior is consistent with the mean being nonstationary, and differencing at least once is proper.

---

*See Box and Jenkins [1, p. 200–201] for an example of a nonstationary process whose estimated acf's decay slowly from approximately $r_1 = 0.5$.

```
+ + + + + + + + + + + + AUTOCORRELATIONS + + + + + + + + + + + +
+ FOR DATA SERIES: AT&T STOCK PRICE                                +
+ DIFFERENCING: 0                        MEAN    =  57.7957        +
+ DATA COUNT = 52                        STD DEV =   3.4136        +
  COEF   T-VAL  LAG _____0_____
  0.93   6.74    1                              [   0>>>>>]>>>>>>>>>>>>>>>>>>>
  0.86   3.75    2                          [       0>>>>>>>>>>>>]>>>>>>>>>>>
  0.81   2.85    3                       [          0>>>>>>>>>>>>>>]>>>>>>>>
  0.75   2.29    4                    [             0>>>>>>>>>>>>>>>>]>>>
  0.68   1.91    5                 [                0>>>>>>>>>>>>>>>>>]
  0.62   1.62    6            [                     0>>>>>>>>>>>>>>>>        ]
  0.55   1.38    7            [                     0>>>>>>>>>>>>>>          ]
  0.49   1.19    8            [                     0>>>>>>>>>>>>            ]
  0.44   1.03    9         [                        0>>>>>>>>>>>                ]
  0.38   0.87   10         [                        0>>>>>>>>>                  ]
  0.29   0.65   11         [                        0>>>>>>>                    ]
  0.22   0.49   12         [                        0>>>>>                      ]
  0.18   0.39   13         [                        0>>>>                       ]
       CHI-SQUARED* =   280.32 FOR DF =   13
```

Figure 7.8 Estimated acf for the stock-price realization in Figure 7.1.

The first differences of the stock-price realization appear in Figure 7.9. The differenced series no longer has a noticeable trend. Instead it fluctuates around a fixed mean of nearly zero (the mean is −0.17). Differencing once appears to have induced a stationary mean, so the nonstationarity in the original series was apparently of the homogeneous variety.

The estimated acf of the differenced stock-price series is shown in Figure 7.10. It dies out to zero quickly, suggesting that the mean of the first differences is stationary. It has no significant t-values; the acf shows neither the decaying pattern suggestive of a pure AR or mixed ARMA model, nor the spike and cutoff pattern associated with MA models. Therefore, the first differences appear to be a white-noise series, suggesting the following ARIMA model for the differenced data:

$$w_t = a_t \qquad (7.5)$$

Substituting $w_t = (1 - B)\tilde{z}_t$ into (7.5) leads to this model for the original series:

$$(1 - B)\tilde{z}_t = a_t \qquad (7.6)$$

In Case 6 in Part II we see that estimation and diagnostic checking confirm model (7.6) as appropriate for the stock-price series.

Example 2. We saw in Figure 7.2 that the parts-availability data also appear to have a nonstationary mean. The estimated acf for this series, shown in Figure 7.11, is consistent with this conclusion. The estimated

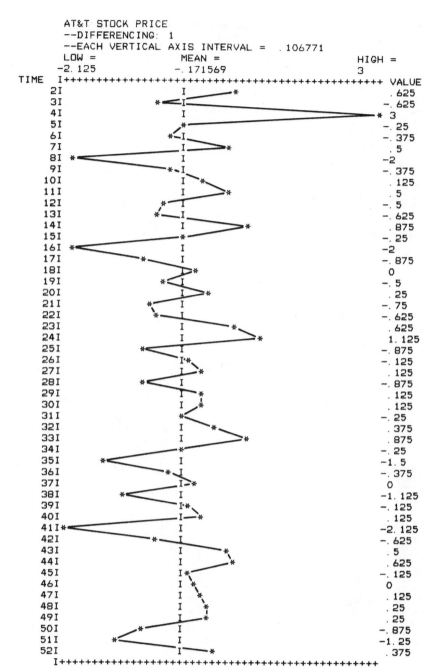

Figure 7.9 First differences of the stock-price realization in Figure 7.1.

```
+ + + + + + + + + + + + AUTOCORRELATIONS + + + + + + + + + + + +
+ FOR DATA SERIES:  AT&T STOCK PRICE                                    +
+ DIFFERENCING: 1                        MEAN    = -.171569            +
+ DATA COUNT =  51                       STD DEV =  .849221            +
  COEF   T-VAL LAG _____O_____
-0.04  -0.31   1                            <<<<O
-0.20  -1.43   2            <<<<<<<<<<<<<<<<<<<<<O
 0.13   0.88   3                            O>>>>>>>>>>>>>
-0.11  -0.71   4                    <<<<<<<<<<<O
-0.07  -0.50   5                      <<<<<<<O
-0.03  -0.17   6                          <<<O
 0.00  -0.02   7                            O
-0.14  -0.96   8                 <<<<<<<<<<<<<O
 0.08   0.50   9                            O>>>>>>>
 0.20   1.33  10                            O>>>>>>>>>>>>>>>>>>>>
-0.17  -1.09  11               <<<<<<<<<<<<<<<O
-0.17  -1.05  12               <<<<<<<<<<<<<<<O
 0.12   0.74  13                            O>>>>>>>>>>>>
     CHI-SQUARED* =    13.81 FOR DF =   13
```

Figure 7.10 Estimated acf for the first differences of the stock-price data in Figure 7.9.

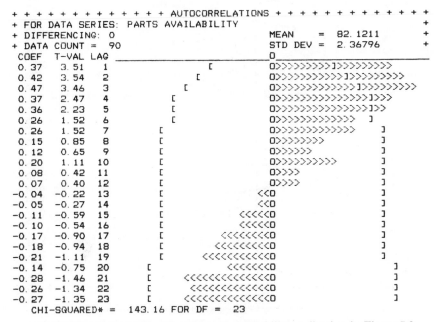

```
+ + + + + + + + + + + + AUTOCORRELATIONS + + + + + + + + + + + +
+ FOR DATA SERIES:  PARTS AVAILABILITY                                  +
+ DIFFERENCING: 0                        MEAN    =  82.1211            +
+ DATA COUNT =  90                       STD DEV =  2.36796            +
  COEF   T-VAL LAG _____O_____
 0.37   3.51   1                    [          O>>>>>>>>>]>>>>>>>>>
 0.42   3.54   2                 [             O>>>>>>>>>>>]>>>>>>>>>
 0.47   3.46   3              [                O>>>>>>>>>>>>]>>>>>>>>>
 0.37   2.47   4           [                   O>>>>>>>>>>>>>>]>>>
 0.36   2.23   5           [                   O>>>>>>>>>>>>]>>
 0.26   1.52   6           [                   O>>>>>>>>>>>  ]
 0.26   1.52   7        [                       O>>>>>>>>>>>     ]
 0.15   0.85   8        [                       O>>>>>>>           ]
 0.12   0.65   9        [                       O>>>>>             ]
 0.20   1.11  10        [                       O>>>>>>>>>>        ]
 0.08   0.42  11        [                       O>>>>              ]
 0.07   0.40  12        [                       O>>>>              ]
-0.04  -0.22  13        [                     <<O                 ]
-0.05  -0.27  14        [                     <<O                 ]
-0.11  -0.59  15        [                  <<<<<O                 ]
-0.10  -0.54  16        [                  <<<<<O                 ]
-0.17  -0.90  17        [               <<<<<<<<O                 ]
-0.18  -0.94  18        [              <<<<<<<<<O                 ]
-0.21  -1.11  19        [            <<<<<<<<<<<O                 ]
-0.14  -0.75  20     [                  <<<<<<<O                     ]
-0.28  -1.46  21     [            <<<<<<<<<<<<<O                     ]
-0.26  -1.34  22     [             <<<<<<<<<<<<O                     ]
-0.27  -1.35  23     [            <<<<<<<<<<<<<O                     ]
     CHI-SQUARED* =   143.16 FOR DF =   23
```

Figure 7.11 Estimated acf for the parts-availability realization in Figure 7.2.

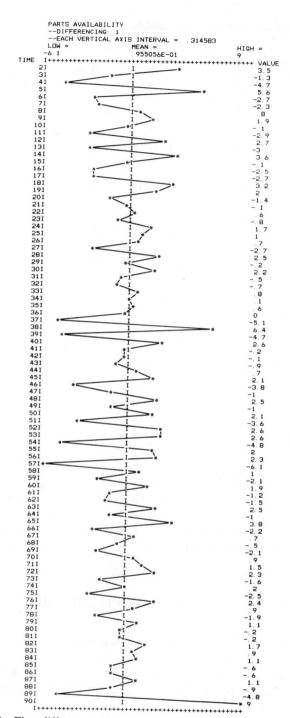

PARTS AVAILABILITY
--DIFFERENCING: 1
--EACH VERTICAL AXIS INTERVAL = .314583

Figure 7.12 First differences of the parts-availability realization in Figure 7.2.

171

```
+ + + + + + + + + + + + AUTOCORRELATIONS + + + + + + + + + + + + +
+ FOR DATA SERIES: PARTS AVAILABILITY                               +
+ DIFFERENCING:  1                           MEAN    = .955056E-01  +
+ DATA COUNT =  89                           STD DEV = 2.56563      +
  COEF   T-VAL  LAG                             0
-0.47   -4.45   1   <<<<<<<<<<<<<<<<[<<<<<<<<<<0              ]
-0.05   -0.42   2              [              <<<0            ]
 0.09    0.72   3              [                 0>>>>>       ]
-0.04   -0.30   4              [              <<0            ]
 0.06    0.46   5              [                 0>>>        ]
-0.11   -0.84   6              [              <<<<<0          ]
 0.12    0.91   7              [                 0>>>>>>      ]
-0.05   -0.40   8        [              <<<0            ]
-0.08   -0.62   9        [              <<<<0           ]
 0.17    1.25  10        [                 0>>>>>>>>>    ]
-0.10   -0.72  11        [              <<<<<0           ]
 0.10    0.72  12        [                 0>>>>>        ]
-0.10   -0.75  13        [              <<<<<0           ]
 0.05    0.39  14        [                 0>>>          ]
-0.04   -0.29  15        [              <<0            ]
 0.03    0.21  16        [                 0>            ]
-0.04   -0.28  17        [              <<0            ]
 0.01    0.06  18        [                 0            ]
-0.07   -0.48  19        [              <<<0            ]
 0.19    1.42  20        [                 0>>>>>>>>>>>  ]
-0.12   -0.88  21        [              <<<<<<0          ]
-0.01   -0.04  22        [                 0            ]
 0.02    0.16  23        [                 0>            ]
     CHI-SQUARED* =   38.93 FOR DF =   23
```

Figure 7.13 Estimated acf for the first differences of the parts-availability data in Figure 7.12.

autocorrelation coefficients actually rise for the first few lags and remain moderately large (absolute t-values > 1.6) until about lag 7.

After differencing once, the parts availability series appears to have a stationary mean. The first differences (plotted in Figure 7.12) seem to fluctuate around a fixed mean of about zero (the mean is 0.096). The estimated acf of the first differences (Figure 7.13) dies out to zero quickly, with only the autocorrelation at lag 1 being significant. This suggests an MA(1) model for the first differences:

$$w_t = (1 - \theta_1 B)a_t \tag{7.7}$$

Since $w_t = (1 - B)\tilde{z}_t$, (7.7) corresponds to an ARIMA(0, 1, 1) model for the original series z_t:

$$(1 - B)\tilde{z}_t = (1 - \theta_1 B)a_t \tag{7.8}$$

Estimation and diagnostic checking of model (7.8), discussed in Case 8 in Part II, show that it provides a good representation of the parts-availability data.

Example 3. The real-estate-loans realization shown in Figure 7.3 appears to change both level and slope, suggesting that differencing twice is needed. The estimated acf of the original data (Figure 7.14) fails to damp out rapidly toward zero, thus confirming the nonstationary character of the realization mean.

The first-differenced data appear in Figure 7.15. This differenced series looks much like the original parts availability realization—its level rises and falls episodically. The estimated acf for the first-differenced data is shown in Figure 7.16. It does not damp out toward zero rapidly, so further differencing seems appropriate.

The twice-differenced data are plotted in Figure 7.17. This series fluctuates around a constant mean of approximately zero (the mean is 0.0176). In Figure 7.18 we see that the estimated acf for this series moves quickly to zero. The significant spike at lag 1, followed by the cutoff to zero, suggests an MA(1) for the second differences:

$$w_t = (1 - \theta_1 B)a_t \tag{7.9}$$

We know that w_t stands for the second differences of z_t: $w_t = (1 - B)^2 \tilde{z}_t$. Therefore, (7.9) implies that the original series z_t follows an ARIMA(0, 2, 1). That is, substitute $(1 - B)^2 \tilde{z}_t$ into (7.9) for w_t to get

$$(1 - B)^2 \tilde{z}_t = (1 - \theta_1 B)a_t \tag{7.10}$$

```
+ + + + + + + + + + + +  AUTOCORRELATIONS + + + + + + + + + + + + +
+ FOR DATA SERIES: REAL ESTATE LOANS                              +
+ DIFFERENCING: 0                        MEAN     =  62.7'        +
+ DATA COUNT =  70                       STD DEV  =  9.42795      +
  COEF  T-VAL LAG _____0_____
  0.93   7.74   1                   [     0>>>>>]>>>>>>>>>>>>>>>>>>>>
  0.85   4.33   2                  [      0>>>>>>>>>]>>>>>>>>>>>>>
  0.78   3.20   3                [        0>>>>>>>>>>>>]>>>>>>>>>
  0.71   2.57   4              [          0>>>>>>>>>>>>>]>>>>
  0.65   2.16   5          [              0>>>>>>>>>>>>>>>]
  0.59   1.85   6          [              0>>>>>>>>>>>>>>>]
  0.54   1.60   7          [              0>>>>>>>>>>>>>> ]
  0.49   1.40   8        [                0>>>>>>>>>>>>       ]
  0.44   1.23   9        [                0>>>>>>>>>>>        ]
  0.39   1.07  10        [                0>>>>>>>>>          ]
  0.35   0.93  11        [                0>>>>>>>>>          ]
  0.30   0.80  12        [                0>>>>>>>>           ]
  0.26   0.69  13      [                  0>>>>>>>               ]
  0.22   0.57  14      [                  0>>>>>>                ]
  0.18   0.48  15      [                  0>>>>>                 ]
  0.15   0.39  16      [                  0>>>>                  ]
  0.12   0.32  17      [                  0>>>                   ]
  0.10   0.26  18      [                  0>>>                   ]
    CHI-SQUARED* =   370.57 FOR DF =   18
```

Figure 7.14 Estimated acf for the loans realization in Figure 7.3.

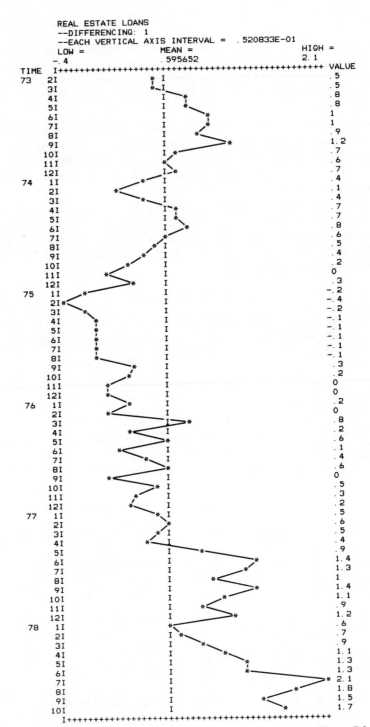

Figure 7.15 First differences of the loans realization in Figure 7.3.

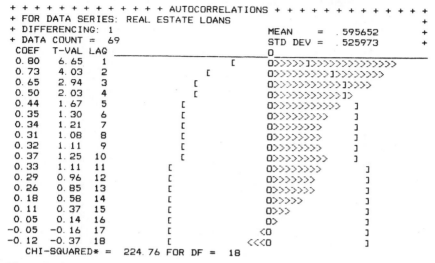

Figure 7.16   Estimated acf for the first differences of the loans data in Figure 7.15.

Estimation and diagnostic-checking results in Part II, Case 7, confirm that (7.10) is an acceptable model for the loans data.

7.2   Nonstationary variance

Some realizations have a variance that changes through time. This occurs most frequently with business and economic data covering a long time span, especially when there is a seasonal element in the data. Such series must be transformed to induce a constant variance before being modeled with the UBJ–ARIMA method. It is possible that no suitable transformation will be found.

Series with a nonstationary variance often have a nonstationary mean also. A series of this type must be transformed to induce a constant variance *and* differenced to induce a fixed mean before being modeled further.

Figure 7.19 is an example of a series whose mean and variance are both nonstationary. These data, which are analyzed in Part II, Case 11, are monthly armed robberies in Boston from 1966 to 1975. The rising trend suggests that the mean is nonstationary, and the variance also seems to get larger as the overall level rises.

The first differences of the original data are shown in Figure 7.20. The first differences appear to fluctuate about a fixed mean which is close to

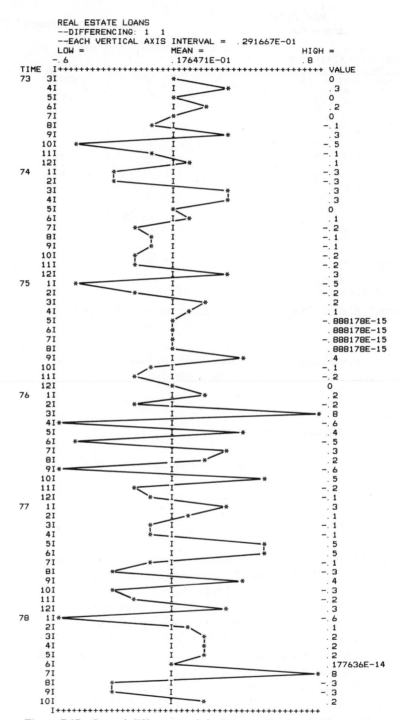

Figure 7.17 Second differences of the loans realization in Figure 7.3.

176

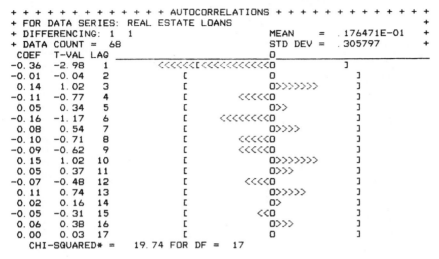

```
+ + + + + + + + + + + + AUTOCORRELATIONS + + + + + + + + + + + +
+ FOR DATA SERIES:  REAL ESTATE LOANS                                +
+ DIFFERENCING:  1   1                          MEAN    =  .176471E-01  +
+ DATA COUNT =  68                              STD DEV =  .305797      +
  COEF   T-VAL LAG                              0
-0.36  -2.98   1          <<<<<<[<<<<<<<<<<<<<0              ]
-0.01  -0.04   2                 [             0             ]
 0.14   1.02   3                 [             0>>>>>>>       ]
-0.11  -0.77   4                 [         <<<<<0             ]
 0.05   0.34   5                 [             0>>            ]
-0.16  -1.17   6                 [        <<<<<<<<0             ]
 0.08   0.54   7                 [             0>>>>          ]
-0.10  -0.71   8                 [         <<<<<0             ]
-0.09  -0.62   9                 [         <<<<<0             ]
 0.15   1.02  10                 [             0>>>>>>>       ]
 0.05   0.37  11                 [             0>>            ]
-0.07  -0.48  12                 [          <<<<0             ]
 0.11   0.74  13                 [             0>>>>>         ]
 0.02   0.16  14                 [             0>             ]
-0.05  -0.31  15                 [            <<0             ]
 0.06   0.38  16                 [             0>>>           ]
 0.00   0.03  17                 [             0             ]
   CHI-SQUARED* =   19.74 FOR DF =   17
```

Figure 7.18 Estimated acf for the second differences of the loans data in Figure 7.17.

zero. However, the variance of the differenced data still seems to be increasing over time.

Logarithmic transformation. Often a series with a nonstationary variance will be stationary in the natural logarithms. This transformation is appropriate if the variance of the original series is proportional to the mean, so that the *percent* fluctuations are constant through time.

The natural logarithms of the armed-robbery realization are plotted in Figure 7.21. This transformation appears to have made the variance of the series stationary. The first differences of the natural logarithms, plotted in Figure 7.22, confirm this conclusion. (Note that we calculated the natural logarithms *before* differencing the data. Differencing first would have caused problems because the differenced series has some negative values, and the natural logarithm of a negative number is undefined.)

We have created a new series, w_t, which is the first differences of the natural logarithms of the original series:

$$w_t = (1 - B)(\ln z_t) \qquad (7.11)$$

We may now model the series w_t using the standard UBJ method. However, our real interest may be in forecasting the original series z_t, not the natural logarithms of z_t. It might seem that we could forecast z_t by merely finding the antilogarithms of forecasts of the logged series. However, there are some complications in this procedure, as discussed in Chapter 10.

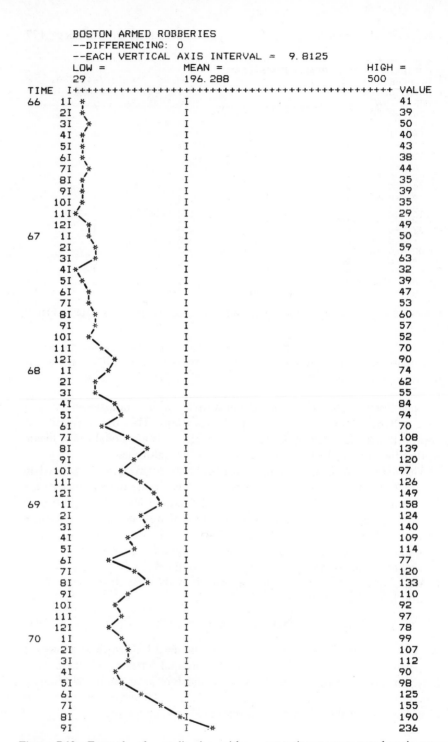

```
          BOSTON ARMED ROBBERIES
          --DIFFERENCING: 0
          --EACH VERTICAL AXIS INTERVAL =   9.8125
          LOW =              MEAN =                      HIGH =
          29                 196.288                     500
  TIME   I+++++++++++++++++++++++++++++++++++++++++++++++++ VALUE
    66    1I *                I                                  41
          2I *                I                                  39
          3I   *              I                                  50
          4I *                I                                  40
          5I *                I                                  43
          6I *                I                                  38
          7I   *              I                                  44
          8I *                I                                  35
          9I *                I                                  39
         10I *                I                                  35
         11I*                 I                                  29
         12I  *               I                                  49
    67    1I   *              I                                  50
          2I     *            I                                  59
          3I    *             I                                  63
          4I*                 I                                  32
          5I *                I                                  39
          6I   *              I                                  47
          7I    *             I                                  53
          8I     *            I                                  60
          9I    *             I                                  57
         10I  *               I                                  52
         11I     *            I                                  70
         12I        *         I                                  90
    68    1I          *       I                                  74
          2I    *             I                                  62
          3I  *               I                                  55
          4I        *         I                                  84
          5I        *         I                                  94
          6I    *             I                                  70
          7I        *         I                                 108
          8I           *      I                                 139
          9I         *        I                                 120
         10I      *           I                                  97
         11I          *       I                                 126
         12I           *      I                                 149
    69    1I            *     I                                 158
          2I         *        I                                 124
          3I          *       I                                 140
          4I       *          I                                 109
          5I        *         I                                 114
          6I   *              I                                  77
          7I        *         I                                 120
          8I         *        I                                 133
          9I      *           I                                 110
         10I   *              I                                  92
         11I    *             I                                  97
         12I  *               I                                  78
    70    1I    *             I                                  99
          2I    *             I                                 107
          3I     *            I                                 112
          4I   *              I                                  90
          5I     *            I                                  98
          6I       *          I                                 125
          7I         *        I                                 155
          8I              *.I                                   190
          9I               I   *                                236
```

Figure 7.19 Example of a realization with a nonstationary mean and variance: Boston armed robberies.

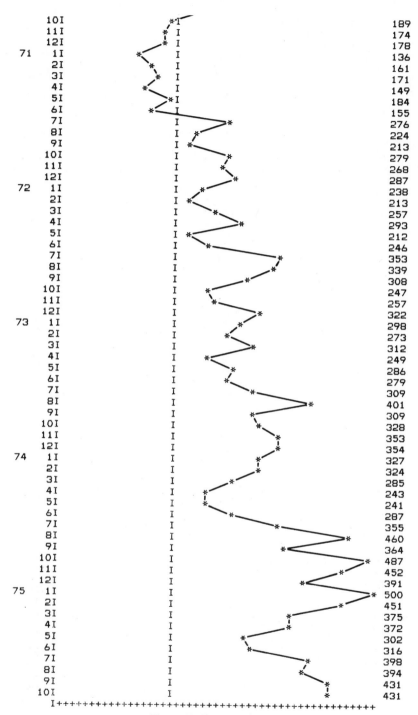

Figure 7.19 (*Continued*)

179

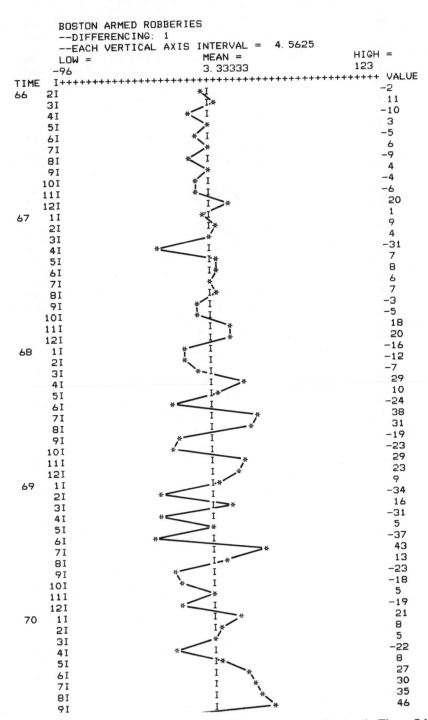

Figure 7.20 First differences of Boston armed-robberies realization in Figure 7.19.

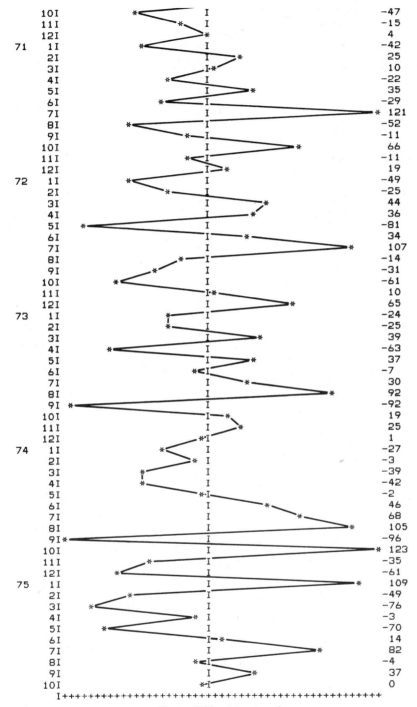

Figure 7.20 (*Continued*)

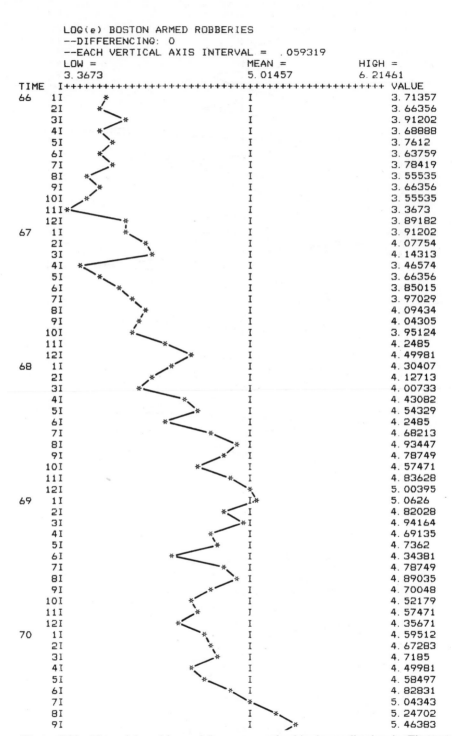

```
        LOG(e) BOSTON ARMED ROBBERIES
        --DIFFERENCING: 0
        --EACH VERTICAL AXIS INTERVAL =  .059319
        LOW =                      MEAN =            HIGH =
        3.3673                     5.01457           6.21461
  TIME  I++++++++++++++++++++++++++++++++++++++++++++++++++ VALUE
  66  1I        *               I                            3.71357
      2I      *                 I                            3.66356
      3I        *               I                            3.91202
      4I      *                 I                            3.68888
      5I      *                 I                            3.7612
      6I      *                 I                            3.63759
      7I       *                I                            3.78419
      8I     *                  I                            3.55535
      9I      *                 I                            3.66356
     10I     *                  I                            3.55535
     11I*                       I                            3.3673
     12I         *              I                            3.89182
  67  1I        *               I                            3.91202
      2I          *             I                            4.07754
      3I            *           I                            4.14313
      4I   *                    I                            3.46574
      5I      *                 I                            3.66356
      6I        *               I                            3.85015
      7I          *             I                            3.97029
      8I           *            I                            4.09434
      9I          *             I                            4.04305
     10I        *               I                            3.95124
     11I           *            I                            4.2485
     12I             *          I                            4.49981
  68  1I           *            I                            4.30407
      2I         *              I                            4.12713
      3I        *               I                            4.00733
      4I             *          I                            4.43082
      5I              *         I                            4.54329
      6I           *            I                            4.2485
      7I               *        I                            4.68213
      8I                *     I                              4.93447
      9I               *        I                            4.78749
     10I             *          I                            4.57471
     11I              *         I                            4.83628
     12I                *       I                            5.00395
  69  1I                I.*                                  5.0626
      2I               *    I                                4.82028
      3I                 *I                                  4.94164
      4I              *       I                              4.69135
      5I              *       I                              4.7362
      6I          *           I                              4.34381
      7I               *      I                              4.78749
      8I                *     I                              4.89035
      9I              *       I                              4.70048
     10I            *         I                              4.52179
     11I             *        I                              4.57471
     12I          *           I                              4.35671
  70  1I             *        I                              4.59512
      2I              *       I                              4.67283
      3I              *       I                              4.7185
      4I           *          I                              4.49981
      5I            *         I                              4.58497
      6I              *       I                              4.82831
      7I                *     I                              5.04343
      8I                 I *                                 5.24702
      9I                 I   *                               5.46383
```

Figure 7.21 Natural logarithms of Boston armed-robberies realization in Figure 7.19.

182

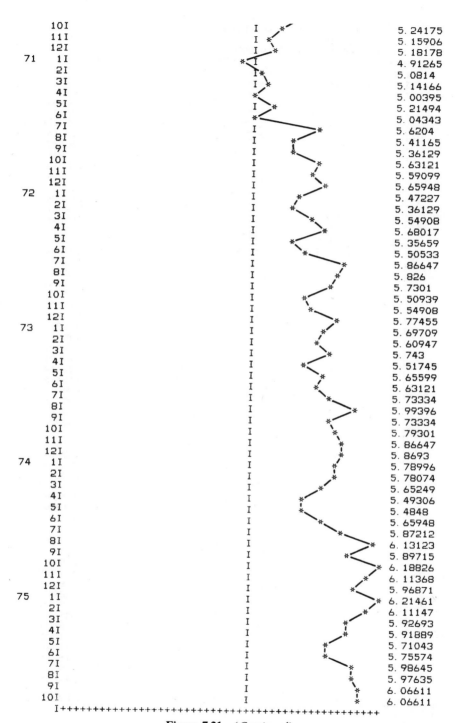

10 I	5. 24175
11 I	5. 15906
12 I	5. 18178
71　1 I	4. 91265
2 I	5. 0814
3 I	5. 14166
4 I	5. 00395
5 I	5. 21494
6 I	5. 04343
7 I	5. 6204
8 I	5. 41165
9 I	5. 36129
10 I	5. 63121
11 I	5. 59099
12 I	5. 65948
72　1 I	5. 47227
2 I	5. 36129
3 I	5. 54908
4 I	5. 68017
5 I	5. 35659
6 I	5. 50533
7 I	5. 86647
8 I	5. 826
9 I	5. 7301
10 I	5. 50939
11 I	5. 54908
12 I	5. 77455
73　1 I	5. 69709
2 I	5. 60947
3 I	5. 743
4 I	5. 51745
5 I	5. 65599
6 I	5. 63121
7 I	5. 73334
8 I	5. 99396
9 I	5. 73334
10 I	5. 79301
11 I	5. 86647
12 I	5. 8693
74　1 I	5. 78996
2 I	5. 78074
3 I	5. 65249
4 I	5. 49306
5 I	5. 4848
6 I	5. 65948
7 I	5. 87212
8 I	6. 13123
9 I	5. 89715
10 I	6. 18826
11 I	6. 11368
12 I	5. 96871
75　1 I	6. 21461
2 I	6. 11147
3 I	5. 92693
4 I	5. 91889
5 I	5. 71043
6 I	5. 75574
7 I	5. 98645
8 I	5. 97635
9 I	6. 06611
10 I	6. 06611

Figure 7.21　(*Continued*)

183

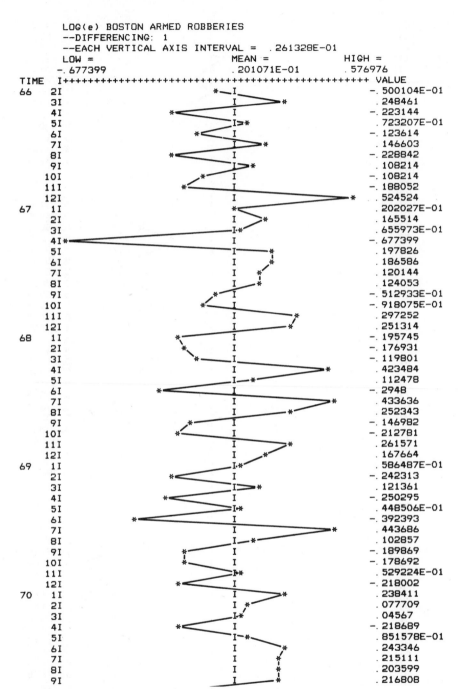

Figure 7.22 First differences of the natural logarithms of Boston armed-robberies data in Figure 7.21.

184

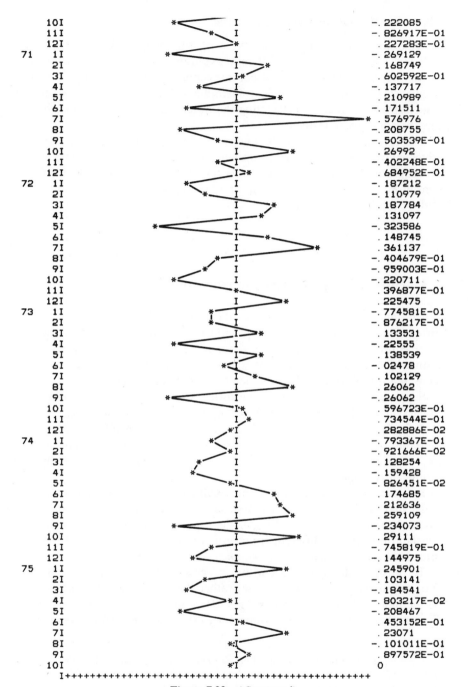

Figure 7.22 (*Continued*)

Other transformations. Sometimes a logarithmic transformation will not induce a stationary variance—it may overtransform or undertransform a series. Many analysts rely on visual inspection of the transformed data to decide whether the logarithmic transformation is adequate. Some other transformations, such as calculating the square roots of the original data, may be appropriate instead.

There is another approach, known as the Box–Cox transformation, which involves *estimating* an appropriate transformation from the data. This procedure is beyond the scope of this text, but the interested reader may consult Box and Jenkins [1, p. 328] for a brief introduction, or Box and Cox [19] for a fuller discussion.

7.3 Differencing and deterministic trends

When the mean $\hat{\mu}$ of an original data series z_t is stationary so that differencing is not required, $\hat{\mu}$ will generally not be zero. Therefore, a model representing such an undifferenced series will generally have a nonzero estimated constant term ($\hat{C}$). As shown in Chapter 5, $\hat{C} = \hat{\mu}\,(1 - \Sigma\hat{\phi}_i)$. Thus, if $\hat{\mu}$ is not statistically zero, then $\hat{C}$ will typically be nonzero.*

Suppose instead that z_t must be differenced ($d > 0$) to achieve a stationary mean. The resulting series w_t often has a mean ($\hat{\mu}_w$) that is not statistically different from zero. A model representing a differenced series therefore often has a constant term of zero. That is, if the estimate of μ_w ($\hat{\mu}_w$) is statistically zero, then $\hat{C} = \hat{\mu}_w\,(1 - \Sigma\hat{\phi}_i)$ will typically also not differ significantly from zero.

But occasionally when $d > 0$ the resulting differenced series w_t has a mean that is significantly different from zero. Then it may be proper to assume that $\mu_w \neq 0$. The resulting model for w_t will therefore usually have a constant term that is different from zero. The corresponding model for the integrated series z_t then has a *deterministic trend* element.

To illustrate this idea, start with a process with no deterministic trend. (For simplicity we refer to processes rather than estimated models.) Let w_t be the first differences of z_t: $w_t = (1 - B)\tilde{z}_t$. Suppose initially that $\mu_w = 0$. For simplicity, let w_t consist of white noise. Then the ARIMA process for w_t

---

*It is possible for $\hat{C}$ to be insignificantly different from zero even though $\hat{\mu}$ differs significantly from zero. This is because the variance of $\hat{C}$ depends not only on the variance of $\hat{\mu}$ but also on the variances and covariances of the estimated ϕ coefficients. In practice, however, if $\hat{\mu}$ is significantly different from zero, $\hat{C}$ is nearly always significant also.

is

$$w_t = a_t \tag{7.12}$$

Substituting $(1 - B)\tilde{z}_t$ for w_t in (7.12), we see that z_t follows a random walk without a constant term:

$$(1 - B)\tilde{z}_t = a_t$$

or

$$z_t = z_{t-1} + a_t \tag{7.13}$$

As a random walk, z_t in (7.13) shows no affinity for a fixed central value. Furthermore, because (7.13) has no constant term, z_t does not move persistently in any particular direction. Instead, z_t moves at random, as dictated by the random shock a_t, starting from the previous value (z_{t-1}).

Now suppose instead that $\mu_w \neq 0$. Then it is proper to write (7.12) with w_t in deviations from its mean:

$$(w_t - \mu_w) = a_t \tag{7.14}$$

Substituting $(1 - B)\tilde{z}_t$ for w_t shows that z_t still follows a random walk, but *with a constant term*:

$$(w_t - \mu_w) = a_t$$

$$(1 - B)\tilde{z}_t - \mu_w = a_t$$

$$z_t = C + z_{t-1} + a_t \tag{7.15}$$

where $C = \mu_w$.

As a random walk (7.15) states that z_t does not tend toward a fixed central value. But unlike (7.13), process (7.15) states that z_t *will* move persistently in a particular direction starting from z_{t-1}. That is, starting from z_{t-1}, z_t will trend upward each time period by amount C if $C > 0$, or downward each period by amount C if $C < 0$. Because C is a constant, this trend component is deterministic.

For models with higher degrees of differencing ($d > 1$) and additional AR and MA terms, the algebraic manipulation and the mathematical nature of the deterministic trend become more complicated than in the preceding example, but the basic conclusions, stated below, remain the same:

1. When the mean of a differenced variable w_t is zero ($\mu_w = 0$), the processes for both w_t and z_t have constant terms of zero. In such cases, any trend element present in forecasts of z_t is stochastic; that is, the trend element depends only on past z values that appear in the

equation because of the differencing and subsequent integration steps. For example, model (7.13) results from integrating model (7.12); neither of these models has a constant term because $\mu_w = 0$. Any trend element in forecasts of z_t depends on the behavior of z_{t-1} on the RHS of (7.13), and z_{t-1} is a stochastic variable.

2. When the mean of a differenced variable is nonzero ($\mu_w \neq 0$), the resulting processes for both w_t and z_t have a nonzero constant term. Forecasts of z_t then have a deterministic trend component in addition to whatever stochastic trend is introduced by the differencing and integration procedures. Thus (7.15) contains a deterministic trend component because C (equal to μ_w in this case) is nonzero. These forecasts may also display a stochastic trend because z_{t-1} appears on the RHS of (7.15).

In practice, when $d > 0$, the UBJ analyst must decide whether or not to include a nonzero constant term (i.e., whether $\mu_w \neq 0$). In business and in economics and other social sciences, $\mu_w = 0$ is often the proper assumption. However, there are several guidelines for making this decision.

1. The most reliable procedure is to include a nonzero mean (and therefore a constant term) in the model at the estimation stage to see if they are statistically nonzero. Some (but not all) computer programs estimate μ simultaneously along with the AR and MA coefficients. These programs usually also provide t-statistics indicating whether $\hat{\mu}$ and $\hat{C}$ are significantly different from zero. If the absolute t-value of the estimated constant term is large (e.g., $|t| > 2.0$), an estimated mean (and therefore a constant term) might be included in this model.

2. Consider the nature of the data. The analyst may know from experience, or from a theoretical understanding of the data source, that the data have a deterministic trend component so that $\mu_w \neq 0$ is a proper assumption. This approach is especially helpful in the physical and engineering sciences where one might conclude that a deterministic element is present from knowledge of the physical or mathematical structure underlying the data.

3. Use a preliminary statistical test (before the estimation stage) to see if $\mu_w = 0$ is a proper assumption. Some computer programs provide rough preliminary tests of the hypothesis $H_0: \mu_w = 0$ based on the sample statistic $\overline{w}$ and its approximate standard error.* If $\overline{w}$ is large compared to its standard error (e.g., $|t| > 2.0$), a nonzero mean might be included in the model.

---

*Box and Jenkins discuss the approximate standard error of $\overline{w}$ in [1, pp. 193–195].

4. Estimate two models, one with a nonzero mean (and constant) and one without, and check the forecasting accuracy of both models.

Finally, however, it is important to remember that models with deterministic trends are relatively uncommon outside the physical sciences. According to Box and Jenkins,

> In many applications, where no physical reason for a deterministic component exists, the mean of w can be assumed to be zero unless such an assumption proves contrary to facts presented by the data. It is clear that, for many applications, the assumption of a stochastic trend is often more realistic than the assumption of a deterministic trend. This is of special importance in forecasting a time series, since a stochastic trend does not necessitate the series to follow the identical pattern which it has developed in the past. [1, pp. 92–93. Quoted by permission.]

(Case 2 in Part II has an example of a model with a statistically significant deterministic trend; this model is rejected because the nature of the data suggests that a deterministic trend makes no sense. Case 15 contains a model with a deterministic trend that can be rationalized.)

Summary

1. The UBJ method applies only to stationary realizations (i.e., those with a mean, variance, and acf that are constant through time.)

2. In practice many realizations are nonstationary. Fortunately, nonstationary realizations can often be transformed into stationary data series.

3. The mean of a realization may change over time. If different parts of a realization behave in a similar fashion except for changes in level and slope, the realization is said to be homogeneously nonstationary.

4. A homogeneously nonstationary realization can be made stationary by differencing. First differencing is needed if the level is changing over time; second differencing is needed if the level and slope are changing over time.

5. Avoid unnecessary differencing. It creates artificial patterns in a data series and reduces forecast accuracy. However, Box and Jenkins suggest differencing when there is a serious question as to whether a stationary or nonstationary model is appropriate.

6. After differencing, we construct an ARMA(p, q) model for the differenced series (w_t).

7.   We may recover the original values (z_t) by integrating a differenced series. Integration involves summing successive values in a differenced series.

8.   An ARMA(p, q) model for a differenced series (w_t) implies an ARIMA(p, d, q) model for the integrated (original, undifferenced) series z_t. The AR and MA coefficients and the constant term are the same for the two models.

9.   The appropriate degree of differencing may be chosen by
   (a)   inspecting the realization visually;
   (b)   examining the estimated acf's of the original realization and of the differenced series;
   (c)   checking any estimated AR coefficients at the estimation stage to see if they satisfy the stationarity conditions stated in Chapter 6.

10.   Some realizations have a variance that changes over time. Such realizations must be transformed to induce a constant variance before the UBJ method may be used. It is possible that no suitable transformation will be found.

11.   Some realizations have both a nonstationary mean and variance. Such realizations must be transformed to induce a constant variance, then differenced to induce a stationary mean.

12.   A common transformation to induce a constant variance involves taking the natural logarithms of the original realization. This is appropriate if the variance of the original realization is proportional to the mean.

13.   If the mean (μ_w) of a differenced series (w_t) is assumed to be zero, the resulting ARIMA model for both w_t and the integrated series z_t has a constant term of zero. Any trend element in forecasts from such a model is stochastic, not deterministic.

14.   If μ_w is assumed to be nonzero, the resulting model for both w_t and z_t has a nonzero constant term. Forecasts from such a model contain a deterministic trend component in addition to any stochastic trend that may be present.

15.   ARIMA models with deterministic trend components are uncommon outside the physical sciences.

Appendix 7A: Integration

A differenced variable w_t is linked deterministically to the original variable z_t by the differencing operator $(1 - B)^d$:

$$w_t = (1 - B)^d z_t \qquad\qquad (7A.1)$$

While w's are differences of the z's, the z's are sums of the w's. We may therefore return to the z's by integrating (summing) the w's. Thus, an ARMA(p, q) model for w_t is an *integrated* ARIMA(p, d, q) model for z_t. This is an important concept because after building an ARIMA model for the stationary series w_t, we often want to forecast the original nonstationary series z_t.

To show that the z's are sums of the w's, consider the case when $d = 1$. Solving (7A.1) for z_t gives

$$z_t = (1 - B)^{-1} w_t \qquad (7A.2)$$

$(1 - B)^{-1}$ can be written as the infinite series $(1 + B + B^2 + B^3 + \cdots)$, so we have

$$z_t = \left(1 + B + B^2 + B^3 + \cdots\right) w_t$$

$$= w_t + w_{t-1} + w_{t-2} + w_{t-3} + \cdots$$

$$= \sum_{i=-\infty}^{t} w_i \qquad (7A.3)$$

If z_t is differenced d times, then the dth difference w_t must be integrated d times to obtain z_t. In the example above, $d = 1$, so z_t results from integrating w_t once. Alternatively, if $d = 2$, w_t is the second difference of z_t and we may solve (7A.1) for z_t to obtain

$$z_t = (1 - B)^{-2} w_t$$

$$= (1 - B)^{-1}(1 - B)^{-1} w_t \qquad (7A.4)$$

To obtain z_t from the second differences, (7A.4) says that we first integrate the second differences w_t to get the first differences (designated x_t):

$$x_t = (1 - B)^{-1} w_t \qquad (7A.5)$$

Substituting this into (7A.4), we then integrate the first differences x_t to obtain the original series z_t:

$$z_t = (1 - B)^{-1} x_t \qquad (7A.6)$$

8

ESTIMATION

Much of Chapters 1 through 7 deals with the first stage of the Box–Jenkins methodology, identification. In this chapter we focus on the second stage, estimation; in Chapter 9 we discuss the third stage, diagnostic checking; and then in Chapter 10 we consider certain elements of forecasting.

Some aspects of ARIMA model estimation involve technical details. Knowledge of these details is not essential for the reader whose primary interest is in the practice of UBJ–ARIMA modeling. Some of these technical matters are treated in two appendixes at the end of this chapter. In the main body of the chapter we focus on the fundamental elements of ARIMA model estimation and on the practical question of how to use estimation results to evaluate a model.

8.1 Principles of estimation

At the identification stage we tentatively select one or more models that seem likely to provide parsimonious and statistically adequate representations of the available data. In making this tentative selection, we calculate a rather large number of statistics (autocorrelation and partial autocorrelation coefficients) to help us. For example, with n observations, we will often estimate about $n/4$ autocorrelation and partial autocorrelation coefficients. Estimating so many parameters is not really consistent with the principle of parsimony. This nonparsimonious procedure is justifiable only as an initial, rough step in analyzing a data series. Our hope is that the broad overview of the data contained in the estimated acf and pacf will get us started in the right direction as we try to identify one or more appropriate models.

By contrast, at the estimation stage we get precise estimates of a small number of parameters. For example, suppose we tentatively choose an MA(2) model, $\tilde{z}_t = (1 - \theta_1 B - \theta_2 B^2)a_t$, at the identification stage based on $n/4$ estimated autocorrelation and partial autocorrelation coefficients. Then at the estimation stage we fit this model to the data to get precise estimates of only three parameters: the process mean μ and the two MA coefficients θ_1 and θ_2.

Although we make more efficient use of the available data at the estimation stage than at the identification stage by estimating fewer parameters, we cannot bypass the identification stage. We need the somewhat crude preliminary analysis at the identification stage to guide us in deciding which model to estimate. But once we have a parsimonious model in mind, we then want to make more efficient use of the available data. That is what we do at the estimation stage: we get accurate estimates of a few parameters as we fit our tentative model to the data.

ARIMA estimation is usually carried out on a computer using a *nonlinear least-squares* (NLS) approach. In the next two sections we introduce the most basic ideas about NLS estimation, bypassing many technical matters. The reader interested in these technical aspects may consult the appendixes at the end of this chapter.*

Maximum likelihood and least-squares estimates.   At the estimation stage the coefficient values must be chosen according to some criterion. Box and Jenkins [1] favor estimates chosen according to the *maximum likelihood* (ML) criterion. Mathematical statisticians frequently prefer the ML approach to estimation problems because the resulting estimates often have attractive statistical properties. It can be shown that the likelihood function (of a correct ARIMA model) from which ML estimates are derived reflects all useful information about the parameters contained in the data.[†]

However, finding exact ML estimates of ARIMA models can be cumbersome and may require relatively large amounts of computer time. For this reason, Box and Jenkins suggest using the *least-squares* (LS) criterion. It can be shown that if the random shocks are Normally distributed (as we suppose they are) then LS estimates are either exactly or very nearly ML estimates.[‡]

---

*Box and Jenkins [1, Chapter 7 and pp. 500–505] provide examples and the estimation algorithm.

[†]See Box and Jenkins' comments on the "likelihood principle" and their references on this matter [1, p. 209].

[‡]If we begin with the conditional likelihood function and the a_t are Normal, the LS method gives exact ML estimates. If we start with the unconditional likelihood function, then the LS method gives very nearly ML estimates if the a_t are Normal and if the sample size is fairly large.

"Least squares" refers to parameter estimates associated with the smallest sum of squared residuals. To explain this idea we first show what is meant by the term *residuals*. Then we illustrate the calculation of the *sum of squared residuals* (SSR). Finally, we consider how we can find the *smallest* SSR.

We will use an AR(1) model to illustrate these ideas, but the same concepts apply to the estimation of any ARIMA model. The AR(1) model we will use is

$$(1 - \phi_1 B)\tilde{z}_t = a_t$$

or

$$z_t = \mu(1 - \phi_1) + \phi_1 z_{t-1} + a_t \qquad (8.1)$$

where $\mu(1 - \phi_1)$ is the constant term.

Residuals.   Suppose for simplicity that we *know* the parameters (μ, ϕ_1) of model (8.1), and suppose we are located at time $t - 1$. Consider how we could predict z_t using our knowledge of the RHS variables in equation (8.1). We cannot observe the random shock a_t during period $t - 1$, but we do know μ, ϕ_1, and z_{t-1} at time $t - 1$. Assign a_t its expected value of zero and use μ, ϕ_1, and z_{t-1} to find the *calculated* value of z_t, designated $\hat{z}_t$:

$$\hat{z}_t = \mu(1 - \phi_1) + \phi_1 z_{t-1} \qquad (8.2)$$

Later, at time t, we can observe z_t. We can then find the random shock a_t by subtracting the calculated value $\hat{z}_t$ [calculated from *known* parameters, equation (8.2)] from the observed value z_t [equation (8.1)]:

$$z_t - \hat{z}_t = a_t \qquad (8.3)$$

So far we have assumed that μ and ϕ_1 are known. In practice we do not know the parameters of ARIMA models; instead, we must estimate them from the data. Designate these estimates in the present case as $\hat{\mu}$ and $\hat{\phi}_1$. Modifying (8.2) accordingly, the calculated value $\hat{z}_t$ now is:

$$\hat{z}_t = \hat{\mu}(1 - \hat{\phi}_1) + \hat{\phi}_1 z_{t-1} \qquad (8.4)$$

When $\hat{z}_t$ is calculated from *estimates* of parameters rather than *known* parameters, (8.3) does not give the exact value of the random shock a_t. Instead, when we subtract (8.4) from (8.1) we get only an estimate of the random shock a_t, denoted $\hat{a}_t$ and called a *residual*:

$$z_t - \hat{z}_t = \hat{a}_t \qquad (8.5)$$

Table 8.1   Calculation of a sum of squared residuals for an AR(1), with $\hat{\phi}_1 = 0.5$

t	$z_t^{\,a}$	$\tilde{z}_t = z_t - \bar{z}^{\,b}$	$\hat{\tilde{z}}_t = \hat{\phi}_1 \tilde{z}_{t-1}$	$\hat{a}_t = z_t - \hat{\tilde{z}}_t$	$\hat{a}_t^{2\,c}$
0	—	—	—	—	—
1	80	20	—	—	—
2	60	0	10	-10	100
3	30	-30	0	-30	900
4	40	-20	-15	-5	25
5	70	10	-10	20	400
6	80	20	5	15	225

$^a \Sigma z_t = 360;\ \hat{\mu} = \bar{z} = \dfrac{1}{n}(\Sigma z_t) = \frac{1}{6}(360) = 60.$

$^b \Sigma \tilde{z}_t = 0.$

$^c \Sigma \hat{a}_t^2 = 1650.$

Equation (8.5) is the definition of a residual for any ARIMA model. In general, $\hat{z}_t$ depends on $\hat{\mu}$ and the estimated AR and MA coefficients (along with their corresponding past z's and past residuals, which are estimated random shocks). Thus our example above was an AR(1), so $\hat{z}_t$ depends on $\hat{\mu}$ and $\hat{\phi}_1$, along with z_{t-1}, as shown in equation (8.4). For an MA(1), $\hat{z}_t$ depends on $\hat{\mu}$ and $\hat{\theta}_1$, along with $\hat{a}_{t-1}$: $\hat{z}_t = \hat{\mu} - \hat{\theta}_1 \hat{a}_{t-1}$. For an ARMA(1, 1), $\hat{z}_t$ depends on $\hat{\mu}$, $\hat{\phi}_1$, and $\hat{\theta}_1$, along with z_{t-1} and $\hat{a}_{t-1}$: $\hat{z}_t = \hat{\mu}(1 - \hat{\phi}_1) + \hat{\phi}_1 z_{t-1} - \hat{\theta}_1 \hat{a}_{t-1}$.

Sum of squared residuals.   Next consider the idea of the sum of squared residuals (SSR). Table 8.1 illustrates the calculation of an SSR.

Suppose we have the realization shown in column 2 of Table 8.1. (A realization with $n = 6$ is much too small to use in practice; we use it only for illustration.) We have tentatively identified an AR(1) model as shown in (8.1) to represent this realization.

We want to find estimates ($\hat{\mu}$ and $\hat{\phi}_1$) of the two unknown parameters (μ and ϕ_1). We could estimate μ and ϕ_1 simultaneously; however, with a large sample it is acceptable to first use the realization mean $\bar{z}$ as an estimate of μ and then proceed to estimate the remaining AR and MA coefficients.* In

---

*When μ and the ϕ's and θ's are estimated simultaneously, the resulting estimate $\hat{\mu}$ is usually very close to $\bar{z}$. The advantage of estimating μ and the ϕ's and θ's simultaneously is that we can then test both $\hat{\mu}$ and the estimated constant $\hat{C}$ to see if they are significantly different from zero. All examples and cases in this text are based on simultaneous estimation of μ with the ϕ's and θ's. The only exception is in this chapter where we first use $\bar{z}$ to estimate μ to simplify the numerical examples.

this example, we find $\bar{z} = 60$ as shown at the bottom of column 2 in Table 8.1.*

Having settled on $\bar{z}$ as our estimate of μ, we remove this element temporarily by expressing the data in deviations from the mean, that is, $\tilde{z}_t = z_t - \bar{z}$. Recall that the $\tilde{z}_t$ series has the same stochastic properties as z_t; we have merely shifted the series so its mean is identically zero. Thus in column 3 of Table 8.1 we calculate $\tilde{z}_t = z_t - \bar{z}$. As expected the $\tilde{z}_t$ series adds to zero and thus has a mean of zero. Therefore, if we rewrite model (8.1) in terms of $\tilde{z}_t$ we have a model with a constant term of zero:

$$\tilde{z}_t = \phi_1 \tilde{z}_{t-1} + a_t \tag{8.6}$$

Now we want to find a set of residuals for this model using the realization in column 3 of Table 8.1. Following our earlier discussion, replace ϕ_1 in (8.6) with its estimated value $\hat{\phi}_1$, and replace a_t with its expected value of zero to obtain this equation for the calculated $\tilde{z}$'s:

$$\hat{\tilde{z}}_t = \hat{\phi}_1 \tilde{z}_{t-1} \tag{8.7}$$

Subtracting (8.7) from (8.6) we obtain this equation for the residual $\hat{a}_t$ (column 5 in Table 8.1):

$$\hat{a}_t = \tilde{z}_t - \hat{\tilde{z}}_t \tag{8.8}$$

where $\tilde{z}_t$ is observed (column 3 in Table 8.1) and $\hat{\tilde{z}}_t$ is calculated using $\hat{\phi}_1$ (column 4 in Table 8.1).

To illustrate the calculation of the SSR, suppose we arbitrarily choose $\hat{\phi}_1 = 0.5$. Later we consider whether this is the best (i.e., least-squares) estimate of ϕ_1. Column 4 in Table 8.1 shows the calculated values $\hat{\tilde{z}}_t$ based on $\hat{\phi}_1 = 0.5$. For example,

$$\hat{\tilde{z}}_2 = \hat{\phi}_1 \tilde{z}_1 = (0.5)(20) = 10$$

$$\hat{\tilde{z}}_3 = \hat{\phi}_1 \tilde{z}_2 = (0.5)(0) = 0$$

$$\vdots \qquad \vdots \qquad \vdots$$

*If the data are differenced ($d > 0$), setting the estimated mean $\bar{w}$ of the differenced series w_t equal to zero is often appropriate. Setting $\bar{w}$ equal to a nonzero value introduces a deterministic trend into the model. This topic is discussed in Chapter 7.

In column 5 of Table 8.1 we find the residuals as shown in equation (8.8) by subtracting each $\hat{\tilde{z}}_t$ from each $\tilde{z}_t$:

$$\hat{a}_2 = \tilde{z}_2 - \hat{\tilde{z}}_2 = 0 - 10 = -10$$

$$\hat{a}_3 = \tilde{z}_3 - \hat{\tilde{z}}_3 = -30 - 0 = -30$$

$$\vdots \qquad \vdots \qquad \vdots$$

Each $\hat{a}_t$ is squared in column 6:

$$\hat{a}_2^2 = (-10)^2 = 100$$

$$\hat{a}_3^2 = (-30)^2 = 900$$

$$\vdots \qquad \vdots$$

Finally, summing column 6 we obtain the sum of squared residuals $\Sigma \hat{a}_t^2 = 1650$. This is the SSR *given* the estimates $\hat{\mu} = 60$ and $\hat{\phi}_1 = 0.5$. If $\hat{\mu}$ and $\hat{\phi}_1$ were different, we would get a different SSR. To get LS estimates of our parameters, we need to find the values of $\hat{\mu}$ and $\hat{\phi}_1$ that give the smallest SSR. In the present example our task is to find the value of $\hat{\phi}_1$ that minimizes the SSR given $\hat{\mu} = \bar{z} = 60$.

8.2   Nonlinear least-squares estimation

In general, least-squares estimation of ARIMA models requires the use of a *nonlinear* least-squares method. Readers may be familiar with the *linear* least-squares (LLS) method (also known as ordinary least squares or classical least squares) since this is the estimation method applied to regression models encountered in introductory statistics texts. The LLS estimator is derived by applying the calculus to the sum of squared residuals function.* This produces a set of linear equations which may be solved simultaneously rather easily. But proceeding in the same fashion with an ARIMA SSR function produces a set of equations which are, in general, highly nonlinear and solvable only with a nonlinear, iterative search technique.†

---

*Wonnacott and Wonnacott [9, Chapter 12] show how this is done.

† The only exception is a pure AR model with no multiplicative seasonal AR terms. (Multiplicative seasonal models are discussed in Chapter 11.)

Grid search. One search technique is the *grid-search* method. This method is not often used, but we present the idea because it provides a simple illustration of an iterative search procedure.

Some computer programs for estimating ARIMA-model parameters proceed as we have by first using $\bar{z}$ to estimate μ; the remaining parameters are then estimated. In the case of the AR(1) model in the last section, for $\bar{z} = 60$, each possible $\hat{\phi}_1$ produces a different SSR since each $\hat{\phi}_1$ gives a different set of $\hat{\bar{z}}_t$ values (column 4, Table 8.1) and therefore a different set of $\hat{a}_t$ values (column 5, Table 8.1). According to the LS criterion, we want to choose the value of $\hat{\phi}_1$ that gives the smallest SSR. We can imagine trying all possible values of $\hat{\phi}_1$ between -1 and $+1$ (the values of $\hat{\phi}_1$ permitted by the stationarity condition $|\hat{\phi}_1| < 1$) and comparing the resulting SSR's. This is the grid-search method.

For example, suppose we arbitrarily choose the series of values $\hat{\phi}_1 = -0.9$, $-0.8, \ldots, 0, \ldots, 0.8, 0.9$. Performing the same sequence of calculations shown in Table 8.1 with each of these $\hat{\phi}_1$ values produces the results shown in Table 8.2. Figure 8.1 is a plot of the pairs of numbers in Table 8.2. The vertical axis in Figure 8.1 shows the SSR corresponding to the values of $\hat{\phi}_1$ shown on the horizontal axis. To read this graph, first find a value of $\hat{\phi}_1$ on the horizontal axis; then find the corresponding SSR on the vertical axis by reading from the SSR function drawn on the graph.* It appears that the smallest SSR occurs somewhere between $\hat{\phi}_1 = 0.2$ and $\hat{\phi}_1 = 0.5$. Further calculations are required to find a more accurate value of $\hat{\phi}_1$. For example, we could now apply the grid-search method to the values $0.20, 0.21, \ldots,$ $0.49, 0.50$. (In fact, the LS estimate of $\hat{\phi}_1$ turns out to be (0.3333.)

The grid-search procedure can be time-consuming because there are so many possible values for $\hat{\phi}_1$. This problem becomes worse when more than one parameter is estimated. For example, estimating an AR(2) model requires considering all possible combinations of $\hat{\phi}_1$ and $\hat{\phi}_2$, not just the possible values of each coefficient separately.

Algorithmic nonlinear least squares. The grid-search method is a nonlinear least-squares (NLS) method but it is rarely used. The commonly used NLS method is similar to the grid-search approach because both involve a trial-and-error search for the least-squares estimates. However, the commonly used NLS method follows an algorithm to converge quickly to the least-squares estimates using a computer.

---

*This function depends on the value of $\hat{\mu}$. If we had used some value other than $\hat{\mu} = \bar{z} = 60$, we would get a different SSR function in Figure 8.1.

Table 8.2 Schedule of sum of squared residuals for various values of $\hat{\phi}_1$ applied to the realization in Table 8.1

$\hat{\phi}_1$	SSR
-0.9	4338
-0.8	3912
-0.7	3522
-0.6	3168
-0.5	2850
-0.4	2568
-0.3	2322
-0.2	2112
-0.1	1938
0	1800
0.1	1698
0.2	1632
0.3	1602
0.4	1608
0.5	1650
0.6	1728
0.7	1842
0.8	1992
0.9	2178

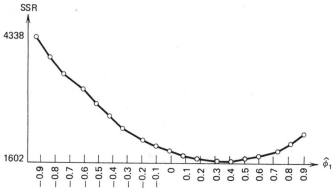

Figure 8.1 Sum of squared residuals as a function of $\hat{\phi}_1$ for the realization in Table 8.1.

The NLS technique most commonly used is a combination of two NLS procedures, Gauss–Newton linearization and the gradient method. This combination is often referred to as "Marquardt's compromise."* The basic idea is that given some initial "guess" values for the coefficients, Marquardt's method selects new coefficients which (i) produce a smaller SSR and which (ii) usually are much closer to the minimum SSR coefficient values. In other words, in choosing new coefficients, this method not only moves in the right direction (moves toward a smaller SSR), it also chooses a good coefficient correction size (moves rapidly to a minimum SSR). Unlike the grid-search method which involves an arbitrary search, Marquardt's method uses a systematic search procedure that makes efficient use of the computer and usually converges quickly to the least-squares estimates. (The reader interested in the details of this method should consult Appendix 8A.)

8.3 Estimation-stage results: have we found a good model?

In Chapter 4 we introduced the following characteristics of a good model:

1. it is parsimonious;
2. it is stationary;
3. it is invertible;
4. it has estimated coefficients of high quality;
5. it has statistically independent residuals;
6. it fits the available data satisfactorily; and
7. it produces sufficiently accurate forecasts.

The typical computer printout of estimation-stage results provides the information needed to evaluate items (2), (3), (4), and (6). In this section we discuss how estimation-stage results are used to evaluate a model according to these four criteria. [We consider item (5) in Chapter 9 where we examine the third stage of UBJ modeling, diagnostic checking. Item (7), forecast accuracy, can be evaluated only by using a model to produce real forecasts. Item (1), parsimony, is an overriding principle in UBJ–ARIMA modeling whose role is illustrated in the case studies.]

As an example to illustrate the use of criteria (2), (3), (4), and (6), consider the estimation results for an ARMA(1, 1) model, shown in Figure 8.2.

---

*This combination is named after Donald W. Marquardt who developed the algorithm for an optimal interpolation between the Gauss–Newton method and the gradient method in [20].

```
+ + + + + + + + +ECOSTAT UNIVARIATE B-J RESULTS+ + + + + + + + +
+ FOR DATA SERIES:   SIMULATED DATA                                +
+ DIFFERENCING:      0                           DF   = 56         +
+ AVAILABLE:            DATA = 60   BACKCASTS = 0  TOTAL = 60       +
+ USED TO FIND SSR: DATA = 59   BACKCASTS = 0  TOTAL = 59          +
+ (LOST DUE TO PRESENCE OF AUTOREGRESSIVE TERMS:            1 )    +

  COEFFICIENT    ESTIMATE      STD ERROR      T-VALUE
  PHI    1       0. 908        0. 070         13. 04
  THETA  1       0. 605        0. 145          4. 18
  CONSTANT       9. 15733

  ADJUSTED RMSE =  . 92857    MEAN ABS % ERR =    0. 71
        CORRELATIONS
          1      2
  1    1. 00
  2    0. 68  1. 00
```

Figure 8.2   Estimation results for an ARMA(1, 1) model.

Checking coefficients for stationarity and invertibility.   The most obvious items included in estimation-stage output are the estimated coefficients. In Figure 8.2 we see that the estimate of ϕ_1 is $\hat{\phi}_1 = 0.908$ and the estimate of θ_1 is $\hat{\theta}_1 = 0.605$. Thus the model may be written in backshift form as $(1 - 0.908B)\tilde{z}_t = (1 - 0.605B)\hat{a}_t$.

The mean of the data series used in this example is $\bar{z} = 99.44$, so our estimate of μ is $\hat{\mu} = \bar{z} = 99.44$. We know from Chapter 5 that the estimated constant term $(\hat{C})$ for this model is $\hat{\mu}(1 - \hat{\phi}_1)$. Therefore, the printed constant term in Figure 8.2 is $\hat{C} = 99.44 \ (1 - 0.908) = 9.15733$. (Hand calculations give slightly different results because of rounding.)

The estimated coefficients can be used to check for both stationarity and invertibility. The stationarity requirement applies only to the autoregressive portion of a model; therefore, the relevant condition for an ARMA(1, 1) is the same as that for an AR(1): $|\phi_1| < 1$. Since ϕ_1 is unknown, we examine $\hat{\phi}_1$ instead. We find the stationarity condition is satisfied since $|\hat{\phi}_1| = 0.908 < 1$. However, we must be careful in reaching this conclusion. Although $|\hat{\phi}_1|$ is less than 1, it is fairly close to 1. In fact, it is only about 1.31 standard errors below 1.* This makes the AR operator for this model, $(1 - 0.908B)$, almost identical to the differencing operator $(1 - B)$. As pointed out in Chapter 7, it is good practice to difference a realization if there is serious doubt about its mean being stationary. We will not pursue the matter further in this case, but this example illustrates how estimation-stage results provide clues about stationarity and how a model might be reformulated.

*This number is obtained by dividing the difference between $\hat{\phi}_1$ and 1 by the estimated standard error of the coefficient, printed in Figure 8.2 as 0.070: $(0.908 - 1)/0.070 = -1.31$.

The invertibility requirement applies only to the moving-average part of a model. The requirement for an ARMA(1, 1) is the same as that for an MA(1): $|\theta_1| < 1$. The estimated coefficient $\hat{\theta}_1 = 0.605$, shown in Figure 8.2, clearly satisfies this requirement.

Coefficient quality: statistical significance.   Included in Figure 8.2 along with each estimated coefficient is its standard error and t-value. Each estimated coefficient has a standard error because it is a statistic based on sample information. A different sample would presumably give different estimates of ϕ_1 and θ_1. Thus each estimated coefficient has a sampling distribution with a certain standard error that is estimated by the computer program.* Most ARIMA estimation routines automatically test the hypothesis that the true coefficient is zero. An approximate t-value to test this hypothesis for each coefficient is calculated in this way:

$$t = \frac{(\text{estimated coefficient}) - (\text{hypothesized coefficient value})}{\text{estimated standard error of the coefficient}}$$

In our example, these calculations are

$$t_{\hat{\phi}_1} = \frac{0.908 - 0}{0.070}$$

$$= 13.04$$

and

$$t_{\hat{\theta}_1} = \frac{0.605 - 0}{0.145}$$

$$= 4.18$$

As a practical rule we should consider excluding any coefficient with an absolute t-value less than 2.0. Any coefficient whose absolute t-value is 2.0 or larger is significantly different from zero at roughly the 5% level. Including coefficients with absolute t-values substantially less than 2.0 tends to produce nonparsimonious models and less accurate forecasts.[†]

Coefficient quality: correlation matrix.   Most ARIMA estimation programs print the correlations between the estimated coefficients. We cannot

---

*It must be emphasized that the estimated standard errors are only approximations. The manner in which these approximations are found is discussed in Appendix 8A.

†Applying t-tests to one coefficient at a time is a very approximate way of gauging the precision of the estimates. The coefficients may also be tested jointly using a chi-squared or F-test. See Box and Jenkins' [1, pp. 224–231] discussion of approximate confidence regions for parameters.

avoid getting estimates that are correlated, but very high correlations between estimated coefficients suggest that the estimates may be of poor quality. When coefficients are highly correlated, a change in one coefficient can easily be offset by a corresponding change in another coefficient with little impact on the SSR. Thus if estimated coefficients are highly correlated, the final coefficient estimates depend heavily on the particular realization used; a slightly different realization could easily produce quite different estimated coefficients. If different realizations from the same process could easily give widely different estimated coefficients, the resulting estimates are of rather poor quality. Under these conditions, estimates based on a given realization could be inappropriate for future time periods unless the behavior of future observations matches the behavior of the given realization very closely.

As a practical rule, one should suspect that the estimates are somewhat unstable when the absolute correlation coefficient between any two estimated ARIMA coefficients is 0.9 or larger. When this happens we should consider whether some alternative models are justified by the estimated acf and pacf. One of these alternatives might provide an adequate fit with more stable parameter estimates. Figure 8.2 shows that the correlation between $\hat{\phi}_1$ and $\hat{\theta}_1$ is 0.68 in the present example. Therefore, the estimated model is satisfactory in this regard.

Coefficient quality: coefficient near-redundancy.   Mixed (ARMA) models are frequently useful, but they sometimes present a problem known as coefficient near-redundancy. A model with near-redundant coefficients presents two problems: it tends to be nonparsimonious, and it is difficult to estimate accurately. The resulting estimated coefficients are typically of low quality.

To understand the idea of coefficient near-redundancy, consider first an ARMA(1, 1) process with $\phi_1 = 0.6$ and $\theta_1 = 0.6$:

$$(1 - 0.6B)\tilde{z}_t = (1 - 0.6B)a_t \qquad (8.9)$$

There is nothing theoretically unacceptable about process (8.9), but note that the AR operator on the LHS $(1 - 0.6B)$ exactly cancels the MA operator on the RHS $(1 - 0.6B)$ leaving $\tilde{z}_t$ as a white-noise process, $\tilde{z}_t = a_t$. The parameters ϕ_1 and θ_1 are perfectly redundant. Thus we could not distinguish (8.9) from a white-noise process based on estimated acf's since they would, on average, yield no significant autocorrelations. This is not a problem in and of itself since the model $\tilde{z}_t = a_t$ requires the estimation of fewer parameters than an ARMA(1, 1) model, yet it should fit a typical realization generated by (8.9) just as well as a less parsimonious ARMA(1, 1) model.

But now consider an ARMA(1, 1) process with $\phi_1 = 0.6$ and $\theta_1 = 0.5$:

$$(1 - 0.6B)\tilde{z}_t = (1 - 0.5B)a_t \qquad (8.10)$$

The term $(1 - 0.6B)$ nearly cancels $(1 - 0.5B)$. The AR and MA coefficients are not perfectly redundant, but they are nearly so. Therefore, estimated acf's based on process (8.10) will, on average, be close to (though not identical to) white-noise acf's. How closely they would approximate white noise would depend on the number of observations. With a relatively large sample size, estimated acf's based on (8.10) would more frequently provide evidence about the existence of an ARMA(1, 1) process.

Unfortunately, if we did identify an ARMA(1, 1) from a realization generated by (8.10) it would be difficult to get good estimates of ϕ_1 and θ_1 with the least-squares method. The reason is there would be a set of near-minimum points on the sum-of-squares surface rather than a clear, well-defined minimum. Figure 8.3 shows an example of this for an ARMA(1, 1) model. The numbers recorded on the contours are the SSR's

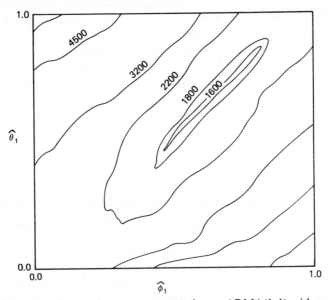

Figure 8.3 Residual sum-of-squares contours for an ARMA(1, 1) with near-redundant coefficients.

associated with the various combinations of $\hat{\phi}_1$ and $\hat{\theta}_1$. We see that a large number of $\hat{\phi}_1$ and $\hat{\theta}_1$ values have nearly identical SSR's in the neighborhood of the minimum SSR. Estimation results in a case like this are quite unstable, heavily dependent on the individual realization. Again, the larger the sample size the better the quality of the estimates, but we might be better off with the more parsimonious $\tilde{z}_t = a_t$ model than with a less parsimonious ARMA(1, 1) with estimated coefficients of poor quality.

Coefficient near-redundancy may not be nearly so obvious as in the last example. Consider the following ARMA(2, 1) process with $\phi_1 = 1.2$, $\phi_2 = -0.32$, and $\theta_1 = 0.5$:

$$(1 - 1.2B + 0.32B^2)\tilde{z}_t = (1 - 0.5B)a_t \qquad (8.11)$$

Factoring the AR operator on the LHS, we get

$$(1 - 0.4B)(1 - 0.8B)\tilde{z}_t = (1 - 0.5B)a_t \qquad (8.12)$$

The AR term $(1 - 0.4B)$ nearly cancels the MA term $(1 - 0.5B)$, so (8.11) is very nearly this AR(1) process:

$$(1 - 0.8B)\tilde{z}_t = a_t \qquad (8.13)$$

Thus we would expect estimated acf's calculated from realizations generated by process (8.11) to look much like AR(1) acf's, especially with moderate sample sizes. Following the usual identification procedures, the analyst might arrive at an AR(1) like (8.13) as an adequate representation of the data. Even if an ARMA(2, 1) were identifiable from an estimated acf, an AR(1) might be preferable, since the latter is more parsimonious, might fit a typical realization about as well, and would not have the unstable coefficient estimates associated with the near-redundant ARMA(2, 1) in (8.11).

The estimated model in Figure 8.2 does not appear to suffer from coefficient near-redundancy. The AR operator $(1 - \hat{\phi}_1 B) = (1 - 0.908B)$ does not come very close to canceling the MA operator $(1 - \hat{\theta}_1 B) = (1 - 0.605B)$. Figure 8.4 shows the residual sum-of-squares contours for this model. Unlike the SSR contours in Figure 8.3, the one in Figure 8.4 has only a limited range of $\hat{\phi}_1$ and $\hat{\theta}_1$ values that are consistent with the minimum SSR.

The practical lesson here is that we should construct a mixed model with great care; avoid including both AR and MA terms in a model without solid evidence that both are needed, and check estimation results for coefficient near-redundancy. This will help produce better forecasts by avoiding non-parsimonious models with unstable estimated coefficients.

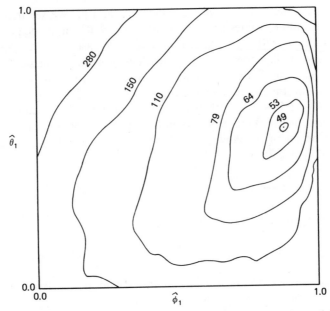

Figure 8.4   Residual sum-of-squares contours for the ARMA(1, 1) model in Figure 8.2.

Closeness of fit: root-mean-squared error.   There is no guarantee that a properly constructed ARIMA model will fit the available data closely. Some data sets have a large amount of statistical "noise" that cannot be removed with AR or MA terms. That is, the variance of the underlying random shocks (σ_a^2) may be large. It could be that the best ARIMA model will not fit the available data well enough to satisfy the UBJ analyst.

Since we cannot observe the random shocks, we cannot measure their variance directly. But we have the estimation-stage residuals ($\hat{a}_t$) and we can use them to estimate the variance of the random shocks with this formula:

$$\hat{\sigma}_a^2 = \frac{1}{n - m} \sum \hat{a}_t^2 \qquad (8.14)$$

where the summation is across all n available squared residuals and m is the number of parameters estimated. By subtracting m from n, we are adjusting $\hat{\sigma}_a^2$ for degrees of freedom.

The square root of $\hat{\sigma}_a^2$ is interpreted as the estimated standard deviation of the random shocks. On the computer printout in Figure 8.2, this statistic is referred to as the adjusted root-mean-squared error (adjusted RMSE). Its value in this instance in 0.92857.

The adjusted RMSE is useful for comparing different models estimated from the same realization. Two or more models could give essentially the same results in most respects. That is, they could be equally parsimonious, equally justifiable based on the estimated acf's and pacf's, and so forth. But if one model has a noticeably lower RMSE, we prefer that one because it fits the available data more closely. And importantly, as we see in Chapter 10, the model with the smaller RMSE tends to have a smaller forecast-error variance.

Closeness of fit: mean absolute percent error.   Another measure of how well a model fits the available data is the mean absolute percent error (MAPE). If a residual is divided by the corresponding observed value, we have a percent residual. The MAPE is simply the mean of the absolute values of these percent residuals:

$$\frac{100}{n} \sum \left| \frac{\hat{a}_t}{z_t} \right| \tag{8.15}$$

where the summation is across all n available absolute percent residuals. Dividing $\hat{a}_t$ by z_t gives the percent residual. The two vertical lines (||) indicate that we are considering the absolute values of the percent residuals. Dividing the sum by n gives the mean absolute percent residual, and multiplying by 100 merely relocates the decimal. Applying (8.15) to the residuals for the model in Figure 8.2 gives a MAPE of 0.71%.

The MAPE generally should not be used for choosing among alternative models that are equivalent in other respects. The adjusted RMSE is used for that purpose since it is related to the forecast error variance. Instead, the MAPE may be useful for conveying the accuracy of a model to managers or other nontechnical users.

Using the example shown in Figure 8.2, we could report that this model fits the available data with an average error of $\pm 0.71\%$. The MAPE suggests, very roughly, the kind of accuracy we could expect from forecasts produced by this model. However, the preferred way of conveying forecast accuracy is to derive confidence intervals for the forecasts. This latter topic is discussed in Chapter 10.

Summary

1. At the identification stage we obtain somewhat rough estimates of many autocorrelation and partial autocorrelation coefficients as a guide to find an appropriate model.

2. At the estimation stage we make more efficient use of the available data by obtaining precise estimates of just a few parameters (the mean and some AR and/or MA coefficients).

3. Box and Jenkins favor choosing coefficient estimates at the estimation stage according to the maximum likelihood (ML) criterion. Assuming a correct model, the likelihood function from which ML estimates are derived reflects all useful information about the parameters contained in the data.

4. Finding exact ML estimates can be computationally burdensome, so Box and Jenkins favor the use of least-squares (LS) estimates. If the random shocks are Normally distributed, LS estimates are computationally easier to find and provide exactly, or very nearly, ML estimates.

5. LS estimates are those which give the smallest sum of squared residuals (SSR $= \Sigma \hat{a}_t^2$).

6. A residual ($\hat{a}_t$) is an estimate of a random shock (a_t). It is defined as the difference between an observed value (z_t) and a calculated value ($\hat{z}_t$). In practice the calculated values are found by inserting estimates of the mean and the AR and MA coefficients into the ARIMA model being estimated, with the current random shock assigned its expected value of zero, and applying these estimates to the available data.

7. Linear least squares (LLS) may be used to estimate only pure AR models without multiplicative seasonal terms. All other models require a nonlinear least-squares (NLS) method.

8. One NLS method is the grid-search procedure. In this approach, each AR and MA coefficient is assigned a series of admissible values and an SSR is found for each combination of these values. The combination of coefficients with the smallest SSR is chosen as the set of LS estimates. This method is not often used because evaluating the sum of squared residuals for each combination of coefficient estimates can be very time-consuming.

9. The most commonly used NLS method is algorithmic in nature. It is a combination of two NLS procedures: Gauss–Newton linearization and the gradient method. This combination, sometimes called "Marquardt's compromise," involves a systematic search for LS estimates. Given some initial coefficient estimates, this algorithm chooses a series of optimal coefficient corrections. This method converges quickly to LS values in most cases.

10. Estimation-stage results may be used to check a model for stationarity and invertibility. The estimated AR and MA coefficients should satisfy the conditions stated in Chapter 6.

11. Most computer programs for estimating ARIMA models provide approximate t-values for each coefficient. A practical rule is to include only estimated coefficients with absolute t-values of about 2.0 or larger.

12. Estimated coefficients are nearly always correlated, but if they are too highly correlated, the estimates are heavily dependent on the particular realization used and tend to be unstable. As a practical rule, we should suspect that the estimates may be of poor quality if the absolute correlation between any two coefficients is 0.9 or larger. If we can find an alternative adequate model with less highly correlated estimates, we should use that alternative since its estimated coefficients will be of higher quality.

13. The adjusted root-mean-squared error (RMSE = $\hat{\sigma}_a$) is an estimate of the standard deviation of the random shocks (σ_a). Other things equal, we prefer a model with a smaller RMSE since it fits the available data better and tends to produce forecasts with a smaller error variance.

14. The mean absolute percent error (MAPE) provides another measure of goodness of fit. It is sometimes used for conveying to nonexperts the approximate accuracy that can be expected from an ARIMA forecasting model. However, the preferred way to convey forecast accuracy (as discussed in Chapter 10) is to derive confidence intervals for the forecasts.

Appendix 8A: Marquardt's compromise*

In the main body of Chapter 8 we said that, in general, ARIMA coefficients (the ϕ's and θ's) must be estimated using a *nonlinear* least-squares (NLS) procedure. While several NLS methods are available, the one most commonly used to estimate ARIMA models is known as "Marquardt's compromise," after Donald W. Marquardt, who wrote an article in which he proved some of the key properties of this method.

In this appendix we set forth the most basic ideas associated with Marquardt's compromise and illustrate some of the relevant calculations. Because the procedure is relatively complicated, we will not present the theory rigorously or give a numerical example of every step. The purpose is to convey the overall structure of the method, along with some supporting calculations. Readers wanting more details about the technique may consult

---

*The material in this appendix is aimed at the reader with a knowledge of calculus, matrix algebra, and linear regression. All variables printed in boldface type represent matrices.

Marquardt's original article [20], or Box and Jenkins' text [1, Chapter 7, and pp. 500–505].

8A.1   Overview

Marquardt's method is called a compromise because it combines two NLS procedures: Gauss–Newton linearization, and the gradient method, also known as the steepest-descent method.

The practical advantage of the Gauss–Newton method is that it tends to converge rapidly to the least-squares (LS) estimates, *if* it converges; the disadvantage is that it may not converge at all. The practical advantage of the gradient method is that, in theory, it will converge to LS estimates; however, it may converge so slowly that it becomes impractical to use.

Marquardt's compromise combines the best of these two approaches: it not only converges to LS estimates (except in rare cases), it also converges relatively quickly. Before working through an algebraic example and some numerical illustrations, we summarize the major steps in the algorithm. Figure 8A.1 shows these steps.

At step 1, the analyst (or the computer program) chooses some starting values for the k ϕ and θ coefficients to be estimated; these initial estimates are entered into the $k \times 1$ vector $\mathbf{B}_0$.* Then the sum of squared residuals (SSR$_0$) associated with these initial values is calculated at step 2. Up to this point, the procedure is essentially the same as the grid-search method illustrated earlier in Section 8.2.

Step 3 is the calculation of numerical derivatives needed for the Gauss–Newton method. We discuss this concept further in the next section. At step 4 equations are formed (using the numerical derivatives found at step 3) that are linear approximations to the nonlinear relationship between the residuals $\hat{a}_t$ and the $\hat{\phi}$ and $\hat{\theta}$ elements in $\mathbf{B}_0$. These linearized equations are then solved (step 5) for the linear least-squares corrections (vector $\mathbf{h}$) that yield the new estimates $\mathbf{B}_1 = \mathbf{B}_0 + \mathbf{h}$.

Since the new estimates $\mathbf{B}_1$ are derived from only linear approximations to the relevant nonlinear equations, they may not give a smaller SSR than SSR$_0$. Thus we must insert the new estimates into the model to see if the SSR is smaller with the $\mathbf{B}_1$ estimates than with the previous $\mathbf{B}_0$ estimates. These are steps 6 and 7, where SSR$_1$ is calculated and then compared with SSR$_0$.

---

*Some programs estimate μ simultaneously with the ϕ's and θ's; in that case, there would be $k + 1$ initial estimates required. For simplicity we will discuss the case where μ is estimated first from the realization mean rather than simultaneously with the ϕ and θ coefficients.

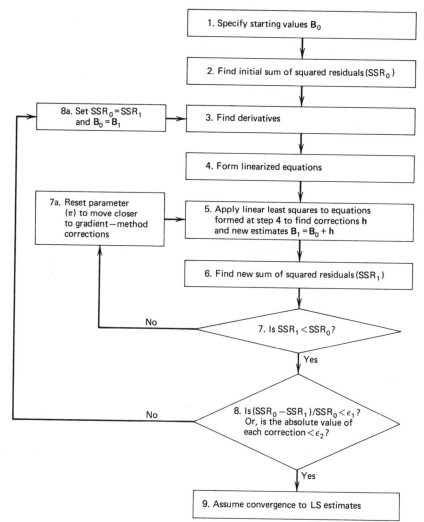

Figure 8A.1 The steps in Marquardt's compromise.

If $SSR_1 < SSR_0$, we test to see if the method has converged to a minimum SSR. The relative reduction in the SSR may be compared with a convergence parameter ϵ_1, or the absolute values of the corrections in **h** may be tested against some convergence parameter ϵ_2. If the relative reduction in the SSR is smaller than ϵ_1, or alternatively, if the absolute values of the corrections in **h** are all smaller than ϵ_2, we assume convergence has occurred

—that is, we assume the last estimates (in $\mathbf{B}_1$) are LS estimates. If parameter ϵ_1 (or ϵ_2) is exceeded, we return to step 3 to find new derivatives after reinitializing (at step 8a) by setting SSR_0 equal to the new (lower) SSR_1, and setting $\mathbf{B}_0$ equal to the new (better) estimates $\mathbf{B}_1$.

As described thus far, Marquardt's compromise is just the Gauss–Newton linearization procedure. But as pointed out above, it is possible that this method will not lead to a reduced SSR at step 7. This is where the gradient method enters. If, at step 7, we find $SSR_1 > SSR_0$, then a parameter π is increased by a predetermined amount and the linear equations (whose contents depend on π) are modified and new corrections are found. As π increases, the corrections move closer to the gradient-method corrections; this means that the absolute values of the corrections will tend to be smaller, but they are more likely to produce a reduced SSR.

8A.2 Application to an MA(1)

In this section we apply the procedure summarized in Figure 8A.1 to an MA(1) model. We present this application primarily in algebraic form, with numerical illustrations for some of the steps. We use the following realization (assumed stationary) as an example:

t	z_t
1	4
2	−5
3	3
4	2
5	−6
6	5
7	−2
8	−1

This realization has far fewer observations than are required in practice. It is used here only for purposes of illustration.

The MA(1) model is

$$\tilde{z}_t = (1 - \theta_1 B)a_t \tag{8A.1}$$

or

$$z_t = \mu + a_t - \theta_1 a_{t-1} \tag{8A.2}$$

Because μ, θ_1, and the random shocks are unknown and must be estimated, we rewrite model (8A.2) as

$$z_t = \hat{\mu} + \hat{a}_t - \hat{\theta}_1 \hat{a}_{t-1} \qquad (8A.3)$$

where the "$\hat{\ }$" sign stands for estimated values.

We use the realization mean $\bar{z}$ to estimate μ. For the realization above, $\bar{z} = 0$:

$$\bar{z} = \frac{\Sigma z_t}{n}$$

$$= \frac{4 - 5 + 3 + 2 - 6 + 5 - 2 - 1}{8}$$

$$= \frac{0}{8}$$

$$= 0$$

Having estimated μ with $\bar{z}$, we remove this nonstochastic element from the data temporarily by expressing the data in deviations from the mean: $\tilde{z}_t = z_t - \bar{z}$. In this case, $\tilde{z}_t = z_t$ because $\bar{z} = 0$. Throughout the remainder of this appendix, we refer to z_t rather than $\tilde{z}_t$, since they are identical in our example and the notation will be less cumbersome. However, bear in mind that *in this appendix, z_t represents deviations from the mean.*

Letting $\hat{\mu} = \bar{z} = 0$, subtracting this value from both sides of (8A.3) to express the model in deviations from the mean, solving the result for $\hat{a}_t$, and recognizing that $\hat{a}_t$, $\hat{a}_{t-1}$, and z_t are vectors gives this expression:

$$\hat{\mathbf{a}}_t = \mathbf{z}_t + \hat{\theta}_1 \hat{\mathbf{a}}_{t-1} \qquad (8A.4)$$

We cannot find the LS estimate of θ_1 directly from (8A.4) by minimizing* the SSR $= \hat{\mathbf{a}}_t' \hat{\mathbf{a}}_t$ with respect to $\hat{\theta}_1$ because we do not know the contents of vector $\hat{\mathbf{a}}_{t-1}$ if $\hat{\theta}_1$ is unknown, and we cannot solve directly for the LS estimate of $\hat{\theta}_1$ if the elements in $\hat{\mathbf{a}}_{t-1}$ are unknown. Thus we must use an iterative search technique.

Initial coefficient values. At step 1 we specify initial values for the contents of $\mathbf{B}_0$, the vector of estimated coefficients. In the present example,

---

*The superscript ($'$) represents matrix transposition.

Table 8A.1   Calculation of the initial sum of squared residuals (SSR_0)

t	z_t	$\hat{a}_{t-1,0}$	$\hat{z}_{t,0} = -\hat{\theta}_{1,0}\hat{a}_{t-1,0}$	$\hat{a}_{t,0} = z_t - \hat{z}_{t,0}$	$\hat{a}_{t,0}^2$ [a]
1	4	0	0	4.0000	16.0000
2	−5	4.0000	−0.4000	−4.6000	21.1600
3	3	−4.6000	0.4600	2.5400	6.4516
4	2	2.5400	−0.2540	2.2540	5.0805
5	−6	2.2540	−0.2254	−5.7746	33.3460
6	5	−5.7746	0.5775	4.4225	19.5585
7	−2	4.4223	−0.4422	−1.5578	2.4267
8	−1	−1.5578	0.1558	−1.1558	1.3359

[a] $\sum \hat{a}_{t,0}^2 = 105.3592$.

$\mathbf{B}_0$ is merely a 1×1 vector containing $\hat{\theta}_1$, the initial estimate of θ_1. Let $\hat{\theta}_{1,0}$ stand for this initial value.

Some computer programs require the user to enter initial values for the estimated coefficients. (Guidelines for choosing initial values are provided in Chapter 12.) Other programs generate their own initial values based on a preliminary analysis of the data.* Still others pick initial values arbitrarily; often these programs use 0.1 as the initial value for all estimated coefficients. In our example, let $\hat{\theta}_{1,0} = 0.1$.

Initial SSR.   At step 2 we use the initial value $\hat{\theta}_{1,0} = 0.1$ to generate an initial set of residuals $\hat{\mathbf{a}}_{t,0}$ and the corresponding initial sum of squared residuals $SSR_0 = \mathbf{a}'_{t,0}\mathbf{a}_{t,0}$. The calculations for finding SSR_0 from a realization for the MA(1) are essentially the same as those shown earlier in Section 8.2 where we calculated an SSR for an AR(1) model to illustrate the idea of the grid-search method.

Table 8A.1 shows the calculation of SSR_0 for the realization presented above (reproduced in column 2), with $\hat{\theta}_{1,0} = 0.1$. Recall that a residual $\hat{a}_t$ is defined as the difference between an observed value z_t and a calculated value $\hat{z}_t$. In this example the elements in both $\hat{z}_t$ and $\hat{a}_t$ depend on the value of $\hat{\theta}_{1,0}$, so we add a zero subscript to these terms:

$$\hat{\mathbf{a}}_{t,0} = \mathbf{z}_t - \hat{\mathbf{z}}_{t,0} \qquad (8A.5)$$

These values are calculated in column 5 of Table 8A.1.

---

*Box and Jenkins discuss this type of preliminary analysis and present the relevant algorithm in [1, pp. 187–192 and 498–500].

The calculated values $\hat{z}_{t,0}$ are found from (8A.3) by letting $\hat{\mu} = \bar{z} = 0$ and $\hat{\theta}_1 = \hat{\theta}_{1,0}$, and assigning the elements in $\hat{\mathbf{a}}_{t,0}$ their expected values of zero:

$$\hat{z}_{t,0} = -\hat{\theta}_{1,0}\hat{a}_{t-1,0} \qquad (8A.6)$$

These values are calculated in column 4 of Table 8A.1.

The calculations are done as follows: First, let $t = 1$. From (8A.5) we see that $\hat{a}_{0,0}$ cannot be calculated because we have no z_0 value available. Thus we assign $\hat{a}_{0,0}$ its expected value of zero and enter this into column 3, row $t = 1$.

From (8A.6) we may now use the value $\hat{a}_{0,0} = 0$ to calculate $\hat{z}_{1,0}$:

$$\hat{z}_{1,0} = -\hat{\theta}_{1,0}\hat{a}_{0,0}$$
$$= (-0.1)(0)$$
$$= 0$$

Enter this value into column 4, row $t = 1$.

Now from (8A.5), find $\hat{a}_{1,0}$ as

$$\hat{a}_{1,0} = z_1 - \hat{z}_{1,0}$$
$$= 4 - 0$$
$$= 4$$

and enter this in column 5 of Table 8A.1, row $t = 1$.

Next, we square each residual. For $t = 1$, we have $\hat{a}_{1,0}^2 = (4)^2 = 16$. Enter this in column 6, row $t = 1$.

Now let $t = 2$. We have already found the next value that belongs in column 3: $\hat{a}_{t-1,0} = \hat{a}_{1,0} = 4$. Next, use (8A.6) to find $\hat{z}_{2,0}$, the next value in column 4:

$$\hat{z}_{2,0} = -\hat{\theta}_{1,0}\hat{a}_{1,0}$$
$$= (-0.1)(4)$$
$$= -0.4$$

The next value in column 5 is $\hat{a}_{2,0}$. From (8A.5),

$$\hat{a}_{2,0} = z_2 - \hat{z}_{2,0}$$
$$= -5 - (-0.4)$$
$$= -4.6$$

The square of this value (21.16) is then entered in column 6.

For $t = 3$, we have already found $\hat{a}_{t-1,0} = \hat{a}_{2,0} = -4.6$. The remaining calculations are simply a continuation of those shown above. The final step is to sum the squared residuals in column 6 to get $SSR_0 = 105.3592$.

Linear approximation with numerical derivatives. Having found SSR_0 corresponding to $\hat{\theta}_{1,0} = 0.1$, we proceed to search for a new value for $\hat{\theta}_1$, designated $\hat{\theta}_{1,1}$ which has a *smaller* SSR than $\hat{\theta}_{1,0}$. Our ultimate goal is to find that value of $\hat{\theta}_1$ which results in a *minimum* SSR.

We approach the problem with a linear approximation by writing a truncated Taylor series expansion of (8A.4):*

$$\hat{a}_t = \hat{a}_{t,0} - \left(\hat{\theta}_{1,1} - \hat{\theta}_{1,0}\right)\left(-\left.\frac{\partial \hat{a}_t}{\partial \hat{\theta}_1}\right|\hat{\theta}_1 = \hat{\theta}_{1,0}\right) \tag{8A.7}$$

Solving (8A.7) for $\hat{a}_{t,0}$:

$$\hat{a}_{t,0} = \left(\hat{\theta}_{1,1} - \hat{\theta}_{1,0}\right)\left(-\left.\frac{\partial \hat{a}_t}{\partial \hat{\theta}_1}\right|\hat{\theta}_1 = \hat{\theta}_{1,0}\right) + \hat{a}_t \tag{8A.8}$$

Forming this linear relationship is step 4 in the algorithm. Equation (8A.8) may be estimated with the method of linear least squares (LLS). We may think of $\hat{a}_{t,0}$ as the "dependent variable"—a set of "observations" generated at step 2, given $\hat{\theta}_1 = \hat{\theta}_{1,0}$. For $\hat{\theta}_{1,0} = 0.1$, these values are shown in column 5 of Table 8A.1. The term $(\hat{\theta}_{1,1} - \hat{\theta}_{1,0})$ in (8A.7) is the coefficient whose value is to be estimated with LLS. The vector $\hat{a}_t$ is the set of residuals whose sum of squares is to be minimized using LLS. The (negative) derivative of $\hat{a}_t$, evaluated initially at $\hat{\theta}_1 = \hat{\theta}_{1,0} = 0.1$ [that is, $(-\partial \hat{a}_t/\partial \hat{\theta}_1|\hat{\theta}_1 = \hat{\theta}_{1,0}]$, may be thought of as the "independent variable" in this linear relationship.

In practice the values of the derivatives in equation (8A.8) are found numerically rather than analytically; that is, we generate these values from the available data. If we increase $\hat{\theta}_1$ slightly from its initial value $\hat{\theta}_{1,0} = 0.1$ up to $\hat{\theta}_{1,0}^* = 0.11$, we can produce a new set of residuals $\hat{a}_{t,0}^*$ corresponding

---

*Let y be a nonlinear function of x: $y = f(x)$. Fix x at x_0. Now we may represent the range of y values around x_0 with a Taylor series expansion:

$$y = f(x_0) - (x_1 - x_0)\left(-\left.\frac{\partial y}{\partial x}\right|x = x_0\right) - \frac{(x_1 - x_0)^2}{2}\left(-\left.\frac{\partial y}{\partial x}\right|x = x_0\right) - \cdots -$$

The first two terms of this expansion are a linear approximation (around x_0) to the nonlinear function $y = f(x)$.

to the new coefficient $\hat{\theta}_{1,0}^* = 0.11$. Then the required set of derivatives is simply the difference between the two sets of residuals:

$$\left(\frac{\partial \hat{a}_t}{\partial \hat{\theta}_1} \middle| \hat{\theta}_1 = \hat{\theta}_{1,0} = 0.1 \right) = \hat{a}_{t,0} - \hat{a}_{t,0}^*$$

Thus, we have one set of residuals $\hat{a}_{t,0}$ produced at step 2 in the algorithm with $\hat{\theta}_1 = \hat{\theta}_{1,0} = 0.1$. These are shown in column 5 of Table 8A.1. We produce another set of residuals $\hat{a}_{t,0}^*$ with $\hat{\theta}_1 = \hat{\theta}_{1,0}^* = 0.11$. The required set of derivatives is then the difference between the two vectors of residuals. This procedure is illustrated numerically in Table 8A.2.

Column 2 of Table 8A.2 merely reproduces the realization we are analyzing. The values in columns 3, 4, and 5 were generated in the same manner as the values in columns 3, 4, and 5 of Table 8A.1. The only difference is that the MA(1) coefficient used in Table 8A.1 is 0.10, while it is 0.11 in Table 8A.2. For an illustration of how these calculations are performed, see the preceding section in this appendix where Table 8A.1 is explained.

Column 6 in Table 8A.2 is the set of derivatives required for equation (8A.8). It is the difference between the residuals in column 5 of Table 8A.1 and those in column 5 of Table 8A.2. The negatives of these values are the "observations" on the "independent variable" in (8A.8).

Finding new estimates. By applying LLS to (8A.8), using the values for $\hat{a}_{t,0}$ in column 5 of Table 8A.1, as the "dependent variable" and the values for the derivatives in column 6 of Table 8A.2 as the "independent variable," we estimate the coefficient $(\hat{\theta}_{1,1} - \hat{\theta}_{1,0})$. This coefficient is the change (the "correction") in $\hat{\theta}_1$ that minimizes the SSR $(\hat{a}_t'\hat{a}_t)$ of (8A.8). This is step 5 in

Table 8A.2   Calculation of numerical derivatives

t	z_t	$\hat{a}_{t-1,0}^*$	$\hat{z}_t = -\hat{\theta}_{1,0}^*\hat{a}_{t-1,0}^*$	$\hat{a}_{t,0}^* = z_t - \hat{z}_t$	$\hat{a}_{t,0} - \hat{a}_{t,0}^*$
1	4	0	0	4.0000	0
2	−5	4.0000	−0.4400	−4.5600	−0.0400
3	3	−4.5600	0.5016	2.4984	0.0416
4	2	2.4984	−0.2748	2.2748	−0.0208
5	−6	2.2748	−0.2502	−5.7498	−0.0248
6	5	−5.7498	0.6325	4.3675	0.0550
7	−2	4.3675	−0.4804	−1.5196	−0.0382
8	−1	−1.5196	0.1672	−1.1672	0.0114

Marquardt's compromise. We will not illustrate this step numerically because it involves complications beyond the scope of our discussion.*

Having used LLS to find the contents of the correction vector $\mathbf{h}$, which in this case consists of the estimated change in $\hat{\theta}_1$ ($\mathbf{h} = \hat{\theta}_{1,1} - \hat{\theta}_{1,0}$), we may easily find the new coefficient $\hat{\theta}_{1,1}$. We know $\mathbf{h} = (\hat{\theta}_{1,1} - \hat{\theta}_{1,0})$ and $\hat{\theta}_{1,0}$, so we solve for $\hat{\theta}_{1,1}$: $\hat{\theta}_{1,1} = \mathbf{h} + \hat{\theta}_{1,0}$. For the realization in our example, the first correction as calculated by the computer program is 0.7263, so $\hat{\theta}_{1,1} = 0.7263 + 0.1000 = 0.8263$.

Testing the new SSR. The new estimated coefficient $\hat{\theta}_{1,1} = 0.8263$ was found by minimizing the sum of squared residuals of (8A.8). However, that equation is only a linear approximation to the relevant nonlinear relationship between $\hat{\theta}_1$ and the sum of squared residuals we want to minimize. It is possible that the correction derived from (8A.8) will not lead to a new estimate $\hat{\theta}_{1,1}$ that reduces the SSR obtained from $\hat{\theta}_{1,0}$. Therefore, we must perform step 6 in the algorithm. This step is identical to step 2 except we now use $\hat{\theta}_{1,1} = 0.8263$ instead of $\hat{\theta}_{1,0} = 0.10$ to generate the sum of squared residuals (SSR_1) corresponding to (8A.4). Having done so, we compare SSR_1 with SSR_0 at step 7. If $SSR_1 < SSR_0$, we assume that $\hat{\theta}_{1,1} = 0.8263$ is a better estimate of θ_1 than $\hat{\theta}_{1,0} = 0.10$ because $\hat{\theta}_{1,1} = 0.8263$ has a smaller SSR. For our example, $SSR_1 = 49.5623$. This is smaller than $SSR_0 = 105.3592$, so we conclude that $\hat{\theta}_{1,1} = 0.8263$ is an improved estimate of θ_1.

Convergence test. If $SSR_1 < SSR_0$, we go to step 8 in the algorithm to decide if we have converged to a minimum SSR. Some computer programs test the relative reduction in the SSR, as shown in step 8 in Figure 8A.1, to see if it is less than some parameter ϵ_1. If it is, convergence to least-squares estimates is assumed. Other programs (including the one used for this example) test the absolute size of the coefficient corrections in vector $\mathbf{h}$. If each absolute correction is smaller than some parameter ϵ_2 (0.001 in the program used here), convergence to LS estimates is assumed. In this example, the correction 0.7263 is much larger than 0.001, so we conclude that we have not yet converged to the least-squares estimate of θ_1.

New starting values. When the convergence parameter (ϵ_1 or ϵ_2) is violated, the estimation procedure begins again at step 8a with $\hat{\theta}_{1,0}$ reset to equal the new, better value $\hat{\theta}_{1,1}$ and with SSR_0 reset to equal SSR_1. New

---

*These complications are related to a transformation of the "data" used in estimating (8A.8). Marquardt [20] points out that the gradient-method results are sensitive to the scaling of the data; therefore, the data are standardized before the LLS estimates of (8A.8) are calculated, and the results are then scaled back again. The scaling procedure is given in Box and Jenkins [1, pp. 504 and 505.]

Table 8A.3 Example of results for iterations of Marquardt's compromise, steps 3–8a

Iteration Number	$\hat{\theta}_1$	π	SSR
0	0.1000	0.01	105.3592
1	0.8263	0.01	49.5623
2	0.9089	1.0	48.3370
3	0.9091	10	48.3370

derivatives are calculated, new linear equations are formed, new corrections found, and so forth. For our example, at step 8a we set $SSR_0 = 49.5623$ and $\hat{\theta}_{1,0} = 0.8263$.

As long as $SSR_1 < SSR_0$ at each iteration, Marquardt's compromise is identical to the Gauss–Newton method. This method tends to produce relatively rapid convergence *if* $SSR_1 < SSR_0$ at each iteration, since it produces relatively large correction values. Thus the program may pass through steps 3–8a only a few times.

Ensuring a reduced SSR. Let us return to step 7 in the algorithm. If $SSR_1 > SSR_0$, we make an adjustment (an increase in a parameter π) to the linear equations formed at step 4. This adjustment produces new coefficient estimates at step 5 which are closer to the gradient-method results; because of this, these new estimates are more likely to lead to a reduced SSR. If these new corrections still do not give a value for SSR_1 that is less than SSR_0, π is increased again. In fact, π is increased until a reduced SSR is induced. In theory, a sufficiently large value for π will ensure corrections that give a reduced SSR, assuming we have not yet converged to a minimum SSR.*

Table 8A.3 shows how π changed during the estimation of θ_1 for our example. Iteration 0 is simply calculation of the initial SSR_0 for the starting value $\hat{\theta}_1 = 0.10$; the starting value of π happens to be 0.01 in this program.

At iteration 1, we achieved a reduced SSR without having to increase π: it remained at 0.01. In order to achieve a reduced SSR at iteration 2, however, π had to be increased gradually up to 1.0 (it was increased two times to reach that level). Then at iteration 3, π had to be increased to 10 to ensure getting an SSR that was not larger than the previous one. Since the absolute value of the coefficient correction at iteration 3 ($\mathbf{h} = 0.0002$) was

---

*In practice, the value of π required to yield a reduced SSR could be so large that machine limits are exceeded and computational errors occur. Fortunately, this rarely happens.

smaller than the relevant convergence parameter ($\epsilon_2 = 0.001$), convergence was assumed and the program halted.

Standard errors of the estimated coefficients. In finding coefficient corrections, we apply LLS to the linearized model, such as equation (8A.8). Let **X** be the matrix of derivatives. Then the variance–covariance matrix of the estimates is

$$V = \hat{\sigma}_a^2 (X'X)^{-1} \tag{8A.9}$$

where $\hat{\sigma}_a^2$ is the estimated residual variance as discussed in Section 8.3 and **X** is the matrix of derivatives calculated at the *last* linearization. Note that the estimated variances of the coefficients derived from (8A.9) are only approximate since they are based on a linear approximation to a nonlinear function. It follows that the *t*-values associated with the estimated coefficients provide only rough tests of the significance of the coefficients.

Appendix 8B: Backcasting

Box and Jenkins distinguish between two estimation procedures: *conditional least squares* (CLS), which is identical in results to the conditional maximum likelihood method, and *unconditional* least squares (ULS). They suggest a practical procedure called "backcasting" or "backforecasting" which gives ULS estimates that are very nearly unconditional maximum likelihood estimates.

Box and Jenkins [1, p. 211] suggest that CLS is satisfactory for estimating models without seasonal elements when the number of observations in the realization is moderate to large. But they emphasize that CLS is generally inferior to ULS for seasonal models. They propose the method of backcasting as a practical way of producing ULS estimates that are very nearly unconditional maximum likelihood estimates.*

8B.1 Conditional least squares

We can get at the idea of CLS by considering how we calculated the SSR's in Tables 8.1 and 8A.1. In Table 8.1 we calculated an SSR for an AR(1)

---

*However, Newbold and Ansley [21] present evidence, based on Monte Carlo methods, that ULS results can deviate significantly from maximum likelihood results when process parameters are close to the nonstationarity or noninvertibility boundaries with small samples.

model. We found residuals ($\hat{a}_t$) for periods 2 through 6, but not for period 1 because there is no value z_0 available to find the calculated value $\hat{z}_1 = \hat{\phi}_1 z_0$ and therefore we could not find $\hat{a}_1$. Thus the SSR calculated there is conditional in the sense that it depended on our using $z_1 = 20$ as the *starting value* of the z_t series.

In Table 8A.1 we calculated an SSR for an MA(1) model. In this case we found residuals for periods 1–8. But to find $\hat{a}_{1,0}$, we had to set $\hat{a}_{0,0}$ equal to its expected value of zero. That allowed us to find the calculated value $\hat{z}_1 = -\hat{\theta}_{1,0} \hat{a}_{0,0} = 0$, and therefore we could find $\hat{a}_{1,0} = z_1 - \hat{z}_1 = 4 - 0 = 4$. Thus the SSR calculated in Table 8A.1 is conditional in the sense that it depended on our using $\hat{a}_{0,0} = 0$ as the starting value for the $\hat{a}_t$ series, and $z_1 = 4$ as the starting value for the z_t series.

8B.2   Unconditional least squares

Consider the sequence of observations $z_1, z_2, z_3, \ldots, z_n$. Now consider some subsequent value z_{n+l+1} with a certain probability relationship to the previous values $z_1, \ldots, z_n$. It can be shown that a value *preceding* the sequence $z_1, \ldots, z_n$ (designated z_{-l}) has the *same* probability relationship to the sequence $z_n, z_{n-1}, z_{n-2}, \ldots, z_1$ as z_{n+l+1} has to the sequence $z_1, z_2, z_3, \ldots, z_n$.

In other words, we can do somewhat better than to confess complete ignorance about the values preceding $z_1, z_2, \ldots, z_n$ if we know something about the probability relationship between z_{t+l+1} and the sequence $z_1, z_2, \ldots, z_n$ (i.e., if we have a tentative ARIMA model in hand for the $z_1, z_2, \ldots, z_n$ series). If we know something about that probability relationship, then we also know something about the probability relationship between the available (reversed) data sequence $z_n, z_{n-1}, \ldots, z_1$ and a value (z_{-l}) that precedes that sequence.

This fact leads Box and Jenkins to propose the following backcasting procedure. First, start with a realization, such as the one in Table 8.1, expressed in deviations from the mean. This $\tilde{z}_t$ sequence is reproduced in column 2 of Table 8B.1. Now, reverse the $\tilde{z}_t$ series in time, that is, $\tilde{z}_1$ becomes the last observation, $\tilde{z}_2$ becomes the next to last, and so forth, and $\tilde{z}_6$ becomes the first observation. This is shown in Table 8B.2 for $t = 6, 5, \ldots, 1$.

Next, "forecast" the reversed series (i.e., backcast the original series) using the most recent values of the estimated coefficients. The example in Table 8.1 is an AR(1) model with $\hat{\phi}_1 = 0.5$. Applying this estimate of $\hat{\phi}_1$ to the $\tilde{z}_t$ series, the forecasting equation is $\hat{\tilde{z}}_t = \hat{\phi}_1 \tilde{z}_{t-1} = 0.5 \tilde{z}_{t-1}$. But for the reversed $\tilde{z}_t$ series, the backcasting equation is $\hat{\tilde{z}}_t = \hat{\phi}_1 \tilde{z}_{t+1} = 0.5 \tilde{z}_{t+1}$. Thus

Table 8B.1   A reproduction of
the realization in Table 8.1

t	$\tilde{z}_t$
1	20
2	0
3	−30
4	−20
5	10
6	20

the backcast for $t = 0$ is $\hat{\tilde{z}}_0 = 0.5\tilde{z}_1 = 0.5(20) = 10$. This value is entered in column 2 of Table 8B.2, in the row where $t = 0$. Then the backcast for $t = -1$ is $\hat{\tilde{z}}_{-1} = 0.5z_0$; since z_0 is not available, we replace it with its backcast value $\hat{\tilde{z}}_0 = 10$. Therefore, $\hat{\tilde{z}}_{-1} = 0.5(10) = 5$. Continue in this manner, each time calculating $\hat{\tilde{z}}_t = 0.5\tilde{z}_{t+1}$ and substituting the backcast value $\hat{\tilde{z}}_{t+1}$ for $\tilde{z}_{t+1}$ when necessary. Backcasting may be halted when a satisfactory number of successive backcasts are sufficiently close to zero (the mean of the $\tilde{z}_t$ series).

Table 8B.2   The realization in Table 8B.1
reversed, with "forecasts"

t	$\tilde{z}_t$ [a]
6	20
5	10
4	−20
3	−30
2	0
1	20
0	10
−1	5
−2	2.5
−3	1.25
−4	0.625
−5	0.3125
⋮	⋮

[a] $\tilde{z}_t$ is reversed in sequence.

Table 8B.3   The Realization in Table 8B.1 with backcasts

t	$\tilde{z}_t{}^a$
$\vdots$	$\vdots$
-5	0.3125
-4	0.625
-3	1.25
-2	2.5
-1	5
0	10
1	20
2	0
3	-30
4	-20
5	10
6	20

$^a\tilde{z}_t$ includes backcasts.

Now we may reverse the reversed series with its "forecasts" to obtain the original $\tilde{z}_t$ series, with some estimated previous values (backcasts) included. This is shown in Table 8B.3, which is simply the series in Table 8B.2 reversed. An estimation procedure is now applied to this $\tilde{z}_t$ series, including its backcasts, to reestimate $\hat{\phi}_1$. This value of $\hat{\phi}_1$ is then used to generate new backcasts, and estimation is reapplied to $\tilde{z}_t$ including the new backcasts. This procedure continues until convergence to LS estimates occurs.

When MA terms are present in a model, backcasting involves more computation because the calculations must start with $t = n$ to produce the required estimation residuals. Nevertheless, the concepts are the same as in the preceding example.

9

DIAGNOSTIC CHECKING

Once we have obtained precise estimates of the coefficients in an ARIMA model, we come to the third stage in the UBJ procedure, diagnostic checking. At this stage we decide if the estimated model is statistically adequate. Diagnostic checking is related to identification in two important ways. First, when diagnostic checking shows a model to be inadequate, we must return to the identification stage to tentatively select one or more other models. Second, diagnostic checking also provides clues about how an inadequate model might be reformulated.

The most important test of the statistical adequacy of an ARIMA model involves the assumption that the random shocks are independent. In Section 9.1 we focus on the residual acf as a device for testing whether that assumption is satisfied. In Section 9.2 we consider several other diagnostic checks. Then in Section 9.3 we discuss how to reformulate an ARIMA model when diagnostic checking suggests it is inadequate.

9.1 Are the random shocks independent?

A statistically adequate model is one whose random shocks are statistically independent, meaning not autocorrelated. In practice we cannot observe the random shocks (a_t), but we do have estimates of them; we have the residuals ($\hat{a}_t$) calculated from the estimated model. At the diagnostic-checking stage we use the residuals to test hypotheses about the independence of the random shocks.

Why are we concerned about satisfying the independence assumption? There is a very practical reason. The random shocks are a component of z_t, the variable we are modeling. Thus, if the random shocks are serially correlated, then there is an autocorrelation pattern in z_t that has not been accounted for by the AR and MA terms in that model. Yet the whole idea in UBJ–ARIMA modeling is to account for any autocorrelation pattern in z_t with a parsimonious combination of AR and MA terms, thus leaving the random shocks as white noise. If the residuals are autocorrelated they are not white noise and we must search for another model with residuals that are consistent with the independence assumption.

When the residuals are autocorrelated we must consider how the estimated ARIMA model could be reformulated. Sometimes this means returning to reexamine the initial estimated acf's and pacf's. However, as we see in Section 9.3, the results at the diagnostic-checking stage can also provide clues about how the model could be improved.

The residual acf. The basic analytical tool at the diagnostic-checking stage is the *residual acf*. A residual acf is basically the same as any other estimated acf. The only difference is that we use the residuals $(\hat{a}_t)$ from an estimated model instead of the observations in a realization (z_t) to calculate the autocorrelation coefficients. In Chapter 2 we stated the commonly used formula for calculating autocorrelation coefficients. To find the residual acf we use the same formula, but we apply it to the estimation-stage residuals:

$$r_k(\hat{a}) = \frac{\sum_{t=1}^{n-k} (\hat{a}_t - \bar{a})(\hat{a}_{t+k} - \bar{a})}{\sum_{t=1}^{n} (\hat{a}_t - \bar{a})^2} \tag{9.1}$$

The $\hat{a}$ in parentheses on the LHS of (9.1) indicates that we are calculating residual autocorrelations. The idea behind the use of the residual acf is this: if the estimated model is properly formulated, then the random shocks (a_t) should be uncorrelated. If the random shocks are uncorrelated, then our estimates of them $(\hat{a}_t)$ should also be uncorrelated on average. Therefore, the residual acf for a properly built ARIMA model will ideally have autocorrelation coefficients that are all statistically zero.

In the last two sentences we use the words "on average" and "ideally" because we cannot expect all residual autocorrelations to be exactly zero, even for a properly constructed model. The reason is that the residuals are calculated from a realization (not a process) using only estimates of the ARIMA coefficients (not their true values). Therefore, we expect that sampling error will cause some residual autocorrelations to be nonzero even if we have found a good model.

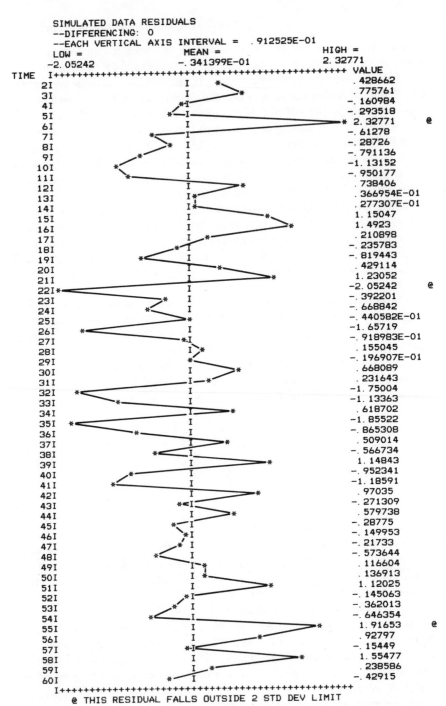

```
        SIMULATED DATA RESIDUALS
        --DIFFERENCING: 0
        --EACH VERTICAL AXIS INTERVAL =  .912525E-01
        LOW =                   MEAN =                   HIGH =
        -2.05242                -.341399E-01             2.32771
TIME    I+++++++++++++++++++++++++++++++++++++++++++++++++ VALUE
   2I                         I        *                    .428662
   3I                         I          *                  .775761
   4I                       *I                             -.160984
   5I                      *  I                            -.293518
   6I                         I                          * 2.32771      @
   7I                  *      I                            -.61278
   8I                    *    I                            -.28726
   9I               *  *      I                            -.791136
  10I           *             I                            -1.13152
  11I           *             I                            -.950177
  12I                         I        *                    .738406
  13I                         I*                            .366954E-01
  14I                         I*                            .277307E-01
  15I                         I                *            1.15047
  16I                         I                  *          1.4923
  17I                       *_I *                           .210898
  18I                     *  I                             -.235783
  19I               *         I                            -.819443
  20I                         I     *                       .429114
  21I                         I                      *      1.23052
  22I*                        I                            -2.05242      @
  23I                     *   I                            -.392201
  24I                  *      I                            -.668842
  25I                        *I                            -.440582E-01
  26I          *              I                            -1.65719
  27I                      *.I                             -.918983E-01
  28I                        I *                            .155045
  29I                      *  I                            -.196907E-01
  30I                         I        *                    .668089
  31I                         I_ *                           .231643
  32I          *              I                            -1.75004
  33I              *          I                            -1.13363
  34I                         I      *                       .618702
  35I         *               I                            -1.85522
  36I                    *    I                            -.865308
  37I                         I    *                        .509014
  38I                     *  I                             -.566734
  39I                         I              *             1.14843
  40I              *          I                            -.952341
  41I            *            I                            -1.18591
  42I                         I             *               .97035
  43I                      *<I                             -.271309
  44I                         I     *                       .579738
  45I                     *   I                            -.28775
  46I                       *I                             -.149953
  47I                       *_I                            -.21733
  48I                     *   I                            -.573644
  49I                         I *                           .116604
  50I                         I *                           .136913
  51I                         I              *             1.12025
  52I                       *I                             -.145063
  53I                     *   I                            -.362013
  54I                  *      I                            -.646354
  55I                         I                      *     1.91653      @
  56I                         I             *               .92797
  57I                       *I                             -.15449
  58I                         I                  *         1.55477
  59I                         I *                           .238586
  60I                     *   I                            -.42915
        I++++++++++++++++++++++++++++++++++++++++++++++++++
        @ THIS RESIDUAL FALLS OUTSIDE 2 STD DEV LIMIT
```

Figure 9.1 Residuals from the ARMA(1, 1) model estimated in Chapter 8.

In Chapter 8 we presented the results of estimating an ARMA(1, 1) model: $(1 - 0.908B)(z_t - 99.44) = (1 - 0.605B)\hat{a}_t$. The residuals from this model are plotted in Figure 9.1. Applying equation (9.1) to this series produces the residual acf in Figure 9.2.

t-tests. Having calculated and plotted the residual autocorrelations, it is important to determine if each is significantly different from zero. We use Bartlett's approximate formula, first introduced in Chapter 3, to estimate the standard errors of the residual autocorrelations. When applied to residual autocorrelations, the formula is

$$s[r_k(\hat{a})] = \left(1 + 2\sum_{j=1}^{k-1} r_j(\hat{a})^2\right)^{1/2} n^{-1/2} \tag{9.2}$$

Having found the estimated standard errors of $r_k(\hat{a})$ from (9.2), we can test the null hypothesis H_0: $\rho_k(a) = 0$ for each residual autocorrelation coefficient. The symbol ρ and the a in parentheses indicate that we are testing a hypothesis about the random shocks in a process. We do not have $\rho_k(a)$ values available, but we have estimates of them in the form of the residual autocorrelations $r_k(\hat{a})$. We test the null hypothesis by calculating how many standard errors (t) away from zero each residual autocorrelation coefficient falls:

$$t = \frac{r_k(\hat{a}) - 0}{s[r_k(\hat{a})]} \tag{9.3}$$

```
++RESIDUAL ACF++
 COEF   T-VAL LAG _____0_____
 0.00    0.03   1                                 0
-0.07   -0.57   2                        <<<<<<<<0
 0.14    1.09   3                                 0>>>>>>>>>>>>>>
-0.03   -0.21   4                              <<<0
-0.04   -0.32   5                             <<<<0
-0.05   -0.38   6                            <<<<<0
-0.08   -0.60   7                         <<<<<<<<0
-0.01   -0.08   8                                <0
 0.14    1.07   9                                 0>>>>>>>>>>>>>>
-0.02   -0.11  10                               <<0
-0.07   -0.49  11                         <<<<<<<0
 0.03    0.25  12                                 0>>>
 0.17    1.24  13                                 0>>>>>>>>>>>>>>>>>
 0.13    0.91  14                                 0>>>>>>>>>>>>>
 0.01    0.10  15                                 0>
    CHI-SQUARED* =      8.07 FOR DF =   12
```

Figure 9.2   Residual acf for the residuals in Figure 9.1.

In practice, if the absolute value of a residual acf t-value is less than (roughly) 1.25 at lags 1, 2, and 3, and less than about 1.6 at larger lags, we conclude that the random shocks at that lag are independent. We could be wrong in this conclusion, of course, but we always run that risk when making decisions based on sample information.

If any residual acf t-value is larger than the critical values suggested above, we tentatively reject the null hypothesis and conclude that the random shocks from the estimated model are correlated and that the estimated model may be inadequate. We then tentatively identify a new model and estimate it to see if our suspicion is justified. We discuss how models are reformulated in Section 9.3.

Unfortunately, there is a potential problem in using Bartlett's formula in testing *residual* autocorrelations: the estimated standard errors are some-times seriously *overstated* when applying Bartlett's formula to residual autocorrelations. This is especially possible at the very short lags (for practical purposes, lags 1 and 2 especially, and perhaps lag 3 also). If the estimated standard errors are overstated, we see from (9.3) that the corre-sponding t-values are understated. Since finding the exact values for the estimated standard errors and t-values for residual autocorrelations is relatively difficult, most computer programs print residual acf t-values calculated using Bartlett's approximation. Therefore, we must be careful in using these printed t-values, especially those at the short lags. This is why we suggest using a warning level for absolute t-values of roughly 1.25 at lags 1, 2, and perhaps 3 in the residual acf.*

Chi-squared test.   There is another way of dealing with the problem of underestimated residual acf t-values. Ljung and Box [24] and Davies et al. [25] suggest a test statistic based on all the residual autocorrelations *as a set.* We are given K residual autocorrelations. We test the following joint null hypothesis about the correlations among the random shocks

$$H_0: \quad \rho_1(a) = \rho_2(a) = \cdots = \rho_K(a) = 0 \qquad (9.4)$$

with this test statistic

$$Q^* = n(n+2) \sum_{k=1}^{K} (n-k)^{-1} r_k^2(\hat{a}) \qquad (9.5)$$

where n is the number of observations used to estimate the model. The

---

*This problem is discussed by Durbin [22] and further analyzed by Box and Pierce [23].

statistic Q^* approximately follows a chi-squared distribution with $(K - m)$ degrees of freedom, where m is the number of parameters estimated in the ARIMA model. This approximate chi-squared test is sometimes referred to as a Ljung–Box test.[†] A table of critical chi-squared values appears at the end of this book. If Q^* is large (significantly different from zero) it says that the residual autocorrelations as a set are significantly different from zero, and the random shocks of the estimated model are probably autocorrelated. We should then consider reformulating the model.

We use the residual acf in Figure 9.2 to illustrate the calculation of a Q^*-statistic. With 15 residual autocorrelations, we have $K = 15$. Apply (9.5) to the $r_k(\hat{a})$ values shown in the COEF column in Figure 9.2:

$$Q^* = n(n + 2) \sum_{k=1}^{K} (n - k)^{-1} r_k^2(\hat{a})$$

$$= 59(61) \sum_{k=1}^{15} (59 - k)^{-1} r_k^2(\hat{a})$$

$$= 3599\left[(1/58)0^2 + (1/57)(-0.07)^2 + \cdots \right.$$

$$\left. + (1/45)(0.13)^2 + (1/44)(0.01)^2\right]$$

$$= 8.07$$

If you perform these calculations by hand you may get a slightly different result due to rounding. The model for which these residual autocorrelations were calculated is an ARMA(1, 1), so $m = 3$. (We have estimated three parameters: ϕ_1, θ_1, and μ.) Therefore, we have $(K - m) = (15 - 3) = 12$ degrees of freedom. The chi-squared statistic and the degrees of freedom (abbreviated df) are both printed beneath the residual acf in Figure 9.2.

According to the chi-squared tables at the end of this book, the critical value with df = 12 at the 10% level is 18.5. Since our calculated chi-squared is less than this critical value, we conclude that the residual autocorrelations in Figure 9.2 are not significantly different from zero as a set, and we accept hypothesis (9.4) that the random shocks are independent. (See Case 12 in

---

[†]Some analysts and computer programs use a statistic suggested by Box and Pierce [23]:

$$Q = n \sum_{k=1}^{K} r_k^2(\hat{a})$$

The Ljung–Box statistic is preferred to the Box–Pierce statistic since its sampling distribution more nearly approximates the chi-squared distribution when the sample size is moderate. All chi-squared statistics in this text are calculated using the Ljung–Box formula (9.5).

Part II for an example of a significant chi-squared statistic that leads to rejection of a model, despite the fact that the residual autocorrelation t-values are only moderately large.)

9.2 Other diagnostic checks

The residual acf, along with the associated t-tests and chi-squared test, is the device most commonly used for diagnostic checking; we make extensive use of it in the case studies in Part II. In this section we discuss several other methods for checking the adequacy of a model.*

Residual plot. The residuals from a fitted model constitute a time series that can be plotted just as the original realization is plotted. Visual analysis of a plot of the residuals is sometimes helpful in detecting problems with the fitted model.

For example, the residuals may display a variance that changes over time, suggesting a logarithmic transformation (or some other transformation) of the original data. In fact, it is sometimes easier to see a changing variance in a plot of the residuals than in a plot of original data. The original realization may contain patterns that interfere with our ability to visualize the variance of the realization. But these patterns are filtered out of the data at the estimation stage, sometimes leaving a more clear picture of the variance of the data in the residuals. The residuals for the ARMA(1, 1) model discussed earlier are plotted in Figure 9.1. Inspection does not suggest that the variance is changing systematically over time.

The residual plot can also be helpful in detecting data errors or unusual events that impact a time series. In Figure 9.1, residuals more than two standard deviations from the mean have an @ note next to them. Of course, we must expect some residuals to be large just by chance. But they might also represent data that were incorrectly recorded, or perturbations to the data caused by identifiable exogenous events. Thus, careful inspection of residuals can sometimes lead to improved accuracy in the data base or insight into the causes of fluctuations in a data series. (Case 2 in Part II shows an example of a large residual that could have arisen because of an identifiable economic policy action by the U.S. Congress.)

Overfitting. Another way of checking a fitted model is to add another coefficient to see if the resulting model is better. This diagnostic check is

---

*One tool not discussed here is the *cumulative periodogram*. Box and Jenkins [1, pp. 294–298] suggest that this device is especially helpful when checking the adequacy of models with seasonal components.

known as *overfitting*. One should have a reason for expanding a model in a certain direction. Otherwise, overfitting is arbitrary and tends to violate the principle of parsimony.

Overfitting is justified especially if the initial estimated acf and pacf are ambiguous. For example, suppose an estimated acf decays toward zero while the pacf has a significant spike at lag 1, suggesting an AR(1) model. But suppose the pacf also has a spike at lag 2 with a *t*-value of 1.8, for example. While this value is not highly significant, it is moderately large. Therefore, an AR(2) is plausible though the evidence favoring it is not overwhelming. According to the principle of parsimony, we should start with an AR(1) model. But using the overfitting strategy, we check our judgment by also trying an AR(2). In this case the moderately large pacf spike at lag 2 gives a clue about the direction in which the model should be expanded.

A special warning is in order: in overfitting be careful not to add coefficients to both sides of the model. That is, do not overfit with both AR and MA terms simultaneously. Doing so not only runs counter to the principle of parsimony but can also lead to serious estimation problems because of coefficient redundancy. This latter problem is discussed in Chapter 8.

Fitting subsets of the data.   Sometimes data continue to be generated by the same type of process [e.g., an ARMA(1, 1)], but the coefficients (ϕ_1 and θ_1) in that process change in value over time. If this happens, forecasts based on a model fitted to the entire data set are less accurate than they could be.

One way to check a model for this problem is to divide the data set in half, for example, and estimate the same model for each half. Then perform a statistical test to see if the coefficients from the two data sets are significantly different.

For example, suppose an AR(1) model has been fitted to both the first (A) and second (B) halves of a realization, with the following results:

$$\hat{\phi}_{1A} = 0.5, \qquad s\left(\hat{\phi}_{1A}\right) = 0.20$$

$$\hat{\phi}_{1B} = 0.7, \qquad s\left(\hat{\phi}_{1B}\right) = 0.25$$

where 0.20 and 0.25 are the standard errors of the two coefficients. Now consider the statistic $\hat{\phi}_{1A} - \hat{\phi}_{1B} = 0.5 - 0.7 = -0.2$. The variance of this difference is the sum of the two variances. Therefore, the estimated standard error of this difference is

$$s\left(\hat{\phi}_{1A} - \hat{\phi}_{1B}\right) = \left[(0.20)^2 + (0.25)^2\right]^{1/2} = 0.32$$

Testing the hypothesis H_0 that $\phi_{1A} = \phi_{1B}$, or

$$\phi_{1A} - \phi_{1B} = 0$$

gives this t-statistic

$$t = \frac{\left(\hat{\phi}_{1A} - \hat{\phi}_{1B}\right) - 0}{s\left(\hat{\phi}_{1A} - \hat{\phi}_{1B}\right)}$$

$$= \frac{-0.2}{0.32}$$

$$= -0.625$$

Since this t-statistic is not significantly different from zero at the 5% level, we conclude that $\phi_{1A} = \phi_{1B}$. In other words, the coefficient ϕ_1 is the same for both halves of the data set.

There is another, less formal check for changing coefficients. We may drop the latter part of the realization (e.g., the last 10% of the observations) and reestimate the same model for this shortened realization. If the resulting coefficients are close to those estimated using the full realization (e.g., within ± 0.1), we conclude that the most recent observations are being generated by the same process as the earlier data.

The first of the two preceding approaches has the advantage of involving a formal statistical test. However, the decision to divide the realization in half is arbitrary. It may be, for example, that the last two-thirds of the realization is generated by coefficients different from those generating the first third. Furthermore, the number of observations must be relatively large before we can consider splitting the data into segments.

The second approach (dropping the latter part of the realization) has the advantage of emphasizing the very recent past. If recent data behave quite differently from the rest of the realization, this raises a serious concern about the ability of the overall model to forecast the near-term future very well. The disadvantage of this check is that it is informal; however, the last 10% or so of a realization is often not a large-enough data set to allow useful formal tests of the change in coefficients, as suggested in the first approach above. (Case 2 in Part II shows an example where estimated coefficients fitted to a subset of the realization are very close to those obtained from the entire data set.)

9.3   Reformulating a model

Suppose we decide tentatively that a model is statistically inadequate because (i) some residual acf t-values exceed the suggested warning values, or (ii) the residual acf chi-squared statistic is too large. According to the UBJ method, we then return to the identification stage to tentatively select one or more other models. There is no guarantee, of course, that we will discover a better model: the residual autocorrelations from the original model could be large just because of sampling error.

One way to reformulate an apparently inadequate model is to reexamine the estimated acf and pacf calculated from the original realization. Because they are based on a realization, estimated acf's and pacf's can give ambiguous evidence about the process generating the data. For example, they might have a pattern that could be interpreted as either a decay or a cutoff to zero, so that either an AR model or an MA model could be justified. Reexamination of the original estimated acf and pacf might suggest one or more alternative models that did not initially seem obvious.

Another way to reformulate a model is to use the residual acf as a guide. For example, suppose the original estimated acf decays toward zero and we fit this AR(1) model to the data initially:

$$(1 - \phi_1 B)\tilde{z}_t = b_t \qquad (9.6)$$

where b_t is a set of autocorrelated shocks. Suppose the residual acf for (9.6) has a spike at lag 1 followed by a cutoff to zero. This suggests an MA(1) model for b_t:

$$b_t = (1 - \theta_1 B)a_t \qquad (9.7)$$

where a_t is not autocorrelated. Use (9.7) to substitute for b_t in (9.6). The result is an ARMA(1, 1) model for z_t:

$$(1 - \phi_1 B)\tilde{z}_t = (1 - \theta_1 B)a_t \qquad (9.8)$$

As an illustration, consider the estimated acf and pacf (based on a simulated realization) in Figure 9.3. Suppose we tentatively identify an AR(1) model for the realization underlying these functions. This model is justified because the estimated acf decays toward zero rather than cutting off to zero, and the estimated pacf has a single spike (at lag 1) with a t-value greater than 2.0.

The top of Figure 9.4 shows the results of fitting model (9.6) to the data. The estimated coefficient $\hat{\phi}_1 = 0.693$ satisfies the stationarity requirement

234   Diagnostic checking

```
+ + + + + + + + + + + + AUTOCORRELATIONS + + + + + + + + + + + +
+ FOR DATA SERIES: SIMULATED DATA                                       +
+ DIFFERENCING:  0                      MEAN    =  99.482               +
+ DATA COUNT =  60                      STD DEV =   2.06206             +
  COEF   T-VAL LAG                      0
  0.68    5.28   1                 [         0>>>>>J>>>>>>>>>>>>
  0.34    1.89   2                   [       0>>>>>>>J
  0.19    1.02   3              [         0>>>>>      J
  0.18    0.95   4              [         0>>>>>      J
  0.08    0.40   5              [         0>>          J
 -0.01   -0.04   6              [         0            J
  0.00    0.02   7              [         0            J
 -0.04   -0.22   8              [        <0            J
 -0.12   -0.62   9              [      <<<0            J
     CHI-SQUARED* =   43.03 FOR DF =   9
```

```
+ + + + + + + + + + + PARTIAL AUTOCORRELATIONS + + + + + + + + + + +
  COEF   T-VAL LAG                        0
  0.68    5.28   1                    [         0>>>>>J>>>>>>>>>>>>
 -0.24   -1.83   2                    [<<<<<0        J
  0.14    1.07   3                    [         0>>>  J
  0.08    0.60   4                    [         0>>   J
 -0.19   -1.47   5                    [<<<<<0        J
  0.06    0.45   6                    [         0>    J
  0.05    0.41   7                    [         0>    J
 -0.20   -1.51   8                    [<<<<<0        J
  0.01    0.04   9                    [         0     J
```

Figure 9.3   Estimated acf and pacf for a simulated realization.

$|\hat{\phi}_1| < 1$. It is also statistically significant at better than the 5% level since its t-value is substantially larger than 2.0.

The residual acf is printed below the estimation results in Figure 9.4. The residual autocorrelation coefficient at lag 1 has a t-value of 1.64. This exceeds the practical warning level of 1.25 suggested earlier for lags 1, 2, and 3 in a residual acf; therefore, we consider modifying the initial AR(1) model.

The residual acf in Figure 9.4 is similar to an MA(1) acf, with a significant spike at lag 1 followed by autocorrelations that are not significantly different from zero. In other words, the residuals ($\hat{b}_t$) of model (9.6) appear to be autocorrelated, following an MA(1) pattern as in (9.7). Using the substitution procedure followed above we arrive at (9.8), an ARMA(1, 1) model, for the original realization z_t.

The results of estimating and checking (9.8) are shown in Figure 9.5. This model satisfies the stationarity requirement $|\hat{\phi}_1| < 1$ and the invertibility requirement $|\hat{\theta}_1| < 1$, and both estimated coefficients have absolute t-values greater than 2.0.

Model (9.8) is better than (9.6) since its adjusted RMSE = 1.43754 (the estimated standard deviation of a_t) is smaller than the estimated standard deviation of b_t, 1.50238. Furthermore, the residual acf at the bottom of

```
+ + + + + + + + +ECOSTAT UNIVARIATE B-J RESULTS+ + + + + + + + +
+ FOR DATA SERIES:   SIMULATED DATA                              +
+ DIFFERENCING:      0                          DF    = 57       +
+ AVAILABLE:         DATA = 60   BACKCASTS = 0   TOTAL = 60       +
+ USED TO FIND SSR:  DATA = 59   BACKCASTS = 0   TOTAL = 59       +
+ (LOST DUE TO PRESENCE OF AUTOREGRESSIVE TERMS:           1)    +
```

COEFFICIENT	ESTIMATE	STD ERROR	T-VALUE
PHI 1	0.693	0.095	7.31
CONSTANT	30.4453	9.43373	3.22728
MEAN	99.2746	.641068	154.858

ADJUSTED RMSE = 1.50238   MEAN ABS % ERR =   1.14

```
        CORRELATIONS
        1      2
1    1.00
2   -0.10   1.00
```

```
++RESIDUAL ACF++
  COEF    T-VAL  LAG                          0
  0.21    1.64    1                           0>>>>>>>>>>>>>>>>>>>>>>>>
 -0.15   -1.13    2       <<<<<<<<<<<<<<<<<0
 -0.13   -0.91    3       <<<<<<<<<<<<<0
  0.12    0.84    4                           0>>>>>>>>>>>>>
 -0.01   -0.09    5                           <0
 -0.15   -1.04    6       <<<<<<<<<<<<<<<<<0
  0.05    0.32    7                           0>>>>>>
  0.03    0.24    8                           0>>>
 -0.11   -0.76    9           <<<<<<<<<<<<0
     CHI-SQUARED* =     8.88 FOR DF =   7
```

Figure 9.4 Estimation and diagnostic-checking results for an AR(1) with a simulated realization.

Figure 9.5 is satisfactory since none of the absolute t-values exceeds the warning levels suggested earlier, and the calculated chi-squared statistic is not significantly different from zero.

The preceding example is not unusual, and it suggests that modifying a model in light of the residual acf is rather straightforward. That is, the initial acf decays to zero, suggesting an AR(1). Then the residual acf has a single spike at lag 1, suggesting the addition of an MA term at lag 1. The resulting model (9.8), in this case, is an obvious composite of the initial model for z_t and the subsequent model for the residuals $\hat{b}_t$. It appears from this example that we can reformulate models by simply adding to the original model the coefficient implied by the residual acf. (See Cases 2 and 5 in Part II for similar examples.) However, the information contained in the residual acf may be less clear than in the preceding illustration.

For example, suppose the initial fitted model is an AR(1):

$$(1 - \phi_1' B)\tilde{z}_t = b_t \qquad (9.9)$$

```
+ + + + + + + + +ECOSTAT UNIVARIATE B-J RESULTS+ + + + + + + + + +
+ FOR DATA SERIES:  SIMULATED DATA                                    +
+ DIFFERENCING:      0                      DF    = 56               +
+ AVAILABLE:              DATA = 60   BACKCASTS = 0   TOTAL = 60      +
+ USED TO FIND SSR:  DATA = 59   BACKCASTS = 0   TOTAL = 59          +
+ (LOST DUE TO PRESENCE OF AUTOREGRESSIVE TERMS:            1)       +

COEFFICIENT    ESTIMATE      STD ERROR      T-VALUE
  PHI    1      0.485         0.154         3.15
  THETA  1     -0.419         0.170        -2.47
  CONSTANT     51.1921       15.306         3.34457

  MEAN         99.3326        .515285     192.772

  ADJUSTED RMSE =  1.43754    MEAN ABS % ERR =   1.14
       CORRELATIONS
        1     2      3
  1    1.00
  2    0.68  1.00
  3   -0.11 -0.08  1.00

++RESIDUAL ACF++
  COEF   T-VAL LAG _____0_____
  0.04   0.30   1                         0>>>>>
  0.04   0.33   2                         0>>>>>
 -0.07  -0.53   3                   <<<<<<<0
  0.17   1.27   4                         0>>>>>>>>>>>>>>>>>>>
  0.00   0.04   5                         0
 -0.14  -1.01   6                <<<<<<<<<<<<<<<0
  0.07   0.49   7                         0>>>>>>>
 -0.02  -0.12   8                       <<0
 -0.04  -0.26   9                     <<<<0
       CHI-SQUARED* =    3.99 FOR DF =  6
```

Figure 9.5 Estimation and diagnostic-checking results for an ARMA(1, 1) with a simulated realization.

where again b_t is a set of autocorrelated shocks. But suppose the residual acf also suggests an AR(1) model for b_t:

$$(1 - \phi_1^* B)b_t = a_t \tag{9.10}$$

where a_t is a set of uncorrelated shocks.

In this case we cannot just add to the original model the coefficient suggested by the residual acf—it is not possible to add an AR coefficient at lag 1 when we already have this coefficient in the model. We can, however, use the same algebraic procedure used in the previous example. Solve (9.10) for b_t:

$$b_t = (1 - \phi_1^* B)^{-1} a_t \tag{9.11}$$

Now substitute (9.11) into (9.9):

$$(1 - \phi_1' B)\tilde{z}_t = (1 - \phi_1^* B)^{-1} a_t \tag{9.12}$$

Next, multiply both sides of (9.12) by $(1 - \phi_1^* B)$:

$$(1 - \phi_1^* B)(1 - \phi_1' B)\tilde{z}_t = a_t \qquad (9.13)$$

Expanding the LHS of (9.13) gives

$$\left(1 - \phi_1' B - \phi_1^* B + \phi_1' \phi_1^* B^2\right)\tilde{z}_t = a_t \qquad (9.14)$$

Combining terms, we get this AR(2) model:

$$\left(1 - \phi_1 B - \phi_2 B^2\right)\tilde{z}_t = a_t \qquad (9.15)$$

where $\phi_1 = \phi_1' + \phi_1^*$ and $\phi_2 = -\phi_1' \phi_1^*$.

This example shows that the information contained in the residual acf can be subtle at times. In particular, it may not be appropriate to simply add to the initial model the coefficients that appear to describe the residual series. Case 4 in Part II illustrates a similar substitution procedure. It also shows that the residual acf can be critically important in finding an adequate model. Case 13 likewise demonstrates how the residual acf is sometimes virtually the only means by which an appropriate model can be found.

Summary

1. At the diagnostic-checking stage we determine if a model is statistically adequate. In particular, we test if the random shocks are independent. If this assumption is not satisfied, there is an autocorrelation pattern in the original series that has not been explained by the ARIMA model. Our goal, however, is to build a model that fully explains any autocorrelation in the original series.

2. In practice we cannot observe the random shocks (a_t) in a process, but we have estimates of them in the form of estimation-stage residuals $(\hat{a}_t)$.

3. To test the hypothesis that the random shocks are independent we construct a residual acf. This acf is like any estimated acf except we construct it using the estimation residuals $\hat{a}_t$ instead of the realization z_t.

4. Approximate *t*-values are calculated for residual autocorrelation coefficients using Bartlett's approximation for the standard error of estimated autocorrelations.

5. If the absolute *t*-values of residual autocorrelations exceed certain warning values, we should consider reformulating the model. The critical values are

Lag	Practical Warning Level
1, 2, 3	1.25
All others	1.6

6.   The warning values above are smaller for the short lags (1, 2, and 3) because using Bartlett's approximation can result in understated residual acf t-values, especially at the short lags.

7.   Another way to deal with potentially underestimated residual acf t-values is to test the residual autocorrelations as a set rather than individually. An approximate chi-squared statistic (the Ljung–Box statistic) is available for this test. If this statistic is significant we should consider reformulating the model.

8.   Other diagnostic checks are to (i) plot the residuals to see if their variance is changing over time, as a clue about incorrectly recorded data, and as a clue about identifiable exogenous events that may perturb the data series; (ii) overfit the model by adding another AR or MA term if there is reason to think another might be called for; (iii) fit the chosen model to subsets of the available realization to see if the estimated coefficients change significantly.

9.   When reformulating a model that seems inadequate in light of the diagnostic checks, it is wise to return to the original estimated acf and pacf to look for further clues to an appropriate model.

10.   The residual acf is an important guide to reformulating a model. At times, we may simply add to the original model the coefficients that are appropriate for the residuals of that model based on the residual acf. At other times, we must algebraically substitute the model that is appropriate for the residuals into the original model to see what new model is implied for the original series.

Questions and Problems

9.1   An estimated ARIMA model with significantly autocorrelated residuals is inadequate. Explain why.

9.2   How does a residual acf differ from an estimated acf calculated from an original realization?

9.3 "A properly constructed ARIMA model has residual autocorrelations that are all zero." Comment on this statement.

9.4 It is suggested that the practical warning level for the absolute values of residual acf t-statistics at lags 1, 2, and 3 is about 1.25. Why is such a small t-value used as the warning level at the short lags in the residual acf?

9.5 What is the motivation for applying a chi-squared test in addition to the t-tests applied to residual autocorrelations?

10

FORECASTING

The ultimate application of UBJ–ARIMA modeling, as studied in this text, is to forecast future values of a time series. In this chapter we first consider how *point* forecasts (single numerical values) are derived algebraically from an estimated ARIMA model. We then discuss how to establish probability limits around point forecasts, thus creating *interval* forecasts. Next, we consider complications that arise in forecasting a series estimated in logarithmic form. Finally, we discuss the sense in which ARIMA forecasts are best, or optimal.

Unless indicated otherwise, throughout this chapter we assume, for simplicity, that any ARIMA model we consider is known; that is, the mean μ, all ϕ and θ coefficients, and all past random shocks are assumed known. Fortunately, the conclusions based on this simplifying assumption are essentially correct in practice if we have properly identified and estimated an ARIMA model using a sufficient number of observations: The properties of ARIMA forecasts are little affected by ordinary sampling error when the sample size is appropriate.*

In the appendix to this chapter we discuss how UBJ–ARIMA methods may be used to complement econometric (regression and correlation) forecasting models.

---

*The robustness of ARIMA forecasts with respect to sampling error in parameter estimates is discussed in Box and Jenkins [1, pp. 306–308].

10.1   The algebra of ARIMA forecasts

Difference-equation form.   The most convenient way to produce point forecasts from an ARIMA model is to write the model in *difference-equation* form. In this section we give several examples of how this is done.

Let t be the current time period. When forecasting we are interested in future values of a time series variable, denoted z_{t+l}, where $l \geq 1$. Period t is called the forecast *origin*, and l is called the forecast *lead time*. In ARIMA analysis, forecasts depend on the available observations on variable z up through period t. Let the information contained in the set of available observations $(z_t, z_{t-1}, \dots)$ be designated I_t. Then the forecast of z_{t+l}, designated $\hat{z}_t(l)$, is the conditional mathematical expectation of z_{t+l}. That is, $\hat{z}_t(l)$ is the mathematical expectation of z_{t+l} given I_t:

$$\hat{z}_t(l) = E(z_{t+l}|I_t) \tag{10.1}$$

where the vertical line means "given."

As an illustration, consider an ARIMA$(1, 0, 1)$ model. We develop the general algebraic form for the first several forecasts from this model. Then we show a numerical example.

The ARIMA$(1, 0, 1)$ model is

$$(1 - \phi_1 B)\tilde{z}_t = (1 - \theta_1 B)a_t$$

or

$$z_t = \mu(1 - \phi_1) + \phi_1 z_{t-1} - \theta_1 a_{t-1} + a_t \tag{10.2}$$

Now let $l = 1$. By altering time subscripts appropriately, use (10.2) to write an expression for z_{t+1}:

$$z_{t+1} = \mu(1 - \phi_1) + \phi_1 z_t - \theta_1 a_t + a_{t+1} \tag{10.3}$$

Applying (10.1) to (10.3), we find that the forecast of z_{t+1} is

$$\hat{z}_t(1) = E(z_{t+1}|I_t)$$

$$= \mu(1 - \phi_1) + \phi_1 z_t - \theta_1 a_t \tag{10.4}$$

Since a_{t+1} is unknown at time t, we assign its expected value of zero. In this example z_t and a_t together constitute I_t. That is, z_t and a_t are all the relevant information about past z's needed to forecast z_{t+1}. (Remember that MA terms are parsimonious algebraic substitutes for AR terms; thus a_t represents a set of past z's.)

Continuing the preceding example with $l = 2$, use (10.2) to write an expression for z_{t+2}. Then the conditional expected value of that expression is the forecast $\hat{z}_t(2)$:

$$\hat{z}_t(2) = E(z_{t+2}|I_t)$$

$$= \mu(1 - \phi_1) + \phi_1 z_{t+1} - \theta_1 a_{t+1} \qquad (10.5)$$

Since z_{t+1} is unknown at origin t it must be replaced by its conditional expectation $\hat{z}_t(1)$ from (10.4). Likewise, a_{t+1} is unknown at origin t and is replaced by its expected value of zero. With these two substitutions, (10.5) becomes

$$\hat{z}_t(2) = \mu(1 - \phi_1) + \phi_1 \hat{z}_t(1) \qquad (10.6)$$

Proceeding as above, we find that each subsequent forecast for this ARIMA(1, 0, 1) is based on the preceding forecast value of z. That is, $\hat{z}_t(3)$ depends on $\hat{z}_t(2)$, $\hat{z}_t(4)$ depends on $\hat{z}_t(3)$, and so on:

$$\hat{z}_t(3) = \mu(1 - \phi_1) + \phi_1 \hat{z}_t(2)$$

$$\hat{z}_t(4) = \mu(1 - \phi_1) + \phi_1 \hat{z}_t(3)$$

$$\hat{z}_t(5) = \mu(1 - \phi_1) + \phi_1 \hat{z}_t(4)$$

$$\vdots \qquad \qquad \vdots$$

In the example above, forecasts for $l > 1$ are called "bootstrap" forecasts because they are based on forecast z's rather than observed z's.

Forecasts from other ARIMA models are found in essentially the same manner as above. In practice, μ is unknown and is replaced by its estimate $\hat{\mu}$. Likewise, the ϕ and θ coefficients are replaced by their estimates, $\hat{\phi}$ and $\hat{\theta}$. As shown above, past z observations are employed when available. They are available up to time t, the forecast origin; thereafter, they must be replaced by their forecast counterparts (their conditional expected values). Past a_t values are replaced by their corresponding estimates, the estimation residuals $\hat{a}_t$, when these residuals are available. But when the time subscript on a random shock exceeds the forecast origin t, that shock is replaced by its expected value of zero. This is what happened as we moved from (10.5) to (10.6) in the case of the ARIMA(1, 0, 1): there is no estimation residual $\hat{a}_{t+1}$ available when we forecast from origin t, so we substitute zero.

Now we consider an *estimated* model as a numerical example. Estimation with $n = 60$ produced these values: $\hat{\mu} = 101.26$, $\hat{\phi}_1 = 0.62$, and $\hat{\theta}_1 = -0.58$. Thus the estimated model can be written as

$$(1 - 0.62B)\tilde{z}_t = (1 + 0.58B)\hat{a}_t$$

where $\tilde{z} = z_t - 101.26$. The estimated constant term is $\hat{C} = \hat{\mu}(1 - \hat{\phi}_1) = 101.26(1 - 0.62) = 38.48$. The last observation of z_t in this data series is $z_{60} = 96.91$. We will show how the first three forecasts from this model are calculated.

With a forecast origin $t = 60$ and a forecast lead time $l = 1$, from (10.4) the forecast for period $t = 61$ is

$$\hat{z}_{61} = \hat{z}_{60}(1) = \hat{C} + \hat{\phi}_1 z_{60} - \hat{\theta}_1 \hat{a}_{60}$$

$$= 38.48 + 0.62(96.91) + 0.58(-1.37)$$

$$= 97.77 \qquad 60.08 \qquad -.7946$$

In the preceding calculations the observation for period $t = 60$ is known ($z_{60} = 96.91$). The random shock for period $t = 60$ is unknown, but we have the estimation residual $\hat{a}_{60} = -1.37$ to put in its place. (Most computer programs for estimating ARIMA models have an option for printing the estimation-stage residuals.)

With a forecast origin $t = 60$ and a forecast lead time $l = 2$, (10.5) gives this forecast for period $t = 62$:

$$\hat{z}_{62} = \hat{z}_{60}(2) = \hat{C} + \hat{\phi}_1 \hat{z}_{61} - \hat{\theta}_1 \hat{a}_{61}$$

$$= 38.48 + 0.62(97.77) + 0.58(0)$$

$$= 99.10$$

In these calculations, z_{61} is unknown at origin $t = 60$; therefore, z_{61} is replaced by its conditional expectation $\hat{z}_{61} = \hat{z}_{60}(1) = 97.77$. The random shock a_{61} is not observable and is replaced by its estimate, the residual $\hat{a}_{61}$. However, since the data extend only through period 60, $\hat{a}_{61}$ is unknown and so is replaced by its expected value of zero.

For origin $t = 60$ and lead time $l = 3$, (10.5) gives

$$\hat{z}_{63} = \hat{z}_{60}(3) = \hat{C} + \hat{\phi}_1 \hat{z}_{62} - \hat{\theta}_1 \hat{a}_{62}$$

$$= 38.48 + 0.62(99.10) + 0.58(0)$$

$$= 99.92$$

Forecasts from other estimated ARIMA models are calculated in the same manner as above. Most computer programs for identifying and estimating ARIMA models also have an option to generate forecasts from any estimated model, so the forecasts need not be produced by hand. However, the necessary calculations are illustrated further in Part II in Cases 3, 9, and 15.

Note in the calculations above that the forecasts are converging toward the mean of the series (101.26). This occurs with the forecasts from all *stationary* models. Cases 1 and 2 in Part II illustrate forecasts that converge to the estimated mean. The convergence may be rapid or slow depending on the model. In general, forecasts from pure MA models converge more rapidly to the mean, since we quickly lose information about past estimated random shocks as we forecast further into the future. As shown above, with pure AR or mixed models we can "bootstrap" ourselves by using forecast z's to replace observed z's; but with a pure MA model we must replace random shocks with the expected value of zero when the forecast lead time exceeds the lag length of a past shock term. When the forecast lead time exceeds q, the maximum lag length of the MA terms, the forecasts from a pure MA model are equal to the estimated constant $\hat{C}$, which is equal to the estimated mean $\hat{\mu}$ in pure MA models.

Forecasts from *nonstationary* models do not converge toward the series mean: a nonstationary series does not fluctuate around a fixed central value, and forecasts for such a series reflect that nonstationary character. (See Cases 5 and 7 in Part II for examples.)

Figure 10.1 shows forecasts generated from an AR(1) model. The letter F represents a forecast value. (The square brackets [ ] represent confidence intervals whose construction and interpretation are discussed in the next section.) Note that the printed values of the forecasts are gravitating toward the estimated mean (6.04) just as the forecasts calculated above from another AR(1) converged toward the estimated mean of that series (101.26).

Figure 10.2 shows forecasts produced from an MA(1) model. The first forecast is $\hat{z}_{101} = \hat{z}_{100}(1) = \hat{C} - \hat{\theta}_1 \hat{a}_{100}$. Since the estimation residual $\hat{a}_{100}$ is available, the forecast reflects not only the estimated constant $\hat{C}$ (equal to the estimated mean $\hat{\mu}$ in a pure MA model), but also the last estimated random shock $\hat{a}_{100}$. But for lead time $l = 2$, the forecast is simply $\hat{z}_{102} = \hat{z}_{100}(2) = \hat{C}$. The estimation residual $\hat{a}_{101}$ is not available so it is replaced by its expected value (zero) and the forecast converges to the estimated mean ($\hat{\mu} = \hat{C} = 99.8344$).

Figure 10.3 shows forecasts derived from a nonstationary model, an ARIMA(0, 1, 1). In difference-equation form this model is $z_t = z_{t-1} - \hat{\theta}_1 \hat{a}_{t-1} + \hat{a}_t$. The forecast origin is the fourth quarter of 1978, designated

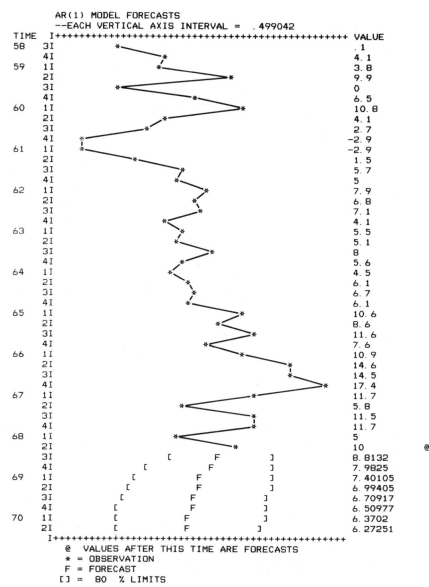

```
        AR(1) MODEL FORECASTS
        --EACH VERTICAL AXIS INTERVAL =  .499042
TIME    I+++++++++++++++++++++++++++++++++++++++++++++++++++ VALUE
58   3I            *——————                                    .1
     4I                  *                                    4.1
59   1I                  *                                    3.8
     2I                       ————————*                       9.9
     3I            *                                          0
     4I                       *                               6.5
60   1I                       ————————*                       10.8
     2I                  *——                                  4.1
     3I               *                                       2.7
     4I       *—————                                          -2.9
61   1I       *                                               -2.9
     2I            *——                                        1.5
     3I                  *                                    5.7
     4I                  *                                    5
62   1I                    *                                  7.9
     2I                  *                                    6.8
     3I                    *                                  7.1
     4I                *                                      4.1
63   1I                 *                                     5.5
     2I                *                                      5.1
     3I                  *                                    8
     4I                *                                      5.6
64   1I                *                                      4.5
     2I                *                                      6.1
     3I                *                                      6.7
     4I                 *                                     6.1
65   1I                     *                                 10.6
     2I                  *                                    8.6
     3I                     *                                 11.6
     4I               *                                       7.6
66   1I                   *                                   10.9
     2I                        *                              14.6
     3I                        *                              14.5
     4I                          *                            17.4
67   1I                     *                                 11.7
     2I             *                                         5.8
     3I                     *                                 11.5
     4I                     *                                 11.7
68   1I            *                                          5
     2I                    *                                  10        @
     3I            [       F       ]                          8.8132
     4I          [        F        ]                          7.9825
69   1I         [          F          ]                       7.40105
     2I           [        F         ]                        6.99405
     3I        [           F           ]                      6.70917
     4I        [           F           ]                      6.50977
70   1I         [          F          ]                       6.3702
     2I        [           F           ]                      6.27251
     I+++++++++++++++++++++++++++++++++++++++++++++++++++
        @   VALUES AFTER THIS TIME ARE FORECASTS
        *  = OBSERVATION
        F  = FORECAST
        [] =   80  % LIMITS
```

Figure 10.1 Forecasts from an AR(1) model.

245

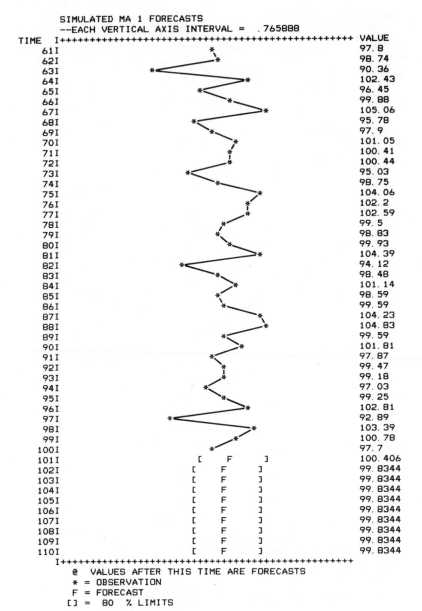

SIMULATED MA 1 FORECASTS
--EACH VERTICAL AXIS INTERVAL = .765888

TIME		VALUE
61		97.8
62		98.74
63		90.36
64		102.43
65		96.45
66		99.88
67		105.06
68		95.78
69		97.9
70		101.05
71		100.41
72		100.44
73		95.03
74		98.75
75		104.06
76		102.2
77		102.59
78		99.5
79		98.83
80		99.93
81		104.39
82		94.12
83		98.48
84		101.14
85		98.59
86		99.59
87		104.23
88		104.83
89		99.59
90		101.81
91		97.87
92		99.47
93		99.18
94		97.03
95		99.25
96		102.81
97		92.89
98		103.39
99		100.78
100		97.7
101	[F]	100.406
102	[F]	99.8344
103	[F]	99.8344
104	[F]	99.8344
105	[F]	99.8344
106	[F]	99.8344
107	[F]	99.8344
108	[F]	99.8344
109	[F]	99.8344
110	[F]	99.8344

@   VALUES AFTER THIS TIME ARE FORECASTS
* = OBSERVATION
F = FORECAST
[] =   80   % LIMITS

Figure 10.2   Forecasts from an MA(1) model.

246

78(4). For $l = 1$, the forecast is

$$\hat{z}_{79(1)} = \hat{z}_{78(4)}(1) = z_{78(4)} - \hat{\theta}_1 \hat{a}_{78(4)}$$

Since both $z_{78(4)}$ and $\hat{a}_{78(4)}$ are available, the forecast includes both of these terms. But with lead time $l = 2$ the forecast becomes $\hat{z}_{79(2)} = \hat{z}_{78(4)}(2) = \hat{z}_{79(1)}$. That is, $z_{79(1)}$ is not observed so it is replaced by its forecast value. The estimation residual $\hat{a}_{79(1)}$ is not available and is replaced by its expected value of zero. By similar reasoning all subsequent forecasts are equal to the preceding forecast, so the forecasts converge to the one-step-ahead forecast $\hat{z}_{78(4)}(1)$.

Note that these forecasts are not converging to the calculated mean of the series (193.3) because the model is nonstationary. Differencing ($d = 1$) has freed the forecasts from a fixed mean. If a series mean is shifting significantly through time, we do not want to tie forecasts to the overall mean of the series; although that value can be calculated, it is not useful for describing the shifting level of the series.

Finally, consider the forecasts in Figure 10.4. These were produced from another nonstationary model, an ARIMA(0, 2, 1). The realization used to identify and estimate this model (not all of which is displayed in Figure 10.4) shows changes in both level and slope. As discussed in Chapter 7, such series require second differencing ($d = 2$) to induce a constant mean. The overall mean of the original series is 62.7. Once again, although we can calculate this single value, it is not helpful in describing the behavior of the series since the level of the data is shifting through time. Clearly, the forecasts in Figure 10.4 are not gravitating toward the overall realization mean of 62.7.

These forecasts are dominated by the differencing element in the model. To see this, consider the difference-equation form of the ARIMA(0, 2, 1) model: $z_t = 2z_{t-1} - z_{t-2} - \hat{\theta}_1 \hat{a}_{t-1} + \hat{a}_t$. The terms $2z_{t-1}$ and z_{t-2} are present because of the differencing operation. The first forecast is

$$\hat{z}_{78(11)} = \hat{z}_{78(10)}(1) = 2z_{78(10)} - z_{78(9)} - \hat{\theta}_1 \hat{a}_{78(10)}$$

The values $z_{78(10)}$ and $z_{78(9)}$ are both available from the realization, and $\hat{a}_{78(10)}$ is available from the estimation residuals. But for lead time $l = 2$, the forecast is $\hat{z}_{78(12)} = \hat{z}_{78(10)}(2) = 2\hat{z}_{78(11)} - z_{78(10)}$. Although observation $z_{78(10)}$ is available, neither the observation $z_{78(11)}$ nor the estimation residual $\hat{a}_{78(11)}$ is available. The former is replaced by its forecast value $\hat{z}_{78(11)}$ and the latter is replaced by its expected value of zero. All subsequent forecasts are entirely bootstrap forecasts. For example, for $l = 3$, we have $\hat{z}_{79(1)} = \hat{z}_{78(10)}(3) = 2\hat{z}_{78(12)} - \hat{z}_{78(11)}$. Thus we see that the forecasts are dominated

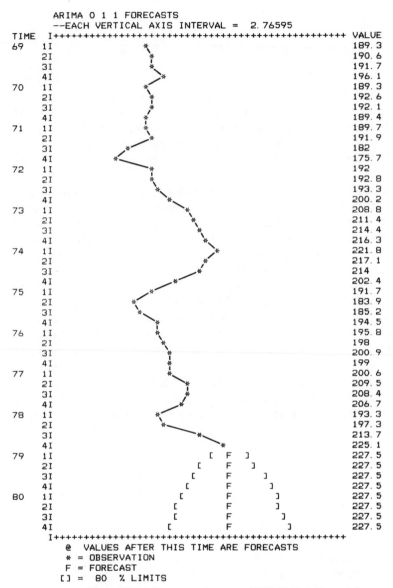

```
          ARIMA 0 1 1 FORECASTS
          --EACH VERTICAL AXIS INTERVAL =   2.76595
TIME   I+++++++++++++++++++++++++++++++++++++++++++++++++++++ VALUE
69      1I                     *                                189.3
        2I                      *                               190.6
        3I                      *                               191.7
        4I                        *                             196.1
70      1I                    *                                 189.3
        2I                     *                                192.6
        3I                     *                                192.1
        4I                    *                                 189.4
71      1I                    *                                 189.7
        2I                     *                                191.9
        3I                 *                                    182
        4I            *                                         175.7
72      1I                 *                                    192
        2I                   *                                  192.8
        3I                   *                                  193.3
        4I                     *                                200.2
73      1I                       *                              208.8
        2I                        *                             211.4
        3I                        *                             214.4
        4I                         *                            216.3
74      1I                          *                           221.8
        2I                         *                            217.1
        3I                       *                              214
        4I                     *                                202.4
75      1I                  *                                   191.7
        2I               *                                      183.9
        3I               *                                      185.2
        4I                 *                                    194.5
76      1I                  *                                   195.8
        2I                   *                                  198
        3I                   *                                  200.9
        4I                  *                                   199
77      1I                   *                                  200.6
        2I                     *                                209.5
        3I                     *                                208.4
        4I                    *                                 206.7
78      1I                  *                                   193.3
        2I                  *                                   197.3
        3I                    *                                 213.7
        4I                      *                               225.1          @
79      1I                    [   F   ]                         227.5
        2I                  [     F   ]                         227.5
        3I                  [     F     ]                       227.5
        4I                 [     F       ]                      227.5
80      1I                   [     F       ]                    227.5
        2I                 [       F       ]                    227.5
        3I                 [       F         ]                  227.5
        4I                 [       F         ]                  227.5
       I+++++++++++++++++++++++++++++++++++++++++++++++++++++
          @   VALUES AFTER THIS TIME ARE FORECASTS
          *  = OBSERVATION
          F  = FORECAST
          [] =   80  % LIMITS
```

Figure 10.3   Forecasts from an ARIMA(0, 1, 1) model.

248

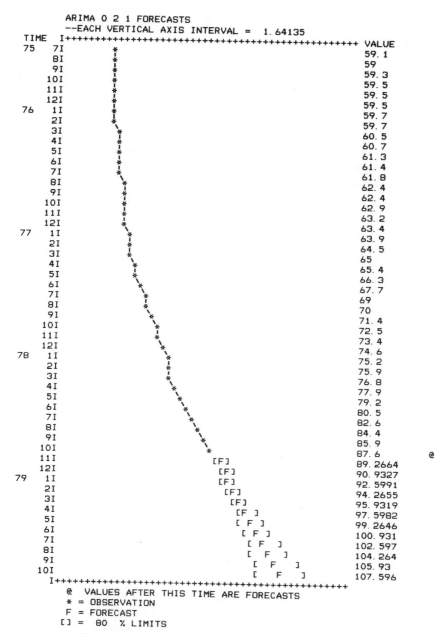

```
            ARIMA 0 2 1 FORECASTS
            --EACH VERTICAL AXIS INTERVAL =  1.64135
   TIME   I+++++++++++++++++++++++++++++++++++++++++++++++++++++  VALUE
    75    7I        *                                            59. 1
          8I        *                                            59
          9I        |                                            59. 3
         10I        *                                            59. 5
         11I        *                                            59. 5
         12I        |                                            59. 5
    76    1I        *                                            59. 7
          2I        *                                            59. 7
          3I         *                                           60. 5
          4I         *                                           60. 7
          5I         *                                           61. 3
          6I         |                                           61. 4
          7I          *                                          61. 8
          8I          *                                          62. 4
          9I          *                                          62. 4
         10I          *                                          62. 9
         11I           *                                         63. 2
         12I           *                                         63. 4
    77    1I           *                                         63. 9
          2I            *                                        64. 5
          3I            *                                        65
          4I             *                                       65. 4
          5I             *                                       66. 3
          6I              *                                      67. 7
          7I              *                                      69
          8I               *                                     70
          9I               *                                     71. 4
         10I                *                                    72. 5
         11I                 *                                   73. 4
         12I                 *                                   74. 6
    78    1I                  *                                  75. 2
          2I                  *                                  75. 9
          3I                   *                                 76. 8
          4I                    *                                77. 9
          5I                     *                               79. 2
          6I                     *                               80. 5
          7I                      *                              82. 6
          8I                       *                             84. 4
          9I                        *                            85. 9
         10I                        *                            87. 6        @
         11I                         [F]                         89. 2664
         12I                         [F]                         90. 9327
    79    1I                         [F]                         92. 5991
          2I                          [F]                        94. 2655
          3I                          [F]                        95. 9319
          4I                          [F ]                       97. 5982
          5I                          [ F ]                      99. 2646
          6I                          [ F ]                      100. 931
          7I                           [ F  ]                    102. 597
          8I                           [  F  ]                   104. 264
          9I                            [  F   ]                 105. 93
         10I                            [   F   ]                107. 596
         I+++++++++++++++++++++++++++++++++++++++++++++++++++++
           @  VALUES AFTER THIS TIME ARE FORECASTS
           *  = OBSERVATION
           F  = FORECAST
           []  =  80  % LIMITS
```

Figure 10.4 Forecasts from an ARIMA(0, 2, 1) model.

249

by the differencing component in the model, and are not tied to any fixed central value.

Random-shock form.   Any ARIMA model can be written in *random-shock* form. That is, we can replace any AR terms with an infinite series of MA terms. A pure MA model is already in random-shock form.

Although the random-shock form is not usually convenient for producing forecasts, it is especially useful for estimating the variance of forecasts and thus for deriving confidence intervals around point forecasts.

The coefficients in the random-shock form are denoted by the symbol ψ_i, with i corresponding to the time lag of the associated past random shock:

$$z_t = \mu + \psi_0 a_t + \psi_1 a_{t-1} + \psi_2 a_{t-2} + \psi_3 a_{t-3} + \cdots \tag{10.7}$$

where $\psi_0 = 1$. If the sequence $\psi_1, \psi_2, \ldots$ is finite, then (10.7) is a pure MA model. If the sequence is infinite, (10.7) represents an AR or mixed model. For a stationary series, μ is simply the mean. For a nonstationary series, μ represents the changing level of the series as determined by the differencing operations.

Any pure MA model is already in random-shock form, with order q. For example, consider an MA(2):

$$(z_t - \mu) = \left(1 - \theta_1 B - \theta_2 B^2\right) a_t$$

or

$$z_t = \mu + a_t - \theta_1 a_{t-1} - \theta_2 a_{t-2} \tag{10.8}$$

Letting $\psi_0 = 1$, $\psi_1 = -\theta_1$, and $\psi_2 = -\theta_2$, we may write (10.8) in random-shock form as

$$z_t = \mu + \psi_0 a_t + \psi_1 a_{t-1} + \psi_2 a_{t-2} \tag{10.9}$$

which is simply a truncated version of (10.7).

AR models can be written in random-shock form by inverting and expanding the AR operator. For example, we showed in Chapter 5 how an AR(1) could be written in MA (random-shock) form. The AR(1) is $(1 - \phi_1 B)(z_t - \mu) = a_t$. Dividing both sides by the AR operator gives $z_t - \mu = (1 - \phi_1 B)^{-1} a_t$. If $|\phi_1| < 1$, $(1 - \phi_1 B)^{-1}$ is equivalent to the convergent infinite series $(1 + \phi_1 B + \phi_1^2 B^2 + \phi_1^3 B^3 + \cdots)$. Thus the AR(1) may be written as

$$z_t - \mu = \left(1 + \phi_1 B + \phi_1^2 B^2 + \phi_1^3 B^3 + \cdots\right) a_t$$

or

$$z_t = \mu + a_t + \phi_1 a_{t-1} + \phi_1^2 a_{t-2} + \phi_1^3 a_{t-3} + \cdots \qquad (10.10)$$

Letting $\psi_0 = 1$, $\psi_1 = \phi_1$, $\psi_2 = \phi_1^2$, $\psi_3 = \phi_1^3$, and so forth, we see that (10.10) is equivalent to (10.7) with an infinite sequence of ψ's.

The values of the ψ coefficients for different ARIMA models vary (except for ψ_0, which is always 1) depending on the degree of differencing and the values of the AR and MA coefficients in the model. It can be cumbersome to find the ψ weights for more complex models by hand; they are usually generated by a computer program. However, we illustrate a method for finding ψ weights using two examples.

It can be shown that the ψ weights are found by equating coefficients of like powers of B in this expansion:*

$$\left(\psi_0 + \psi_1 B + \psi_2 B^2 + \cdots\right)\left(1 - \phi_1 B - \phi_2 B^2 - \cdots - \phi_p B^p\right)(1 - B)^d$$

$$= \left(1 - \theta_1 B - \theta_2 B^2 - \cdots - \theta_q B^q\right) \qquad (10.11)$$

Consider again the AR(1) model. The relevant version of (10.11) then is

$$\left(\psi_0 + \psi_1 B + \psi_2 B^2 + \cdots\right)(1 - \phi_1 B) = 1$$

or

$$\psi_0 + (\psi_1 - \phi_1 \psi_0)B + (\psi_2 - \phi_1 \psi_1)B^2 + (\psi_3 - \phi_1 \psi_2)B^3 + \cdots = 1$$

Now set the coefficients of the various powers of B on the LHS equal to the coefficients of the same powers of B on the RHS. For B^0 we find $\psi_0 = 1$. For B^1 we get $\psi_1 - \phi_1 \psi_0 = 0$, or $\psi_1 = \phi_1$. For B^2 we have $\psi_2 - \phi_1 \psi_1 = 0$, or $\psi_2 = \phi_1^2$. With B^3 we get $\psi_3 - \phi_1 \psi_2 = 0$, or $\psi_3 = \phi_1^3$. We see that this method produces the same result for the AR(1) obtained in (10.10) above.

Next consider an ARIMA$(1, 0, 1)$. In this case (10.11) is

$$\left(\psi_0 + \psi_1 B + \psi_2 B^2 + \psi_3 B^3 + \cdots\right)(1 - \phi_1 B) = (1 - \theta_1 B)$$

---

*This expansion is found as follows: Define $\phi(B)$ as the AR operator, $\theta(B)$ as the MA operator, ∇^d as the differencing operator, and $\psi(B)$ as the ψ-weight operator. Then the ARIMA model for z_t is $\phi(B)\nabla^d \tilde{z}_t = \theta(B)a_t$, where the ψ-weight form is $\tilde{z}_t = \psi(B)a_t$. Write the ψ-weight form as $a_t = \psi(B)^{-1}\tilde{z}_t$; substitute this for a_t in the ARIMA model and rearrange to get $\psi(B)\phi(B)\nabla^d \tilde{z}_t = \theta(B)\tilde{z}_t$. Dividing by $\tilde{z}_t$ and writing the operators in long form gives (10.11).

or

$$\psi_0 + (\psi_1 - \phi_1\psi_0)B + (\psi_2 - \phi_1\psi_1)B^2 + (\psi_3 - \phi_1\psi_2)B^3 + \cdots$$
$$= 1 - \theta_1 B$$

Now equate the coefficients of like powers of B on both sides of the equation. This leads to the following results:

j	B^j	ψ_j
0	B^0	$\psi_0 = 1$
1	B^1	$\psi_1 = \phi_1 - \theta_1$
2	B^2	$\psi_2 = \phi_1(\phi_1 - \theta_1)$
3	B^3	$\psi_3 = \phi_1^2(\phi_1 - \theta_1)$
$\vdots$	$\vdots$	$\vdots$

We see from the pattern that, in general, $\psi_j = \phi_1^{j-1}(\phi_1 - \theta_1)$ for the ARIMA(1, 0, 1).

Apply this result to the ARIMA(1, 0, 1) estimation results presented in the last section. There we had $\hat{\phi}_1 = 0.62$ and $\hat{\theta}_1 = -0.58$. Inserting these estimated coefficients into the above expression for ψ_j, we get

$$\hat{\psi}_0 = 1$$

$$\hat{\psi}_1 = \hat{\phi}_1 - \hat{\theta}_1 \qquad = 0.62 + 0.58 \ = 1.20$$

$$\hat{\psi}_2 = \hat{\phi}_1(\hat{\phi}_1 - \hat{\theta}_1) = 0.62(1.20) \qquad = 0.74$$

$$\hat{\psi}_3 = \hat{\phi}_1^2(\hat{\phi}_1 - \hat{\theta}_1) = (0.62)^2(1.20) = 0.46$$

$$\vdots \qquad\qquad \vdots \qquad\qquad \vdots$$

We will use these results in the next section to illustrate how the ψ weights are used to construct confidence intervals around point forecasts.

10.2 The dispersion of ARIMA forecasts

Using the difference-equation form of an ARIMA model, we can produce a series of point forecasts, where "point" means the forecast is a single value rather than a range. Using the random-shock form, we can find the variance of the forecast errors. This allows us to construct approximate confidence intervals around our forecasts, thus providing some information on how reliable forecasts may be. In this section we find general expressions for the variance and standard deviation of ARIMA forecast errors. Then we show how confidence intervals are constructed, and we illustrate the relevant calculations using the ARIMA$(1, 0, 1)$ model presented earlier.

Forecast-error variance and standard deviation.   First, define a forecast error for origin t and lead time l, designated $e_t(l)$, as the observed z for period $t + l$ minus the forecast z for that period:

$$e_t(l) = z_{t+l} - \hat{z}_t(l) \tag{10.12}$$

Use (10.7) to write z_{t+l} in random-shock form as

$$z_{t+l} = \mu + \psi_0 a_{t+l} + \psi_1 a_{t-1+l} + \psi_2 a_{t-2+l} + \cdots \tag{10.13}$$

The corresponding forecast value $\hat{z}_t(l)$, which is the conditional mathematical expectation $E(z_{t+l}|I_t)$, is found from (10.13) to be

$$\hat{z}_t(l) = E(z_{t+l}|I_t)$$

$$= \mu + \psi_l a_t + \psi_{l+1} a_{t-1} + \psi_{l+2} a_{t-2} + \cdots \tag{10.14}$$

That is, the information set I_t is defined as information about the series z only through period t. Thus (10.14) contains random shocks only from period t or earlier since any random shock after period t is unknown at time t. (We are assuming for simplicity that shock terms at time t or earlier are observable. Of course, in practice they must be estimated from the estimation-stage residuals. Shock terms after time t are not only unknown at time t, they cannot be estimated at time t.)

For example, let $l = 1$. We want to find the expectation of (10.13) given I_t. The first shock term on the RHS of (10.13) is $\psi_0 a_{t+l} = \psi_0 a_{t+1}$. Since a_{t+1} is unknown (and cannot be estimated) at origin t we assign this term its expected value of zero. The next shock term is $\psi_1 a_{t-1+l} = \psi_1 a_t$. The value a_t is known (or may be estimated) at origin t, so we include $\psi_1 a_t = \psi_1 a_t$ in the expectation (10.14). The next shock term in (10.13) is $\psi_2 a_{t-2+l} = \psi_2 a_{t-1}$. The value a_{t-1} is available at time t so the term $\psi_{l+1} a_{t-1} = \psi_2 a_{t-1}$ appears in (10.14). By the same reasoning all subsequent shock terms in (10.13) are known at origin t and therefore appear in (10.14). The reader is encouraged to proceed as above to see how finding the expectation of (10.13) leads to (10.14) for $l = 2$. The result is $\hat{z}_t(2) = \mu + \psi_2 a_t + \psi_3 a_{t-1} + \psi_4 a_{t-2} + \dots,$

where the terms $\psi_0 a_{t+2}$ and $\psi_1 a_{t+1}$ in (10.13) have an expected value of zero since a_{t+2} and a_{t+1} are unknown (and cannot be estimated) at origin t.

Substituting (10.13) and (10.14) into (10.12), we find

$$e_t(l) = \psi_0 a_{t+l} + \psi_1 a_{t-1+l} + \cdots + \psi_{l-1} a_{t+1} \qquad (10.15)$$

That is, (10.13) contains all random-shock terms up through period $t + l$, whereas (10.14) contains only those up through period t. Subtracting (10.14) from (10.13) leaves the random-shock terms from period $t + l$ back through period $t + 1$.

Now using (10.15) we find that the (conditional) variance of $e_t(l)$ is

$$\sigma^2[e_t(l)] = E\{e_t(l) - E[e_t(l)]|I_t\}^2$$

$$= E[e_t(l)]^2$$

$$= \sigma_a^2(1 + \psi_1^2 + \psi_2^2 + \cdots + \psi_{l-1}^2) \qquad (10.16)$$

and therefore the standard deviation of $e_t(l)$ is

$$\sigma[e_t(l)] = \sigma_a(1 + \psi_1^2 + \psi_2^2 + \cdots + \psi_{l-1}^2)^{1/2} \qquad (10.17)$$

The variance (10.16) is found by squaring (10.15). {Note in (10.15) that $E[e_t(l)] = 0$, and recall that $\psi_0 = 1$.} All cross-product terms have an expected value of zero since the random shocks are assumed to be independent. The expected value of each remaining squared shock term is, by assumption, the constant σ_a^2.

In practice, $\sigma[e_t(l)]$ must be estimated, since σ_a is unknown and is replaced by the RMSE ($\hat{\sigma}_a$) and since the coefficients are unknown and are replaced by estimates ($\hat{\psi}_i$) calculated from the estimated ARIMA coefficients ($\hat{\phi}$'s and $\hat{\theta}$'s). The resulting forecast-error variances (and forecast confidence intervals) are therefore only approximate.

Consider the sequence of estimated ψ coefficients calculated in the last section for an estimated ARIMA(1, 0, 1). They were

$$\psi_0 = 1$$

$$\hat{\psi}_1 = 1.20$$

$$\hat{\psi}_2 = 0.74$$

$$\hat{\psi}_3 = 0.46$$

$$\vdots$$

The estimated standard deviation of the shocks for this model is $\hat{\sigma}_a = 1.60$. Use these values and (10.17) to find the estimated standard deviation of the forecast errors for lead times $l = 1, 2,$ and 3:

$$\hat{\sigma}[e_t(1)] = \hat{\sigma}_a(1)^{1/2}$$

$$= 1.60$$

$$\hat{\sigma}[e_t(2)] = \hat{\sigma}_a(1 + \hat{\psi}_1^2)^{1/2}$$

$$= 1.6[1 + (1.20)^2]^{1/2}$$

$$= 2.50$$

$$\hat{\sigma}[e_t(3)] = \hat{\sigma}_a(1 + \hat{\psi}_1^2 + \hat{\psi}_2^2)^{1/2}$$

$$= 1.6[1 + (1.20)^2 + (0.74)^2]^{1/2}$$

$$= 2.77$$

Forecast confidence intervals. If the random shocks are Normally distributed (as we assume they are) and if we have estimated an appropriate ARIMA model with a sufficiently large sample, forecasts from that model are approximately Normally distributed. Using (10.17) we can therefore construct confidence intervals around each point forecast using a table of probabilities for standard Normal deviations. Thus an approximate 95% confidence interval is given by

$$\hat{z}_t(l) \pm 1.96\hat{\sigma}[e_t(l)]$$

and an approximate 80% confidence interval is

$$\hat{z}_t(l) \pm 1.28\hat{\sigma}[e_t(l)]$$

Earlier we presented forecasts for lead times $l = 1, 2,$ and 3 for an estimated ARIMA(1, 0, 1) model. These point forecasts were

$$\hat{z}_{60}(1) = 97.77$$

$$\hat{z}_{60}(2) = 99.10$$

$$\hat{z}_{60}(3) = 99.92$$

The estimated standard deviations of the forecast errors calculated above are used to produce approximate 95% confidence intervals around the point forecasts as follows:

$$\hat{z}_{60}(1) \pm 1.96\hat{\sigma}[e_{60}(1)]$$

$$97.77 \pm 1.96(1.60)$$

$$97.77 \pm 3.14$$

or

$$(94.63, 100.91)$$

$$\hat{z}_{60}(2) \pm 1.96\hat{\sigma}[e_{60}(2)]$$

$$99.10 \pm 1.96(2.50)$$

$$99.10 \pm 4.90$$

or

$$(94.20, 104.00)$$

$$\hat{z}_{60}(3) \pm 1.96\hat{\sigma}[e_{60}(3)]$$

$$99.92 \pm 1.96(2.77)$$

$$99.92 \pm 5.43$$

or

$$(94.49, 105.35)$$

These intervals are interpreted in the usual way. For example, the last interval is interpreted in this way: We can be 95% confident that the interval $(99.49, 105.35)$ will contain the observed value $z_{t+l} = z_{60+3} = z_{63}$.

10.3   Forecasting from data in logarithmic form

In Chapter 7 we said if the variance of a data series changes in proportion to its mean, then building a model of the natural logarithms of the series is appropriate. (Cases 9 and 11 in Part II are examples of such a series.) However, usually we are interested in forecasting the original data rather than the log values. It might be tempting merely to calculate the antilogs of the logarithmic forecasts to get the forecasts of the original series. But doing this creates a problem: if the random shocks of the log series are Normally distributed, then the shocks of the original series (and the forecasts of this series) follow a *log-Normal* distribution.*

Let a log series be denoted by z'_t, where z_t is the original series. Then it can be shown that the forecast for z_{t+l} depends on both the forecast and the forecast-error variance of z'_{t+l} in this way:

$$\hat{z}_t(l) = \exp\{\hat{z}'_t(l) + \tfrac{1}{2}\sigma^2[e'_t(l)]\} \tag{10.18}$$

Thus, we should not simply find the antilog of $\hat{z}'_t(l)$ to find $\hat{z}_t(l)$. Instead, we must take into account the variance of the logarithmic forecast as shown in (10.18). However, the upper and lower confidence limits around $\hat{z}_t(l)$ are found by taking the antilogs of the limits around $\hat{z}'_t(l)$. That is, if U and L are the upper and lower limits of an α-percent confidence interval around $\hat{z}'_t(l)$, then exp(U) and exp(L) are the α-percent upper and lower limits around $\hat{z}_t(l)$. It follows that the interval around $\hat{z}_t(l)$ is not symmetrical since the interval around $\hat{z}'_t(l)$ is symmetrical.

Finally, note that forecasts of z'_t may be interpreted in terms of z_t without finding antilogs because the change of a log value is the percent change of the corresponding antilog value. For example, suppose the following forecasts for z'_{t+l} are generated from origin t:

l	$\hat{z}'_t(l)$	90% Confidence Values ($\pm$)
1	3.7866	0.014
2	3.8084	0.015
3	3.8209	0.017

Let the log of the last available observation (z'_t) be 3.7525. Then we have these forecast log *changes*.

l	$\Delta\hat{z}'_t(l)$
1	$0.0341 = \hat{z}'_t(1) - z'_t\ \ \ = 3.7866 - 3.7525$
2	$0.0218 = \hat{z}'_t(2) - \hat{z}'_t(1) = 3.8084 - 3.7866$
3	$0.0125 = \hat{z}'_t(3) - \hat{z}'_t(2) = 3.8209 - 3.8084$

These log changes are interpreted directly as forecast percent changes for z_{t+l}, and the interval values above are interpreted as percent intervals. That

*The log-Normal distribution is discussed by Olkin et al. [26, pp. 299–302]. Nelson [27, pp. 161–165] discusses the log-Normal distribution in the context of ARIMA models.

is, multiply the forecast log changes and the 90% confidence values by 100 to get

l	Percent $\Delta\hat{z}_t(l)$	90% Confidence Values ($\pm$)
1	3.41%	1.4%
2	2.18%	1.5%
3	1.25%	1.7%

Thus, the original series is forecast to rise by 3.41% from period t to $t + 1$, then it is forecast to rise by another 2.18% from period $t + 1$ to $t + 2$, and by 1.25% from period $t + 2$ to $t + 3$.

10.4 The optimality of ARIMA forecasts

Forecasts from ARIMA models are said to be *optimal* forecasts. This means that no other univariate forecasts have a smaller mean-squared forecast error (abbreviated MSE). That is, let an ARIMA l-step-ahead forecast be $\hat{z}_t(l)$, with corresponding forecast error $e_t(l)$, and let I_t be the information about all available z's through period t. Then given I_t, $\hat{z}_t(l)$ is optimal because the conditional mathematical expectation of the squared ARIMA forecast error, $E[e_t(l)|I_t]^2$, is smaller than for any other univariate forecast.* It also follows that ARIMA forecasts give the minimum forecast-error variance since $E[e_t(l)|I_t]^2$ is that variance.

Several points must be clarified. First, optimality refers to the mathematical expectation of $[e_t(l)]^2$, not to any particular $e_t(l)$. That is, some other (non-ARIMA) univariate model forecast could have a smaller squared forecast error than a properly constructed ARIMA-model forecast in a particular instance, but not on average.

Second, optimality applies strictly only if the particular ARIMA model being considered is the correct one. Thus, ARIMA forecasts are minimum MSE forecasts in practice only if the strategy of identification, estimation, and diagnostic checking is adequate to the problem at hand, and only if that strategy has been properly employed.

Third, we are comparing ARIMA forecasts only with other univariate forecasts. That is, I_t contains information about past z's only. If I_t were expanded to include information about other relevant variables (giving a

---

*Box and Jenkins [1, pp. 127–128] demonstrate that the ARIMA forecast is the minimum MSE forecast of z_{t+l}.

multiple-series model), we could get forecasts with a smaller MSE than ARIMA forecasts.

Fourth, we are considering only univariate models that are linear combinations of past z's, with fixed coefficients. "Linear combination" means that each past z is multiplied by some coefficient, and the resulting terms are then added. Consider that any ARIMA model can be written, by inverting and expanding the MA operator, as an AR model of infinitely high order:*

$$z_t = C + a_t + \pi_1 z_{t-1} + \pi_2 z_{t-2} + \cdots \tag{10.19}$$

where each π_i is some combination of ϕ and θ coefficients. It should be clear that (10.19) is a linear combination of past z's. Now, it is possible that a nonlinear combination of z's could produce forecasts with a smaller MSE than linear ARIMA forecasts.

Furthermore, the π coefficients in (10.19) do not have time subscripts: they are fixed through time because they are composed of ϕ and θ coefficients which are assumed to be fixed. Univariate models with time-varying coefficients could, at times, produce smaller MSE forecasts than the fixed-coefficient ARIMA models we have considered. The theory and practice of nonlinear ARIMA models and time-varying parameter ARIMA models is not well-developed at present.

These conditions on the optimality of ARIMA forecasts might seem quite restrictive. But keep in mind that linear, fixed-coefficient univariate models are often very useful in practice. It is helpful, therefore, to know that forecasts from ARIMA models are optimal within this larger class of useful models.

Summary

1.   Point forecasts (single numerical values) from an ARIMA model are calculated most easily by writing the model in difference equation form.

2.   To find point forecasts from an ARIMA model using the difference-equation form, write the model in common algebraic form and solve for z_t. Insert the estimates of C and the ϕ and θ coefficients and assign a_t its expected value of zero. Now insert the appropriate values for any past observations (past z terms) and past random shocks (past a terms). In practice we must use estimation-stage residuals in place of past random shocks, or the expected value of zero if the forecast lead time l exceeds the

---

*We showed in Chapter 5 how an MA(1) model, for example, can be written in AR form.

lag length of the MA term in question. Likewise, we use forecast z values in place of observed z values when the forecast lead time l exceeds the lag length of the AR term in question.

3. While point forecasts are most conveniently calculated from the difference-equation form of an ARIMA model, in creating confidence intervals around these point forecasts it is convenient to start with the random-shock form of a model.

4. The random-shock form of an ARIMA model is its MA form. That is, by inverting and expanding the AR operator, we replace any AR terms with an infinite series of MA terms.

5. A forecast error for lead time l, $e_t(l)$, is defined as the difference between an observed z_t and its forecast counterpart $\hat{z}_t(l)$:

$$e_t(l) = z_t - \hat{z}_t(l)$$

This forecast error has variance $\sigma^2[e_t(l)]$ given by

$$\sigma_a^2\left(1 + \psi_1^2 + \psi_2^2 + \cdots + \psi_{l-1}^2\right)$$

where the ψ_i coefficients are the coefficients in the random-shock form of the model.

6. If the random shocks are Normally distributed and if we have an appropriate ARIMA model, then our forecasts and the associated forecast errors are approximately Normally distributed.

7. The forecast-error variance for a given ARIMA model is estimated from the available realization. Let $\hat{\sigma}[e_t(l)]$ be the square root of this estimated variance. This estimate may be used to construct a confidence interval around any forecast:

$$\hat{z}_t(l) \pm Z\hat{\sigma}\left[e_t(l)\right]$$

where Z is the standard Normal deviation associated with the desired degree of confidence.

8. If the variance of a realization is made stationary by transformation of the original data into natural log values (z_t'), we may not forecast the original series (z_t) by merely finding the antilogs of the log forecasts. Instead, we must take into account the variance of the log forecasts in this way:

$$\hat{z}_t(l) = \exp\{\hat{z}_t'(l) + \tfrac{1}{2}\sigma^2\left[e_t'(l)\right]\}$$

9. Forecast log changes are interpreted as forecast percentage changes for the original series.

10. ARIMA forecasts are said to be optimal univariate forecasts: the mean-squared forecast error, given the information (I_t) about the z observations available through period t, designated $E[e_t(l)|I_t]^2$, is smaller than for any other univariate forecast. Note that optimality refers to the mean-squared forecast error, not to any particular squared forecast error.

11. ARIMA forecasts are optimal only if we have found an appropriate ARIMA model and only among forecasts from univariate, linear, fixed-coefficient models. A multivariate model, or a model with a nonlinear combination of past z's, or a model with time-varying coefficients might give forecasts with a smaller mean-squared forecast error.

Appendix 10A: The complementarity of ARIMA models and econometric models

This appendix is aimed primarily at the reader with a background in econometrics. However, it should be useful to any reader who knows the fundamentals of regression analysis.

We have noted that ARIMA models are a special class of univariate models: they use only the information contained in past observations of the variable being analyzed. In this appendix we discuss how ARIMA models may be used in conjunction with econometric models, a common class of multiple-series models based on standard regression and correlation methods. While there are some important differences between ARIMA and econometric forecasting models, both have the same purpose: finding statistical relationships that are reliable enough to produce useful forecasts.

A single-equation econometric model specifies how a dependent variable (y) is functionally related to one or more independent variables ($x_1, x_2, \ldots, x_m$) *other than* past values of y. (Sometimes econometric models have past values of y among the "independent" variables, but other variables are also present.) If one or more of the independent variables ($x_1, \ldots, x_m$) is also logically dependent on y, the econometric model may consist of several equations.

A single-equation econometric model might be written as follows:

$$y_t = \alpha + \beta_1 x_{1_t} + \beta_2 x_{2_t} + \epsilon_t \tag{10A.1}$$

where α, β_1, and β_2 are parameters, t is a time subscript, and ϵ is a probabilistic shock element usually assumed to be a set of Normally, independently, and identically distributed random variables with a mean of zero.

In econometric modeling the analyst is guided by some theory when selecting independent variables. This theory may involve human behavior or a technological relationship, for example, but one should have a reason besides mere statistical patterns for choosing the variables to include in an econometric model. By contrast, in UBJ–ARIMA modeling we emphasize statistical appearances (correlation as shown in estimated acf's and pacf's) rather than theories about why one variable is related to another.

UBJ–ARIMA analysis may be used to complement econometric analysis in at least four ways, as discussed below.

Forecasting independent variables.   Econometric models are often used for forecasting time-series data, and they can be very useful for this purpose. However, one must first forecast the values of any independent variables that are contemporaneous with the dependent variable (y_t) before forecasting y_t. That is, if y_t depends on x_{1_t} and x_{2_t}, as in equation (10A.1), we must forecast the future values $x_{1_{t+1}}$ and $x_{2_{t+1}}$ in order to forecast y_{t+1}.

ARIMA models are convenient for producing forecasts of independent variables whenever an independent variable is contemporaneous with the dependent variable in an econometric model. These forecasts can be generated without gathering additional data (assuming enough observations on the independent variable are available initially) since UBJ–ARIMA models are univariate models.

Analyzing residuals.   ARIMA analysis can be applied to the estimated residuals (estimates of the ϵ_t terms, designated $\hat{\epsilon}_t$) to see if they satisfy the standard independence assumption. ARIMA analysis can detect patterns in the $\hat{\epsilon}_t$ terms that might be missed by traditional econometric tests. For example, a common way of testing regression equation shock terms for independence is with the Durbin–Watson statistic (d):*

$$d = \frac{\sum_{t=2}^{n} (\hat{\epsilon}_t - \hat{\epsilon}_{t-1})^2}{\sum_{t=1}^{n} \hat{\epsilon}_t^2} \tag{10A.2}$$

As suggested by the form of the numerator in (10A.2), this statistic tests only for correlation between estimation residuals separated by one time period ($k = 1$), whereas with UBJ–ARIMA analysis one routinely examines residuals separated by various time periods ($k = 1, 2, 3, \dots$) in a residual acf.

---

*For a brief introduction to the Durbin–Watson statistic, see Mansfield [8, Chapter 12].

Combining forecasts.   A forecast which is a weighted average of two or more individual forecasts often has a smaller error variance than any of the individual forecasts. This is especially true when the individual forecasts are based on different information sets and/or different methods.

It may be worthwhile, therefore, to average econometric forecasts and ARIMA forecasts. This is particularly appealing since an econometric model contains information (the independent variables) that a univariate ARIMA model does not contain, and ARIMA models may contain information (past values of the dependent variable) not contained in an econometric model.

This approach to forecasting must be used with care. For example, the weights assigned to the individual forecasts must be chosen properly. Furthermore, if combined forecasts are superior, this suggests that another forecasting method may be called for—one that meshes the individual approaches. For an introductory discussion about combining forecasts, see Granger [28, pp. 157–164]. For a more advanced treatment, along with bibliographic references, see Granger and Newbold [17, Chapter 8].

Checking for misspecification.   The mathematical structures of some econometric models logically imply ARIMA models for the endogenous (dependent) variables. This means that econometric and ARIMA models are, under certain circumstances, alternative ways of expressing the same mathematical model.

The consequences of this are quite interesting. In particular, if an econometric model logically implies an ARIMA model for an endogenous variable which is quite inconsistent with the estimated acf and pacf of that variable, this is strong evidence that the econometric model is incorrectly specified. Thus ARIMA models can aid in the construction of better econometric models.

The logical relationship between econometric models and ARIMA models is a relatively recent area of research. The interested reader may consult two articles by Zellner [29, 30], both of which contain additional references.

Questions and Problems

10.1   Write the following in difference-equation form:

(a) $(1 - \phi_1 B - \phi_2 B^2)\tilde{z}_t = (1 - \theta_1 B)a_t$

(b) $\tilde{z}_t = (1 - \theta_1 B - \theta_2 B^2)a_t$

(c) $(1 - \phi_1 B)(1 - B)\tilde{z}_t = a_t$

(d) $(1 - B)\tilde{z}_t = (1 - \theta_1 B)a_t$

(e) $(1 - B)^2 \tilde{z}_t = (1 - \theta_1 B - \theta_2 B^2)a_t$

(f) $(1 - \phi_1 B)\tilde{z}_t = (1 - \theta_1 B - \theta_2 B^2)a_t$

10.2 Use the following information to forecast for lead times $l = 1, 2,$ and 3 for each of the models in question 10.1.

(a) $n = 100, z_{99} = 53, z_{100} = 56, \hat{a}_{100} = 1.4, \hat{\phi}_1 = 1.4, \hat{\phi}_2 = -0.7,$ $\hat{\theta}_1 = 0.3, \hat{\mu} = 50$

(b) $n = 100, \hat{a}_{99} = 1.3, \hat{a}_{100} = -2.6, \hat{\theta}_1 = 0.7, \hat{\theta}_2 = -0.5, \hat{\mu} = 100$

(c) $n = 100, z_{99} = 217, z_{100} = 232, \hat{\phi}_1 = 0.3$

(d) $n = 100, z_{100} = 28, \hat{\theta}_1 = 0.5, \hat{a}_{100} = -0.7$

(e) $n = 100, z_{99} = 97, z_{100} = 102, \hat{\theta}_1 = 0.3, \hat{\theta}_2 = 0.2$

(f) $n = 100, z_{100} = 103, \hat{\phi}_1 = 0.6, \hat{\theta}_1 = 0.8, \hat{\theta}_2 = -0.3, \hat{\mu} = 100$

10.3 Find the values of the first three $\hat{\psi}$ weights for each of the models in problem 10.2 using expansion (10.11). Present both the algebraic form and numerical values.

10.4 Find the estimated forecast-error variances and standard deviations for each of the forecasts produced in problem 10.2 using the following information:

(a) $\hat{\sigma}_a^2 = 3.3$
(b) $\hat{\sigma}_a^2 = 2.5$
(c) $\hat{\sigma}_a^2 = 8$
(d) $\hat{\sigma}_a^2 = 6.7$
(e) $\hat{\sigma}_a^2 = 1.2$
(f) $\hat{\sigma}_a^2 = 2.5$

10.5 Construct 80% and 95% confidence intervals for each of the forecasts produced in problem 10.2.

11

SEASONAL AND OTHER
PERIODIC MODELS

Time-series data often display *periodic* behavior. A periodic series has a pattern which repeats every s time periods, where $s > 1$. Experience has shown that ARIMA models often produce good forecasts of periodic data series.

One of the most common types of periodic behavior is *seasonal* variation. This is why we use the letter s to stand for the length of periodicity. In this chapter we focus on seasonal models, but everything said here also applies to other types of periodic models.

ARIMA models for seasonal time series are built using the same iterative modeling procedure used for nonseasonal data: identification, estimation, and diagnostic checking. With seasonal data we must often difference the observations by length s. This involves calculating the periodic differences $z_t - z_{t-s}$. We also give special attention to estimated autocorrelation and partial autocorrelation coefficients at multiples of lag s, $(s, 2s, 3s, \ldots)$. Likewise, at the estimation stage we obtain estimates of selected AR and MA coefficients appearing at multiples of lag s. And at the diagnostic-checking stage we focus on residual autocorrelation coefficients at multiples of lag s.

This attention to coefficients at multiples of lag s is *in addition to* our usual concern about nonseasonal patterns in the data. This is why analyzing seasonal series is so challenging—most seasonal series also have a nonseasonal pattern. Distinguishing these two patterns to achieve a parsimonious and statistically adequate representation of a realization can be difficult, especially for the beginning analyst.

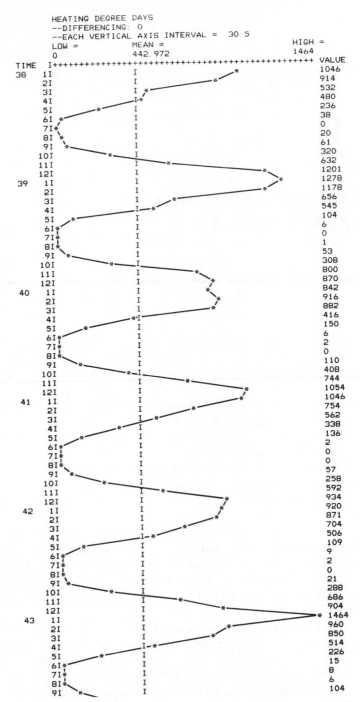

Figure 11.1 Heating degree days, Columbus, Ohio, 1938–1949.

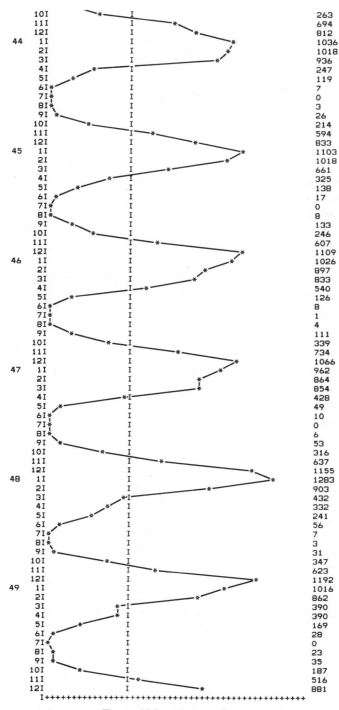

Figure 11.1 (*Continued*)

In this chapter we first discuss the nature of periodic and seasonal data. Next, we examine the theoretical acf's and pacf's associated with seasonal models. Then we consider seasonal differencing. We conclude with a critically important topic—how to build models for data series that have both seasonal and nonseasonal ARIMA patterns.

11.1 Periodic data

As an example of a periodic series, consider the plot of monthly heating degree days (abbreviated HDD) for Columbus, Ohio shown in Figure 11.1. The winter months' values are regularly higher than those in other months within the same year, while the summer months' values are regularly lower. This suggests that HDD values in any given month are similar to HDD values in the corresponding month in other years; that is, the January value in one year is similar to January values in other years, the July value in one year is similar to the July values in other years, and so forth for each month.

In any periodic series we expect observations separated by multiples of s to be similar: z_t should be similar to $z_{t \pm i(s)}$, where $i = 1, 2, 3, \ldots$. In the case of the monthly HDD data, one time period is one-twelfth of a year. This gives a pattern that repeats every 12 observations, so $s = 12$. Therefore, we expect HDD in a given month (z_t) to be related to HDD in the same month one year earlier (z_{t-12}), the same month one year later (z_{t+12}), the same month two years earlier (z_{t-24}), the same month two years later (z_{t+24}), and so forth.

The frequency with which data are recorded determines the value assigned to s, the length of the periodic interval. The monthly HDD data show a similarity between observations twelve periods apart, so $s = 12$. But if the data were recorded quarterly, we would expect a given quarter's value to be similar to values in the same quarter in other years. Thus similar observations would be four periods apart and we would have $s = 4$.

Figure 11.2 is another example of periodic data. It shows the number of students passing through a turnstile as they enter a university library. The building is open seven days a week, and the data in Figure 11.2 display a weekly (seven-day) periodicity. The first observation in each week is a Monday; the data reach a peak value near the middle of each week and drop to a low each Saturday. In this example observations seven periods apart are similar, so $s = 7$. Therefore, we expect z_t to be related to $z_{t \pm i(7)}$ for $i = 1, 2, 3, \ldots$.

Seasonal data. The most common type of periodic data in economics and business is data with seasonal variation, meaning variation within a year. The HDD series in Figure 11.1 is an example of a seasonal series—the

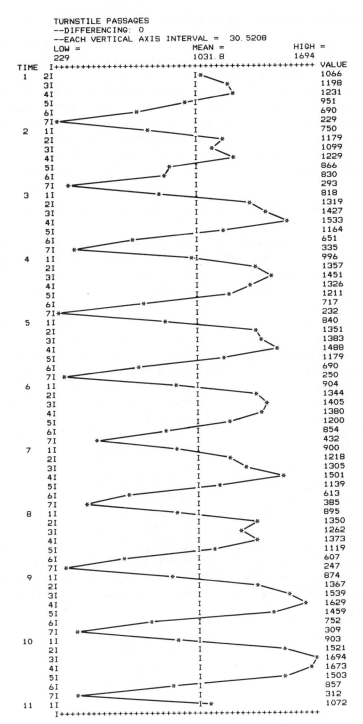

Figure 11.2 Library turnstile passages.

within-year pattern is similar from year to year. The turnstile data are periodic but not seasonal; that is, the repeating patterns in this series occur from week to week rather than from year to year.

Seasonal patterns reflect physical forces, such as changes in the weather, or institutional factors, such as social customs. For example, people may buy more ice cream as the temperature rises during the summer but buy less during the winter as the temperature drops. Other data such as liquor sales show repeated peaks (high values) and troughs (low values) partly because an extended holiday season comes in the late fall and early winter.

11.2   Theoretical acf's and pacf's for seasonal processes

Theoretical and estimated acf's and pacf's play the same role in the construction of seasonal ARIMA models as in the building of nonseasonal models. At the identification stage estimated acf's and pacf's are calculated from the available data. These are compared with some common, known theoretical acf's and pacf's and a tentative model is chosen based on this comparison. The parameters of this model are estimated and the estimation-stage residuals ($\hat{a}_t$) are then analyzed with a residual acf to see if they are consistent with the hypothesis that the random shocks (a_t) are independent. If we reject this hypothesis, the structure within the residual acf may help us tentatively identify another model.

The fundamental fact about seasonal time-series data is that observations s time periods apart ($z_t, z_{t-s}, z_{t+s}, z_{t-2s}, z_{t+2s}, \ldots$) are similar. We therefore expect observations s periods apart to be correlated. Thus, acf's and pacf's for seasonal series should have nonzero coefficients at one or more multiples of lag s ($s, 2s, 3s, \ldots$).

In Chapter 6 we discussed the theoretical acf's and pacf's for five common (stationary) nonseasonal models: AR(1), AR(2), MA(1), MA(2), and ARMA(1, 1). The ideas presented there carry over to the analysis of seasonal data with one exception: the coefficients appearing at lags $1, 2, 3, \ldots$ in *non*seasonal acf's and pacf's appear at lags $s, 2s, 3s, \ldots$ in purely seasonal acf's and pacf's. For example, a stationary nonseasonal AR(1) process with $\phi_1 = 0.7$ has a theoretical acf that decays exponentially in this manner (where k is the lag length and ρ_k represents the autocorrelation coefficient):

k	ρ_k
1	$\rho_1 = 0.7$
2	$\rho_2 = 0.49$
3	$\rho_3 = 0.34$
$\vdots$	$\vdots$

A stationary seasonal process with one seasonal AR coefficient and with $s = 4$, for example, also has a theoretical acf that decays exponentially, but at the *seasonal lags* $(4, 8, 12, \dots)$ which are multiples of 4:

k	ρ_k
1	$\rho_1 = 0$
2	$\rho_2 = 0$
3	$\rho_3 = 0$
4	$\rho_4 = 0.7$
5	$\rho_5 = 0$
6	$\rho_6 = 0$
7	$\rho_7 = 0$
8	$\rho_8 = 0.49$
9	$\rho_9 = 0$
10	$\rho_{10} = 0$
11	$\rho_{11} = 0$
12	$\rho_{12} = 0.34$
$\vdots$	$\vdots$

This parallel between nonseasonal and seasonal acf's and pacf's simplifies the analysis of seasonal data. The reader who is thoroughly familiar with the nonseasonal acf's and pacf's in Chapter 6 should be able to picture the same patterns occurring at multiples of lag s. Because of the similarity between nonseasonal and purely seasonal acf's and pacf's, we examine here only two of the more common seasonal processes.

Consider a purely seasonal process with one autoregressive coefficient at lag s. This is written

$$z_t = C + \Phi_s z_{t-s} + a_t$$

or

$$(1 - \Phi_s B^s)\tilde{z}_t = a_t \tag{11.1}$$

(Upper-case greek letters are used for seasonal coefficients.) Equation (11.1) says that z_t is related to its own past value s periods earlier, z_{t-s}.

A purely seasonal moving-average process with one coefficient at lag s is written

$$z_t = C - \Theta_s a_{t-s} + a_t$$

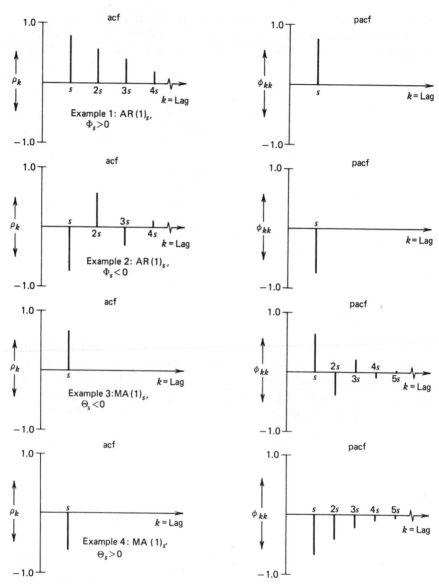

Figure 11.3 Theoretical acf's and pacf's for four stationary seasonal processes.

or

$$\tilde{z}_t = (1 - \Theta_s B^s) a_t \qquad (11.2)$$

Here z_t is related to the random shock s periods earlier, a_{t-s}.

Figure (11.3) shows the theoretical acf's and pacf's for these two purely seasonal processes under various assumptions about the signs of Φ_s and Θ_s. These diagrams are identical to the nonseasonal AR(1) and MA(1) acf's and pacf's except the coefficients for the seasonal processes occur at multiples of lag s $(s, 2s, 3s, \dots)$ instead of at lags $1, 2, 3, \dots$. The theoretical acf for process (11.1) decays exponentially at lags $s, 2s, 3s, \dots$ either all on the positive side or alternating in sign starting from the negative side. The theoretical acf for process (11.2) has a spike at lag s followed by a cutoff to zero at lags $2s, 3s, \dots,$.

In practice, identifying seasonal models from estimated acf's and pacf's can be more difficult than is suggested by the preceding discussion. In particular, many realizations with seasonal variation also contain nonseasonal patterns. The estimated acf and pacf for a combined seasonal–nonseasonal realization reflect both of these elements. Visually separating the seasonal and nonseasonal parts in estimated acf's and pacf's can be difficult. We consider combined seasonal–nonseasonal models in Sections 11.4–11.6.

A seasonal process with a nonstationary mean has an acf similar to the acf for a nonstationary, nonseasonal process. In Chapters 2, 6, and 7 we saw that the acf for a process with a nonstationary mean fails to damp out quickly to zero. A seasonal process with a nonstationary mean has acf spikes at lags $s, 2s, 3s, \dots$ that do not damp out rapidly to zero. Figure 11.4 shows a hypothetical example. These autocorrelations need not be large to

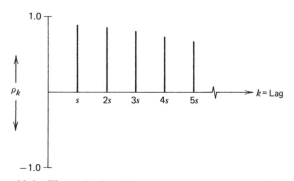

Figure 11.4   Theoretical acf for a nonstationary seasonal process.

indicate a nonstationary mean. The key point is that they do not quickly damp out to zero. When a realization produces an estimated acf similar to the one in Figure 11.4, seasonal differencing is warranted.

11.3 Seasonal differencing

The mean of a realization may shift significantly from period to period because of strong seasonal variation. Nevertheless, the observations for a *given* season may all fluctuate around a constant mean.

Seasonal differencing is similar to regular differencing (introduced in Chapter 2) because both involve calculating the changes in a data series. For regular differencing we calculate the period-to-period changes $z_t - z_{t-1}$. But to perform seasonal differencing, we calculate the change from the last corresponding season $z_t - z_{t-s}$.

For example, consider the HDD data in Figure 11.1. The mean of the series seems to shift *within* each year: while January values tend to lie above those for other months, July values are regularly lower than those for most other months. It is as if the January values are drawn from one probability distribution with a certain mean, February values are drawn from another probability distribution with a different mean, and so forth for other months. The estimated acf for the HDD data in Figure 11.5 confirms the nonstationary character of the seasonal variation in those data: the autocorrelations at the seasonal lags (12, 24, 36) decay slowly. When the mean of a realization shifts according to a seasonal pattern, *seasonal differencing* often induces a constant mean.

To find the seasonal differences (w_t) of the HDD data, subtract from each observation the observation occurring 12 periods earlier:

$$w_{13} = z_{13} - z_1 = 1278 - 1046 = 232$$

$$w_{14} = z_{14} - z_2 = 1178 - 914 = 264$$

$$w_{15} = z_{15} - z_3 = 656 - 532 = 124$$

$$w_{16} = z_{16} - z_4 = 545 - 480 = 65$$

$$w_{17} = z_{17} - z_5 = 104 - 236 = -132$$

$$w_{18} = z_{18} - z_6 = 6 - 38 = -32$$

$$\vdots \qquad \vdots \qquad \vdots$$

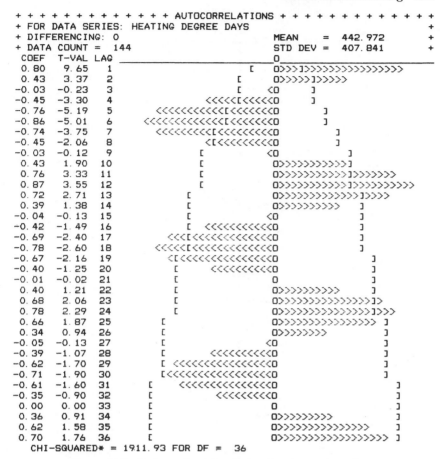

Figure 11.5 Estimated acf for the heating-degree-days data.

The first seasonal difference we can find (w_{13}) is for January 1939. This is because there is no z_0 value available to subtract from z_{12} (December 1938), no z_{-1} value to subtract from z_{11} (November 1938), and so forth. Therefore, we lose 12 observations due to seasonal differencing of length 12.

The seasonally differenced data are plotted in Figure 11.6 and the estimated acf and pacf for this series are shown in Figure 11.7. Inspection of Figure 11.6 suggests that seasonal differencing has removed the obvious peak–trough seasonal variation appearing in the original data in Figure 11.1. The estimated acf in Figure 11.7 now drops quickly to small values at lags 24 and 36 following the spike at lag 12. Since the estimated acf also

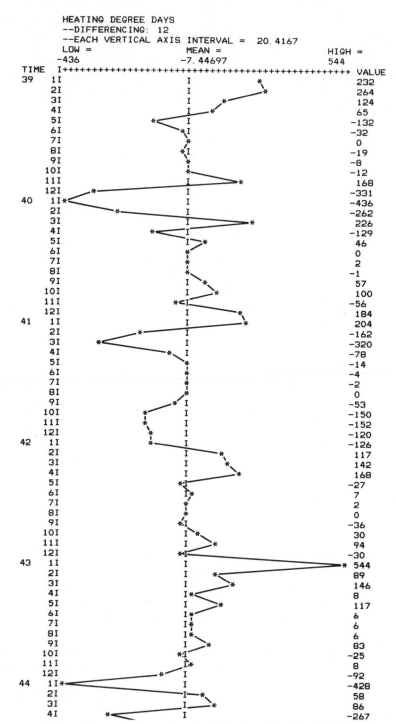

Figure 11.6 Seasonal differences of the heating-degree-days data.

276

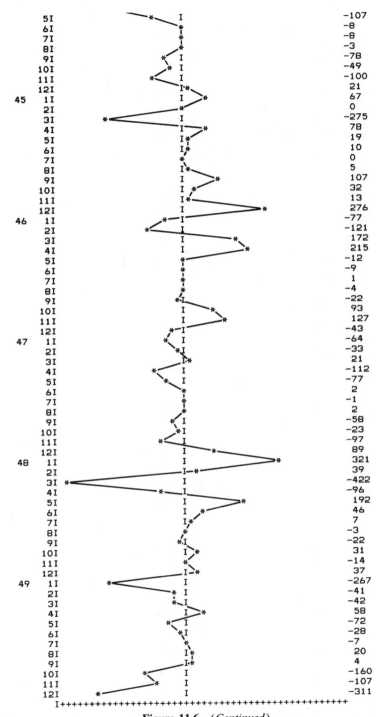

Figure 11.6 (*Continued*)

277

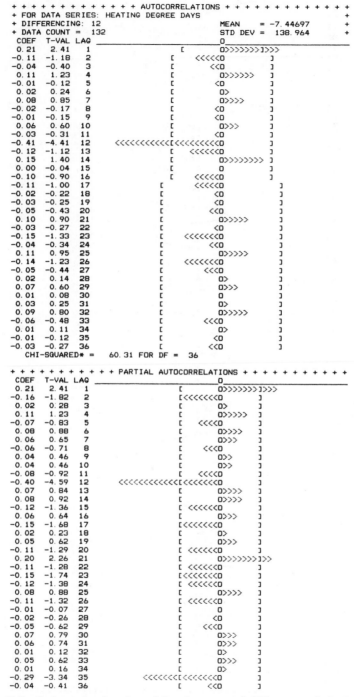

Figure 11.7 Estimated acf and pacf for the seasonal differences of the heating-degree-days data.

moves quickly to statistically insignificant values at the shorter nonseasonal lags, we conclude that the seasonally differenced series is stationary.

The seasonal differencing performed with the HDD series illustrates *first-degree* seasonal differencing which is seasonal differencing performed once. Letting D stand for the degree of seasonal differencing, in this case $D = 1$. We could also calculate the seasonal differences of the seasonal first differences $(w_t - w_{t-12}) - (w_{t-12} - w_{t-24})$. This is called *second-degree* seasonal differencing ($D = 2$). In practice, second-degree (or higher) seasonal differencing is virtually never needed.

The backshift operator B and the differencing operator ∇ introduced in Chapter 5 are useful for writing compact expressions to show the degree of seasonal differencing. Recall that B is defined such that $z_t B^k = z_{t-k}$. Letting $k = s$, the Dth difference (length s) of the series z_t can be written

$$\nabla_s^D z_t = (1 - B^s)^D z_t \qquad (11.3)$$

For example, let $D = 1$. Then the seasonal first differences of z_t are written $\nabla_s z_t = (1 - B^s)z_t$. To show that this is the expression for seasonal first differences, expand this expression and apply the definition of B^k where $k = s$:

$$\nabla_s z_t = (1 - B^s)z_t$$

$$= z_t - z_t B^s$$

$$= z_t - z_{t-s} \qquad (11.4)$$

We get the same result whether we apply the operator $(1 - B^s)$ to z_t or $\tilde{z}_t$ (the deviations of z_t from its mean μ) because the μ terms add to zero:

$$\nabla_s \tilde{z}_t = (1 - B^s)(z_t - \mu)$$

$$= z_t - z_t B^s - \mu + \mu B^s$$

$$= z_t - z_{t-s} - \mu + \mu$$

$$= z_t - z_{t-s} \qquad (11.5)$$

As discussed in Chapter 7, having the μ terms add to zero after differencing is the algebraic counterpart to the fact that a differenced realization (w_t) often has a mean ($\hat{\mu}_w$) that is statistically zero. If $\hat{\mu}_w$ is significantly different from zero, implying that the true mean μ_w is nonzero, we may include a deterministic trend in a model for w_t by expressing w_t in deviations from its mean: $w_t - \hat{\mu}_w$. (See Section 7.3 for a review of the topic of deterministic trends.) Case 15 in Part II is an example of a series that contains a deterministic trend after seasonal differencing.

11.4   Seasonal–nonseasonal multiplicative models

Many realizations contain both seasonal and nonseasonal patterns. (In the remainder of this chapter we use the letters S–NS to stand for "seasonal–nonseasonal.") Building ARIMA models for S–NS realizations is challenging because the estimated acf's and pacf's reflect both the seasonal and nonseasonal elements. We must attempt to separate these two parts visually and mentally.

Box and Jenkins [1] suggest a certain model type as a useful starting place for representing S–NS realizations. In this type of model the seasonal and nonseasonal elements are multiplied by each other.

To arrive at the form of these multiplicative S–NS models, consider our two earlier seasonal processes, equations (11.1) and (11.2). In a more general seasonal process we allow for seasonal differencing and for any number of AR and MA seasonal terms:

$$
\left(1 - \Phi_s B^s - \Phi_{2s} B^{2s} - \cdots - \Phi_{Ps} B^{Ps}\right)\left(1 - B^s\right)^D \tilde{z}_t
$$
$$
= \left(1 - \Theta_s B^s - \Theta_{2s} B^{2s} - \cdots - \Theta_{Qs} B^{Qs}\right) a_t
$$

(11.6)

Now write the AR seasonal operator $(1 - \Phi_s B^s - \Phi_{2s} B^{2s} - \cdots - \Phi_{Ps} B^{Ps})$ compactly as $\Phi_P(B^s)$, the seasonal differencing operator as ∇_s^D, and the MA seasonal operator $(1 - \Theta_s B^s - \Theta_{2s} B^{2s} - \cdots - \Theta_{Qs} B^{Qs})$ as $\Theta_Q(B^s)$. Substituting these terms into (11.6) gives

$$
\Phi_P(B^s)\nabla_s^D \tilde{z}_t = \Theta_Q(B^s) a_t
$$
(11.7)

We have used the symbol a_t to stand for a set of uncorrelated random shocks. Now if (11.7) represents only the seasonal part of a process that also contains nonseasonal patterns, then the shocks in (11.7) are autocorrelated and we should not represent them with the a_t term. Let b_t stand for a set of (nonseasonally) autocorrelated shocks whose behavior is not explained by (11.7). Then (11.7) becomes

$$
\Phi_P(B^s)\nabla_s^D \tilde{z}_t = \Theta_Q(B^s) b_t
$$
(11.8)

If the behavior of the autocorrelated shocks b_t is described by a nonseasonal ARIMA model, we may write a general model for b_t as

$$\phi_p(B)\nabla^d b_t = \theta_q(B)a_t \qquad (11.9)$$

where the random shocks a_t are not autocorrelated. Solving (11.9) for b_t gives $b_t = [\phi_p(B)\nabla^d]^{-1}\theta_q(B)a_t$. Substitute this expression into (11.8) and rearrange terms to get

$$\phi_p(B)\Phi_P(B^s)\nabla^d\nabla_s^D\tilde{z}_t = \Theta_Q(B^s)\theta_q(B)a_t \qquad (11.10)$$

where $\phi_p(B)$ is the nonseasonal AR operator, $\theta_q(B)$ is the nonseasonal MA operator, $\Phi_P(B^s)$ is the seasonal AR operator, $\Theta_Q(B^s)$ is the seasonal MA operator, and $\nabla^d\nabla_s^D$ are the differencing operators.

Thus (11.8) and (11.9) together imply (11.10) for z_t, where z_t has both seasonal and nonseasonal components, and where z_t is differenced d times (length one) and D times (length s). Furthermore, the seasonal and nonseasonal AR elements are multiplied by each other as are the seasonal and nonseasonal MA elements.

Process (11.10) is referred to as an ARIMA(p, d, q)(P, D, Q)$_s$ process. The lower-case letters (p, d, q) indicate the nonseasonal orders and the upper-case letters (P, D, Q) denote the seasonal orders of the process. The parentheses mean that the seasonal and nonseasonal elements are multiplied as shown in (11.10).

As an example consider an ARIMA$(0, 0, 1)(0, 1, 1)_4$ process. Realizations generated by this process have a pattern with a periodicity of four, since $s = 4$. Because $D = 1$, z_t is differenced once by length four. With $d = 0$ there is no nonseasonal differencing. There is one seasonal MA term at lag 4 ($Q = 1$) and one nonseasonal MA term at lag 1 ($q = 1$). Furthermore, the two MA operators are multiplied by each other. In backshift form this model is

$$(1 - B^4)\tilde{z}_t = (1 - \Theta_4 B^4)(1 - \theta_1 B)a_t \qquad (11.11)$$

As another example consider an ARIMA$(1, 0, 0)(1, 0, 1)_{12}$. This process would produce realizations with a periodicity of length 12, since $s = 12$. In this case z_t is not differenced at all because $d = 0$ and $D = 0$. The seasonal part of the process is mixed ($P = 1$ and $Q = 1$) with one AR and one MA coefficient at lag 12. Then there is a nonseasonal AR term at lag 1 ($p = 1$). The two AR operators are multiplied, giving this model in backshift form:

$$(1 - \phi_1 B)(1 - \Phi_{12}B^{12})\tilde{z}_t = (1 - \Theta_{12}B^{12})a_t \qquad (11.12)$$

11.5   An example of a seasonal–nonseasonal multiplicative model

In this section we go through the full cycle of identification, estimation, and diagnostic checking using the HDD realization in Figure 11.1. We will also discuss the stationarity and invertibility conditions for seasonal models and present the forecast profile for our HDD model.

Identification.   The estimated acf for the HDD realization appears in Figure 11.5. The autocorrelations at the seasonal lags (12, 24, 36) fail to die out quickly. This confirms the nonstationary character of the seasonal pattern and calls for seasonal differencing.

Note that the autocorrelations at the seasonal lags in Figure 11.5 are surrounded by other large autocorrelations (especially lags 10, 11, 13, 23, and 25). This is not unusual when a strong seasonal pattern is present. It is wise to ignore these surrounding values for the time being. Most of them usually disappear after the seasonal component is properly modeled. In this example seasonal differencing is sufficient to remove all these large surrounding values as shown by the estimated acf for the seasonally differenced series (Figure 11.7).

Note also in Figure 11.5 that there are other large autocorrelations gathering around the half-seasonal lags (6, 18, 30). Strong seasonal variation can sometimes produce large (and misleading) autocorrelations at fractional multiples of the seasonal lag. These values can be misleading because they often become statistically insignificant after the realization is differenced by length s, or (in the residual acf) when AR or MA coefficients are estimated at the seasonal lags. The estimated acf in Figure 11.7 for the seasonally differenced data shows that differencing clears up the waves of significant values surrounding the half-seasonal lags. (See Case 12 in Part II for an example where seasonality induces large autocorrelations not only at the half-seasonal lags, but also at the quarter-seasonal lags.)

Now focus on the estimated acf and pacf in Figure 11.7. Seasonal differencing has created a stationary series since the estimated acf falls quickly to zero at both the short lags (1, 2, 3) and the seasonal lags (12, 24, 36). We are now ready to identify a tentative model.

Consider the significant spike at lag 12. It is followed by a cutoff to very small values at lags 24 and 36. According to the theoretical acf's in Figure 11.3, this calls for a seasonal MA term and we expect $\hat{\Theta}_{12}$ to be positive. The decaying pattern at lags 12, 24, and 36 in the estimated pacf (Figure 11.7) confirms that an MA term is appropriate at the seasonal lag.

Now focus on the nonseasonal pattern in Figure 11.7. We see a spike at lag 1 in the acf followed immediately by a cutoff to insignificant values. This indicates an MA(1) for the nonseasonal part of our model. The alternating decay in the pacf at lags 1, 2, and 3 confirms this nonseasonal MA(1) pattern.

This analysis leads us to tentatively choose an ARIMA$(0, 0, 1)(0, 1, 1)_{12}$ model:

$$(1 - B)^{12} \ddot{z}_t = (1 - \Theta_{12} B^{12})(1 - \theta_1 B) a_t \qquad (11.13)$$

Estimation.   The estimation of purely seasonal or S–NS models is not fundamentally different from the estimation of nonseasonal models as discussed in Chapter 8. The parameters of seasonal models are usually estimated according to a least-squares criterion using a nonlinear routine such as Marquardt's compromise (see Appendix 8A). However, the computational burden is often greater for models with seasonal components since there are more parameters to be estimated. This computational burden is increased further when the technique of backcasting is employed, a technique that is important when seasonal elements are present. (See Appendix 8B.)

The estimation results for model (11.13) in Figure 11.8 show that both $\hat{\Theta}_{12}$ and $\hat{\theta}_1$ are significant with absolute t-values greater than 2.0. The adjusted RMSE indicates that the standard deviation of the residuals is about 104 heating degree days. Since the estimated coefficients are not highly correlated (their correlation is only -0.02) we can be confident that they are not unstable.

Diagnostic checking.   A statistically adequate model satisfies the assumption that the random shocks are independent. We cannot observe the random shocks a_t but we have estimation residuals $\hat{a}_t$ from model (11.13) which are estimates of the random shocks. If the residuals are independent we accept the hypothesis that the shocks are independent.

The residual acf is used to test the hypothesis that the shocks are independent. The residual acf for (11.13) is shown at the bottom of Figure 11.8. The practical warning levels for residual autocorrelations are (i) about 1.25 at the short lags (1, 2, and perhaps 3); (ii) 1.25 at seasonal lags (12, 24, and 36 in this case since $s = 12$); and (iii) 1.6 elsewhere.

Only the residual autocorrelations at lags 3, 15, and 23 exceed these warning levels. While we could pursue a more complex model by including terms at these lags, out of 36 autocorrelations we expect a few to be

```
+ + + + + + + + +ECOSTAT UNIVARIATE B-J RESULTS+ + + + + + + + +  +
+ FOR DATA SERIES:   HEATING DEGREE DAYS                                +
+ DIFFERENCING:      12                              DF   = 130         +
+ AVAILABLE:              DATA = 132    BACKCASTS = 13   TOTAL = 145     +
+ USED TO FIND SSR: DATA = 132    BACKCASTS = 13   TOTAL = 145          +
+ (LOST DUE TO PRESENCE OF AUTOREGRESSIVE TERMS:               0)       +

COEFFICIENT    ESTIMATE      STD ERROR      T-VALUE
THETA   1       -0.233         0.083         -2.80
THETA* 12        0.932         0.037         25.40

   ADJUSTED RMSE =   104.367    MEAN ABS % ERR =   55.53
          CORRELATIONS
          1      2
   1   1.00
   2  -0.02   1.00


++RESIDUAL ACF++
 COEF   T-VAL LAG _____O_____
-0.04  -0.51   1        [             <<<<O                    ]
-0.05  -0.62   2        [             <<<<O                    ]
-0.15  -1.66   3        [    <<<<<<<<<<<<<<<O                  ]
 0.12   1.30   4        [                   O>>>>>>>>>>>>     ]
-0.08  -0.87   5        [            <<<<<<<<O                 ]
 0.05   0.52   6        [                   O>>>>>             ]
 0.04   0.45   7        [                   O>>>>              ]
 0.03   0.34   8        [                   O>>>               ]
-0.04  -0.45   9        [             <<<<O                    ]
 0.04   0.42  10        [                   O>>>>              ]
-0.06  -0.64  11        [            <<<<<<O                   ]
 0.09   1.01  12        [                   O>>>>>>>>>         ]
-0.03  -0.29  13        [               <<<O                   ]
 0.05   0.58  14        [                   O>>>>>             ]
-0.15  -1.63  15        [    <<<<<<<<<<<<<<<O                  ]
-0.01  -0.08  16        [                  <O                  ]
-0.11  -1.21  17        [       <<<<<<<<<<<<O                  ]
 0.01   0.09  18    [                      O>                      ]
 0.00  -0.01  19    [                      O                       ]
-0.01  -0.11  20    [                     <O                       ]
 0.06   0.65  21    [                      O>>>>>                  ]
 0.01   0.07  22    [                      O>                      ]
-0.21  -2.16  23  <[<<<<<<<<<<<<<<<<<<<<<<O                        ]
-0.04  -0.38  24    [                  <<<<O                       ]
 0.09   0.90  25    [                      O>>>>>>>>>             ]
-0.05  -0.48  26    [                  <<<<O                       ]
-0.09  -0.93  27    [              <<<<<<<<<O                      ]
 0.03   0.26  28    [                      O>>>                    ]
 0.03   0.29  29    [                      O>>>                    ]
-0.01  -0.14  30    [                     <O                       ]
-0.02  -0.21  31    [                    <<O                       ]
 0.06   0.59  32    [                      O>>>>>>                 ]
-0.03  -0.32  33    [                   <<<O                       ]
 0.07   0.66  34    [                      O>>>>>>>               ]
-0.03  -0.32  35    [                   <<<O                       ]
 0.00  -0.01  36    [                      O                       ]
   CHI-SQUARED* =    28.40 FOR DF =   34
```

Figure 11.8 Estimation and diagnostic-checking results for model (11.3).

significant just by chance. While some analysts might try alternative models and track their forecasting records, we can be satisfied that model (11.13) fits the available data adequately. This is confirmed by the chi-squared statistic printed below the residual acf. It is smaller than the critical value (40.3) for 30 degrees of freedom (df) at the 10% level, and thus it is also insignificant for df = 34.

Stationarity and invertibility conditions. One advantage of the multiplicative form is that it simplifies the checking of stationarity and invertibility conditions. With a multiplicative model these conditions apply *separately* to the seasonal and nonseasonal coefficients. For example, consider an ARIMA$(2, 0, 1)(1, 0, 2)_s$. In backshift form this is

$$\left(1 - \phi_1 B - \phi_2 B^2\right)\left(1 - \Phi_s B^s\right)\tilde{z}_t = \left(1 - \Theta_s B^s - \Theta_{2s} B^{2s}\right)\left(1 - \theta_1 B\right)a_t$$

$$(11.14)$$

The stationarity requirement applies only to the AR coefficients and we treat the nonseasonal and seasonal AR components separately since they are multiplied.* Thus the stationarity conditions for the nonseasonal part of (11.14) are the same as for an AR(2) as discussed in Chapter 6: $|\phi_2| < 1$, $\phi_2 - \phi_1 < 1$, and $\phi_2 + \phi_1 < 1$.

Then there is a separate stationarity condition on the AR seasonal part of (11.14): it is the same condition as for a nonseasonal AR(1) model discussed in Chapter 6, except in this case we have an AR(1)$_s$ component; thus the condition is $|\Phi_s| < 1$.

Invertibility applies only to the MA part of (11.14) and we treat the nonseasonal and seasonal components separately. The condition on the nonseasonal part presented in Chapter 6 is $|\theta_1| < 1$. The conditions on the seasonal part are the same as for a nonseasonal MA(2) as discussed in Chapter 6, except we have an MA(2)$_s$ component. Thus the joint conditions are $|\Theta_{2s}| < 1$, $\Theta_{2s} - \Theta_s < 1$, and $\Theta_{2s} + \Theta_s < 1$.

Model (11.13) for the differenced HDD series is stationary since it contains no AR terms. The invertibility requirements are $|\Theta_{12}| < 1$ and $|\theta_1| < 1$. These are satisfied since $|\hat{\Theta}_{12}| = 0.932 < 1$ and $|\hat{\theta}_1| = 0.233 < 1$.

Forecasting. In Chapter 10 we showed how forecasts are produced from the difference-equation form of a model. Expanding (11.13) we get the

---

*The formal stationarity condition for a multiplicative model is that the roots of $\phi_p(B)\Phi_P(B^s)$ = 0 lie outside the unit circle. This is equivalent to the joint condition that the roots of both $\phi_p(B) = 0$ and $\Phi_P(B^s) = 0$ lie outside the unit circle.

difference equation

$$z_t = z_{t-12} - \theta_1 a_{t-1} - \Theta_{12} a_{t-12} + \theta_1 \Theta_{12} a_{t-13} + a_t \qquad (11.15)$$

Inserting the estimated values for θ_1 and Θ_{12} gives the forecast form

$$\hat{z}_t = z_{t-12} + 0.233\hat{a}_{t-1} - 0.932\hat{a}_{t-12} - 0.217\hat{a}_{t-13} \qquad (11.16)$$

where the $\hat{a}$ terms are estimation-stage residuals. Forecasts are generated in the same manner illustrated in Chapter 10.

Equation (11.16) shows how our model accounts for the nonstationary character of the seasonal pattern: each forecast $\hat{z}_t$ starts from the value 12

Table 11.1   Forecasts from equation (11.16).

Time		Forecast Values	80% Confidence Limits		Future Observed Values	Percent Forecast Errors
			Lower	Upper		
50	1	1053.2834	919.6939	1186.8729	904.0000	−16.51
	2	915.0873	777.9187	1052.2560	1147.0000	20.22
	3	631.0462	493.8775	768.2148	943.0000	33.08
	4	413.7132	276.5446	550.8819	392.0000	−5.54
	5	153.8229	16.6543	290.9916	218.0000	29.44
	6	19.4635	−117.7052	156.6321	23.0000	15.38
	7	1.3907	−135.7780	138.5593	7.0000	80.13
	8	9.4737	−127.6950	146.6423	0.0000	n.a.[a]
	9	59.9529	−77.2158	197.1215	112.0000	46.47
	10	270.8146	133.6460	407.9833	88.0000	−207.74
	11	626.1457	488.9771	763.3144	733.0000	14.58
	12	981.1140	843.9454	1118.2827	983.0000	0.19
51	1	1071.6088	934.1364	1209.0813	1324.0000	19.06
	2	915.0873	777.5983	1052.5763	939.0000	2.55
	3	631.0462	493.5572	768.5352	648.0000	2.62
	4	413.7132	276.2243	551.2022	306.0000	−35.20
	5	153.8229	16.3339	291.3119	155.0000	0.76
	6	19.4635	−118.0255	156.9524	3.0000	−548.78
	7	1.3907	−136.0983	138.8796	0.0000	n.a.
	8	9.4737	−128.0153	146.9627	2.0000	−373.68
	9	59.9529	−77.5361	197.4419	45.0000	−33.23
	10	270.8146	133.3257	408.3036	380.0000	28.73
	11	626.1457	488.6568	763.6347	510.0000	−22.77
	12	981.1140	843.6251	1118.6030	885.0000	−10.86

[a]n.a. = not available.

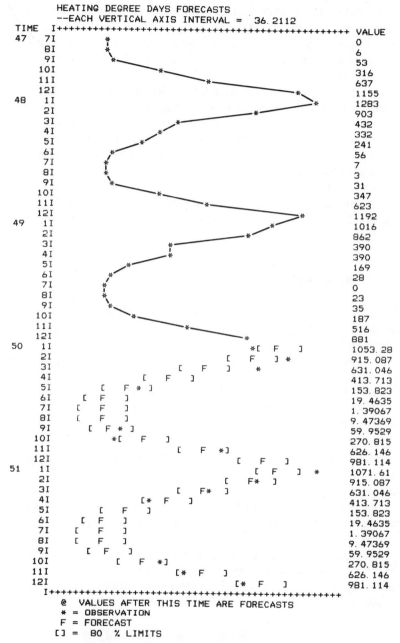

```
    HEATING DEGREE DAYS FORECASTS
    --EACH VERTICAL AXIS INTERVAL =  36.2112
TIME  I++++++++++++++++++++++++++++++++++++++++++++++++++  VALUE
47  7I        *                                              0
    8I        *-.                                            6
    9I         *                                             53
   10I             *                                         316
   11I                 *                                     637
   12I                         *                             1155
48  1I                           *                           1283
    2I                      *                                903
    3I              *                                        432
    4I            *                                          332
    5I          *                                            241
    6I       *                                               56
    7I      *                                                7
    8I      *-                                               3
    9I       *                                               31
   10I            *                                          347
   11I                 *                                     623
   12I                           *                           1192
49  1I                         *                             1016
    2I                      *                                862
    3I            *                                          390
    4I            .*                                         390
    5I         *                                             169
    6I       *                                               28
    7I      *                                                0
    8I      *                                                23
    9I       *                                               35
   10I         *                                             187
   11I              *                                        516
   12I                    *                                  881
50  1I                            *[    F    ]               1053.28
    2I                          [   F    ] *                 915.087
    3I                    [    F    ]    *                   631.046
    4I              [    F    ]                              413.713
    5I          [    F * ]                                   153.823
    6I      [  F    ]                                        19.4635
    7I      [   F    ]                                       1.39067
    8I      [   F    ]                                       9.47369
    9I       [  F * ]                                        59.9529
   10I          *[    F    ]                                 270.815
   11I                [   F  *]                              626.146
   12I                       [    F    ]                     981.114
51  1I                          [   F    ]  *               1071.61
    2I                        [   F*    ]                    915.087
    3I                 [    F*    ]                          631.046
    4I             [*  F    ]                                413.713
    5I         [    F    ]                                   153.823
    6I      [   F    ]                                       19.4635
    7I      [   F    ]                                       1.39067
    8I      [   F    ]                                       9.47369
    9I       [  F    ]                                       59.9529
   10I          [    F *]                                    270.815
   11I                [*  F    ]                             626.146
   12I                       [*  F    ]                      981.114
    I++++++++++++++++++++++++++++++++++++++++++++++++++
        @   VALUES AFTER THIS TIME ARE FORECASTS
        *  = OBSERVATION
        F  = FORECAST
        [] =   80  % LIMITS
```

Figure 11.9 Forecasts from equation (11.16).

287

periods earlier, z_{t-12}. Then a further adjustment is made for the remainder of the seasonal pattern represented by the term $-0.932\hat{a}_{t-12}$, for the period-to-period pattern represented by the term $0.233\hat{a}_{t-1}$, and for the interaction between the seasonal and nonseasonal pattern represented by the term $-0.217\hat{a}_{t-13}$.

Table 11.1 shows the forecasts for 24 months ahead along with 80% confidence limits and the future observed values. These results are plotted in Figure 11.9. These forecasts track the powerful seasonal pattern in this series rather well. The forecasts pick up the extreme seasonal undulations in the observed series, and only five of the 24 confidence intervals fail to contain the observed future values.

You should now be ready to read Cases 9–15 in Part II, all of which involve data containing seasonal variation. Consult the practical rules in Chapter 12 frequently as you study the cases and as you attempt to build your own UBJ–ARIMA models.

11.6   Nonmultiplicative models*

Multiplicative models are an excellent starting place for modeling S–NS realizations, and they often give good results. Most computer programs for estimating UBJ–ARIMA models provide at least an option for multiplicative seasonality; some simply assume that seasonal models are multiplicative.

Occasionally, the assumption that the seasonal and nonseasonal elements are multiplicative is inappropriate. In this section we discuss alternatives to multiplicative models.

Additive models.   Note that (11.15), the difference-equation form of (11.13), contains an implicit MA coefficient at lag $s + 1 = 13$, $\theta_{13} = \theta_1\Theta_{12}$. Other multiplicative models contain coefficients that are the product of seasonal and nonseasonal coefficients. The particular compound coefficients that occur depend on the orders of the seasonal and nonseasonal parts of the model.

Now suppose the implicit coefficient θ_{13} in (11.13) is not, in fact, equal to $\theta_1\Theta_{12}$ for a certain data series. Forcing θ_{13} to equal $\theta_1\Theta_{12}$ by using the multiplicative form (11.13) would tend to reduce forecast accuracy. It might be, instead, that θ_{13} is zero. Then the following additive model is ap-

---

*This section involves some complicated details and may be omitted at the first reading of this chapter. The reader is encouraged to first work through Cases 9–13 in Part II.

propriate.

$$(1 - B^{12})\tilde{z}_t = (1 - \theta_1 B - \Theta_{12}B^{12})a_t \tag{11.17}$$

Another possibility is that the parameter θ_{13} is neither zero nor equal to $\theta_1\Theta_{12}$. In this case θ_{13} could be estimated "freely." Rather than forcing $\theta_{13} = 0$ as in (11.17), or forcing $\theta_{13} = \theta_1\Theta_{12}$ as in (11.13), we could estimate this additive model:

$$(1 - B^{12})\tilde{z}_t = (1 - \theta_1 B - \Theta_{12}B^{12} - \theta_{13}B^{13})a_t \tag{11.18}$$

Theoretical acf's. The theoretical acf's for S–NS models can be quite complicated. But the S–NS models that occur commonly in practice are of relatively low order, and their theoretical acf's have distinguishing characteristics. In this section we illustrate how the theoretical acf's for multiplicative models differ from those of additive models by examining the theoretical acf's of models (11.13), (11.17), and (11.18).

Let w_t represent a realization that has been made stationary by suitable differencing and other appropriate transformations (such as a logarithmic transformation), $w_t = (1 - B)^d(1 - B^s)^D z_t'$. Suppose the stationary series w_t is generated by a multiplicative process, an ARMA(0, 1)(0, 1)$_s$ shown earlier as (11.13):

$$w_t = (1 - \Theta_s B^s)(1 - \theta_1 B)a_t \tag{11.19}$$

Process (11.19) has these theoretical autocorrelations:

$$\rho_1 = \frac{-\theta_1(1 + \Theta_s^2)}{(1 + \theta_1^2)(1 + \Theta_s^2)}$$

$$\rho_s = \frac{-\Theta_s(1 + \theta_1^2)}{(1 + \theta_1^2)(1 + \Theta_s^2)} \tag{11.20}$$

$$\rho_{s-1} = \frac{\theta_1\Theta_s}{(1 + \theta_1^2)(1 + \Theta_s^2)}$$

$$\rho_{s+1} = \rho_{s-1}$$

The most notable characteristic of these autocorrelations is that the spike at lag s is flanked by two nonzero autocorrelations of equal value ($\rho_{s+1} =$

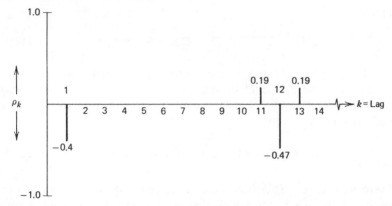

Figure 11.10   Theoretical acf for process (11.19) with $s = 12$, $\theta_1 = 0.5$, and $\Theta_{12} = 0.7$.

$\rho_{s-1} \neq 0$). In fact it can be deduced from (11.20) that $\rho_{s+1} = \rho_{s-1} = \rho_1 \rho_s$ for this process. Figure 11.10 shows an example of a theoretical acf for this process with $s = 12$, $\theta_1 = 0.5$, and $\Theta_{12} = 0.7$. The value of θ_{13} in this case is $\theta_{13} = \theta_1 \Theta_{12} = (0.5)(0.7) = 0.35$. Using (11.20), we find $\rho_1 = -0.4$, $\rho_s = -0.47$, and $\rho_{s-1} = \rho_{s+1} = \rho_1 \rho_s = 0.19$.

This example illustrates a common result for S–NS models: the seasonal autocorrelations are often surrounded by nonzero autocorrelations that are symmetrical (in estimated acf's, roughly symmetrical).

For comparison, consider the additive process appropriate when θ_{s+1} is equal to neither $\theta_1 \Theta_s$ nor zero:

$$w_t = \left(1 - \theta_1 B - \Theta_s B^s - \theta_{s+1} B^{s+1}\right) a_t \qquad (11.21)$$

The theoretical acf for this process differs from the acf of process (11.19) notably in that the two autocorrelations flanking the autocorrelation at lag s are no longer equal, although both are nonzero. Figure 11.11 shows an example of an acf for process (11.21) where $s = 12$, $\theta_1 = 0.5$, $\Theta_{12} = 0.7$, and $\theta_{13} = 0.6$.*

Finally, suppose $\theta_{s+1} = 0$ as in model (11.17). Then we have

$$w_t = \left(1 - \theta_1 B - \Theta_s B^s\right) a_t \qquad (11.22)$$

---

*The autocovariances for processes (11.21) and (11.22) are given in Box and Jenkins [1, p. 332].

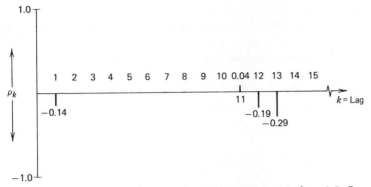

Figure 11.11   Theoretical acf for process (11.21) with $s = 12$, $\theta_1 = 0.5$, $\Theta_{12} = 0.7$, and $\theta_{13} = 0.6$.

In this case, unlike the previous two examples, $\rho_{s+1} = 0$ although ρ_{s-1} is not. Figure 11.12 is an example of a theoretical acf for (11.22) with $s = 12$, $\theta_1 = 0.5$, and $\Theta_{12} = 0.7$.

The preceding discussion suggests that an estimated acf can be helpful in deciding whether a multiplicative or nonmultiplicative model is called for. However, while the differences between the theoretical acf's in Figures 11.10, 11.11, and 11.12 are fairly clear, the evidence in estimated acf's is often ambiguous, especially when the sample size is modest. For this reason the beginning analyst is encouraged not to pay too much attention to estimated autocorrelations near the seasonal lags or at fractional seasonal

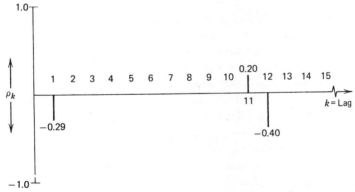

Figure 11.12   Theoretical acf for process (11.22) with $s = 12$, $\theta_1 = 0.5$, and $\Theta_{12} = 0.7$.

lags. The inexperienced modeler is likely to obtain more satisfactory results by focusing on the first few nonseasonal lags and on the obvious seasonal lags $(s, 2s, 3s)$ in the estimated acf, and by working with common multiplicative models. Nevertheless, bear in mind that multiplicative models are occasionally inappropriate, and experimentation with nonmultiplicative models may be called for. Case 14 in Part II is an example showing that a nonmultiplicative model is sometimes superior.

Summary

1. Realizations with periodic patterns (those repeating every s time period, where $s > 1$) can often be forecast well using ARIMA models. Such models are built with the same iterative three-stage procedure as nonseasonal models—identification, estimation, and diagnostic checking.

2. One of the most common types of periodic behavior is seasonal variation. A seasonal pattern occurs within a year, and each season shows behavior similar to the corresponding season in other years. Thus a given observation z_t may be related to other observations that are $i(s)$ periods $(i = 1, 2, 3, \ldots)$ into the past or future $(z_{t-s}, z_{t+s}, z_{t-2s}, z_{t+2s}, \ldots)$.

3. In analyzing seasonal data, we focus on autocorrelation coefficients, partial autocorrelation coefficients, and AR and MA coefficients at multiples of lag s.

4. Seasonal data must sometimes be differenced by length s $(z_t - z_{t-s})$ to achieve a stationary mean. This is because the observations from each season may gather around a mean that is different from the means of other seasons.

5. Theoretical acf's and pacf's for purely seasonal processes are identical to those for their nonseasonal counterparts except the autocorrelations appear at lags that are multiples of s. For example, an MA(1) has a spike at lag 1 while a seasonal process with one MA coefficient at lag s has a spike at lag s. A stationary AR(1) has decaying autocorrelations at lags $1, 2, 3, \ldots$, while a seasonal process with one AR coefficient at lag s has decaying autocorrelations at lags $s, 2s, 3s, \ldots$.

6. A strong seasonal pattern may produce large estimated autocorrelations at lags surrounding the seasonal lags and at fractional multiples of the seasonal interval. For example, it is not unusual to find significant autocorrelations at lags $0.5s, 1.5s, 2.5s, \ldots$ in the initial estimated acf. It is best to ignore these autocorrelations in the early stages when building a model since they frequently become insignificant after seasonal differencing or (in the residual acf) after AR or MA coefficients are estimated at the seasonal lag.

7. When a realization contains both seasonal and nonseasonal patterns, it is wise to start with a multiplicative model. This means the seasonal AR operator is multiplied by the nonseasonal AR operator, and the seasonal MA operator is multiplied by the nonseasonal MA operator. The general form of such models is

$$\phi_p(B)\Phi_P(B^s)\nabla^d\nabla_s^D\tilde{z}_t = \Theta_Q(B^s)\theta_q(B)a_t$$

8. Multiplicative models may be written in ARIMA(p, d, q)(P, D, Q)$_s$ notation, where s is the length of the periodic interval; P is the order of the seasonal AR portion of the model; D is the number of times the series is differenced by length s to achieve a stationary mean; and Q is the order of the seasonal MA portion of the model.

9. Identification of models with both seasonal and nonseasonal elements can be difficult, since the two patterns are mixed together in estimated acf's and pacf's.

10. Stationarity and invertibility conditions for multiplicative seasonal models apply separately to the seasonal and nonseasonal components. For example, the stationarity requirements for an ARIMA(1, 0, 0)(1, 0, 0)$_s$ model are $|\phi_1| < 1$ and $|\Phi_s| < 1$.

11. The practical warning level for absolute t-values at seasonal lags in estimated (including residual) acf's is about 1.25. When a seasonal-lag autocorrelation coefficient has an absolute t-value greater than 1.25, we should consider including a seasonal AR or MA coefficient in our model.

12. Occasionally an additive seasonal model is more satisfactory than a multiplicative model. In these cases the implied multiplicative coefficients (those appearing at lags that are products of the seasonal and nonseasonal lags) may be set equal to zero or estimated freely as additive terms.

Questions and Problems

11.1 A police reporter records the number of crimes committed each hour for 240 hours. What length periodicity would you expect to see in these data?

11.2 Explain the logic behind seasonal differencing. When is seasonal differencing performed?

11.3 Calculate the seasonal first differences for the data in Chapter 1, problem 1.1.

11.4 Write the following in backshift form:
- **(a)** ARIMA$(0, 1, 1)(0, 1, 1)_s$
- **(b)** ARIMA$(1, 1, 0)(0, 1, 1)_4$
- **(c)** ARIMA$(2, 0, 0)(0, 0, 1)_{12}$
- **(d)** ARIMA$(0, 2, 2)(2, 1, 2)_{12}$
- **(e)** ARIMA$(0, 1, 0)(0, 1, 1)_7$

11.5 Construct the theoretical acf's for these processes:
- **(a)** $\tilde{z}_t = (1 - 0.8B^{12})(1 + 0.6B)a_t$
- **(b)** $\tilde{z}_t = (1 - 0.9B^4)(1 - 0.7B)a_t$

11.6 What are the numerical values of the implicit coefficients at lag $s + 1$ for the two processes in question 11.5?

11.7 Which of the following estimated models are stationary and invertible? Explain.
- **(a)** $(1 - 0.8B)(1 - B^4)\tilde{z}_t = (1 - 0.8B^4)\hat{a}_t$
- **(b)** $w_t = (1 - 0.4B^{12} - 0.3B^{24})(1 - 0.5B^2)\hat{a}_t$
- **(c)** $(1 - 1.2B + 0.5B^2)(1 - 0.5B^{12})(1 - B)\tilde{z}_t = \hat{a}_t$

Part II
THE ART OF
ARIMA MODELING

12

PRACTICAL RULES

In this chapter we state some practical rules for building proper UBJ–ARIMA models with relative ease. Many of these rules have been stated in earlier chapters; this chapter is intended to be a compact and convenient summary for easy reference. You should be familiar with the essential concepts underlying UBJ–ARIMA models before reading this chapter.

Following this chapter are 15 case studies illustrating the use of the practical rules presented in this chapter and showing how to build UBJ–ARIMA models in a step-by-step manner. By studying these cases carefully and experimenting with several dozen data sets, you should be able to build most UBJ models within about 30 minutes using an interactive computer program.

Here are the practical rules for building UBJ–ARIMA models:

1. Forecasts from an ARIMA model are only as good as the accuracy of the data and the judgment of the analyst. All data should be checked for accuracy before any analysis is performed.

2. Ideally, one should have a minimum of about 50 observations to build an ARIMA model. This usually allows sufficient degrees of freedom for adequate identification and estimation even if one loses observations due to differencing. When seasonal variation is present, it is desirable to have more than 50 observations. Occasionally, some analysts may use less than 50 observations, but the results

must be used cautiously. (See Case 15 for an example of a seasonal model based on 42 observations.) With less than 50 observations the analyst should consider alternatives to an ARIMA model.

3. An important preliminary step is visual inspection of a plot of the original realization. Inspection of the data is most important in deciding if the variance of the realization is stationary. While there are some formal statistical tests available, Granger and Newbold [17] argue that informal inspection of the realization is as useful as any other procedure. The logarithmic transformation is the most common one for data in economics and business. This transformation is appropriate when the variance of a realization is proportional to the mean. (See Cases 9 and 11 for examples.)

4. Inspection of the realization may also help you form a preliminary opinion about whether the mean of the realization is stationary. If the mean of the series seems to change over time, look for confirmation in the estimated acf as discussed below.

5. Inspection of the realization may also help you form an initial impression about the presence of a seasonal pattern. There may be obvious seasonal variation (see Cases 9, 13, and 14), mild seasonality, or perhaps no seasonal pattern will be apparent. The final decision about including seasonal elements in a model must rest on autocorrelation analysis and estimation-stage results, but preliminary inspection of the data can be a helpful supplement. (See Cases 10 and 11 where seasonality is detected only with autocorrelation analysis.)

6. The number of useful estimated autocorrelations is about $n/4$, that is, about one-fourth of the number of observations.

7. The majority of the data series in economics and business show seasonal patterns. Even seasonally adjusted data may still show a seasonal pattern if the adjustment is insufficient or excessive.

8. There are three things to examine in deciding on the degree of differencing to achieve a stationary mean:
   (a) *A plot of the data.* This should not be the sole criterion for differencing but in conjunction with other tools it provides clues. A realization with major changes in level (especially a strong "up" or "down" trend) or slope is a candidate for differencing. Significant changes in level require nonseasonal first differencing (see Case 6), while slope changes require nonseasonal second differencing (see Case 7). Strong seasonal variation usually calls for no more than seasonal first differencing.

(b) *The estimated acf.* If the estimated autocorrelation coefficients decline slowly at longer lags (they neither cut off to zero nor damp out to zero rapidly), the mean of the data is probably nonstationary and differencing is needed. This applies separately to autocorrelations at seasonal lags (e.g., multiples of 4 for quarterly data, multiples of 12 for monthly data). When considering seasonal-length differencing you must mentally suppress all autocorrelations except those at multiples of the length of seasonality, temporarily treating the seasonal lags as a separate structure.

(c) *The signs and sizes of estimated AR coefficients.* These must satisfy the formal stationarity conditions discussed in Chapter 6 and restated in Table 12.1. These conditions apply separately to nonseasonal and seasonal AR coefficients when estimating a multiplicative AR seasonal model. If the evidence for a nonstationary mean from the estimated acf is strong and clear, one may difference on that basis alone without estimating any AR coefficients for the undifferenced data. (See Cases 7, 9, and 10 for examples showing how estimated AR coefficients confirm the need for differencing.)

9. Nonseasonal first differencing ($d = 1$) is required more often if data are measured in current dollar values rather than in physical units

Table 12.1 Summary of stationarity and invertibility conditions

Model Type	Stationarity Conditions	Invertibility Conditions
Pure AR	Depends on p; see below	Always invertible
Pure MA	Always stationary	Depends on q; see below
AR(1)	$|\phi_1| < 1$	Always invertible
AR(2)	$|\phi_2| < 1$ $\phi_2 + \phi_1 < 1$ $\phi_2 - \phi_1 < 1$	Always invertible
MA(1)	Always stationary	$|\theta_1| < 1$
MA(2)	Always stationary	$|\theta_2| < 1$ $\theta_2 + \theta_1 < 1$ $\theta_2 - \theta_1 < 1$
ARMA(p, q)	Depends on p; see above	Depends on q; see above

or base period dollars. Nonseasonal second differencing ($d = 2$) is relatively unusual; it is used only in Case 7.

10. Seasonal-length differencing is required frequently. Doing so once ($D = 1$) is virtually always sufficient.

11. Needless differencing produces less satisfactory models and should be avoided. It creates artificial patterns in a data series and causes the forecast-error variance to be unnecessarily large. However, one should difference when in doubt; doing so in borderline cases usually yields superior forecasts because they are not tied to a fixed mean. (See Case 5 for a doubtful situation that is resolved in favor of differencing.)

12. The correlation matrix of estimated parameters may provide evidence about a nonstationary mean. That is, some computer programs estimate the mean simultaneously along with the AR and MA coefficients rather than just using $\bar{z}$ as the estimated mean. Some estimated models satisfy the formal stationarity conditions on the AR coefficients, but one or more of the estimated AR coefficients will be highly correlated (absolute $r > 0.9$) with the estimated mean. In these cases the estimate of the mean can be highly inaccurate, and the wisest course is to difference the data. (Cases 7 and 9 illustrate this phenomenon.)

13. *The following cannot be overemphasized: best results are usually achieved with parsimonious, common models.* The great majority of data series in economics and business can be adequately represented, after proper differencing, with ARMA models of order two or less. First-order models are typically adequate for the seasonal portion. By focusing on common, parsimonious models, you will discover that many estimated autocorrelations at lags other than 1, 2 and the first two seasonal lags ($s, 2s$), while significant at the identification stage, become insignificant in the residual acf. Higher-order models are best tried only after a previous effort has been made to build a satisfactory model of order two or less. Many of the following case studies show how common models with just a few estimated parameters account for a large number of significant autocorrelation coefficients in the original estimated acf.

14. Better results are sometimes achieved by identifying and estimating the seasonal element first, then examining the residual acf for a nonseasonal pattern. This is especially true when seasonality is strong enough to dominate the original estimated acf. (Cases 13–15 illustrate how useful this practical rule can be.)

15. It is wise to examine the estimated acf of the first differences even if differencing does not seem necessary to induce a stationary mean

because the seasonal pattern is often clearer in the acf of the first differences than in the acf of the original series. (This point is well-illustrated in Case 10.)

16. Sometimes strong seasonality makes it difficult for the inexperienced analyst to build an adequate UBJ–ARIMA model. It is permissible to first deseasonalize the data using one of the standard methods, build a UBJ–ARIMA model of the deseasonalized series, then apply the seasonal adjustment factors to the deseasonalized forecasts. However, it is much preferable to incorporate the seasonal pattern into the ARIMA model.

17. The standard multiplicative seasonal model (discussed in Chapter 11) usually produces satisfactory results. When in doubt, the analyst should compare the results of the multiplicative and additive forms. An advantage of the multiplicative form is that the stationarity and invertibility conditions are more easily checked because they apply separately to the seasonal and nonseasonal parts of the model. (Case 14 is an example of an additive seasonal model.)

18. Most seasonal processes in economics are autoregressive or mixed (ARMA) in the seasonal component. The autoregressive element

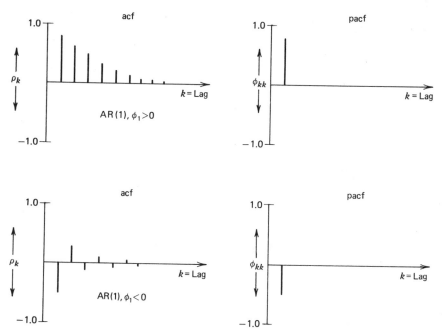

Figure 12.1   Examples of theoretical acf's and pacf's for stationary AR processes.

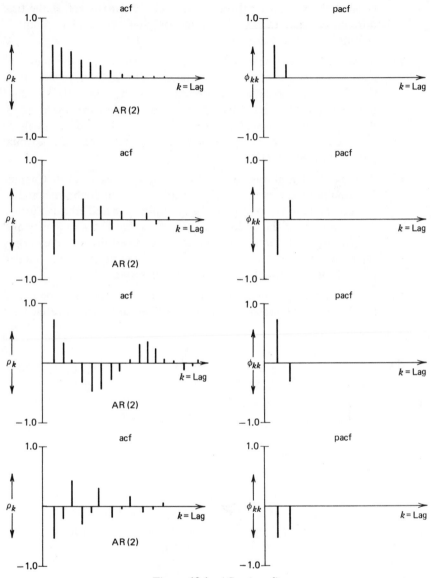

Figure 12.1 (*Continued*)

often appears only in the form of seasonal differencing which forces the AR coefficient at the seasonal lag to equal 1.0.

19. A stationary AR process has a theoretical acf showing exponential decay or a damped sine wave. The corresponding theoretical pacf has spikes followed by a cutoff to zero; the lag length of the last significant pacf spike equals the AR order of the process. [Figure 12.1 shows examples of theoretical acf's and pacf's for stationary AR(1) and AR(2) processes.]

20. An MA process has a theoretical acf with spikes followed by a cutoff to zero. The lag length of the last acf spike equals the MA order of the process. The corresponding theoretical pacf has exponential decay or a damped sine wave. [Figure 12.2 shows examples of theoretical acf's and pacf's for MA(1) and MA(2) processes.]

21. A stationary ARMA process has a theoretical acf and pacf both of which tail off toward zero. The acf tails off after the first $q - p$ lags with either exponential decay or a damped sine wave. The pacf tails off after the first $p - q$ lags. [Figure 12.3 shows examples of theoretical acf's and pacf's for stationary ARMA(1, 1) processes.]

22. A seasonal AR or MA process has the same acf and pacf characteristics as the corresponding nonseasonal process, but the autocorrelations and partial autocorrelations for the seasonal process occur at

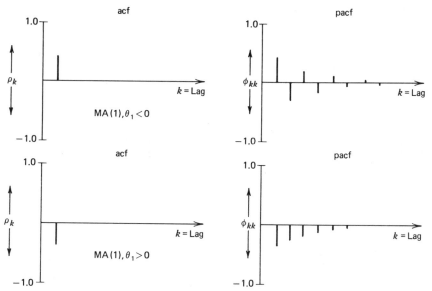

Figure 12.2 Examples of theoretical acf's and pacf's for MA processes.

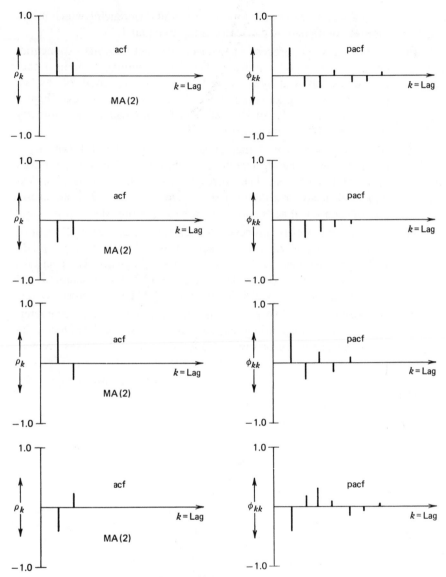

Figure 12.2 (*Continued*)

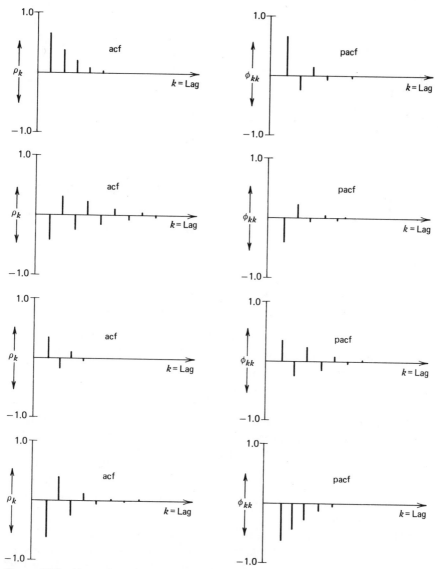

Figure 12.3 Examples of theoretical acf's and pacf's for stationary ARMA(1, 1) processes.

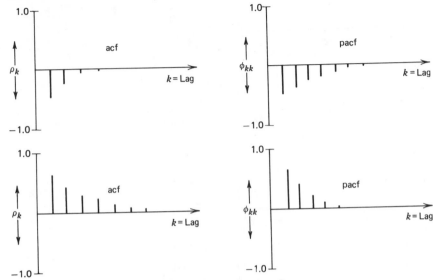

Figure 12.3 (*Continued*)

lags which are multiples of the seasonal length. For example, a stationary AR seasonal process for quarterly data has a theoretical acf with exponential decay at lags $4, 8, 12, \ldots$, whereas an MA seasonal process for quarterly data has a theoretical acf with a spike at lag 4, but zero at subsequent seasonal lags $(8, 12, 16, \ldots)$.

23. It is sometimes difficult to identify a parsimonious mixed model at the initial identification stage. Cases 2, 4, and 5 illustrate this difficulty. It is sometimes helpful to first estimate a pure AR model, then add coefficients based on the residual acf. This procedure can make it easier to find a parsimonious model and can help avoid coefficient near-redundancy. (Case 4 illustrates the usefulness of this technique.)

24. The order of a model determines only the maximum number of coefficients to be estimated, not the total number. For example, an MA(2) model always contains a θ_2 coefficient, but it may or may not contain a θ_1 coefficient. Thus both of the following are MA(2) models:

$$\tilde{z}_t = \left(1 - \theta_1 B - \theta_2 B^2\right) a_t \tag{12.1}$$

$$\tilde{z}_t = \left(1 - \theta_2 B^2\right) a_t \tag{12.2}$$

[Case 2 shows an example of a model similar to (12.2).]

Unfortunately, some computer programs for estimating ARIMA models automatically estimate all coefficients up to and including the one whose subscript is equal to the order of the model. For example, if the user specifies an MA(2) model, a program of this type automatically estimates model (12.1). Such programs should be avoided because they may force us to estimate and forecast with nonparsimonious models.

25. Some computer programs require the user to enter initial estimates for the AR and MA coefficients. Using 0.1 for all initial estimates gives good results in many cases. Better initial estimates, as shown below, may be obtained from the estimated acf and pacf used to identify the model. Remember that r_k is the estimated autocorrelation coefficient at lag k, while $\hat{\phi}_{kk}$ is the estimated partial autocorrelation coefficient at lag k. Note that initial MA estimates are the negatives of the corresponding r_k values.

Model	Coefficient Requiring Initial Estimate	Initial Estimate
AR(1)	$\hat{\phi}_1$	r_1
AR(2)	$\hat{\phi}_1$	$\hat{\phi}_{11}$
	$\hat{\phi}_2$	$\hat{\phi}_{22}$
MA(1)	$\hat{\theta}_1$	$-r_1$
MA(2)	$\hat{\theta}_1$	$-r_1$
	$\hat{\theta}_2$	$-r_2$

26. At the estimation stage one should be wary of retaining an estimated coefficient whose absolute t-value is much less than 2.0, especially if it occurs at a lag other than 1, 2, or the seasonal lag.

27. With differenced data include a constant term (let the mean of the differenced series be nonzero) only if there is reason to think the series has a deterministic trend. After proper differencing, data outside the physical sciences usually have a mean that is not significantly different from zero; when the mean is zero, the resulting model has a constant term of zero and any trend in the forecasts will be purely stochastic.

When the program estimates the mean simultaneously with the AR and MA coefficients, the decision may be made empirically by examining the statistical significance of the estimated mean and constant. Ideally, a deterministic trend is also interpretable. (See Cases 6 and 15 for models with significant deterministic trends. In

Case 6 this result gives a nonsense model, the statistical significance of the estimated constant notwithstanding; in Case 15 the presence of a deterministic trend is defensible.)

28. Table 12.2 summarizes the practical warning levels for absolute *t*-values in estimated acf's.

(a) At the *identification* stage pay attention to *nonseasonal autocorrelations* with absolute *t*-values in excess of about 1.6. Coefficients at these lags often prove to be statistically significant at the estimation phase.

(b) At the *identification and diagnostic-checking* stages pay attention to *seasonal autocorrelations* with absolute *t*-values in excess of about 1.25. The corresponding estimated AR or MA seasonal coefficients are often highly significant at the estimation stage. If the residual acf shows statistical zeros at the seasonal lags $(s, 2s, \dots)$, the same *t*-value warning level (absolute value > 1.25) also applies to half-seasonal lag residual autocorrelations (lags $0.5s, 1.5s, \dots$) and to residual autocorrelations contiguous to the seasonal lags (lags $s + 1, s - 1, 2s + 1, 2s - 1, \dots$). (See Case 12 for an example of an estimated half-seasonal MA coefficient that is significant despite its corresponding residual autocorrelation being insignificant at the 5% level.)

(c) At the *diagnostic-checking* stage when examining *residual autocorrelations* consider estimating coefficients at the short lags (1, 2,

Table 12.2 Practical warning levels for absolute *t*-values in estimated acf's

acf Lag	Identification Stage (initial estimated acf)	Diagnostic-Checking Stage (residual acf)
Short (1, 2, perhaps 3)	1.6	1.25
Seasonal $(s, 2s, \dots)$	1.25	1.25
Near-seasonal[a] $(s - 1, s + 1, 2s - 1, 2s + 1, \dots)$ and half-seasonal $(0.5s, 1.5s, \dots)$	—	1.25
All others	1.6	1.6

[a]*Note*: Focus on the near-seasonal or half-seasonal lags only after the seasonal lag $(s, 2s, \dots)$ autocorrelations are rendered insignificant by seasonal differencing or by including seasonal AR or MA coefficients in the model.

and perhaps 3) if the corresponding absolute t-values in the residual acf exceed about 1.25. The short-lag residual autocorrelation t-values are sometimes seriously underestimated by the standard formula (Bartlett's approximation).

(d) In all cases the absolute t-value warning level for partial autocorrelations is 2.0.

29. When reformulating a model based on the residual acf, add one coefficient at a time to achieve more-parsimonious results. This procedure also helps avoid coefficient near-redundancy, a potential problem when adding AR and MA coefficients simultaneously. Furthermore, adding just one coefficient will sometimes clear up many other residual autocorrelations. (Case 13 illustrates how useful this practical rule can be.)

30. Examine a plot of the residuals from any final ARIMA model. This can give clues about misrecorded data, or it can lead to greater insight into the causes of variation in the data. (Case 2 illustrates the latter phenomenon.)

31. Be wary of using an ARIMA model to forecast if the estimated coefficients have absolute correlations of 0.9 or higher. In such cases the coefficient estimates tend to be unstable: they are highly dependent on the particular realization used, and slight changes in the data pattern can lead to widely differing estimates.

32. The ultimate test of a model is its ability to forecast. While we should give preference to common, low-order models, an unusual model should be used if it regularly forecasts better.

CASE STUDIES: INTRODUCTION

The following 15 case studies are designed to show how to construct ARIMA models by following the practical rules summarized in Chapter 12. You are encouraged to review that chapter frequently as you work through the cases. You might also try building an ARIMA model yourself for each series before reading the case study. Then you can compare your own efforts with the procedures and results presented here.

Most of the data in the cases are taken from economics and business situations. But the methods of analysis presented here also apply to time-series data that arise in other contexts such as engineering, chemistry, or sociology, for example. The first four cases (Group A) involve stationary, nonseasonal data series. The next four (Group B) illustrate nonstationary models, but still without seasonal variation. The last seven cases (Group C) all involve realizations that are both nonstationary and seasonal. Within each group the cases move from easier to more complicated ones.

Some of the data in the first eight cases have been seasonally adjusted prior to analysis. This has been done *only* to create series that would be relatively easy to analyze for the beginning modeler. It must be emphasized that it is preferable in practice to account for seasonal patterns within an ARIMA model rather than removing the seasonal element first.

Most of the series analyzed here are relatively short ($n < 100$). This was a conscious choice since there are so many practical situations when the analyst must contend with a moderate sample size.

Some readers may delve into the case studies before completing Chapters 1–12. Although each case contains material drawn from later chapters, many cases can be read with benefit after the reading of selected chapters.

Here is a suggested schedule:

1. After Chapters 1–4, read Cases 1–4.
2. After Chapter 6, review Cases 1–4.
3. After Chapter 7, read Cases 5–8.
4. After Chapter 9, review Cases 1–8.
5. After Chapter 11, read Cases 9–15.

Since some readers may get into Cases 1–4 before reading Chapter 5 on backshift notation, models in those cases are presented in both backshift and common algebraic form. In Cases 5–15 all models are written in backshift form only.

Any realization can be modeled in more than one way. No claim is made here that the final models in the case studies are the only defensible ones. You may discover alternatives that also provide good representations of the data. The author will appreciate hearing from readers who find superior alternatives.

Group A

STATIONARY, NONSEASONAL MODELS

CASE 1.   CHANGE IN BUSINESS INVENTORIES

In this case we analyze the quarterly change in business inventories, stated at annual rates in billions of dollars. We examine 60 observations covering the period from the first quarter of 1955 through the fourth quarter of 1969.*

For pedagogical reasons the data used in this case study have been seasonally adjusted. As noted in the Introduction to the case studies, this has been done in some of the first eight cases only to create series that are relatively easy to analyze. The better practice is to include seasonal terms directly in the ARIMA model as we do in Cases 9–15.

Figure C1.1 is a plot of the seasonally adjusted data. A casual examination suggests that the series is stationary. The observations seem to fluctuate around a fixed mean, and the variance seems to be constant over time. However, we must withhold judgment about stationarity of the mean until we examine the estimated acf and perhaps some estimated AR coefficients.

Identification.   Figure C1.2 is the estimated acf and pacf for the undifferenced series. Box and Jenkins [1] suggest that the most autocorrelations we may safely examine is about one-fourth of the number of observations. With 60 observations we may calculate $60/4 = 15$ autocorrelations. The computer program used for this analysis limits the number of partial

---

*The original series is found in *Business Conditions Digest*, November 1979, p. 97.

315

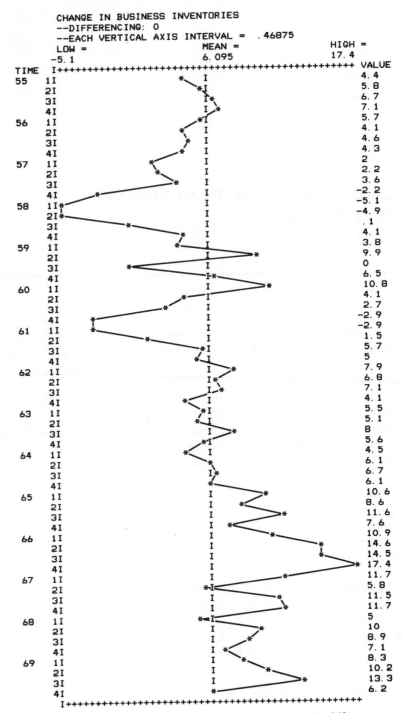

Figure C1.1 Change in business inventories, 1955–1969.

autocorrelations to the number of autocorrelations, so we also find 15 partials. Fifteen partial autocorrelations should be more than adequate since usually only the first few partials are helpful, especially when we have data without seasonal variation. Partials are useful primarily for identifying the AR order of a model (the value of p for a nonseasonal model, or p and P for a model with a seasonal component).

Consider the estimated acf in Figure C1.2. Only the first three autocorrelations are significantly different from zero at about the 5% level: only the first three spikes in the acf extend beyond the square brackets. The position of those brackets is based on Bartlett's approximation for the standard error of estimated autocorrelations as discussed in Chapter 3. The brackets are placed about two standard errors above and below zero.

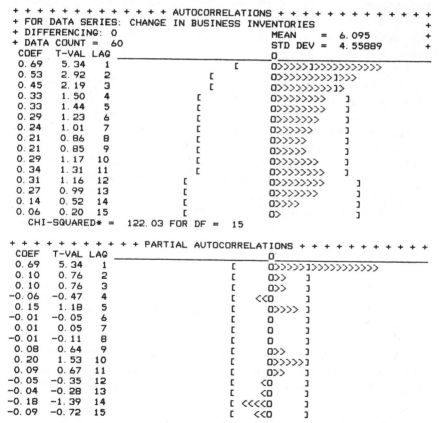

Figure C1.2 Estimated acf and pacf for the realization in Figure C1.1.

The autocorrelations decay to statistical insignificance rather quickly. We conclude that the mean of the series is probably stationary. An AR model seems appropriate because the acf *decays* toward zero rather than cutting off sharply to zero. If the acf cuts off to zero, it suggests a moving-average model. (See Chapters 3, 6, and 12 for examples of theoretical acf's and pacf's.)

A decaying acf is also consistent with a mixed (ARMA) model. But starting with a mixed model is often unwise for three reasons. First, it is often difficult to correctly identify a mixed model initially. The mixed nature of a realization frequently becomes more clear at the diagnostic-checking stage. Second, the principle of parsimony suggests that we try a simple AR or MA model before considering a less-parsimonious mixed model. This is especially important since the estimated acf and pacf are rather crude guides to model choice. Finally, starting immediately with a mixed model may lead to coefficient redundancy as discussed in Chapter 8.

We have tentatively selected an AR model to represent the data. But an AR model of what order? Experience and the principle of parsimony suggest a low-order model, with $p = 1$ or $p = 2$. The estimated pacf should help us make the decision.

The estimated pacf is useful primarily for choosing the order of an AR model. An AR(1) is associated with a single spike in the pacf followed by a cutoff to zero. An AR(2) has two spikes in the pacf, then a cutoff to zero. The estimated pacf in Figure C1.2 suggests an AR(1). It has one spike at lag 1 which is significantly different from zero at about the 5% level, then it cuts off to zero. The 5% significance level is shown by the square brackets on the pacf. Only the partial autocorrelation at lag 1 extends past the brackets.

Based on the preceding analysis, we tentatively choose an AR(1) model to represent the available data. This model is written as*

$$(1 - \phi_1 B)\tilde{z}_t = a_t \qquad\qquad (C1.1)$$

or

$$z_t = C + \phi_1 z_{t-1} + a_t$$

Estimation. Figure C1.3 shows the results of estimating model (C1.1). The computer program estimates μ (the process mean) and ϕ_1 simultaneously. When the mean is estimated in this way, the result is usually not

---

*Recall that $\tilde{z}_t$ is the deviation of z_t from μ: $\tilde{z}_t = (z_t - \mu)$.

much different from the arithmetic mean of the realization ($\bar{z}$). In this case $\bar{z} = 6.095$ while the estimate from the program is $\hat{\mu} = 6.19155$. The advantage of estimating μ simultaneously with the other parameters is that we may then test the mean and the constant term for statistical significance. As discussed in Chapter 8, ϕ_1 and μ are estimated using the least-squares criterion.

In Figure C1.3 we see that $\hat{\phi}_1 = 0.690$ and $\hat{\mu} = 6.19155$. Then the estimated constant is found to be $\hat{C} = \hat{\mu}(1 - \hat{\phi}_1) = 6.19155\ (1 - 0.690) = 1.92091$.

This model satisfies the stationarity requirement $|\hat{\phi}_1| < 1.0$. $\hat{\phi}_1$ is also significantly different from zero at better than the 5% level since its absolute t-value (7.21) is greater than 2.0. As discussed in Chapter 8, we find this

```
+ + + + + + + + +ECOSTAT UNIVARIATE B-J RESULTS+ + + + + + + + + +
+ FOR DATA SERIES:   CHANGE IN BUSINESS INVENTORIES                +
+ DIFFERENCING:      0                          DF      = 57       +
+ AVAILABLE:         DATA = 60    BACKCASTS = 0  TOTAL = 60         +
+ USED TO FIND SSR: DATA = 59    BACKCASTS = 0  TOTAL = 59         +
+ (LOST DUE TO PRESENCE OF AUTOREGRESSIVE TERMS:            1)     +

COEFFICIENT    ESTIMATE      STD ERROR     T-VALUE
PHI      1     0.690         0.096         7.21
CONSTANT       1.92091       .731404       2.62633

MEAN           6.19155       1.4181        4.3661

ADJUSTED RMSE =   3.37903    MEAN ABS % ERR =   73.50
         CORRELATIONS
         1      2
1     1.00
2     0.02   1.00

++RESIDUAL ACF++
COEF    T-VAL  LAG _____0_____
-0.07   -0.51   1              <<<<<<<0
-0.01   -0.07   2                     <0
 0.13    1.02   3                     0>>>>>>>>>>>>>
-0.10   -0.72   4          <<<<<<<<<<0
 0.11    0.79   5                     0>>>>>>>>>>>
 0.07    0.48   6                     0>>>>>>>
 0.02    0.18   7                     0>>
-0.01   -0.09   8                     <0
-0.07   -0.55   9              <<<<<<<0
 0.11    0.79  10                     0>>>>>>>>>>>
 0.14    1.04  11                     0>>>>>>>>>>>>>>>
 0.05    0.37  12                     0>>>>>
 0.16    1.11  13                     0>>>>>>>>>>>>>>>>>
-0.02   -0.16  14                    <<0
-0.07   -0.51  15              <<<<<<<0
    CHI-SQUARED* =   8.53 FOR DF =  13
```

Figure C1.3   Estimation and diagnostic-checking results for model (C1.1).

t-statistic by testing the null hypothesis $H_0: \phi_1 = 0$ as follows:

$$t = \frac{\hat{\phi}_1 - 0}{s(\hat{\phi}_1)}$$

$$= \frac{0.690}{0.096}$$

$$= 7.21$$

where $s(\hat{\phi}_1)$ is the estimated standard error of $\hat{\phi}_1$ taken from the printout in Figure C1.3.

The mean absolute percent error for this model is quite large (73.50%). This is partly because the original observations are small in absolute value. Each residual $(\hat{a}_t)$ from the estimated equation is being divided by a number (z_t) that is relatively close to zero.

Diagnostic checking.   To determine if model (C1.1) is statistically adequate, we test the random shocks a_t for independence using the residuals $\hat{a}_t$ from the estimated equation. The residuals are estimates of the random shocks, and these shocks are assumed to be statistically independent. We use the estimated acf of the residuals to test whether the shocks are independent. With 59 residuals we may examine about 59/4 or 15 residual autocorrelations. (See Chapter 9 for a discussion of the residual acf.)

The residual acf appears below the estimation results in Figure C1.3. None of the residual autocorrelations has an absolute t-value exceeding the warning levels summarized in Chapter 12 (1.25 at lags 1, 2, and 3 and 1.6 elsewhere). Furthermore, according to the chi-squared test the residual autocorrelations are not significantly different from zero *as a set*. The estimated chi-squared statistic shown at the bottom of the residual acf is not significant. For 13 degrees of freedom, this statistic would have to exceed 19.81 to indicate statistical dependence in the random shocks at the 10% level. (The chi-squared statistic used here is based on the Ljung–Box statistic as discussed in Chapter 9.)

Model (C1.1) is satisfactory: $\hat{\phi}_1$ meets the stationarity requirement and is statistically different from zero, $\hat{C}$ is significant and the shocks appear to be independent according to both the t-tests and the chi-squared test. We have found a good model according to the first five criteria summarized in Table 4.1. Therefore, we may move to the forecasting stage.

Forecasting.   Forecasts for lead times $l = 1, 2, 3,$ and 4 from origin $t = 60$ appear in Table C1.1 along with 80% confidence intervals for the true, but unobserved, changes in business inventories. The last two columns have n.a. (not available) because we used all 60 observations at the estimation and forecasting stages. The forecasts are for the first quarter of 1970 through the fourth quarter of 1970, but our last available observation is the fourth quarter of 1969. Therefore, the differences between the forecast and the observed values are not available after 1969.

The forecasts are gradually converging toward the estimated mean (6.19155). As discussed in Chapter 10, this happens with all ARIMA forecasts of stationary series.

Additional checks.   There are some further checks we can perform to determine how much confidence we might place in the forecasting ability of model (C1.1). These checks are informal, but can be quite helpful.

One check as discussed in Chapter 9 is to examine a plot of the estimation-stage residuals. A model could fit an entire data set rather well, but fit the distant past or the recent past very poorly. If a model fits the data well on average but fits the early part of the realization poorly, the early values may have been generated by a different process. We might then drop the early values and return to the identification stage. Alternatively, if a model fits poorly immediately before the time when forecasts must be made, we should be wary. The statistics describing the overall fit of the model (the RMSE and mean absolute percent error) may overstate the forecast accuracy we can expect if the fit has deteriorated over the recent past.

The residuals from model (C1.1) are plotted in Figure C1.4. Neither the very early ones nor the very recent ones seem any larger, on average, than we might expect based on the adjusted RMSE and mean absolute percent error.

Table C1.1   Forecasts from model (C1.1)

Time		Forecast Values	80% Confidence Limits		Future Observed Values	Percent Forecast Errors
			Lower	Upper		
70	1	6.1974	1.8722	10.5225	n.a.[a]	n.a.
	2	6.1956	0.9413	11.4498	n.a.	n.a.
	3	6.1943	0.5515	11.8371	n.a.	n.a.
	4	6.1935	0.3749	12.0120	n.a.	n.a.

[a]n.a. = not available.

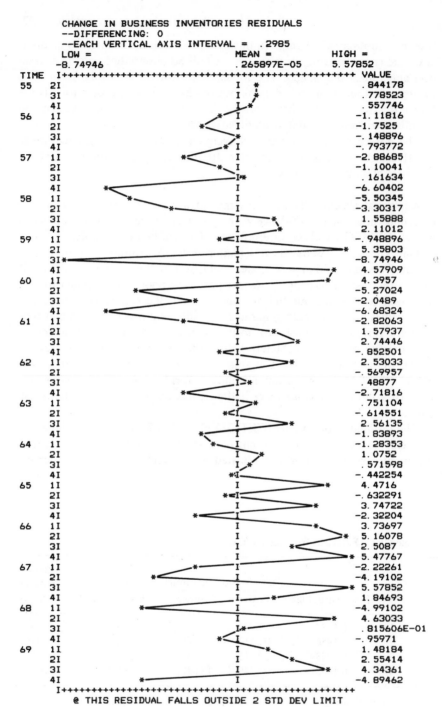

CHANGE IN BUSINESS INVENTORIES RESIDUALS
--DIFFERENCING: 0
--EACH VERTICAL AXIS INTERVAL = .2985

| | LOW = | MEAN = | HIGH = |
| | -8.74946 | .265897E-05 | 5.57852 |

TIME			VALUE
55	2I		.844178
	3I		.778523
	4I		.557746
56	1I		-1.11816
	2I		-1.7525
	3I		-.148896
	4I		-.793772
57	1I		-2.88685
	2I		-1.10041
	3I		.161634
	4I		-6.60402
58	1I		-5.50345
	2I		-3.30317
	3I		1.55888
	4I		2.11012
59	1I		-.948896
	2I		5.35803
	3I		-8.74946
	4I		4.57909
60	1I		4.3957
	2I		-5.27024
	3I		-2.0489
	4I		-6.68324
61	1I		-2.82063
	2I		1.57937
	3I		2.74446
	4I		-.852501
62	1I		2.53033
	2I		-.569957
	3I		.48877
	4I		-2.71816
63	1I		.751104
	2I		-.614551
	3I		2.56135
	4I		-1.83893
64	1I		-1.28353
	2I		1.0752
	3I		.571598
	4I		-.442254
65	1I		4.4716
	2I		-.632291
	3I		3.74722
	4I		-2.32204
66	1I		3.73697
	2I		5.16078
	3I		2.5087
	4I		5.47767
67	1I		-2.22261
	2I		-4.19102
	3I		5.57852
	4I		1.84693
68	1I		-4.99102
	2I		4.63033
	3I		.815606E-01
	4I		-.95971
69	1I		1.48184
	2I		2.55414
	3I		4.34361
	4I		-4.89462

@ THIS RESIDUAL FALLS OUTSIDE 2 STD DEV LIMIT

Figure C1.4 Residuals from model (C1.1).

322

Another check is to drop the last few observations (e.g., the last 10% of the data set), reestimate the model, check the reestimated coefficients for stability, and "forecast history." Figure C1.5 shows the results of reestimating (C1.1) with only the first 56 observations.

This check gives satisfactory results in three ways. First, $\hat{\phi}_1$ is virtually unchanged. A useful rule of thumb is that a coefficient is stable if the reestimated value falls within about 0.1 of the original estimate. The reestimation check in this case is satisfactory according to this guideline.

Second, the residual acf in Figure C1.5 shows no drastic change from the one in Figure C1.3. If one or more residual autocorrelations become highly significant just because a few observations are dropped, we might doubt the stability of the model.

```
+ + + + + + + + + +ECOSTAT UNIVARIATE B-J RESULTS+ + + + + + + + + +
+ FOR DATA SERIES:   CHANGE IN BUSINESS INVENTORIES                    +
+ DIFFERENCING:      0                              DF    = 53         +
+ AVAILABLE:         DATA = 56    BACKCASTS = 0     TOTAL = 56         +
+ USED TO FIND SSR:  DATA = 55    BACKCASTS = 0     TOTAL = 55         +
+ (LOST DUE TO PRESENCE OF AUTOREGRESSIVE TERMS:                1)     +

COEFFICIENT     ESTIMATE        STD ERROR       T-VALUE
  PHI    1       0.698           0.098            7.10
  CONSTANT       1.81072         .730834          2.4776

  MEAN           5.99151         1.50087          3.99203

  ADJUSTED RMSE =  3.36176    MEAN ABS % ERR =   78.59
        CORRELATIONS
          1      2
  1     1.00
  2     0.04   1.00

++RESIDUAL ACF++
  COEF   T-VAL LAG _____0_____
 -0.07  -0.51   1                       <<<<<<<O
  0.00  -0.02   2                             O
  0.15   1.09   3                             O>>>>>>>>>>>>>>>
 -0.13  -0.91   4               <<<<<<<<<<<<<O
  0.10   0.69   5                             O>>>>>>>>>>
  0.12   0.85   6                             O>>>>>>>>>>>>
 -0.04  -0.31   7                         <<<<O
 -0.02  -0.12   8                           <<O
 -0.01  -0.08   9                            <O
  0.07   0.47  10                             O>>>>>>>
  0.07   0.50  11                             O>>>>>>>
  0.05   0.34  12                             O>>>>>
  0.14   0.98  13                             O>>>>>>>>>>>>>>
 -0.01  -0.07  14                            <O
 -0.05  -0.33  15                       <<<<<O
      CHI-SQUARED* =   6.80 FOR DF =   13
```

Figure C1.5 Estimation and diagnostic-checking results for model (C1.1) using the first 56 observations.

Third, the model based on the shorter data set is able to forecast history fairly well. Since we did not use the last four observations in our reestimation, we can compare the forecasts from this model with the four available "future" observations. The percent forecast errors in the last column of Table C1.2 are all smaller than the overall mean absolute percent error, and the 80% confidence limits established around the forecast values contain three of the four observed values.

A digression on seasonality.   A confession is in order: to simplify this case study, we passed over an important practical step. It is wise to examine the acf of the nonseasonal first differences even if differencing does not seem necessary to induce stationarity. This often allows a seasonal pattern to show through more clearly. (See Chapter 11 for a discussion of seasonal models.) This step is especially valuable when the original estimated acf has a decaying (AR or ARMA) pattern. Even when data are seasonally adjusted there may be a seasonal factor remaining in the data if the adjustment is faulty.

Figure C1.6 is the acf and pacf for the first differences of the seasonally adjusted business-inventory data. With quarterly data the seasonal lags are multiples of 4, that is, $4, 8, 12, \ldots$. The spike at lag 4 has an absolute t-value greater than 1.25 (the practical warning value at seasonal lags), so there might be some seasonal variation remaining in the adjusted data. A seasonal $MA(1)_4$ pattern is implied for the seasonal element since the acf cuts off to zero at lag 8, while the pacf decays from lags 4 to 8. Estimation results (not shown) did not produce a significant seasonal coefficient. This result is consistent with the residual acf for model (C1.1) in Figure C1.3 which shows no seasonal pattern.

Table C1.2   Forecasts from model (C1.1) using the first 56 observations

Time		Forecast Values	80% Confidence Limits		Future Observed Values	Percent Forecast Errors
			Lower	Upper		
69	1	6.7650	2.4619	11.0681	8.3000	18.49
	2	6.5312	1.2841	11.7783	10.2000	35.97
	3	6.3681	0.7182	12.0181	13.3000	52.12
	4	6.2543	0.4183	12.0903	6.2000	−0.88

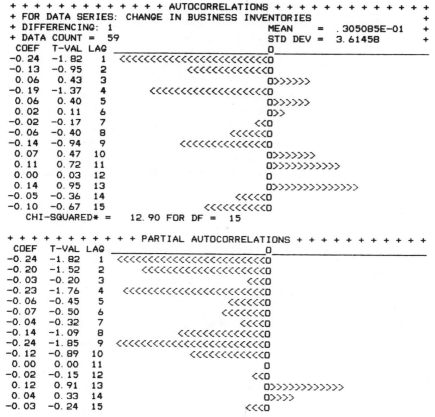

Figure C1.6 Estimated acf and pacf for the first differences of the realization in Figure C1.1.

Final comments.   We now make the following points:

1. The analyst must yet decide if model (C1.1) gives sufficiently accurate forecasts. This depends on the purposes of the forecasts. The analyst wanting more accurate forecasts should choose other forecasting methods and compare their accuracy with the results of the ARIMA model.

2. As pointed out in Chapter 1, ARIMA models are especially suited to short-term forecasting. In general, an ARIMA model should be updated as new data become available. Ideally, this means repeating

the entire cycle of identification, estimation, and diagnostic checking. Often this cycle can be repeated quickly with new data because the original model provides a good guide.

3. The technique of backcasting (discussed in Appendix 8B) is important when estimating models with seasonal coefficients. For models with no seasonal component, the advantages of backcasting are modest and may not be worth the additional computational costs.

   We estimated (C1.1) without using backcasting. Figure C1.7 shows the results of using backcasting to estimate the same model. As shown at the top of this figure, 15 backcasts were produced, thus providing 15 additional "observations." The estimated coefficient $\hat{\phi}_1$,

```
+ + + + + + + + +ECOSTAT UNIVARIATE B-J RESULTS+ + + + + + + + +
+ FOR DATA SERIES:   CHANGE IN BUSINESS INVENTORIES               +
+ DIFFERENCING:     0                              DF    = 58     +
+ AVAILABLE:          DATA = 60    BACKCASTS = 15   TOTAL = 75     +
+ USED TO FIND SSR: DATA = 60    BACKCASTS = 14   TOTAL = 74     +
+ (LOST DUE TO PRESENCE OF AUTOREGRESSIVE TERMS:            1 )   +

COEFFICIENT    ESTIMATE       STD ERROR      T-VALUE
  PHI    1      0. 691         0. 095          7. 29
  CONSTANT     1. 86442        . 69199        2. 69429

  MEAN         6. 04071        1. 2624        4. 7851

  ADJUSTED RMSE =   3. 352    MEAN ABS % ERR =   73. 49
        CORRELATIONS
          1      2
  1    1. 00
  2    0. 00   1. 00

++RESIDUAL ACF++
  COEF   T-VAL LAG _____0_____
 -0. 07  -0. 53   1              <<<<<<<O
 -0. 01  -0. 08   2                    <O
  0. 13   1. 01   3                    O>>>>>>>>>>>>>
 -0. 10  -0. 73   4           <<<<<<<<<<O
  0. 11   0. 79   5                    O>>>>>>>>>>>
  0. 06   0. 48   6                    O>>>>>>
  0. 02   0. 17   7                    O>>
 -0. 01  -0. 09   8                    <O
 -0. 08  -0. 55   9              <<<<<<<O
  0. 11   0. 79  10                    O>>>>>>>>>>
  0. 14   1. 03  11                    O>>>>>>>>>>>>>>
  0. 05   0. 37  12                    O>>>>>
  0. 16   1. 10  13                    O>>>>>>>>>>>>>>>>
 -0. 02  -0. 16  14                   <<O
 -0. 07  -0. 52  15              <<<<<<<O
    CHI-SQUARED* =     8. 52 FOR DF =   13
```

Figure C1.7 Estimation and diagnostic-checking results for model (C1.1) with backcasting.

the estimated constant, and the residual acf are all nearly the same as those obtained without backcasting (Figure C1.3).

The first eight case studies in Part II involve data with no seasonal pattern or data from which the seasonal pattern has been removed. Models for these data sets are estimated without using backcasting. The models in cases 9–15 are estimated with backcasting because they involve data with seasonal variation.

CASE 2. SAVING RATE

The saving rate is personal saving as a percent of disposable personal income. Some economists believe shifts in this rate contribute to business fluctuations. For example, when people save more of their income they spend less for goods and services. This reduction in total demand for output may cause national production to fall and unemployment to rise.

In this case we analyze 100 quarterly observations of the U.S. saving rate for the years 1955–1979. The data are seasonally adjusted prior to publication by the U.S. Department of Commerce.* Figure C2.1 is a plot of the data. Visual inspection suggests that the variance is approximately constant through time.

Identification. The estimated acf and pacf for the undifferenced data appear in Figure C2.2. About 25 autocorrelations is a safe number to examine since that is one-fourth of the number of observations. The computer program then limits us to the same number of partial autocorrelations.

The estimated acf and pacf together suggest two things: (i) the undifferenced data have a stationary mean, and (ii) an AR(1) model is a good first choice to try at the estimation stage.

A stationary mean is implied because the autocorrelations fall quickly to statistical insignificance: they are not statistically different from zero after

---

*The series is found in the Commerce Department publication *Business Conditions Digest*, November 1978, p. 103 and July 1980, p. 83.

lag 3 or 4. Only the first three spikes extend past the square brackets representing the 5% significance level, and only the first four have absolute *t*-values exceeding the practical warning level of 1.6.

The decaying pattern in the acf suggests an AR model. An MA model would be implied if the acf cut off sharply to zero rather than decaying. A mixed model is also possible since mixed processes also have decaying acf's. But unless the evidence for a mixed model is strong and clear at the initial identification stage, we should start with a pure AR model.

The best choice is an AR(1) rather than a higher-order AR model for two reasons. First, the autocorrelations decline in approximately the manner we would expect for an AR(1). For an AR(1) process with $\phi_1 = 0.77$, the theoretical autocorrelations are 0.77, $(0.77)^2 = 0.59$, $(0.77)^3 = 0.46$, $(0.77)^4 = 0.35$, and so forth. The estimated autocorrelations in Figure C2.2 follow this pattern rather closely. Of course, we cannot expect an estimated acf to be identical to a theoretical acf. In Chapter 3 we showed five estimated acf's and pacf's constructed from simulated realizations generated by a known AR(1) process. Those results illustrate that estimated acf's and pacf's only approximate their theoretical counterparts because of sampling error.

Second, the estimated pacf supports the choice of an AR(1) model. The estimated partials cut off (rather than decaying) to statistical insignificance, so we should consider an AR model. Because the cutoff occurs after lag 1, with all absolute *t*-values after lag 1 being less than 2.0, we should entertain an AR model of order one. [Theoretical acf's and pacf's for AR(1) processes are presented in Chapters 3, 6, and 12.] At the estimation stage we estimate this AR(1) model:

$$(1 - \phi_1 B)\tilde{z}_t = a_t \qquad\qquad (C2.1)$$

or

$$z_t = C + \phi_1 z_{t-1} + a_t$$

Estimation and diagnostic checking.   Figure C2.3 shows the results of estimating (C2.1). Stationarity is confirmed since $\hat{\phi}_1 = 0.81$; this satisfies the condition $|\hat{\phi}_1| < 1$. The large *t*-value attached to $\hat{\phi}_1$ indicates this term should be kept in the model. (An estimated coefficient with an absolute *t*-value equal to or greater than 2.0 is significantly different from zero at about the 5% level.) As discussed in Chapter 5, the estimated constant is $\hat{C} = \hat{\mu}(1 - \hat{\phi}_1)$. Inserting the estimated values $\hat{\mu}$ and $\hat{\phi}_1$ we have 6.12259(1 − 0.810) = 1.16398. We should retain the constant term since its *t*-value is substantially larger than 2.0. Thus far (C2.1) is satisfactory. We are ready for some diagnostic checking.

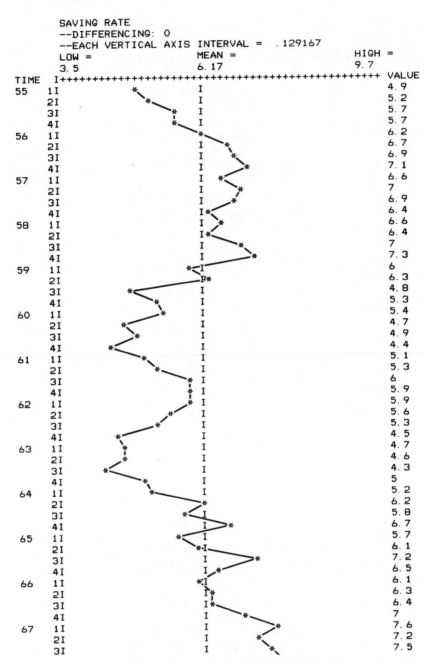

```
SAVING RATE
--DIFFERENCING: 0
--EACH VERTICAL AXIS INTERVAL =  .129167
      LOW =                    MEAN =                        HIGH =
      3. 5                     6. 17                         9. 7
TIME    I++++++++++++++++++++++++++++++++++++++++++++++++++++   VALUE
55   1I              *        I                                 4. 9
     2I                *      I                                 5. 2
     3I                  *    I                                 5. 7
     4I                 *     I                                 5. 7
56   1I                     * I                                 6. 2
     2I                       I  *                              6. 7
     3I                       I    *                            6. 9
     4I                       I      *                          7. 1
57   1I                       I  *                              6. 6
     2I                       I     *                           7
     3I                       I    *                            6. 9
     4I                       I*                                6. 4
58   1I                       I  *                              6. 6
     2I                       I*                                6. 4
     3I                       I     *                           7
     4I                       I        *                        7. 3
59   1I                    *  I                                 6
     2I                     I*                                  6. 3
     3I             *         I                                 4. 8
     4I                 *     I                                 5. 3
60   1I                  *    I                                 5. 4
     2I             *         I                                 4. 7
     3I              *        I                                 4. 9
     4I           *           I                                 4. 4
61   1I                *      I                                 5. 1
     2I                 *     I                                 5. 3
     3I                   *   I                                 6
     4I                  *    I                                 5. 9
62   1I                  *    I                                 5. 9
     2I                *  *   I                                 5. 6
     3I               *       I                                 5. 3
     4I             *         I                                 4. 5
63   1I            *          I                                 4. 7
     2I           *           I                                 4. 6
     3I         *             I                                 4. 3
     4I             *         I                                 5
64   1I                *      I                                 5. 2
     2I                     * I                                 6. 2
     3I                 *     I                                 5. 8
     4I                       I   *                             6. 7
65   1I                *      I                                 5. 7
     2I                     *I                                  6. 1
     3I                       I      *                          7. 2
     4I                       I *                               6. 5
66   1I                     *I                                  6. 1
     2I                      I*                                 6. 3
     3I                      I*                                 6. 4
     4I                       I     *                           7
67   1I                       I         *                       7. 6
     2I                       I       *                         7. 2
     3I                       I         *                       7. 5
```

Figure C2.1   U.S. saving rate, 1955–1979.

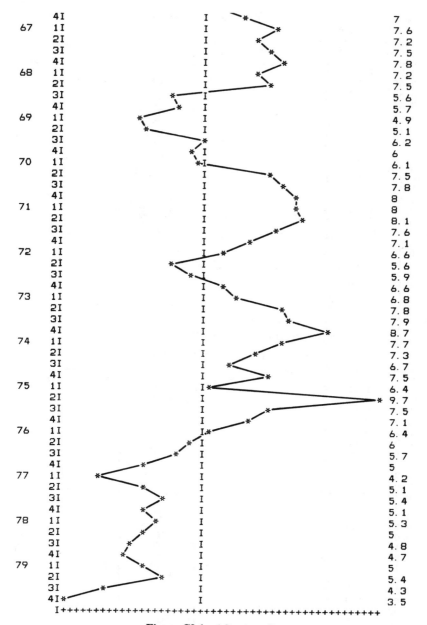

Figure C2.1 (*Continued*)

```
+ + + + + + + + + + + + AUTOCORRELATIONS + + + + + + + + + + + + +
+ FOR DATA SERIES: SAVING RATE                                         +
+ DIFFERENCING: 0                          MEAN      =   6. 17          +
+ DATA COUNT =   100                       STD DEV =   1. 14083         +
  COEF   T-VAL LAG _____O_____
  0. 77   7. 65   1                                 [       0>>>>>]>>>>>>>>>>>>>>
  0. 65   4. 43   2                               [       0>>>>>>>]>>>>>>>>
  0. 49   2. 79   3                               [       0>>>>>>>]>>>>
  0. 34   1. 84   4                          [       0>>>>>>>>>]
  0. 24   1. 24   5                          [       0>>>>>>>      ]
  0. 16   0. 82   6                          [       0>>>>         ]
  0. 11   0. 54   7                          [       0>>>          ]
  0. 13   0. 68   8                          [       0>>>          ]
  0. 13   0. 65   9                          [       0>>>          ]
  0. 16   0. 82  10                          [       0>>>>         ]
  0. 15   0. 75  11                          [       0>>>          ]
  0. 07   0. 36  12                          [       0>>           ]
  0. 04   0. 20  13                          [       0>            ]
 -0. 01  -0. 06  14                          [       0             ]
 -0. 05  -0. 25  15                          [      <0             ]
 -0. 09  -0. 43  16                          [     <<0             ]
 -0. 17  -0. 82  17                          [   <<<<0             ]
 -0. 18  -0. 87  18                          [   <<<<0             ]
 -0. 15  -0. 73  19                          [   <<<<0             ]
 -0. 14  -0. 67  20                          [    <<<0             ]
 -0. 14  -0. 68  21                          [   <<<<0             ]
 -0. 12  -0. 56  22                          [    <<<0             ]
 -0. 12  -0. 58  23                          [    <<<0             ]
 -0. 16  -0. 76  24                          [   <<<<0             ]
 -0. 17  -0. 83  25                          [   <<<<0             ]
       CHI-SQUARED* =   190. 61 FOR DF =   25

+ + + + + + + + + + + PARTIAL AUTOCORRELATIONS + + + + + + + + + + +
  COEF   T-VAL LAG _____O_____
  0. 77   7. 65   1                                 [       0>>>>>]>>>>>>>>>>>>>>>
  0. 16   1. 63   2                          [       0>>>> ]
 -0. 14  -1. 43   3                          [ <<<<0       ]
 -0. 09  -0. 88   4                          [   <<0       ]
  0. 01   0. 10   5                          [     0       ]
  0. 01   0. 15   6                          [     0       ]
  0. 00   0. 02   7                          [     0       ]
  0. 16   1. 63   8                          [     0>>>>   ]
  0. 00   0. 02   9                          [     0       ]
  0. 04   0. 41  10                          [     0>      ]
 -0. 05  -0. 52  11                          [    <0       ]
 -0. 20  -1. 98  12                          [<<<<<0       ]
  0. 03   0. 26  13                          [     0>      ]
  0. 01   0. 12  14                          [     0       ]
  0. 00  -0. 04  15                          [     0       ]
 -0. 05  -0. 45  16                          [    <0       ]
 -0. 16  -1. 60  17                          [ <<<<0       ]
  0. 00  -0. 01  18                          [     0       ]
  0. 10   1. 03  19                          [     0>>>    ]
  0. 00   0. 03  20                          [     0       ]
 -0. 12  -1. 22  21                          [   <<<0      ]
  0. 07   0. 68  22                          [     0>>     ]
 -0. 02  -0. 18  23                          [     0       ]
 -0. 20  -2. 00  24                          [<<<<<0       ]
  0. 02   0. 22  25                          [     0>      ]
```

Figure C2.2 Estimated acf and pacf for the realization in Figure C2.1.

The chief tool for diagnostic checking is the residual acf shown at the bottom of Figure C2.3. Recall from Chapters 1, 4, and 9 that this acf is calculated using the estimation-stage residuals from model (C2.1). These residuals $(\hat{a}_t)$ are estimates of the unobservable random shocks (a_t) in model (C2.1). As discussed in Chapter 3, these random shocks are assumed to be statistically independent. We use the estimation residuals to test the hypothesis that the shocks of model (C2.1) are independent by constructing

```
+ + + + + + + + +ECOSTAT UNIVARIATE B-J RESULTS+ + + + + + + + + +
+ FOR DATA SERIES:   SAVING RATE                                    +
+ DIFFERENCING:      0                        DF     = 97           +
+ AVAILABLE:           DATA = 100  BACKCASTS = 0  TOTAL = 100        +
+ USED TO FIND SSR:  DATA = 99    BACKCASTS = 0  TOTAL = 99          +
+ (LOST DUE TO PRESENCE OF AUTOREGRESSIVE TERMS:           1)        +

COEFFICIENT      ESTIMATE        STD ERROR      T-VALUE
PHI       1       0.810          0.063          12.78
CONSTANT          1.16398        .398547         2.92056

MEAN              6.12259        .372094        16.4544

ADJUSTED RMSE =    702547    MEAN ABS % ERR =    8.45
       CORRELATIONS
       1      2
1    1.00
2   -0.06   1.00

++RESIDUAL ACF++
  COEF   T-VAL  LAG _____0_____
 -0.14   -1.40   1              [   <<<<<<<O          ]
  0.25    2.46   2              [          O>>>>>>>>>>]>>>
 -0.02   -0.19   3              [          <O         ]
 -0.04   -0.34   4              [          <<O        ]
  0.00   -0.03   5              [          O          ]
 -0.01   -0.10   6              [          <O         ]
 -0.14   -1.28   7              [   <<<<<<<O          ]
  0.07    0.68   8          [              O>>>>       ]
 -0.04   -0.34   9          [              <<O         ]
  0.12    1.04  10          [              O>>>>>>     ]
  0.13    1.14  11          [              O>>>>>>     ]
 -0.05   -0.40  12          [              <<O         ]
  0.09    0.77  13          [              O>>>>       ]
  0.00   -0.03  14          [              O           ]
  0.02    0.16  15          [              O>          ]
  0.08    0.69  16          [              O>>>>       ]
 -0.15   -1.27  17          [    <<<<<<<O              ]
 -0.08   -0.65  18          [      <<<<O               ]
 -0.12   -1.06  19          [    <<<<<<O               ]
  0.06    0.49  20          [              O>>>         ]
 -0.14   -1.17  21          [    <<<<<<<O              ]
  0.07    0.59  22          [              O>>>>       ]_
  0.05    0.39  23          [              O>>          ]
 -0.04   -0.30  24          [              <<O         ]
 -0.09   -0.76  25          [      <<<<O               ]
   CHI-SQUARED* =   27.42 FOR DF =  23
```

Figure C2.3 Estimation and diagnostic-checking results for model (C2.1).

the residual acf. If the residual autocorrelations are statistically zero, both individually and as a set, we conclude that (C2.1) is adequate.

The most obvious characteristic of the residual acf is the significant spike at lag 2 ($t = 2.46$). It suggests that (C2.1) is not adequate because the residuals are significantly correlated. Significant residual autocorrelations are especially important when they occur at the short lags (1, 2, and perhaps 3) and at the seasonal lags. This calls for a return to the identification stage.

Further identification. It is wise to return to the original acf and pacf when diagnostic checking shows a tentatively identified model to be inadequate. Sometimes this reexamination brings to light characteristics that seemed obscure earlier. In this case hindsight analysis of Figure C2.2 is not very helpful. We do not seem to have overlooked anything obvious in our earlier analysis. Perhaps we can now see a slight wavelike decaying pattern in the pacf across the first four or five lags. Along with the decaying acf this suggests a mixed model, but the pacf still seems to cut off rather than tail off. Therefore, we use the residual acf in Figure C2.3 as our main guide to further identification.

We have at least three choices in dealing with the residual acf spike at lag 2. First, we might ignore it, believing that one significant autocorrelation out of 24 could occur just because of sampling error. But in this case we will not ignore it. It occurs at a very short lag and therefore merits special attention. Furthermore, its absolute t-value is well in excess of the residual acf short-lag warning value of 1.25.

Second, we might alter model (C2.1) to include an AR coefficient at lag 2. This is not an attractive alternative. If the residuals from (C2.1) had an AR structure, we would expect the residual acf in Figure C2.3 to have a decaying pattern. No such pattern appears, so estimating a ϕ_2 coefficient is not required. Furthermore, the pacf in Figure C2.2 does not show a significant spike at lag 2. Thus we have two pieces of evidence leading us to reject the addition of a ϕ_2 coefficient.

Third, we might estimate a θ_2 coefficient in addition to estimating ϕ_1. The residual autocorrelation at lag 2 in Figure C2.3 is significant, but the subsequent autocorrelations cut off to statistical zeros. This suggests an MA term at lag 2 rather than an AR term.

This choice is reinforced when we employ the substitution procedure discussed in Chapter 9. Let b_t represent the serially correlated random shocks in model (C2.1):

$$(1 - \phi_1 B)\tilde{z}_t = b_t \qquad (C2.2)$$

The single spike at lag 2 in Figure C2.1 suggests an MA(2) model (with θ_1

constrained to zero) for b_t,

$$b_t = \left(1 - \theta_2 B^2\right) a_t \qquad\qquad (\text{C2.3})$$

where a_t is not serially correlated. Substituting (C2.3) into (C2.2) gives an ARMA(1, 2) model for z_t with θ_1 constrained to zero:

$$\left(1 - \phi_1 B\right)\tilde{z}_t = \left(1 - \theta_2 B^2\right) a_t \qquad\qquad (\text{C2.4})$$

or

$$z_t = C + \phi_1 z_{t-1} - \theta_2 a_{t-2} + a_t$$

Two other alternative models deserve comment before we proceed to estimate (C2.4). Look again at the residual acf in Figure C2.3. The autocorrelation at lag 1 has an absolute t-value (1.40) larger than the relevant warning value (1.25). Thus we might consider an ARMA(1, 1) model, or an ARMA(1, 2) with θ_1 not constrained to zero.

For the moment we choose (C2.4) over these alternatives for two reasons. First, the t-value of the residual autocorrelation at lag 2 in Figure C2.3 is far larger than the t-value at lag 1. If these two t-values were much closer in absolute size, we might prefer to try a θ_1 coefficient before trying a θ_2 coefficient. Second, as pointed out in Chapter 3, estimated autocorrelations can be highly correlated with each other. One significant autocorrelation can cause nearby autocorrelations to be fairly large also. In this case the residual autocorrelation at lag 2 is so large that it could cause the autocorrelation at lag 1 to be fairly large also.

Further estimation and diagnostic checking.   Estimation results for model (C2.4) are shown at the top of Figure C2.4. The stationarity requirement is satisfied since $|\hat{\phi}_1| < 1$, and the invertibility condition is met because $|\hat{\theta}_2| < 1$. These conditions are discussed in Chapter 6. Furthermore, both $\hat{\phi}_1$ and $\hat{\theta}_2$ have absolute t-values greater than 2.0, so we conclude that both coefficients are significantly different from zero at better than the 5% level. Thus far (C2.4) is satisfactory, so we proceed to the diagnostic-checking stage.

The residual acf appears at the bottom of Figure C2.4. For (C2.4) to be adequate, the residual acf should be consistent with the hypothesis that the random shocks (a_t) are independent. We cannot observe the shocks directly, but the residuals $(\hat{a}_t)$ from fitting model (C2.4) to the data are estimates of the shocks. None of the residual autocorrelations has an absolute t-value greater than our practical warning levels, and the chi-squared (Ljung–Box) statistic discussed in Chapter 9 is not significant at the 10% level.

```
+ + + + + + + + + +ECOSTAT UNIVARIATE B-J RESULTS+ + + + + + + + + +
+ FOR DATA SERIES:   SAVING RATE                                    +
+ DIFFERENCING:      0                           DF    = 96         +
+ AVAILABLE:          DATA = 100  BACKCASTS = 0   TOTAL = 100        +
+ USED TO FIND SSR: DATA = 99   BACKCASTS = 0   TOTAL = 99         +
+ (LOST DUE TO PRESENCE OF AUTOREGRESSIVE TERMS:            1)      +
```

COEFFICIENT	ESTIMATE	STD ERROR	T-VALUE
PHI 1	0.733	0.078	9.35
THETA 2	-0.352	0.107	-3.29
CONSTANT	1.63065	.492003	3.3143
MEAN	6.1133	.341582	17.897

ADJUSTED RMSE = .674023   MEAN ABS % ERR =   8.31

```
     CORRELATIONS
      1      2      3
1    1.00
2    0.34   1.00
3   -0.05   0.01   1.00
```

```
++RESIDUAL ACF++
 COEF   T-VAL LAG _____0_____
-0.03  -0.28   1         [            <<<O                     ]
 0.00   0.01   2         [             O                       ]
 0.04   0.37   3         [             O>>>>                    ]
 0.00   0.05   4         [             O                       ]
 0.06   0.55   5         [             O>>>>>>                  ]
 0.01   0.06   6         [             O>                      ]
-0.11  -1.10   7         [    <<<<<<<<<<<O                      ]
 0.05   0.46   8         [             O>>>>>                   ]
-0.01  -0.12   9         [            <O                       ]
 0.13   1.30  10         [             O>>>>>>>>>>>>>>          ]
 0.12   1.15  11         [             O>>>>>>>>>>>>>           ]
-0.06  -0.58  12       [            <<<<<<O                       ]
 0.05   0.47  13       [             O>>>>>                       ]
 0.00  -0.04  14       [             O                           ]
 0.05   0.47  15       [             O>>>>>                       ]
 0.11   1.02  16       [             O>>>>>>>>>>>>                ]
-0.14  -1.33  17       [   <<<<<<<<<<<<<<O                        ]
-0.13  -1.19  18       [    <<<<<<<<<<<<<O                        ]
-0.05  -0.47  19       [            <<<<<O                        ]
 0.03   0.29  20       [             O>>>                         ]
-0.16  -1.44  21       [  <<<<<<<<<<<<<<<<O                       ]
 0.05   0.41  22       [             O>>>>>                       ]
 0.10   0.91  23       [             O>>>>>>>>>>>                 ]
-0.06  -0.48  24       [            <<<<<<O                       ]
-0.09  -0.81  25     [              <<<<<<<<<O                      ]
 CHI-SQUARED* =   19.94 FOR DF =  22
```

Figure C2.4 Estimation and diagnostic-checking results for model (C2.4).

Finally, our estimated coefficients are not too highly correlated as shown by the correlation matrix in Figure C2.4. Their absolute correlation (0.34) is less than the practical warning level of 0.9, so we are fairly confident that our coefficient estimates are stable. (The problem of correlated estimates is discussed in Chapter 8.) We conclude that (C2.4) is a good model according to the first five criteria summarized in Table 4.1.

The residual autocorrelation at lag 1 in Figure C2.4 is now much smaller than it was in Figure C2.3. Apparently in model (C2.1) the residual autocorrelation at lag 1 was correlated with the residual autocorrelation at lag 2. When we accounted for the residual autocorrelation at lag 2 with model (C2.4), we also removed nearly all of the residual autocorrelation at lag 1.

Forecasting.   Forecasts for lead times $l = 1, 2, \ldots, 8$ from origin $t = 100$ appear in Table C2.1. Several points deserve emphasis. First, the first several forecasts are less risky statistically than the later forecasts. This is suggested by the smaller width of the 80% confidence intervals associated with the earlier forecasts. After the first forecast we are producing "bootstrap" forecasts as discussed in Chapter 10. This means they are based at least partly on previous forecasts. For example, when forecasting for the first quarter of 1980 (period 101), we multiply $\hat{\phi}_1$ by the last observed z value (period 100) to get the AR portion of the forecast. When forecasting two periods ahead (for period 102), we would like to multiply $\hat{\phi}_1$ by the observed value z_{101}. But all we have available for that period is the forecast value $\hat{z}_{101}$, so we use it instead. All forecasts from (C2.4) from forecast origin $t = 100$ with a lead time greater than 1 are bootstrap forecasts. This illustrates why it is desirable to reestimate UBJ–ARIMA models as new data become available. (See Chapter 10 for illustrations of how forecasts are calculated.)

Table C2.1   Forecasts from model (C2.4)

Time		Forecast Values	80% Confidence Limits		Future Observed Values	Percent Forecast Errors
			Lower	Upper		
80	1	3.7335	2.8708	4.5962	n.a.[a]	n.a.
	2	3.8876	2.8178	4.9574	n.a.	n.a.
	3	4.4813	3.1644	5.7981	n.a.	n.a.
	4	4.9166	3.4845	6.3487	n.a.	n.a.
81	1	5.2358	3.7453	6.7263	n.a.	n.a.
	2	5.4699	3.9490	6.9907	n.a.	n.a.
	3	5.6415	4.1045	7.1785	n.a.	n.a.
	4	5.7673	4.2217	7.3129	n.a.	n.a.

[a] n.a. = not available.

Second, the forecasts are gravitating towards the estimated mean ($\hat{\mu} = 6.1133$). As pointed out in Chapter 10, this occurs with forecasts from all stationary models. How rapidly this convergence occurs depends on the form of the model. A model with AR terms tends to converge less rapidly to the estimated mean than a pure MA model.

Additional checks.   It is wise to examine the residuals from the fitted model. If the very early residuals are unusually large, we should consider dropping the early part of the realization and returning to the identification stage. If our model fits the recent past quite poorly, we should use our forecasts cautiously.

The estimation residuals are shown in Figure C2.5. Neither the very early ones nor the most recent ones as a set are unusually large. However, the last two are noticeably negative, with the last one falling more than two standard deviations below the mean of the residuals. While the fit of the model does not deteriorate sharply over the last 10 to 15 observations, the model clearly does not explain very well the decline of the saving rate during the last half of 1979. The most notable residual occurs in the second quarter of 1975. We return to a discussion of this particular residual at the end of this case study.

Another check is to exclude some of the later observations and reestimate the model to see if it is stable over time. A rule of thumb is to drop about the last 10% of the data set. But note in the plot of the data (Figure C2.1) that the observation for the second quarter of 1975 deviates sharply from the surrounding observations; we have also seen that the residual for this period is quite large. A single deviant observation can sometimes have a powerful effect on estimation results. Therefore, in our reestimation we exclude that observation and all subsequent ones, dropping a total of 30 observations. This is a large number to exclude. But it is a bit unusual to find a significant MA coefficient at lag 2 without also finding one at lag 1, so we have reason to be concerned about whether we have identified an appropriate model. Excluding 30 observations should provide a good test of the stability of this model.

Figure C2.6 shows the results of reestimating model (C2.4) after dropping the last 30 observations. Both $\hat{\phi}_1$ and $\hat{\theta}_2$ are still highly significant. They are also remarkably close to the estimates shown in Figure C2.4 considering that we have excluded 30% of the available data. Both the reestimated $\hat{\phi}_1$ and the reestimated $\hat{\theta}_2$ fall within 0.1 of their initial estimates. The residual acf in Figure C2.6 does not suggest that (C2.4) is inadequate for the shortened data set. The residual autocorrelations at lags 7 and 8 have t-values that are somewhat large, but they are not too disturbing. (The square brackets are not shown after lag 7 because they lie beyond the largest value accommodated by the scale of this graph.) The results in Figure C2.6 suggest that model (C2.4) is adequate for at least one major subset of our data as well as for the entire set.

The ultimate practical test of a model is its ability to forecast. To put (C2.4) to this test, we drop the last eight observations, reestimate, and compare forecasts with observed values. See Table C2.2 for the results. The

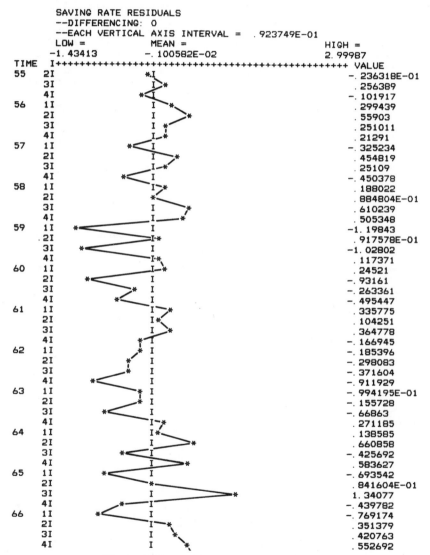

Figure C2.5   Residuals from model (C2.4).

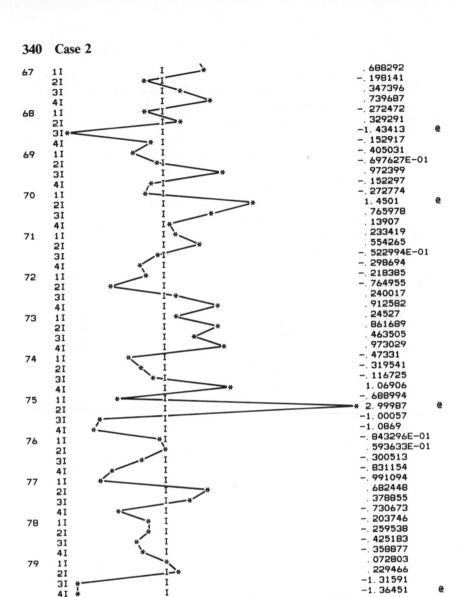

Figure C2.5 (*Continued*)

forecasts are fairly accurate for one and two periods ahead but their accuracy declines substantially thereafter as measured by the percent forecast errors. This is evidence that UBJ–ARIMA models tend to perform best when used for short-term forecasting. Of course, we must remember that the observed values of the saving-rate series are relatively small numbers, so the percent forecast errors can occasionally be quite large. A better check of

```
+ + + + + + + + +ECOSTAT UNIVARIATE B-J RESULTS+ + + + + + + + +
+ FOR DATA SERIES:   SAVING RATE                                    +
+ DIFFERENCING:      0                              DF    = 66      +
+ AVAILABLE:              DATA = 70   BACKCASTS = 0   TOTAL = 70    +
+ USED TO FIND SSR: DATA = 69   BACKCASTS = 0   TOTAL = 69          +
+ (LOST DUE TO PRESENCE OF AUTOREGRESSIVE TERMS:              1)    +
```

COEFFICIENT		ESTIMATE	STD ERROR	T-VALUE
PHI	1	0.712	0.091	7.82
THETA	2	-0.427	0.121	-3.51
CONSTANT		1.79085	.569269	3.14588
MEAN		6.20916	.328539	18.8993

```
ADJUSTED RMSE =  .557597   MEAN ABS % ERR =   7.08
      CORRELATIONS
      1      2      3
1    1.00
2    0.37   1.00
3    0.04   0.02   1.00
```

```
++RESIDUAL ACF++
 COEF   T-VAL  LAG _____0_____
 0.02    0.21   1  [                       0>>                         ]
-0.03   -0.25   2  [                     <<<0                          ]
 0.10    0.80   3  [                       0>>>>>>>>>>>                 ]
-0.06   -0.48   4  [                 <<<<<<0                            ]
 0.13    1.08   5  [                       0>>>>>>>>>>>>>>              ]
-0.08   -0.67   6  [                <<<<<<<<0                           ]
-0.20   -1.61   7  [  <<<<<<<<<<<<<<<<<<<<<0                            ]
-0.20   -1.54   8   <<<<<<<<<<<<<<<<<<<<<0
 0.00   -0.01   9                         0
 0.18    1.33  10                         0>>>>>>>>>>>>>>>>>>>
 0.04    0.26  11                         0>>>>
-0.04   -0.30  12                     <<<<0
 0.14    1.01  13                         0>>>>>>>>>>>>>>
 0.01    0.06  14                         0>
 0.09    0.61  15                         0>>>>>>>>>
 0.07    0.47  16                         0>>>>>>>
-0.16   -1.12  17       <<<<<<<<<<<<<<<<0
-0.15   -1.05  18        <<<<<<<<<<<<<<<0
      CHI-SQUARED* =   19.44 FOR DF =   15
```

Figure C2.6 Estimation and diagnostic-checking results for model (C2.4) using the first 70 observations.

forecast accuracy is to see if the estimated 80% confidence intervals contain the observed values. Only the intervals for forecasts seven and eight periods ahead fail to contain the observed value. We have already noted from the residuals in Figure C2.5 that even when we include the last two observations at the estimation stage, the model does not explain their behavior very well.

Checking for adequacy of seasonal adjustment. To simplify the first eight case studies, we based them on data which initially lack seasonal variation or which have been seasonally adjusted. As noted at the beginning

Table C2.2   Forecasts from model (C2.4) using the first 92 observations

Time		Forecast Values	80% Confidence Limits		Future Observed Values	Percent Forecast Errors
			Lower	Upper		
78	1	5.6327	4.7791	6.4863	5.3000	− 6.28
	2	5.5283	4.5017	6.5549	5.0000	− 10.57
	3	5.7871	4.5334	7.0407	4.8000	− 20.56
	4	5.9599	4.6172	7.3026	4.7000	− 26.81
79	1	6.0754	4.6948	7.4560	5.0000	− 21.51
	2	6.1526	4.7554	7.5498	5.4000	− 13.94
	3	6.2042	4.7996	7.6087	4.3000	− 44.28
	4	6.2386	4.8308	7.6464	3.5000	− 78.25

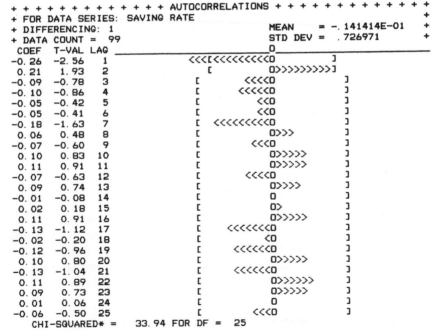

```
+ + + + + + + + + + + + AUTOCORRELATIONS + + + + + + + + + + + + +
+ FOR DATA SERIES:  SAVING RATE                                     +
+ DIFFERENCING:  1                      MEAN     = -. 141414E-01    +
+ DATA COUNT =  99                      STD DEV =  .726971          +
  COEF   T-VAL LAG _____0_____
 -0. 26  -2. 56   1              <<<[<<<<<<<<<<0          ]
  0. 21   1. 93   2                 [          0>>>>>>>>>>]
 -0. 09  -0. 78   3                 [      <<<<0          ]
 -0. 10  -0. 86   4                 [     <<<<<0          ]
 -0. 05  -0. 42   5                 [       <<0          ]
 -0. 05  -0. 41   6                 [       <<0          ]
 -0. 18  -1. 63   7                 [  <<<<<<<<<0          ]
  0. 06   0. 48   8                 [          0>>>         ]
 -0. 07  -0. 60   9                 [       <<<0          ]
  0. 10   0. 83  10                 [          0>>>>>       ]
  0. 11   0. 91  11                 [          0>>>>>       ]
 -0. 07  -0. 63  12                 [       <<<0          ]
  0. 09   0. 74  13                 [          0>>>>        ]
 -0. 01  -0. 08  14                 [          0          ]
  0. 02   0. 18  15                 [          0>          ]
  0. 11   0. 91  16                 [          0>>>>>       ]
 -0. 13  -1. 12  17                 [    <<<<<<<0          ]
 -0. 02  -0. 20  18                 [         <0          ]
 -0. 12  -0. 96  19                 [     <<<<<<0          ]
  0. 10   0. 80  20                 [          0>>>>>       ]
 -0. 13  -1. 04  21                 [    <<<<<<0          ]
  0. 11   0. 89  22                 [          0>>>>>>      ]
  0. 09   0. 73  23                 [          0>>>>>       ]
  0. 01   0. 06  24                 [          0          ]
 -0. 06  -0. 50  25                 [       <<<0          ]
       CHI-SQUARED* =   33. 94 FOR DF =   25
```

Figure C2.7   Estimated acf of the first differences for the realization in Figure C2.1.

of the present case, the saving-rate data are seasonally adjusted before publication by the U.S. Commerce Department. Nevertheless, we will consider the possibility of seasonal variation since there is no guarantee that the seasonal adjustment is adequate.

Even when first differencing does not seem necessary, it is wise to examine the estimated acf of the first differences to check for seasonal variation. (Generally, this should be done at the initial identification stage. We bypassed this step in the present case to avoid unnecessary complications since this is only the second case study.) Often the estimated acf of the first differences gives a clearer picture of seasonality since the differencing step filters out much of the nonseasonal variation, especially when the original acf has an AR pattern.

The estimated acf for the first differences of the saving-rate data appears in Figure C2.7. With quarterly data the seasonal interval is $s = 4$. Therefore, we focus on autocorrelations at lags that are multiples of 4 (4, 8, 12, ...) to check for a seasonal pattern. None of the autocorrelations at the seasonal lags in Figure C2.7 have absolute t-values exceeding the relevant practical warning value of 1.25, so we conclude that the U.S. Commerce Department's seasonal adjustment procedure was adequate. This conclusion is reinforced by the residual acf for the complete model (Figure C2.4). It contains no significant residual autocorrelations at the seasonal lags.

An alternative model.   In this section we present an alternative to model (C2.4). Suppose after fitting (C2.1) we had tried an AR(2) instead of model (C2.4). Keep in mind that (C2.4) is more defensible than an AR(2) based on our analysis of the residual acf in Figure C2.3, and based on the lack of a significant spike at lag 2 in the pacf in Figure C2.2. But beginning analysts often have a difficult time choosing between AR and MA terms. It may therefore be instructive to examine the results of estimating this model:

$$\left(1 - \phi_1 B - \phi_2 B^2\right)\tilde{z}_t = a_t \qquad (C2.5)$$

or

$$z_t = C + \phi_1 z_{t-1} + \phi_2 z_{t-2} + a_t$$

Figure C2.8 shows that the results are not as satisfactory as those for model (C2.4). The t-value for $\hat{\phi}_2$ is less than 2.0 and the RMSE is larger than the one produced by model (C2.4). More important, the residual acf has a large t-value remaining at lag 2.

Updating and forecast accuracy.   We have emphasized that UBJ–ARIMA models are best suited to short-term forecasting and that we should

```
+ + + + + + + + +ECOSTAT UNIVARIATE B-J RESULTS+ + + + + + + + +
+ FOR DATA SERIES:  SAVING RATE                                          +
+ DIFFERENCING:      0                              DF    = 95           +
+ AVAILABLE:          DATA = 100  BACKCASTS = 0    TOTAL = 100          +
+ USED TO FIND SSR:  DATA = 98   BACKCASTS = 0    TOTAL = 98           +
+ (LOST DUE TO PRESENCE OF AUTOREGRESSIVE TERMS:          2)            +

COEFFICIENT    ESTIMATE     STD ERROR     T-VALUE
  PHI    1      0.668        0.102         6.53
  PHI    2      0.183        0.103         1.77
  CONSTANT     .906688      .424774       2.13452

  MEAN          6.08625      .477098      12.7568

  ADJUSTED RMSE =  .698449    MEAN ABS % ERR =    8.62
        CORRELATIONS
        1      2      3
  1    1.00
  2   -0.79   1.00
  3   -0.03  -0.05   1.00
```

```
++RESIDUAL ACF++
  COEF   T-VAL LAG                          0
  0.04   0.37    1         [              0>>>>                ]
  0.18   1.82    2         [              0>>>>>>>>>>>>>>>>>>  ]
 -0.03  -0.26    3         [            <<<0                  ]
 -0.09  -0.88    4         [        <<<<<<<<0                 ]
 -0.05  -0.48    5       [            <<<<<0                   ]
 -0.07  -0.70    6       [           <<<<<<<0                  ]
 -0.17  -1.58    7       [   <<<<<<<<<<<<<<<<<0                ]
  0.02   0.23    8       [              0>>                    ]
 -0.02  -0.20    9       [             <<0                     ]
  0.13   1.24   10       [              0>>>>>>>>>>>>>         ]
  0.14   1.30   11       [              0>>>>>>>>>>>>>>        ]
 -0.02  -0.20   12       [             <<0                     ]
  0.08   0.69   13       [              0>>>>>>>>              ]
  0.01   0.13   14       [              0>                     ]
  0.04   0.39   15       [              0>>>>                  ]
  0.07   0.65   16       [              0>>>>>>>               ]
 -0.14  -1.27   17       [    <<<<<<<<<<<<<<0                  ]
 -0.12  -1.06   18     [      <<<<<<<<<<<<0                     ]
 -0.11  -0.97   19     [       <<<<<<<<<<<0                     ]
  0.03   0.23   20     [              0>>>                      ]
 -0.11  -0.91   21     [       <<<<<<<<<<<0                     ]
  0.08   0.70   22     [              0>>>>>>>>                 ]
  0.09   0.72   23     [              0>>>>>>>>                 ]
 -0.03  -0.21   24     [            <<<0                        ]
 -0.09  -0.71   25     [        <<<<<<<<0                       ]
  CHI-SQUARED* =    25.01 FOR DF =   22
```

Figure C2.8 Estimation and diagnostic-checking results for model (C2.5).

reestimate regularly as new data become available. This can reduce or eliminate the use of bootstrap forecasts and improve forecast accuracy.

Let us return to the forecasts in Table C2.2. They were generated using the first 92 observations. The forecast origin for all these forecasts is $t = 92$. The last seven forecasts shown there are bootstrap forecasts. Suppose instead we forecast only one period ahead, then reestimate by including the next available observation (z_{93}), forecast one more period ahead, then reestimate again with one more observation (z_{94}), and so forth. We may then compare the percent forecast errors from this procedure with the column of percent errors in Table C2.2. The results are shown in Table C2.3.

The first percent error in Table C2.3 is the same as the first percent error shown in Table C2.2. In both cases this is the one-period-ahead percent forecast error for period $t = 93$ (the first quarter of 1978). The rest of the percent forecast errors in Table C2.3 are based on reestimating model (C2.4), adding one observation each time and forecasting only one period ahead. In contrast, the last seven forecasts in Table C2.2 are for more than one period ahead.

All the percent forecast errors in Table C2.3 are smaller than the percent forecast errors in Table C2.2 (except for the first one, of course, which is identical). This illustrates how advantageous it can be to avoid bootstrap forecasts by reestimating when a new observation is available and forecasting only one period ahead. It also emphasizes that UBJ–ARIMA models generally produce better forecasts when the lead time is short.

Large residuals and intervention analysis. In Chapter 9 we said that examination of residuals can sometimes lead to insight into the causes of

Table C2.3   One-period-ahead percent forecast errors for model (C2.4)

Number of Observations Used	One-Period-Ahead Percent Forecast Errors
92	− 6.28
93	− 6.14
94	− 10.18
95	− 9.66
96	0.13
97	3.48
98	− 31.75
99	− 41.41

variations in a realization. In Figure C2.5 we saw that the residual for the second quarter of 1975 was especially large. A review of economic policy actions in 1975 shows that the U.S. Congress had formulated a tax cut by March 1975. This produced a windfall increase in disposable personal income beginning in April 1975. If the bulk of this increase in income was not spent on goods and services, the saving rate would have risen. This is what we see in the data.

When we can identify a specific event that might be responsible for producing one or more large residuals, we can modify our univariate model accordingly by including a variable to represent that event. The result is a multivariate model often called an *intervention model*. This type of analysis is beyond the scope of our discussion, but it represents an important extension of univariate Box–Jenkins analysis. The interested reader may consult Box and Tiao [31], McCleary and Hay [32], or Cleary and Levenbach [33].

CASE 3. COAL PRODUCTION

In Chapter 3 we distinguished a process from a model. A process is the true, but unknown, generating mechanism that could produce all possible observations in a time series. A model is merely a way of representing the behavior of a certain realization. We hope that any model we build can mimic the underlying process, but it is unlikely to be identical to the process. A good model is one which adequately fits the available data with a small number of estimated parameters.

It is sometimes possible to find several good models based on a single realization. Estimated acf's and pacf's can be ambiguous and may suggest two or more different models we might reasonably entertain. Furthermore, estimation and diagnostic-checking results may not show that one model is clearly superior to another: two or more models may fit the data equally well. Our two ultimate guides in choosing a model are the principle of parsimony, and the forecasting ability of the alternative models. In this case study we find several reasonable models, each providing a good fit to the available data.

The data series in this case study is monthly bituminous coal production in the United States from January 1952 through December 1959, a total of 96 observations. The data have been seasonally adjusted, but only to simplify the analysis since this is one of the early case studies.* Remember

*The original, unadjusted data are found on page 263 of the 1973 edition of *Business Statistics*, published by the U.S. Department of Commerce.

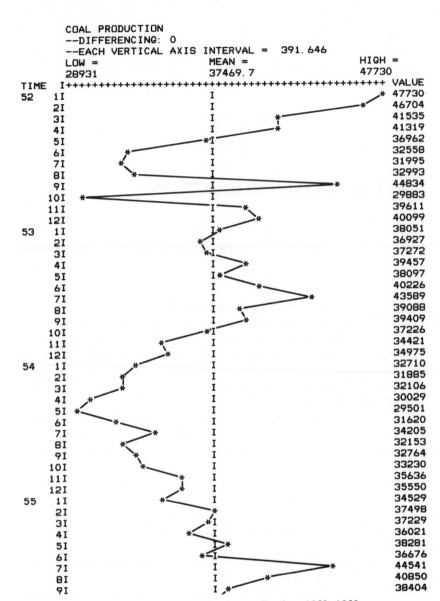

COAL PRODUCTION
--DIFFERENCING: 0
--EACH VERTICAL AXIS INTERVAL = 391.646

	LOW = 28931	MEAN = 37469.7	HIGH = 47730

TIME		VALUE
52	1	47730
	2	46704
	3	41535
	4	41319
	5	36962
	6	32558
	7	31995
	8	32993
	9	44834
	10	29883
	11	39611
	12	40099
53	1	38051
	2	36927
	3	37272
	4	39457
	5	38097
	6	40226
	7	43589
	8	39088
	9	39409
	10	37226
	11	34421
	12	34975
54	1	32710
	2	31885
	3	32106
	4	30029
	5	29501
	6	31620
	7	34205
	8	32153
	9	32764
	10	33230
	11	35636
	12	35550
55	1	34529
	2	37498
	3	37229
	4	36021
	5	38281
	6	36676
	7	44541
	8	40850
	9	38404

Figure C3.1 Coal-production realization, 1952–1959.

348

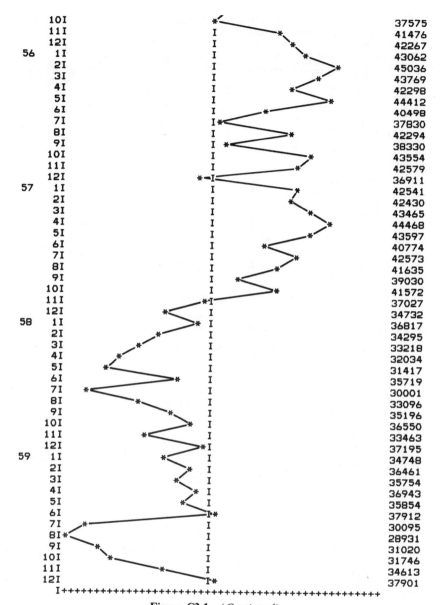

Figure C3.1 (*Continued*)

```
+ + + + + + + + + + + AUTOCORRELATIONS + + + + + + + + + + + + +
+ FOR DATA SERIES: COAL PRODUCTION                                       +
+ DIFFERENCING: 0                         MEAN    =  37469. 7            +
+ DATA COUNT =  96                        STD DEV =  4444. 92            +
  COEF   T-VAL LAG                               0
  0. 67   6. 53   1                        [      0>>>>>]>>>>>>>>>>>>>>>>>>
  0. 60   4. 26   2                     [         0>>>>>>>>>]>>>>>>>>>>
  0. 52   3. 13   3                     [         0>>>>>>>>>]>>>>>>>
  0. 38   2. 09   4                  [            0>>>>>>>>>>>]>
  0. 31   1. 65   5                  [            0>>>>>>>>>>   ]
  0. 26   1. 33   6                  [            0>>>>>>>>>    ]
  0. 24   1. 21   7            [                  0>>>>>>>>          ]
  0. 21   1. 07   8            [                  0>>>>>>>           ]
  0. 10   0. 50   9            [                  0>>>               ]
  0. 11   0. 52  10            [                  0>>>>              ]
  0. 05   0. 23  11            [                  0>>                ]
  0. 04   0. 19  12            [                  0>                 ]
 -0. 02  -0. 08  13            [                 <0                 ]
  0. 00   0. 00  14            [                  0                 ]
  0. 03   0. 13  15            [                  0>                 ]
 -0. 01  -0. 07  16            [                  0                 ]
  0. 00   0. 00  17            [                  0                 ]
 -0. 04  -0. 19  18            [                 <0                 ]
 -0. 08  -0. 40  19            [               <<<0                 ]
 -0. 11  -0. 55  20            [              <<<<0                 ]
 -0. 22  -1. 08  21            [           <<<<<<<0                 ]
 -0. 23  -1. 11  22            [           <<<<<<<<0                 ]
 -0. 28  -1. 35  23            [          <<<<<<<<<0                 ]
 -0. 35  -1. 62  24            [        <<<<<<<<<<<<0                 ]
         CHI-SQUARED* =  193. 57 FOR DF =  24
```

```
+ + + + + + + + + + + PARTIAL AUTOCORRELATIONS + + + + + + + + + + + +
  COEF   T-VAL LAG                               0
  0. 67   6. 53   1                        [      0>>>>>]>>>>>>>>>>>>>>>>>>
  0. 28   2. 71   2                        [      0>>>>>]>>>
  0. 08   0. 83   3                        [      0>>>      ]
 -0. 11  -1. 08   4                        [ <<<<0         ]
 -0. 02  -0. 16   5                        [    <0         ]
  0. 03   0. 26   6                        [     0>        ]
  0. 08   0. 79   7                        [     0>>>      ]
  0. 03   0. 26   8                        [     0>        ]
 -0. 19  -1. 81   9                        [<<<<<0         ]
  0. 03   0. 25  10                        [     0>        ]
 -0. 02  -0. 24  11                        [    <0         ]
  0. 07   0. 64  12                        [     0>>       ]
 -0. 10  -0. 93  13                        [ <<<0          ]
  0. 04   0. 39  14                        [     0>        ]
  0. 07   0. 67  15                        [     0>>       ]
 -0. 04  -0. 41  16                        [    <0         ]
  0. 01   0. 11  17                        [     0         ]
 -0. 11  -1. 06  18                        [ <<<<0         ]
 -0. 05  -0. 52  19                        [   <<0         ]
 -0. 06  -0. 54  20                        [   <<0         ]
 -0. 16  -1. 62  21                        [<<<<<0         ]
 -0. 06  -0. 61  22                        [   <<0         ]
 -0. 07  -0. 69  23                        [   <<0         ]
 -0. 09  -0. 92  24                        [  <<<0         ]
```

Figure C3.2 Estimated acf and pacf for the realization in Figure C3.1.

that it is desirable to account for seasonality within the ARIMA model. (Seasonal models are discussed in Chapter 11 and illustrated in Cases 9–15.)

The adjusted data are plotted in Figure C3.1. The variance of the series appears to be stationary. The mean also seems to be stationary since the series has no persistent trend. However, a series could have several statistically significant changes in its overall level without showing a clear trend. Our tentative view that the mean is stationary must be confirmed with autocorrelation analysis and possibly with estimates of some AR coefficients.

Identification. The estimated acf and pacf for the undifferenced data appear in Figure C3.2. With 96 observations we may safely examine about $96/4 = 24$ autocorrelation coefficients. The computer program then limits us to the same number of partial autocorrelation coefficients.

The acf in Figure C3.2 indicates that the mean of the data is stationary since the autocorrelations drop to zero fairly rapidly. Only the first four exceed zero at about the 5% significance level as indicated by the square brackets. Furthermore, only the first five autocorrelations have absolute t-values exceeding the practical warning level of 1.6. While subsequent analysis may prove us wrong, we conclude for now that the original series has a stationary mean.

Nevertheless, we will also examine the estimated acf of the first (nonseasonal) differences. We calculate the acf for the series $w_t = z_t - z_{t-1}$, where z_t represents the original observations shown in Figure C3.1. Our purpose is to obtain a better picture of any seasonal pattern that might remain in the data. Although the data have been seasonally adjusted, the adjustment might be incomplete or excessive. If the seasonal adjustment is adequate, we should not find significant autocorrelations at the seasonal lags. With monthly data, observations 12 periods apart may be related because they occur in the same season, though in separate years. The length of seasonality is therefore $s = 12$, and the seasonal lags are multiples of 12 (12, 24, ...). See Chapter 11 for a fuller discussion of seasonal patterns.

In the acf for the undifferenced data in Figure C3.2, the autocorrelation at lag 12 is quite small, but the coefficient at lag 24 has an absolute t-value greater than the seasonal-lag practical warning level (1.25). These seasonal-lag values may be dominated by the nonseasonal pattern and may not clearly show the seasonal pattern. The acf of the first (nonseasonal) differences should provide a clearer image of any remaining seasonal pattern. This acf appears in Figure C3.3. Only the coefficients at lags 12 and 24 are relevant for present purposes. They both have absolute t-values well below the warning level of 1.25. We conclude that the seasonal adjustment procedure was adequate and we do not concern ourselves with identifying a seasonal pattern.

```
+ + + + + + + + + + + + AUTOCORRELATIONS + + + + + + + + + + + +
+ FOR DATA SERIES: COAL PRODUCTION                                    +
+ DIFFERENCING: 1                              MEAN    = -103.463     +
+ DATA COUNT =   95                            STD DEV =  3493.65      +
  COEF   T-VAL  LAG _____0_____
 -0.42   -4.14   1      <<<<<<<<<<<<<<[<<<<<<<<<<<<<0                ]
  0.07    0.56   2                    [             0>>>              ]
  0.09    0.78   3                    [             0>>>>>            ]
 -0.08   -0.65   4                    [          <<<<0               ]
  0.02    0.13   5                    [             0>               ]
 -0.05   -0.44   6                    [          <<<0                ]
  0.00    0.02   7                    [             0                ]
  0.04    0.32   8                    [             0>>              ]
 -0.06   -0.49   9                    [          <<<0                ]
  0.02    0.14  10                    [             0>               ]
 -0.09   -0.71  11                    [          <<<<0               ]
  0.09    0.75  12                    [             0>>>>>           ]
 -0.10   -0.84  13                    [         <<<<<0               ]
 -0.02   -0.17  14                    [            <0                ]
  0.09    0.74  15                    [             0>>>>>           ]
 -0.08   -0.63  16                    [          <<<<0               ]
  0.07    0.53  17                    [             0>>>             ]
 -0.03   -0.20  18                    [            <0                ]
  0.02    0.14  19                    [             0>               ]
  0.13    1.00  20                    [             0>>>>>>>         ]
 -0.14   -1.12  21                    [       <<<<<<<0               ]
  0.09    0.72  22                    [             0>>>>>           ]
  0.01    0.10  23                    [             0>               ]
 -0.05   -0.39  24                    [          <<<0                ]
      CHI-SQUARED* =    31.85 FOR DF =   24
```

Figure C3.3   Estimated acf for the first differences of the realization in Figure C3.1.

We now return to the estimated acf and pacf in Figure C3.2. Should we consider fitting an AR model, an MA model, or a mixed model to the undifferenced data? The decaying pattern in the estimated acf suggests that we start with an AR model or a mixed model. Because the autocorrelations do not cut off sharply to statistical zeros, we have no evidence that a pure MA model is appropriate. Since proper identification of a mixed model at this stage of the analysis is often difficult, we begin with a pure AR model.

What order AR model should we entertain? Use the estimated pacf for a clue. There we find two significant spikes at lags 1 and 2 followed by a cutoff to zero. This pacf pattern is consistent with an AR(2) model. [See Chapters 6 and 12 for examples of theoretical AR(2) acf's and pacf's.]

We have tentatively identified an AR(2) model for the undifferenced data:

$$\left(1 - \phi_1 B - \phi_2 B^2\right)\tilde{z}_t = a_t \tag{C3.1}$$

or

$$z_t = C + \phi_1 z_{t-1} + \phi_2 z_{t-2} + a_t$$

Estimation and diagnostic checking.   At the estimation stage we obtain least-squares estimates of C, ϕ_1, and ϕ_2. The results are shown at the top of Figure C3.4. Both $\hat{\phi}_1$ and $\hat{\phi}_2$ are significantly different from zero at about the 5% level since both t-values exceed 2.0.

The estimated coefficients also satisfy the three stationarity conditions for an AR(2) presented in Chapter 6. The first requirement is $|\hat{\phi}_2| < 1$; this

```
+ + + + + + + + +ECOSTAT UNIVARIATE B-J RESULTS+ + + + + + + + +
+ FOR DATA SERIES:   COAL PRODUCTION                              +
+ DIFFERENCING:      0                          DF    = 91        +
+ AVAILABLE:        DATA = 96      BACKCASTS = 0   TOTAL = 96      +
+ USED TO FIND SSR: DATA = 94      BACKCASTS = 0   TOTAL = 94      +
+ (LOST DUE TO PRESENCE OF AUTOREGRESSIVE TERMS:            2)    +

COEFFICIENT      ESTIMATE          STD ERROR      T-VALUE
  PHI    1         0.432            0.099          4.35
• PHI    2         0.311            0.097          3.22
  CONSTANT       9461.01          2913.8          3.24697

  MEAN          36822.4           1237.62        29.7527

  ADJUSTED RMSE =   3047.82     MEAN ABS % ERR =    6.32
          CORRELATIONS
          1      2      3
  1    1.00
  2   -0.69   1.00
  3   -0.03  -0.09   1.00

++RESIDUAL ACF++
  COEF   T-VAL  LAG _____0_____
 -0.02  -0.19   1          [              <<0                   ]
  0.01   0.08   2          [              0>                    ]
  0.18   1.74   3          [              0>>>>>>>>>>>>>>>>>> ]
  0.01   0.12   4     [              0>                             ]
  0.04   0.36   5     [              0>>>>                          ]
  0.01   0.10   6     [              0>                             ]
  0.03   0.26   7     [              0>>>                           ]
  0.06   0.55   8     [              0>>>>>>                        ]
 -0.03  -0.25   9     [            <<<0                             ]
 -0.03  -0.30  10     [            <<<0                             ]
 -0.06  -0.59  11     [          <<<<<<0                            ]
  0.04   0.38  12     [              0>>>>                          ]
 -0.10  -0.97  13     [      <<<<<<<<<<0                            ]
 -0.03  -0.29  14     [            <<<0                             ]
  0.07   0.65  15     [              0>>>>>>>                       ]
 -0.04  -0.40  16     [            <<<<0                            ]
  0.02   0.20  17     [              0>>                            ]
 -0.01  -0.11  18     [             <0                             ]
  0.04   0.38  19     [              0>>>>                          ]
  0.09   0.81  20     [              0>>>>>>>>>                     ]
 -0.11  -0.99  21     [      <<<<<<<<<<<0                           ]
  0.04   0.39  22     [              0>>>>                          ]
  0.01   0.12  23     [              0>                             ]
 -0.09  -0.81  24     [        <<<<<<<<<0                           ]
       CHI-SQUARED* =   10.80 FOR DF =   21
```

Figure C3.4   Estimation and diagnostic-checking results for model (C3.1).

is satisfied since $\hat{\phi}_2 = 0.311$. The second condition is $\hat{\phi}_1 + \hat{\phi}_2 < 1$; this is met because the two coefficients sum to 0.743. The third requirement is $\hat{\phi}_2 - \hat{\phi}_1 < 1$; this is also satisfied since $\hat{\phi}_2 - \hat{\phi}_1 = -0.121$.

Next, use the estimation residuals $(\hat{a}_t)$ of model (C3.1) to test the hypothesis that the shocks of this model (a_t) are statistically independent. To do so we calculate autocorrelation coefficients using the residuals from the estimated model. The residual acf is shown below the estimation results in Figure C3.4.

Now we apply t-tests to the individual residual autocorrelation coefficients and a chi-squared test to the residual autocorrelations as a set. The only residual autocorrelation which appears at all troublesome is at lag 3; its t-value exceeds the practical warning level of 1.25. (Recall from Chapter 9 that absolute t-values at very short lags in the *residual* acf can sometimes be badly underestimated by the standard formula. The warning value of 1.25 for residual acf t-values at short lags is a practical rule to compensate for this occasional underestimation.) All other residual acf t-values in Figure C3.4 are satisfactory, and the chi-squared statistic printed at the bottom of the residual acf is not significant at the 10% level for 21 degrees of freedom.

We now must make a judgment about the residual autocorrelation at lag 3. We might decide that one residual autocorrelation out of 24 with a t-value as large as 1.74 could easily occur by chance. Then we would consider model (C3.1) to be adequate. Alternatively, we might be especially concerned about the residual autocorrelation at lag 3 since its t-value exceeds the practical warning level of 1.25 by a wide margin. And autocorrelations at the short lags (1, 2, perhaps 3) and the seasonal lags deserve more attention than those at other lags. We will pursue some additional models to see if we can account for the residual autocorrelation at lag 3.

Further identification and estimation.   The best modification to model (C3.1) is the addition of an MA term at lag 3. The original pacf in Figure C3.1 shows no evidence that an AR(3) model is appropriate: the spike at lag 3 is not significant. Furthermore, the residual acf in Figure C3.4 does not show a decaying pattern as it should if an AR coefficient is called for. The one argument in favor of adding a ϕ_3 coefficient is it would be consistent with the overfitting strategy.

While estimating a θ_3 coefficient seems preferable, we first estimate an AR(3) model for comparison:

$$\left(1 - \phi_1 B - \phi_2 B^2 - \phi_3 B^3\right)\tilde{z}_t = a_t \qquad \text{(C3.2)}$$

or

$$z_t = C + \phi_1 z_{t-1} + \phi_2 z_{t-2} + \phi_3 z_{t-3} + a_t$$

The estimation results for this model (not shown) indicate that $\hat{\phi}_3$ is not significantly different from zero: its t-value is only 0.85. Therefore, we proceed to estimate a model with an MA term at lag 3 in addition to two AR terms at lags 1 and 2. This new model is an ARMA(2, 3) with θ_1 and θ_2 constrained to zero:

$$\left(1 - \phi_1 B - \phi_2 B^2\right)\tilde{z}_t = \left(1 - \theta_3 B^3\right)a_t \tag{C3.3}$$

or

$$z_t = C + \phi_1 z_{t-1} + \phi_2 z_{t-2} - \theta_3 a_{t-3} + a_t$$

The results for this model in Figure C3.5 indicate that the absolute t-value for $\hat{\theta}_3$ is 1.80. Although $\hat{\theta}_3$ is more significant than was $\hat{\phi}_3$, experience suggests it is usually wiser to exclude coefficients with absolute t-values less than 2.0 at the estimation stage. But this is only a practical rule of thumb, and some analysts might choose to use model (C3.3). As pointed out in Chapter 8 and Appendix 8A, t-values are only approximate tests of significance, especially when MA terms are present in a model.

We conclude that (C3.1) is an adequate representation of the available data, though we should monitor carefully the effect of any new data on the adequacy of model (C3.1). Model (C3.3) may become preferable as more data appear. In fact, there is nothing wrong with accepting both models and using both to forecast future values. We could then choose the one that performs better. For now we use model (C3.1) to forecast.

```
+ + + + + + + + + +ECOSTAT UNIVARIATE B-J RESULTS+ + + + + + + + + +
+ FOR DATA SERIES:   COAL PRODUCTION                                +
+ DIFFERENCING:     0                             DF    = 90        +
+ AVAILABLE:             DATA = 96    BACKCASTS = 0   TOTAL = 96     +
+ USED TO FIND SSR  DATA = 94    BACKCASTS = 0   TOTAL = 94         +
+ (LOST DUE TO PRESENCE OF AUTOREGRESSIVE TERMS:              2)    +
```

COEFFICIENT		ESTIMATE	STD ERROR	T-VALUE
PHI	1	0.392	0.102	3.82
PHI	2	0.302	0.097	3.11
THETA	3	-0.202	0.112	-1.80
CONSTANT		11304.6	3376.53	3.34799
MEAN		36929.6	1222.67	30.204

ADJUSTED RMSE = 3008.2   MEAN ABS % ERR =   6.23

```
     CORRELATIONS
     1      2      3      4
1   1.00
2  -0.60   1.00
3   0.23   0.06   1.00
4  -0.04  -0.08  -0.07   1.00
```

Figure C3.5 Estimation results for model (C3.3).

Table C3.1   Forecasts from model (C3.1)

Time		Forecast Values	80% Confidence Limits		Future Observed Values	Percent Forecast Errors
			Lower	Upper		
60	1	36600.4000	32699.2000	40501.6000	n.a.[a]	n.a.
	2	37062.3000	32813.0000	41311.6000	n.a.	n.a.
	3	36856.9000	32184.9000	41528.8000	n.a.	n.a.
	4	36912.0000	32045.3000	41778.6000	n.a.	n.a.
	5	36871.8000	31861.1000	41882.5000	n.a.	n.a.
	6	36871.6000	31773.7000	41969.6000	n.a.	n.a.
	7	36859.1000	31702.3000	42015.9000	n.a.	n.a.
	8	36853.6000	31658.7000	42048.4000	n.a.	n.a.
	9	36847.3000	31627.1000	42067.4000	n.a.	n.a.
	10	36842.9000	31606.1000	42079.6000	n.a.	n.a.
	11	36839.0000	31591.2000	42086.7000	n.a.	n.a.
	12	36835.9000	31580.9000	42091.0000	n.a.	n.a.

[a] n.a. = not available.

Forecasting.   Forecasts from origin $t = 96$ for lead times up to $l = 12$ are shown in Table C3.1 along with an 80% confidence interval around each forecast. The further into the future we forecast, the more uncertain our forecasts become, as indicated by the widening of the confidence intervals at the longer lead times.

Starting with a forecast lead time $l = 3$, we have strictly bootstrap forecasts: they are based completely on previous forecasts, not on previous observed values. The first forecast $\hat{z}_{97}$ is based on the last two observed values (z_{95}, z_{96}). That is, let period 97 be January 1960, period 98 be February 1960, and so forth. Then the forecast, with lead time $l = 1$ for period 97 from origin $t = 96$, based on model (C3.1) is calculated from the difference-equation form as discussed in Chapter 10:

$$\hat{z}_{97} = \hat{z}_{96}(1) = \hat{C} + \hat{\phi}_1 z_{96} + \hat{\phi}_2 z_{95}$$

Substituting observed values for z_{96} and z_{95}, and substituting the estimated values for $\hat{C}$, $\hat{\phi}_1$, and $\hat{\phi}_2$, we get

$$36,600 = 9461 + 0.432(37,901) + 0.311(34,613)$$

(If you perform the calculations shown on the right-hand side of this equation you will not get exactly 36,600 because $\hat{\phi}_1$ and $\hat{\phi}_2$ are rounded.)

The second forecast with lead time $l = 2$ and origin $t = 96$ is a partial bootstrap forecast since it is based partly on $\hat{z}_{97}$:

$$\hat{z}_{98} = \hat{z}_{96}(2) = \hat{C} + \hat{\phi}_1 \hat{z}_{97} + \hat{\phi}_2 z_{96}$$

$$37{,}062 = 9461 + 0.432(36{,}600) + 0.311(37{,}901)$$

In place of the observed value for period 97, which we do not know, we use the forecast value found above, 36,600. The third forecast is entirely a bootstrap forecast since it is based only on previous forecast values:

$$\hat{z}_{99} = \hat{z}_{96}(3) = \hat{C} + \hat{\phi}_1 \hat{z}_{98} + \hat{\phi}_1 \hat{z}_{97}$$

$$36{,}857 = 9461 + 0.432(37{,}062) + 0.311(36{,}600)$$

All other forecasts are entirely bootstrap forecasts.

Our estimated AR coefficients are potentially correlated. As discussed in Chapter 8, if they are too highly correlated, small shifts in the data could produce large changes in the coefficients; we should then consider our estimates to be of poor quality. Forecasts from such a model may be less reliable than is suggested by how well the model fits the past. A practical rule is to be wary of forecasting with an ARIMA model when absolute correlations among estimated coefficients exceed about 0.9. In the present example we need not be concerned about this problem since the correlation between $\hat{\phi}_1$ and $\hat{\phi}_2$ is -0.69. This statistic is found in Figure C3.4 just below the RMSE. Row 1 and column 1 in the correlation matrix refer to the first estimated coefficient ($\hat{\phi}_1$), row 2 and column 2 refer to the second estimated coefficient ($\hat{\phi}_2$), and so forth. The last row and column always refer to the estimated mean if there is a constant term in the model.

Additional checks. As suggested in Case 1, some informal checks are frequently helpful in deciding how reliable a model's forecasts may be.

One check is to examine the estimation-stage residuals to see if the model fits the distant past and the recent past as well as it fits the full data set. If the fit during the distant past is markedly worse than the overall fit, we might remove those values from our realization. If the fit over the recent past is especially poor, we might not want to use this model to forecast, or we may want to make a subjective adjustment to the ARIMA forecasts. The residuals for model (C3.1) are shown in Figure C3.6. While there are a few large residuals in both the early and the late segment, on average they do not suggest that the model fits these segments poorly.

Another informal check is to drop the last few observations (perhaps the last 10% or 20% of the data set) and reestimate the model. The results of reestimating (C3.1) after dropping the last 12 observations appear in Figure C3.7.

This check reinforces our earlier conclusion that model (C3.1) is acceptable. First, the reestimated coefficients are well within 0.1 of the original

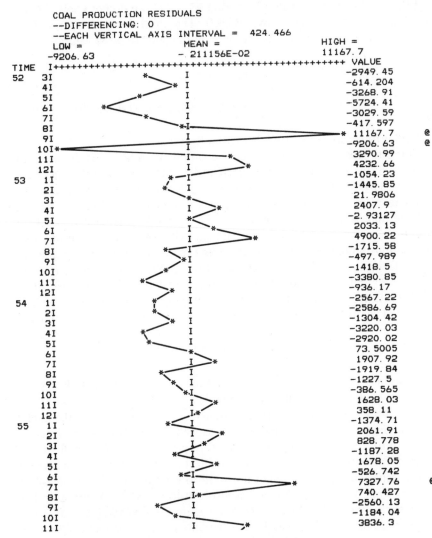

Figure C3.6   Residuals from model (C3.1).

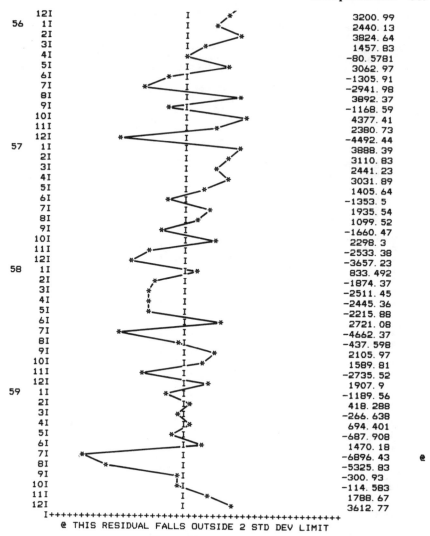

	12I	3200. 99
56	1I	2440. 13
	2I	3824. 64
	3I	1457. 83
	4I	-80. 5781
	5I	3062. 97
	6I	-1305. 91
	7I	-2941. 98
	8I	3892. 37
	9I	-1168. 59
	10I	4377. 41
	11I	2380. 73
	12I	-4492. 44
57	1I	3888. 39
	2I	3110. 83
	3I	2441. 23
	4I	3031. 89
	5I	1405. 64
	6I	-1353. 5
	7I	1935. 54
	8I	1099. 52
	9I	-1660. 47
	10I	2298. 3
	11I	-2533. 38
	12I	-3657. 23
58	1I	833. 492
	2I	-1874. 37
	3I	-2511. 45
	4I	-2445. 36
	5I	-2215. 88
	6I	2721. 08
	7I	-4662. 37
	8I	-437. 598
	9I	2105. 97
	10I	1589. 81
	11I	-2735. 52
	12I	1907. 9
59	1I	-1189. 56
	2I	418. 288
	3I	-266. 638
	4I	694. 401
	5I	-687. 908
	6I	1470. 18
	7I	-6896. 43 @
	8I	-5325. 83
	9I	-300. 93
	10I	-114. 583
	11I	1788. 67
	12I	3612. 77

@ THIS RESIDUAL FALLS OUTSIDE 2 STD DEV LIMIT

Figure C3.6 (*Continued*)

estimates shown in Figure C3.4. Second, the residual acf in Figure C3.7 is not strikingly different from the residual acf based on the full data set. Although the residual autocorrelation at lag 3 is now more statistically significant, its value has not changed drastically. Third, as shown in Table C3.2 the model forecasts rather well, at least for the first six periods ahead. The first six percent forecast errors in the right-most column are all smaller

```
+ + + + + + + + +ECOSTAT UNIVARIATE B-J RESULTS+ + + + + + + + + +
+ FOR DATA SERIES:   COAL PRODUCTION                                  +
+ DIFFERENCING:      0                         DF    = 79             +
+ AVAILABLE:         DATA = 84    BACKCASTS = 0  TOTAL = 84           +
+ USED TO FIND SSR:  DATA = 82    BACKCASTS = 0  TOTAL = 82           +
+ (LOST DUE TO PRESENCE OF AUTOREGRESSIVE TERMS:          2)          +
```

COEFFICIENT	ESTIMATE	STD ERROR	T-VALUE
PHI 1	0.402	0.105	3.81
PHI 2	0.338	0.103	3.28
CONSTANT	9652.81	3207.23	3.0097
MEAN	37134.9	1330.34	27.9139

ADJUSTED RMSE = 3074.73   MEAN ABS % ERR =   6.45

```
       CORRELATIONS
        1     2     3
1    1.00
2   -0.67  1.00
3   -0.04 -0.11  1.00
```

```
++RESIDUAL ACF++
  COEF  T-VAL LAG _____0_____
 -0.04 -0.33   1    [                  <<<<0                        ]
 -0.01 -0.06   2    [                    <0                         ]
  0.23  2.06   3    [                    0>>>>>>>>>>>>>>>>>>>>>>>1>
  0.06  0.52   4    [                    0>>>>>>                    ]
  0.08  0.65   5    [                    0>>>>>>>>                  ]
 -0.02 -0.13   6    [                   <<0                         ]
  0.05  0.44   7    [                    0>>>>>                     ]
  0.04  0.36   8    [                    0>>>>                      ]
 -0.04 -0.32   9    [                  <<<<0                        ]
 -0.02 -0.17  10    [                   <<0                         ]
 -0.05 -0.42  11    [                  <<<<<0                       ]
 -0.01 -0.10  12    [                    <0                         ]
 -0.12 -1.05  13    [            <<<<<<<<<<<<0                      ]
 -0.05 -0.45  14    [                  <<<<<0                       ]
  0.03  0.28  15    [                    0>>>                       ]
 -0.07 -0.62  16    [                <<<<<<<0                       ]
  0.00  0.00  17    [                    0                          ]
 -0.02 -0.21  18    [                   <<0                         ]
  0.04  0.34  19    [                    0>>>>                      ]
  0.06  0.51  20    [                    0>>>>>>                    ]
 -0.10 -0.82  21    [             <<<<<<<<<<0                       ]
        CHI-SQUARED* =    10.63 FOR DF =   18
```

Figure C3.7 Estimation and diagnostic-checking results for model (C3.1) with the first 84 observations.

than the mean absolute percent error for the fitted model (6.45%), and the 80% confidence intervals for the first six forecasts all contain the observed values. The relatively large forecast errors starting at lag 7 could probably be reduced if the model were reestimated sequentially by adding one observation at a time so that each forecast was only a one-step-ahead forecast.

Table C3.2   Forecasts from model (C3.1) using the first 84 observations

Time		Forecast Values	80% Confidence Limits		Future Observed Values	Percent Forecast Errors
			Lower	Upper		
59	1	35918.3000	31982.6000	39853.9000	34748.0000	−3.37
	2	36666.0000	32424.0000	40907.9000	36461.0000	−0.56
	3	36535.2000	31859.6000	41210.8000	35754.0000	−2.18
	4	36735.3000	31875.4000	41595.1000	36943.0000	0.56
	5	36771.6000	31766.3000	41776.8000	35854.0000	−2.56
	6	36853.8000	31762.9000	41944.6000	37912.0000	2.79
	7	36899.1000	31749.1000	42049.1000	30095.0000	−22.61
	8	36945.1000	31757.0000	42133.1000	28931.0000	−27.70
	9	36978.9000	31765.3000	42192.4000	31020.0000	−19.21
	10	37008.0000	31777.6000	42238.4000	31746.0000	−16.58
	11	37031.2000	31789.6000	42272.8000	34613.0000	−6.99
	12	37050.3000	31801.3000	42299.3000	37901.0000	2.24

Alternative models.   Before leaving this case let us return to the identification stage to consider a model other than (C3.1). This exercise will demonstrate the value of the overfitting strategy and the principle of parsimony. (Overfitting is discussed in Chapter 9.)

Suppose a less careful analysis of the acf and pacf in Figure C3.2 had led us to start with an AR(1) model rather than an AR(2). We might have entertained this model:

$$(1 - \phi_1 B)\tilde{z}_t = a_t \tag{C3.4}$$

or

$$z_t = C + \phi_1 z_{t-1} + a_t$$

The results of estimating this model are shown in Figure C3.8. The estimated coefficient $\hat{\phi}_1$ satisfies the stationarity condition since its absolute value is less than one. It is also significantly different from zero at better than the 5% level since its absolute t-value is larger than 2.0. But the residual acf shows that (C3.4) is not adequate. This is indicated especially by the large residual autocorrelation at lag 1, but those at lags 2 and 3 also have absolute t-values exceeding the short-lag residual acf warning level of 1.25.

```
+ + + + + + + + +ECOSTAT UNIVARIATE B-J RESULTS+ + + + + + + + + +
+ FOR DATA SERIES:   COAL PRODUCTION                                    +
+ DIFFERENCING:      0                              DF    = 93          +
+ AVAILABLE:          DATA = 96   BACKCASTS = 0     TOTAL = 96          +
+ USED TO FIND SSR:  DATA = 95   BACKCASTS = 0     TOTAL = 95          +
+ (LOST DUE TO PRESENCE OF AUTOREGRESSIVE TERMS:                1)      +
```

```
COEFFICIENT    ESTIMATE      STD ERROR     T-VALUE
  PHI    1      0.666         0.073          9.09
  CONSTANT     12398.5       2764.15        4.48548

MEAN           37155.1        983.625       37.7737

ADJUSTED RMSE = 3193.31    MEAN ABS % ERR =   6.49
        CORRELATIONS
         1       2
  1    1.00
  2   -0.06    1.00
```

```
++RESIDUAL ACF++
  COEF   T-VAL  LAG                                O
-0.23   -2.26    1    <<<[<<<<<<<<<<<<<<<<<<<<<<<O                  ]
 0.17    1.59    2    [                    O>>>>>>>>>>>>>>>>>       ]
 0.18    1.59    3    [                    O>>>>>>>>>>>>>>>>>>      ]
 0.02    0.18    4    [                    O>>                      ]
 0.10    0.87    5    [                    O>>>>>>>>>>              ]
 0.03    0.29    6    [                    O>>>                     ]
 0.07    0.62    7    [                    O>>>>>>>                 ]
 0.05    0.46    8   [                     O>>>>>                    ]
 0.00    0.01    9   [                     O                        ]
 0.03    0.26   10   [                     O>>>                      ]
-0.07   -0.60   11   [              <<<<<<<O                         ]
 0.08    0.70   12   [                     O>>>>>>>>                 ]
-0.09   -0.74   13   [            <<<<<<<<<O                         ]
-0.02   -0.14   14   [                  <<O                         ]
 0.07    0.64   15   [                     O>>>>>>>                  ]
-0.07   -0.59   16   [              <<<<<<<O                         ]
 0.04    0.38   17   [                     O>>>>                     ]
-0.05   -0.43   18   [                <<<<<O                         ]
-0.01   -0.08   19   [                    <O                         ]
 0.07    0.60   20   [                     O>>>>>>>                  ]
-0.16   -1.38   21   [          <<<<<<<<<<<<<<<O                     ]
 0.04    0.30   22   [                     O>>>>                     ]
-0.04   -0.35   23   [                  <<<<O                        ]
-0.10   -0.82   24   [              <<<<<<<<<O                       ]
  CHI-SQUARED* =    22.95 FOR DF =   22
```

Figure C3.8   Estimation and diagnostic-checking results for model (C3.4).

Ideally, we would now return to Figure C3.2 and discover that an AR(2) model is indicated by the estimated acf and pacf there. But the beginning analyst often experiments a great deal and it may be helpful if we examine the results of several experiments. Letting the overfitting strategy guide us, we could extend (C3.4) and estimate an AR(2). This takes us immediately to model (C3.1), which we have already seen is an adequate representation of the available data.

Another approach is to add MA terms to (C3.4) at the short-lag lengths in an attempt to clean up the residual autocorrelations. Suppose we attack the most significant residual autocorrelation in Figure C3.8 by postulating an ARMA(1, 1) model:

$$(1 - \phi_1 B)\tilde{z}_t = (1 - \theta_1 B)a_t \qquad (C3.5)$$

or

$$z_t = C + \phi_1 z_{t-1} - \theta_1 a_{t-1} + a_t$$

We have stumbled onto model (C3.5) by pretending to experiment as a novice modeler might. But note that (C3.5) also is defensible as an initial model based on the original acf and pacf in Figure C3.2; that acf decays, and we could also interpret the pacf as decaying if we stretch our imaginations. When the estimated acf and pacf both decay, a mixed model is called for. Identifying the exact order of a mixed model from the initial acf and pacf is often difficult, but a good starting place is the common ARMA(1, 1) model (C3.5).

Estimation results for this model are shown in Figure C3.9. Both estimated coefficients are statistically significant and they satisfy the stationarity and invertibility requirements since their absolute values are less than one. Based on the adjusted RMSE's and chi-squared statistics (C3.5) is not quite as good as (C3.1), but we can live with it.

We might still be suspicious of the adequacy of model (C3.5) because the absolute t-values of the residual autocorrelation at lag 3 exceeds the warning level of 1.25. Consider this ARMA(1, 3) model with θ_2 constrained to zero:

$$(1 - \phi_1 B)\tilde{z}_t = \left(1 - \theta_1 B - \theta_3 B^3\right)a_t \qquad (C3.6)$$

or

$$z_t = C + \phi_1 z_{t-1} - \theta_1 a_{t-1} - \theta_3 a_{t-3} + a_t$$

Estimation of this model (results not shown) produced a t-value for $\hat{\theta}_3$ of only -1.40, while the residual acf t-value at lag 2 rose to 1.39. Adding a θ_2 coefficient to model (C3.5) gives this ARMA(1, 2):

$$(1 - \phi_1 B)\tilde{z}_t = \left(1 - \theta_1 B - \theta_2 B^2\right)a_t \qquad (C3.7)$$

or

$$z_t = C + \phi_1 z_{t-1} - \theta_1 a_{t-1} - \theta_2 a_{t-2} + a_t$$

```
+ + + + + + + + + +ECOSTAT UNIVARIATE B-J RESULTS+ + + + + + + + + +
+                                                                      +
+ FOR DATA SERIES:   COAL PRODUCTION                                   +
+ DIFFERENCING:      0                           DF    = 92            +
+ AVAILABLE:         DATA = 96    BACKCASTS = 0   TOTAL = 96            +
+ USED TO FIND SSR:  DATA = 95    BACKCASTS = 0   TOTAL = 95           +
+ (LOST DUE TO PRESENCE OF AUTOREGRESSIVE TERMS:              1)       +
```

COEFFICIENT		ESTIMATE	STD ERROR	T-VALUE
PHI	1	0.841	0.067	12.51
THETA	1	0.356	0.130	2.73
CONSTANT		5828.66	2530.44	2.30341
MEAN		36738.1	1321.35	27.8035

```
ADJUSTED RMSE =  3059.23    MEAN ABS % ERR =   6.31
     CORRELATIONS
        1      2      3
  1   1.00
  2   0.66   1.00
  3  -0.25  -0.18   1.00
```

```
++RESIDUAL ACF++
  COEF  T-VAL LAG                          0
 -0.06  -0.60   1        [              <<<<<<0                    ]
  0.11   1.11   2        [              0>>>>>>>>>>>               ]
  0.15   1.42   3        [              0>>>>>>>>>>>>>>>           ]
  0.00  -0.01   4     [                 0                          ]
  0.04   0.40   5     [                 0>>>>                      ]
  0.00   0.02   6     [                 0                          ]
  0.03   0.30   7     [                 0>>>                       ]
  0.04   0.35   8     [                 0>>>>                      ]
 -0.04  -0.35   9     [              <<<<0                         ]
 -0.02  -0.18  10     [               <<0                         ]
 -0.09  -0.80  11     [           <<<<<<<<<0                       ]
  0.04   0.33  12     [                 0>>>>                      ]
 -0.11  -0.99  13     [            <<<<<<<<<<<0                    ]
 -0.03  -0.28  14     [               <<<0                        ]
  0.07   0.60  15     [                 0>>>>>>>                   ]
 -0.05  -0.44  16     [             <<<<<0                         ]
  0.04   0.40  17     [                 0>>>>                      ]
  0.00  -0.03  18     [                 0                          ]
  0.04   0.34  19     [                 0>>>>                      ]
  0.10   0.91  20     [                 0>>>>>>>>>>                ]
 -0.10  -0.91  21     [            <<<<<<<<<<0                      ]
  0.06   0.52  22     [                 0>>>>>>                    ]
  0.00   0.02  23     [                 0                          ]
 -0.07  -0.67  24     [             <<<<<<<0                       ]
     CHI-SQUARED* =   11.66 FOR DF =   21
```

Figure C3.9 Estimation and diagnostic-checking results for model (C3.5).

Estimation results for this experiment are shown in Figure C3.10. This model is quite satisfactory. The reader should verify that it is stationary and invertible and that all estimated coefficients are significant at about the 5% level. The overall fit is slightly better than for model (C3.1) based on the smaller RMSE, and the chi-squared statistic is insignificant at the 10% level for 20 degrees of freedom.

```
+ + + + + + + + +ECOSTAT UNIVARIATE B-J RESULTS+ + + + + + + + +
+ FOR DATA SERIES:   COAL PRODUCTION                                    +
+ DIFFERENCING:      0                              DF    = 91          +
+ AVAILABLE:            DATA = 96    BACKCASTS = 0   TOTAL = 96          +
+ USED TO FIND SSR:     DATA = 95    BACKCASTS = 0   TOTAL = 95          +
+ (LOST DUE TO PRESENCE OF AUTOREGRESSIVE TERMS:              1)        +
```

COEFFICIENT	ESTIMATE	STD ERROR	T-VALUE
PHI 1	0. 801	0. 080	10. 00
THETA 1	0. 423	0. 125	3. 38
THETA 2	-0. 225	0. 114	-1. 97
CONSTANT	7363. 16	3013. 68	2. 44324
MEAN	36910. 6	1268. 84	29. 0901

ADJUSTED RMSE = 3016. 31   MEAN ABS % ERR =   6. 24

```
        CORRELATIONS
        1      2      3      4
1    1. 00
2    0. 56   1. 00
3    0. 40  -0. 03   1. 00
4   -0. 20  -0. 12  -0. 10   1. 00
```

```
++RESIDUAL ACF++
  COEF   T-VAL. LAG _____0_____
  0. 02   0. 22    1    [                 0>>                       ]
 -0. 04  -0. 37    2    [              <<<<0                        ]
  0. 10   0. 95    3    [                 0>>>>>>>>>>>              ]
  0. 02   0. 17    4    [                 0>>                       ]
  0. 03   0. 33    5    [                 0>>>                      ]
  0. 01   0. 07    6    [                 0>                        ]
  0. 04   0. 42    7    [                 0>>>>                     ]
  0. 05   0. 51    8    [                 0>>>>>                    ]
 -0. 02  -0. 19    9    [               <<0                        ]
 -0. 03  -0. 27   10    [              <<<0                        ]
 -0. 06  -0. 56   11    [            <<<<<<0                       ]
  0. 03   0. 31   12    [                 0>>>                     ]
 -0. 10  -0. 95   13    [        <<<<<<<<<<0                       ]
 -0. 03  -0. 24   14   [               <<<0                       ]
  0. 07   0. 67   15   [                 0>>>>>>>                  ]
 -0. 03  -0. 29   16   [               <<<0                       ]
  0. 02   0. 14   17   [                 0>>                       ]
 -0. 01  -0. 13   18   [                <0                        ]
  0. 04   0. 40   19   [                 0>>>>                     ]
  0. 08   0. 72   20   [                 0>>>>>>>>                 ]
 -0. 11  -1. 00   21   [          <<<<<<<<<<<0                     ]
  0. 03   0. 31   22   [                 0>>>                      ]
  0. 00   0. 02   23   [                 0                        ]
 -0. 09  -0. 86   24   [           <<<<<<<<<0                      ]
  CHI-SQUARED* =      8. 01 FOR DF =  20
```

Figure C3.10 Estimation and diagnostic-checking results for model (C3.7).

Which model is preferable—(C3.1) or (C3.7)? They appear to fit the data about equally well: the RMSE's are nearly identical and the chi-squared statistics for the residual autocorrelations are also quite similar, though both of these criteria favor model (C3.7) slightly. The important principle of parsimony would lead us to select model (C3.1) since it contains one less parameter. If we are going to use a model with three estimated parameters, we might as well choose (C3.3) since it has an RMSE even smaller than (C3.7).

We need not be excessively concerned about the choice between (C3.1) and (C3.7) since these two models are *nearly equivalent mathematically* as well as in their ability to fit the past. To see this, consider that (C3.7) can also be written as

$$\frac{(1 - \phi_1 B)}{\left(1 - \theta_1 B - \theta_2 B^2\right)} \tilde{z}_t = a_t$$

Substituting the estimated coefficients in Figure C3.10 gives

$$\frac{(1 - 0.801B)}{\left(1 - 0.423B + 0.225B^2\right)} \tilde{z}_t = \hat{a}_t \qquad \text{(C3.8)}$$

If we can show that the coefficient of $\tilde{z}_t$ in (C3.8) is virtually the same as the coefficient of $\tilde{z}_t$ in (C3.1), we will have shown that our two estimated models are nearly the same mathematically. The easiest way to do this is to equate the coefficient of $\tilde{z}_t$ in (C3.8) with the coefficient of $\tilde{z}_t$ in (C3.1), and see if this creates a gross contradiction:

$$\frac{(1 - 0.801B)}{(1 - 0.423B + 0.225B^2)} \overset{?}{=} (1 - 0.432B - 0.311B^2) \qquad \text{(C3.9)}$$

Multiplying through by the denominator on the LHS of (C3.9) gives

$$(1 - 0.801B) \overset{?}{=} (1 - 0.855B + 0.097B^2 + 0.035B^3 - 0.070B^4)$$

$$\text{(C3.10)}$$

The last three RHS terms are relatively small, while the coefficients of B on both sides are fairly close in value. Therefore, (C3.1) and (C3.7), while not identical, are mathematically quite similar. Comparison of Tables C3.1 and C3.3 shows that these two models produce similar forecasts. In this case we would be guided by the principle of parsimony and use (C3.1) for forecasting, or, as mentioned earlier, we might use both models to forecast and see which one turns in the better performance.

With the more careful analysis we applied earlier to the estimated acf and pacf in Figure C3.2, we could avoid having to choose between two similar models. Nevertheless, even without such careful analysis we see that use of the overfitting strategy or application of the principle of parsimony leads us to model (C3.1), though by a more indirect path.

Table C3.3 Forecasts from model (C3.7)

Time		Forecast Values	80% Confidence Limits		Future Observed Values	Percent Forecast Errors
			Lower	Upper		
60	1	36673.2000	32812.3000	40534.1000	n.a.[a]	n.a.
	2	37531.4000	33404.0000	41658.7000	n.a.	n.a.
	3	37407.5000	32805.9000	42009.2000	n.a.	n.a.
	4	37308.4000	32427.0000	42189.9000	n.a.	n.a.
	5	37229.1000	32176.5000	42281.7000	n.a.	n.a.
	6	37165.5000	32006.2000	42324.8000	n.a.	n.a.
	7	37114.7000	31888.2000	42341.2000	n.a.	n.a.
	8	37074.0000	31804.8000	42343.1000	n.a.	n.a.
	9	37041.4000	31745.1000	42337.7000	n.a.	n.a.
	10	37015.3000	31701.7000	42328.9000	n.a.	n.a.
	11	36994.4000	31669.7000	42319.1000	n.a.	n.a.
	12	36977.7000	31645.9000	42309.5000	n.a.	n.a.

[a]n.a. = not available.

Final comments. We make the following points:

1. The experimental approach to modeling in the last section is not recommended as a fundamental strategy, especially if it is a haphazard substitute for careful thinking. Ideally, the analyst engages in enough careful thought at the initial identification stage to keep experimentation at a minimum. Even if diagnostic checking shows a model to be inadequate, it is wise to return to the original acf and pacf to rethink their implications instead of just adding terms to the original model.

However, a certain amount of experimentation is both necessary and useful. It is necessary when the acf's and pacf's used for identification are too ambiguous to allow clear choices. And it can be instructive, especially for the beginning modeler who needs to learn the consequences of estimating alternative models. Above all, experimentation should follow sound guidelines, such as those stated in Chapter 12. The single most important guideline is the principle of parsimony: when two or more models provide equivalent fits to the data, choose the one with the least number of estimated parameters.

Of course, the ultimate test of a model is its ability to forecast. Not surprisingly, (C3.1) and (C3.7) are able to forecast history with about equal accuracy. When model (C3.7) was refitted after dropping the last 12

Table C3.4   Forecasts from model (C3.7) using the first 84 observations

Time		Forecast Values	80% Confidence Limits		Future Observed Values	Percent Forecast Errors
			Lower	Upper		
59	1	35912.8000	32048.1000	39777.5000	34748.0000	−3.35
	2	36445.4000	32404.0000	40486.8000	36461.0000	0.04
	3	36572.9000	32066.1000	41079.7000	35754.0000	−2.29
	4	36677.0000	31885.3000	41468.6000	36943.0000	0.72
	5	36761.9000	31789.7000	41734.2000	35854.0000	−2.53
	6	36831.2000	31742.3000	41920.1000	37912.0000	2.85
	7	36887.7000	31722.6000	42052.9000	30095.0000	−22.57
	8	36933.9000	31718.6000	42149.2000	28931.0000	−27.66
	9	36971.5000	31723.1000	42219.9000	31020.0000	−19.19
	10	37002.2000	31731.9000	42272.6000	31746.0000	−16.56
	11	37027.3000	31742.4000	42312.2000	34613.0000	−6.98
	12	37047.7000	31753.2000	42342.3000	37901.0000	2.25

observations, it produced the percent forecast errors shown in the right-most column of Table C3.4. Comparison of these percent errors with those in Table C3.2 shows that (C3.1) and (C3.7) forecast with nearly identical accuracy.

2. Our success in estimating model (C3.7) illustrates how residual acf t-values can, at times, be badly underestimated by the standard formula, as discussed in Chapter 9. Although the t-value for the residual autocorrelation at lag 2 in Figure C3.8 is only 1.11, the $\hat{\theta}_2$ coefficient in model (C3.7) has an absolute t-value nearly equal to 2.0. Using about 1.25 as a practical warning level for t-values at lags 1, 2, and 3 in residual acf's was helpful in this instance. Keep in mind, however, that 1.25 is only a warning value; there is no guarantee that associated coefficients will prove to have absolute t-values greater than 2.0 at the estimation stage.

CASE 4. HOUSING PERMITS

Activity in the housing industry typically leads the rest of the economy. The number of housing permits issued usually declines before the economy moves into a recession, and rises before a recession ends. Therefore, the level of activity in the housing industry is of interest not only to those within that industry, but also to others following broader economic trends.

In this case study we develop a model to forecast the index of new private housing units authorized by local building permits. These are quarterly, seasonally adjusted data covering the years 1947–1967. Figure C4.1 is a plot of the 84 observations we will analyze.* This is an especially challenging series to model.

Identification. Our first task is to decide if the data are stationary. Inspection of Figure C4.1 indicates that the variance is approximately constant. Figure C4.2 shows the estimated acf and pacf of the undifferenced data. (Why is 21 the proper maximum number of coefficients to calculate?) If the estimated acf would not drop quickly to zero we would suspect a nonstationary mean, but the estimated autocorrelations in Figure C4.2 fall to zero very rapidly. Only the first two autocorrelations extend past the 5% significance level indicated by the square brackets, and no absolute t-values after lag 2 exceed the practical warning level of 1.6. There is no evidence here that differencing is required to achieve a stationary mean.

---

*The data are found in the U.S. Department of Commerce publication *Business Conditions Digest*, July 1978, p. 98.

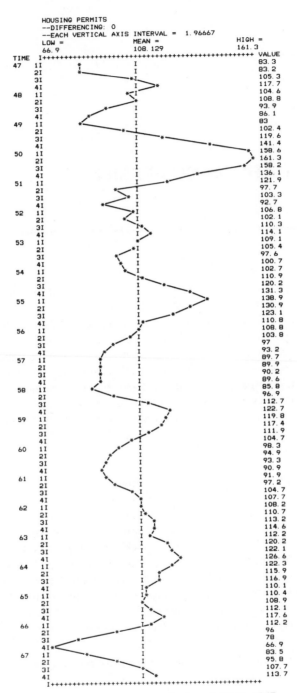

Figure C4.1   Housing permits issued, 1947–1967.

```
+ + + + + + + + + + + + AUTOCORRELATIONS + + + + + + + + + + + + +
+ FOR DATA SERIES:  HOUSING PERMITS                                    +
+ DIFFERENCING:  0                         MEAN    =   108. 129        +
+ DATA COUNT =   84                        STD DEV =   17. 1176        +
   COEF   T-VAL  LAG                              0
   0. 84   7. 65   1                        [        0>>>>>]>>>>>>>>>>>>>>>>>
   0. 54   3. 19   2                        [     0>>>>>>>]>>>>>
   0. 19   1. 03   3                  [        0>>>>>         ]
  -0. 07  -0. 35   4                  [       <<0            ]
  -0. 23  -1. 21   5                  [    <<<<<<0           ]
  -0. 27  -1. 38   6                  [   <<<<<<<0           ]
  -0. 23  -1. 15   7                  [    <<<<<<0           ]
  -0. 18  -0. 90   8                  [     <<<<<0           ]
  -0. 15  -0. 72   9                  [      <<<<0           ]
  -0. 16  -0. 80  10                  [      <<<<0           ]
  -0. 19  -0. 94  11                  [     <<<<<0           ]
  -0. 23  -1. 12  12                  [    <<<<<<0           ]
  -0. 21  -1. 01  13                  [     <<<<<0           ]
  -0. 15  -0. 71  14                  [      <<<0            ]
  -0. 04  -0. 17  15                  [        <0            ]
   0. 08   0. 38  16                  [        0>>           ]
   0. 18   0. 85  17                  [        0>>>>>        ]
   0. 24   1. 09  18                  [        0>>>>>>       ]
   0. 23   1. 06  19              [        0>>>>>>           ]
   0. 18   0. 82  20              [        0>>>>>            ]
   0. 10   0. 44  21              [        0>>>              ]
         CHI-SQUARED* =   152. 05 FOR  DF =   21


+ + + + + + + + + + + PARTIAL AUTOCORRELATIONS + + + + + + + + + + +
   COEF   T-VAL  LAG                              0
   0. 84   7. 65   1                        [        0>>>>>]>>>>>>>>>>>>>>>>>
  -0. 52  -4. 81   2              <<<<<<<<[<<<<<0           ]
  -0. 25  -2. 27   3                       [<<<<<0          ]
   0. 15   1. 41   4                       [      0>>>>     ]
  -0. 09  -0. 82   5                       [    <<0         ]
   0. 07   0. 66   6                       [      0>>       ]
  -0. 04  -0. 32   7                       [     <0         ]
  -0. 21  -1. 89   8                       [<<<<<0          ]
   0. 02   0. 21   9                       [      0>        ]
  -0. 18  -1. 69  10                       [<<<<<0          ]
  -0. 02  -0. 18  11                       [      0         ]
  -0. 04  -0. 39  12                       [     <0         ]
   0. 07   0. 63  13                       [      0>>       ]
  -0. 03  -0. 27  14                       [     <0         ]
   0. 08   0. 73  15                       [      0>>       ]
  -0. 01  -0. 11  16                       [      0         ]
   0. 01   0. 13  17                       [      0         ]
   0. 05   0. 48  18                       [      0>        ]
  -0. 04  -0. 36  19                       [     <0         ]
  -0. 02  -0. 15  20                       [      0         ]
  -0. 01  -0. 09  21                       [      0         ]
```

Figure C4.2 Estimated acf and pacf for the realization in Figure C4.1.

Although the series is adjusted for seasonal variation before publication, it might still have a seasonal pattern if the adjustment procedure is flawed. To check for this possibility, look for significant autocorrelations at lags which are multiples of the length of seasonality (lags $s, 2s, 3s, \ldots$). With $s = 4$ (for quarterly data) we look at lags $4, 8, 12, \ldots$.

The acf in Figure C4.2 shows no significant autocorrelations at the seasonal lags: none of the absolute t-values at those lags exceeds the practical warning level of 1.25. Sometimes a seasonal pattern shows up more clearly in the estimated acf for the first differences (see Figure C4.3). The autocorrelation at lag 4 now has a large absolute t-value (2.49), but it is difficult to tell if this represents a seasonal pattern or if it merely fits in with the overall wavelike decay starting from lag 1. In this case differencing has not removed the nonseasonal pattern sufficiently to allow a better judgment regarding the possible seasonal pattern. Since the data have been seasonally adjusted, and since there is no clear evidence that a seasonal pattern remains, we assume for the time being that the seasonal adjustment is adequate. After modeling the series, we can inspect the residual acf for evidence of a seasonal pattern.

Our next concern is to tentatively identify one or more models whose theoretical acf's and pacf's are similar to the estimated ones. Before you

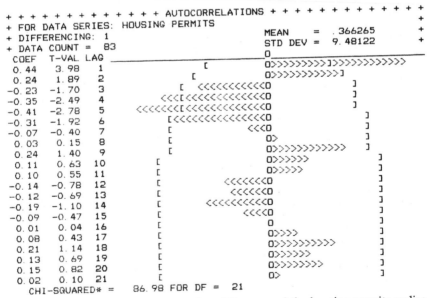

Figure C4.3 Estimated acf for the first differences of the housing-permits realization.

read further, study the acf and pacf in Figure C4.2 carefully. (We have decided that differencing is not needed to achieve a stationary mean, so Figure C4.2 is the relevant one.) Based on your analysis of these graphs, write down one or more ARMA models that seem to be good candidates to explain the data. Then compare your reasoning with the discussion in the next several paragraphs.

First, consider a pure AR model. This is not a very good choice. The acf in Figure C4.2 does not decay as it should if an AR(p) model is called for. If it decays it certainly does so very rapidly. Instead, the autocorrelations seem to cut off after lag 2. Although we might describe the acf in Figure C4.2 as decaying in wavelike fashion, the autocorrelations after lag 2 are all quite insignificant. None of the absolute t-values after lag 2 exceeds the practical warning level of 1.6 much less the 5% standard-error level indicated by the square brackets. It does not seem justified to interpret this estimated acf as decaying either exponentially or in a damped sine-wave pattern. Furthermore, the estimated pacf in Figure C4.2 appears to decay rather than cut off to zero. This is further evidence that a pure AR model is not an attractive alternative.

What about a pure MA model? The estimated acf cuts off to zero at lag 3, suggesting an MA(2). The pacf damps out with sign changes along the first several lags; this is also consistent with an MA(2). [It may help you to review the examples of MA(2) acf's and pacf's in Chapter 6 or 12.] A pure MA model of order two is clearly preferable to a pure AR model.

Finally, consider a mixed model. Properly identifying a mixed model at the initial identification stage can be difficult. Nevertheless, we might at least consider whether a common mixed model, the ARMA(1, 1), is consistent with the estimated acf and pacf in Figure C4.2.

Return again to the examples of theoretical acf's and pacf's in Chapter 6 or 12. Are the present estimated acf and pacf consistent with an ARMA(1, 1) model? In general, both the acf and pacf of a mixed model tail off toward zero rather than cutting off to zero. Although the estimated pacf in Figure C4.2 decays, we have already argued that the estimated acf cuts off after lag 2. Thus an ARMA(1, 1) model is less preferable than an MA(2).

Based on the preceding analysis, the model of choice is the MA(2):

$$\tilde{z}_t = \left(1 - \theta_1 B - \theta_2 B^2\right) a_t \qquad \text{(C4.1)}$$

or

$$z_t = C - \theta_1 a_{t-1} - \theta_2 a_{t-2} + a_t$$

Estimation and diagnostic checking. Estimation results for model (C4.1) appear in Figure C4.4 along with the residual acf and pacf for diagnostic checking. The two estimated coefficients ($\hat{\theta}_1$ and $\hat{\theta}_2$) are significantly

```
+ + + + + + + + + +ECOSTAT UNIVARIATE B-J RESULTS+ + + + + + + + +
+ FOR DATA SERIES:   HOUSING PERMITS                                    +
+ DIFFERENCING:      0                          DF    = 81              +
+ AVAILABLE:         DATA = 84    BACKCASTS = 0  TOTAL = 84              +
+ USED TO FIND SSR:  DATA = 84    BACKCASTS = 0  TOTAL = 84              +
+ (LOST DUE TO PRESENCE OF AUTOREGRESSIVE TERMS:              0)        +

COEFFICIENT    ESTIMATE      STD ERROR     T-VALUE
  THETA  1      -1.133         0.072        -15.73
  THETA  2      -0.771         0.072        -10.70
  CONSTANT     106.837        2.5749        41.4915

  MEAN         106.837        2.5749        41.4915

  ADJUSTED RMSE =  8.55591    MEAN ABS % ERR =   6.11
        CORRELATIONS
          1      2      3
   1    1.00
   2    0.60   1.00
   3    0.16   0.08   1.00
```

```
++RESIDUAL ACF++                                   0
 COEF   T-VAL LAG _____
 0.21   1.95   1                        [        0>>>>>>>>>>]>
 0.29   2.54   2                     [           0>>>>>>>>>>>>]>>
 0.17   1.37   3                     [           0>>>>>>>>    ]
-0.15  -1.23   4                     [    <<<<<<<0            ]
-0.17  -1.37   5                     [    <<<<<<<<0           ]
-0.13  -1.02   6                [       <<<<<<<0              ]
-0.15  -1.11   7                [      <<<<<<<0               ]
-0.09  -0.70   8                [         <<<<<0              ]
-0.01  -0.07   9                [              0              ]
-0.12  -0.87  10                [         <<<<<0              ]
-0.05  -0.36  11                [            <<0              ]
-0.18  -1.36  12                [      <<<<<<<<0              ]
-0.07  -0.50  13                [           <<<0              ]
-0.14  -0.99  14                [       <<<<<<0               ]
 0.01   0.08  15                [              0>             ]
 0.05   0.36  16                [              0>>>           ]
 0.09   0.63  17                [              0>>>>          ]
 0.18   1.24  18                [              0>>>>>>>>>     ]
 0.12   0.83  19                [              0>>>>>         ]
 0.09   0.63  20                [              0>>>>          ]
 0.08   0.56  21                [              0>>>>          ]
      CHI-SQUARED* =    38.36 FOR DF =  18
+RESIDUAL  PACF+                                   0
 COEF   T-VAL LAG _____
 0.21   1.95   1                        [        0>>>>>>>>>>]>
 0.26   2.34   2                        [        0>>>>>>>>>>]>>>
 0.08   0.70   3                        [        0>>>>       ]
-0.30  -2.72   4          <<<<<[<<<<<<<<<<0                  ]
-0.21  -1.96   5             <[<<<<<<<<<<<0                  ]
 0.03   0.32   6                        [        0>>         ]
 0.08   0.69   7                        [        0>>>>       ]
-0.03  -0.26   8                        [       <0           ]
-0.04  -0.37   9                        [       <<0          ]
-0.17  -1.56  10                        [ <<<<<<<<0          ]
-0.06  -0.52  11                        [     <<<0           ]
-0.16  -1.51  12                        [ <<<<<<<<0          ]
 0.05   0.41  13                        [        0>>         ]
-0.06  -0.59  14                        [     <<<0           ]
 0.06   0.58  15                        [        0>>>        ]
 0.00   0.04  16                        [        0           ]
 0.00   0.01  17                        [        0           ]
 0.05   0.47  18                        [        0>>>        ]
 0.02   0.19  19                        [        0>          ]
-0.02  -0.14  20                        [       <0           ]
 0.04   0.35  21                        [        0>>         ]
```

Figure C4.4 Estimation and diagnostic-checking results for model (C4.1).

different from zero and satisfy the invertibility requirements for an MA(2). However, the residual acf is not good at all: the absolute t-values at lags 1 and 2 are far larger than the residual acf short-lag warning level of 1.25. This suggests that the data have an AR pattern since we already have MA coefficients at lags 1 and 2.

The residual pacf is not very helpful in suggesting which AR coefficients we should try. It has 4 spikes that are significant (or nearly so)—at lags 1, 2, 4, and 5. It would violate the principle of parsimony to add four AR terms to model (C4.1). Furthermore, if AR terms are called for, we may not need θ_1 and θ_2 in the model. The wisest choice is to reevaluate the original estimated acf and pacf in Figure C4.2.

Further identification, estimation, and diagnostic checking. One point seems clear as we reconsider Figure C4.2: we should expect to end up with a mixed model. The diagnostic-checking results for model (C4.1) indicate that AR terms are needed, while the estimated pacf in Figure C4.2 decays rather than cutting off to zero. If it cut off to zero it would suggest a pure AR model. Thus it seems we are now forced to conclude that the acf in Figure C4.2 decays since mixed models have acf's and pacf's that tail off. Beyond this there is little additional information we can extract from Figure C4.2 unless we want to derive theoretical acf's and pacf's for a large number of mixed models and compare these with our initial estimated acf and pacf. Instead, a good choice is to try an ARMA(1, 1), a common, parsimonious mixed model:

$$(1 - \phi_1 B)\tilde{z}_t = (1 - \theta_1 B)a_t \qquad (C4.2)$$

or

$$z_t = C + \phi_1 z_{t-1} - \theta_1 a_{t-1} + a_t$$

Figure C4.5 shows the estimation and diagnostic-checking results for model (C4.2). Both $\hat{\phi}_1$ and $\hat{\theta}_1$ have absolute t-values greater than 2.0 and they satisfy the applicable stationarity and invertibility conditions: $|\hat{\phi}_1| < 1$ and $|\hat{\theta}_1| < 1$. Unfortunately, the residual acf is miserable: the first six absolute t-values all approach or exceed the relevant practical warning levels (1.25 at the first two or three lags and 1.6 thereafter). The residual acf has a rough decaying pattern implying that another AR term is needed. The residual pacf also seems to decay, however crudely. If they both decay, the residuals from model (C4.2) follow a mixed model and we need to add both AR and MA terms to model (C4.2).

We have not had much success in finding either a completely satisfactory model or one which seems good except for some simple and obvious

```
+ + + + + + + + +ECOSTAT UNIVARIATE B-J RESULTS+ + + + + + + + +
+ FOR DATA SERIES:   HOUSING PERMITS                                 +
+ DIFFERENCING:      0                           DF    = 80          +
+ AVAILABLE:          DATA = 84    BACKCASTS = 0   TOTAL = 84         +
+ USED TO FIND SSR: DATA = 83    BACKCASTS = 0   TOTAL = 83          +
+ (LOST DUE TO PRESENCE OF AUTOREGRESSIVE TERMS:            1)       +

COEFFICIENT    ESTIMATE      STD ERROR      T-VALUE
PHI    1        0.778         0.077          10.06
THETA  1       -0.333         0.119          -2.80
CONSTANT       24.2868        8.44199        2.8769

MEAN           109.509        5.59856        19.5603

ADJUSTED RMSE =  8.47442    MEAN ABS % ERR =   6.34
     CORRELATIONS
      1      2      3
1    1.00
2    0.46   1.00
3    0.10   0.05   1.00

++RESIDUAL ACF++
 COEF  T-VAL LAG _____0_____
 0.13   1.19   1            [           0>>>>>>>  ]
 0.35   3.10   2            [           0>>>>>>>>>>>>]>>>>>
-0.22  -1.74   3        [<<<<<<<<<<<<0              ]
-0.18  -1.43   4        [  <<<<<<<<<0              ]
-0.30  -2.26   5      <[<<<<<<<<<<<<<0                ]
-0.26  -1.85   6      [<<<<<<<<<<<<<<0                ]
-0.02  -0.16   7        [            <0             ]
-0.09  -0.64   8        [        <<<<<0             ]
 0.22   1.50   9        [           0>>>>>>>>>>>   ]
-0.03  -0.19  10        [           <0             ]
 0.13   0.90  11        [           0>>>>>>>       ]
-0.16  -1.08  12      [     <<<<<<<0                ]
-0.01  -0.08  13      [            <0               ]
-0.15  -0.98  14      [     <<<<<<<0                ]
-0.02  -0.16  15      [           <0                ]
 0.04   0.24  16      [           0>>               ]
 0.03   0.16  17      [           0>                ]
 0.21   1.38  18      [           0>>>>>>>>>>       ]
 0.04   0.22  19      [           0>>               ]
 0.17   1.06  20      [           0>>>>>>>          ]
-0.02  -0.11  21      [           <0                ]
     CHI-SQUARED* =   53.83 FOR DF =  18
+RESIDUAL  PACF+
 COEF  T-VAL LAG _____0_____
 0.13   1.19   1            [         0>>>>>>>  ]
 0.33   3.04   2            [         0>>>>>>>>>]>>>>>>>
-0.33  -3.02   3    <<<<<<<[<<<<<<<<<<0           ]
-0.29  -2.61   4     <<<<[<<<<<<<<<<0           ]
-0.06  -0.57   5          [      <<<0           ]
-0.14  -1.25   6          [   <<<<<<<0          ]
 0.08   0.73   7          [         0>>>>       ]
-0.11  -1.00   8          [      <<<<<0         ]
 0.07   0.60   9          [         0>>>        ]
-0.11  -0.98  10          [      <<<<<0         ]
-0.08  -0.75  11          [       <<<<0         ]
-0.16  -1.42  12          [ <<<<<<<<0           ]
-0.01  -0.12  13          [         <0          ]
-0.04  -0.41  14          [        <<0          ]
-0.04  -0.39  15          [        <<0          ]
 0.01   0.13  16          [         0>          ]
-0.06  -0.52  17          [       <<<0          ]
 0.09   0.83  18          [         0>>>>>      ]
 0.01   0.06  19          [         0           ]
 0.00   0.01  20          [         0           ]
 0.09   0.81  21          [         0>>>>       ]
```

Figure C4.5 Estimation and diagnostic-checking for model (C4.2).

modifications. We have determined that an ARMA(p, q) model is appropriate, but we have encountered difficulty in identifying the orders of the model (the values of p and q.)

One of the practical rules in Chapter 12 is that identification of mixed models is sometimes easier if we start with a pure AR model, then modify based on the residual acf. Starting with AR terms tends to give a cleaner residual acf from which appropriate MA terms may be chosen. Starting with just MA terms tends to remove only a few significant autocorrelations, often leaving a messy decaying or damped sine-wave pattern in the residual acf. Starting with AR terms eliminates a large number of significant autocorrelations since AR terms are associated with decaying or damped sine-wave patterns; it is often relatively easy then to spot the few remaining problem areas in the residual acf.

To apply the strategy suggested above, start by fitting an AR(1) model to the data:

$$(1 - \phi_1 B)\tilde{z}_t = a_t \tag{C4.3}$$

or

$$z_t = C + \phi_1 z_{t-1} + a_t$$

Figure C4.6 shows the estimation results and the residual acf. $\hat{\phi}_1$ is highly significant and satisfies the stationarity condition $|\hat{\phi}_1| < 1$, so we next examine the residual acf to see if just a few MA terms could improve this model. The residual acf in Figure C4.6 has a decaying, wavelike pattern rather than cutting off quickly to zero: the first six residual autocorrelations all have absolute t-values larger than the relevant warning levels. Apparently, the residuals from model (C4.3) can be described with an AR or ARMA model; in either case, there is an autoregressive element remaining in these residuals. The residual pacf in Figure C4.6 suggests that the pattern in the residuals is ARMA rather than just AR since the pacf decays. Our present strategy is to focus on the AR portion of the model until we can isolate the MA part.

The autoregressive element in the residuals for model (C4.3) could be AR(1), AR(2), or even AR(3); the damped sine-wave appearance of the residual acf in Figure (C4.6) is more consistent with an AR(2) or AR(3), but sometimes an AR(1) will produce an *estimated* acf that displays waves rather than a simple decay.

If the AR element in the residuals of model (C4.3) is AR(2), it suggests that the original series z_t requires an AR(3) model. Alternatively, if the residuals follow an AR(1) pattern, it suggests that z_t requires an AR(2). This

```
+ + + + + + + + +ECOSTAT UNIVARIATE B-J RESULTS+ + + + + + + + + +
+ FOR DATA SERIES:   HOUSING PERMITS                                    +
+ DIFFERENCING:      0                           DF    = 81             +
+ AVAILABLE:              DATA = 84   BACKCASTS = 0   TOTAL = 84         +
+ USED TO FIND SSR:  DATA = 83   BACKCASTS = 0   TOTAL = 83             +
+ (LOST DUE TO PRESENCE OF AUTOREGRESSIVE TERMS:              1)        +

COEFFICIENT    ESTIMATE      STD ERROR     T-VALUE
  PHI    1      0.836         0.058         14.31
  CONSTANT     18.0683       6.39536        2.82522

  MEAN        110.297        6.19162        17.814

ADJUSTED RMSE =  9.16348    MEAN ABS % ERR =    6.87
        CORRELATIONS
           1      2
  1     1.00
  2     0.13   1.00


++RESIDUAL ACF++                        0
  COEF   T-VAL LAG _____0_____
  0.47   4.25   1            [            0>>>>>>>>>]>>>>>>>>>>>>>>>
  0.27   2.07   2           [       0>>>>>>>>>>>>>]
 -0.19  -1.37   3           [      <<<<<<<<<0             ]
 -0.33  -2.35   4       <<<[<<<<<<<<<<<<<<<0             ]
 -0.41  -2.70   5     <<<<[<<<<<<<<<<<<<<<<<0          ]
 -0.32  -1.98   6       [<<<<<<<<<<<<<<<<0          ]
 -0.10  -0.58   7         [           <<<<<0            ]
 -0.01  -0.06   8         [           0            ]
  0.20   1.16   9         [           0>>>>>>>>>>         ]
  0.09   0.52  10         [           0>>>>>         ]
  0.09   0.50  11         [           0>>>>         ]
 -0.12  -0.68  12         [        <<<<<<0         ]
 -0.10  -0.57  13         [        <<<<<0         ]
 -0.16  -0.92  14         [       <<<<<<<<0         ]
 -0.06  -0.36  15         [           <<<0         ]
  0.02   0.14  16         [           0>         ]
  0.08   0.47  17         [           0>>>>         ]
  0.21   1.18  18         [           0>>>>>>>>>>>         ]
  0.13   0.72  19         [           0>>>>>>>         ]
  0.16   0.86  20         [           0>>>>>>>>         ]
  0.03   0.14  21         [           0>         ]
     CHI-SQUARED* =   84.89 FOR DF =   19
+RESIDUAL  PACF+                        0
  COEF   T-VAL LAG _____0_____
  0.47   4.25   1            [            0>>>>>>>>>]>>>>>>>>>>>>>>>
  0.07   0.64   2            [            0>>>      ]
 -0.44  -3.97   3  <<<<<<<<<<<<<<[<<<<<<<<<<0      ]
 -0.18  -1.61   4            [<<<<<<<<<0      ]
 -0.07  -0.62   5            [        <<<0      ]
 -0.11  -1.01   6            [     <<<<<<0      ]
  0.07   0.62   7            [            0>>>      ]
 -0.11  -1.04   8            [     <<<<<<0      ]
  0.07   0.63   9            [            0>>>      ]
 -0.15  -1.36  10            [     <<<<<<<0      ]
 -0.08  -0.74  11            [        <<<<0      ]
 -0.14  -1.30  12            [     <<<<<<<0      ]
  0.01   0.05  13            [            0      ]
 -0.08  -0.70  14            [        <<<<0      ]
 -0.03  -0.25  15            [          <0      ]
 -0.01  -0.08  16            [            0      ]
 -0.04  -0.37  17            [          <<0      ]
  0.09   0.85  18            [            0>>>>>      ]
 -0.03  -0.27  19            [          <0      ]
  0.04   0.33  20            [            0>>      ]
  0.07   0.60  21            [            0>>>      ]
```

Figure C4.6 Estimation and diagnostic-checking results for model (C4.3).

may be seen by using the substitution procedure described in Chapter 9. First write model (C4.3) with b_t in place of a_t, where b_t is autocorrelated, and write the AR coefficient as ϕ_1^* to distinguish it from other AR coefficients at lag 1 encountered in this procedure:

$$(1 - \phi_1^* B)\tilde{z}_t = b_t \tag{C4.4}$$

Now suppose b_t is described by an AR(2) model:

$$\left(1 - \phi_1' B - \phi_2' B^2\right)b_t = a_t \tag{C4.5}$$

where a_t is not autocorrelated. Solving (C4.5) for b_t and substituting for b_t in (C4.4) gives

$$\left(1 - \phi_1' B - \phi_2' B^2\right)(1 - \phi_1^* B)\tilde{z}_t = a_t \tag{C4.6}$$

Expanding the LHS of (C4.6) we have

$$\left(1 - \phi_1 B - \phi_2 B^2 - \phi_3 B^3\right)\tilde{z}_t = a_t \tag{C4.7}$$

where

$$\phi_1 = \phi_1^* + \phi_1'$$

$$\phi_2 = \phi_2' - \phi_1^* \phi_1'$$

$$\phi_3 = -\phi_1^* \phi_2'$$

We see that (C4.4), an AR(1) with autocorrelated residuals b_t, together with (C4.5), an AR(2) describing the behavior of b_t, imply (C4.7), an AR(3) model for z_t.

Alternatively, suppose b_t follows an AR(1) pattern. Using the same substitution procedure produces an AR(2) for z_t:

$$\left(1 - \phi_1 B - \phi_2 B^2\right)\tilde{z}_t = a_t \tag{C4.8}$$

or

$$z_t = C + \phi_1 z_{t-1} + \phi_2 z_{t-2} + a_t$$

Hewing to the principle of parsimony, we entertain (C4.8) rather than (C4.7) for the time being.

Estimation results and the accompanying residual acf for model (C4.8) appear in Figure C4.7. Both $\hat{\phi}_1$ and $\hat{\phi}_2$ have absolute t-values greater than

2.0, and together they satisfy the stationarity conditions for an AR(2). The residual acf has cleared up considerably. The wavelike pattern has disappeared and only the first three autocorrelations have absolute *t*-values exceeding the relevant residual acf warning levels (1.25 at the first two or three lags, 1.6 thereafter for nonseasonal data).

The residual autocorrelation at lag 2 is so large that we should focus on it. The absolute *t*-values at lags 1 and 3 exceed the practical warning values, but those autocorrelation coefficients may merely be negatively correlated with the dominant autocorrelation at lag 2. Adding a single MA coefficient

```
+ + + + + + + + +ECOSTAT UNIVARIATE B-J RESULTS+ + + + + + + + +
+ FOR DATA SERIES:  HOUSING PERMITS                                    +
+ DIFFERENCING:     0                           DF    = 79             +
+ AVAILABLE:            DATA = 84   BACKCASTS = 0    TOTAL = 84         +
+ USED TO FIND SSR: DATA = 82   BACKCASTS = 0    TOTAL = 82            +
+ (LOST DUE TO PRESENCE OF AUTOREGRESSIVE TERMS:              2)       +

COEFFICIENT    ESTIMATE      STD ERROR    T-VALUE
  PHI    1      1.305         0.093        13.97
  PHI    2     -0.552         0.092        -5.99
  CONSTANT     26.9703        5.62062       4.79846

  MEAN        109.204         3.44334      31.7146

  ADJUSTED RMSE =  7.68362   MEAN ABS % ERR =   5.70
        CORRELATIONS
         1      2      3
  1    1.00
  2   -0.85   1.00
  3   -0.01   0.05   1.00
```

```
++RESIDUAL ACF++
 COEF   T-VAL  LAG _____0_____
-0.18   -1.60   1          [    <<<<<<<<<<O                ]
 0.36    3.20   2          [              O>>>>>>>>>>>>]>>>>>>
-0.20   -1.54   3          [    <<<<<<<<<<O                ]
 0.04    0.33   4     [              O>>                   ]
-0.14   -1.04   5     [         <<<<<<<O                   ]
-0.16   -1.18   6     [         <<<<<<<O                   ]
 0.04    0.31   7     [              O>>                   ]
-0.17   -1.27   8     [         <<<<<<<<O                  ]
 0.20    1.47   9     [              O>>>>>>>>>>           ]
-0.14   -1.01  10     [         <<<<<<O                    ]
 0.15    1.08  11     [              O>>>>>>>              ]
-0.20   -1.39  12     [    <<<<<<<<<<O                     ]
 0.09    0.58  13     [              O>>>>                 ]
-0.13   -0.89  14     [         <<<<<<<O                   ]
 0.04    0.24  15   [                O>>                   ]
 0.02    0.15  16   [                O>                    ]
-0.01   -0.06  17   [                O                     ]
 0.18    1.21  18   [                O>>>>>>>>>            ]
-0.04   -0.26  19   [              <<O                     ]
 0.14    0.94  20   [                O>>>>>>>              ]
-0.04   -0.26  21   [              <<O                     ]
        CHI-SQUARED* =   45.58 FOR DF =   18
```

Figure C4.7   Estimation and diagnostic-checking results for model (C4.8).

at lag 2 to model (C4.8) might produce an acceptable model. To find out, we postulate the following ARMA(2, 2) with θ_1 constrained to zero:

$$\left(1 - \phi_1 B - \phi_2 B^2\right)\tilde{z}_t = \left(1 - \theta_2 B^2\right)a_t \qquad (C4.9)$$

or

$$z_t = C + \phi_1 z_{t-1} + \phi_2 z_{t-2} - \theta_2 a_{t-2} + a_t$$

```
+ + + + + + + + + +ECOSTAT UNIVARIATE B-J RESULTS+ + + + + + + + + +
+ FOR DATA SERIES:  HOUSING PERMITS                                      +
+ DIFFERENCING:     0                            DF     = 78             +
+ AVAILABLE:           DATA = 84    BACKCASTS = 0  TOTAL = 84             +
+ USED TO FIND SSR:  DATA = 82    BACKCASTS = 0  TOTAL = 82             +
+ (LOST DUE TO PRESENCE OF AUTOREGRESSIVE TERMS:            2)           +

COEFFICIENT    ESTIMATE        STD ERROR      T-VALUE
  PHI    1       1.203          0.104          11.57
  PHI    2      -0.530          0.093          -5.67
  THETA  2      -0.408          0.123          -3.33
  CONSTANT      35.7761         7.19599         4.97167

  MEAN         109.271          3.36689        32.4546

ADJUSTED RMSE =   7.12817    MEAN ABS % ERR =    5.30
        CORRELATIONS
        1       2       3       4
1     1.00
2    -0.78    1.00
3     0.42   -0.15    1.00
4    -0.01    0.06    0.02    1.00

++RESIDUAL ACF++
 COEF   T-VAL  LAG _____0_____
-0.01  -0.05    1      [                    <0                        ]
 0.07   0.60    2      [                    0>>>>>>>                   ]
-0.07  -0.59    3      [              <<<<<<<0                         ]
 0.11   1.01    4      [                    0>>>>>>>>>>>               ]
-0.10  -0.86    5      [             <<<<<<<<<<0                       ]
-0.17  -1.51    6      [       <<<<<<<<<<<<<<<<<0                      ]
-0.02  -0.19    7    [                      <<0                         ]
-0.13  -1.11    8    [              <<<<<<<<<<<<<0                       ]
 0.11   0.92    9    [                      0>>>>>>>>>>>                ]
-0.09  -0.73   10    [                <<<<<<<<<0                         ]
 0.05   0.44   11    [                      0>>>>>                       ]
-0.15  -1.26   12    [            <<<<<<<<<<<<<<<0                       ]
 0.05   0.41   13    [                      0>>>>>                       ]
-0.07  -0.58   14    [                <<<<<<<0                           ]
 0.03   0.21   15    [                      0>>>                         ]
 0.01   0.06   16    [                      0>                           ]
 0.01   0.11   17    [                      0>                           ]
 0.16   1.31   18    [                      0>>>>>>>>>>>>>>>>>           ]
-0.01  -0.04   19                           <0
 0.09   0.72   20                           0>>>>>>>>>
-0.01  -0.07   21                           <0
        CHI-SQUARED* =    16.06 FOR DF =   17
```

Figure C4.8 Estimation and diagnostic-checking results for model (C4.9).

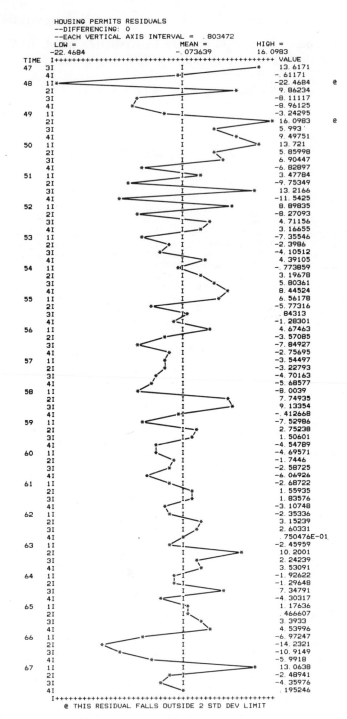

```
HOUSING PERMITS RESIDUALS
--DIFFERENCING: 0
--EACH VERTICAL AXIS INTERVAL =  .803472
        LOW =                  MEAN =              HIGH =
        -22.4684              -.073639            16.0983
TIME  I++++++++++++++++++++++++++++++++++++++++++++++++++ VALUE
47    3I                              I                 *   13.6171
      4I                          *-I                      -.61171
48    1I*                            I                      -22.4684      @
      2I                            I            *         9.86234
      3I             *              I                      -8.11117
      4I           *                I                      -8.96125
49    1I         *        *-I                              -3.24295
      2I                            I                    * 16.0983       @
      3I                            I   *                  5.993
      4I                            I          *           9.49751
50    1I                            I              *       13.721
      2I                            I        *             5.85998
      3I                            I      *               6.90447
      4I          *                 I                      -6.82897
51    1I                            I*  *                  3.47784
      2I               *            I                      -9.75349
      3I               *            I              *       13.2166
      4I         *                  I                      -11.5425
52    1I                            I         *            8.89835
      2I          *                 I                      -8.27093
      3I                            I    *                 4.71156
      4I                            I  *                   3.16655
53    1I            *               I                      -7.35546
      2I                        *   I                      -2.3986
      3I                    *       I                      -4.10512
      4I                            I *                    4.39105
54    1I                          *-I                      -.773859
      2I                            I   *                  3.19678
      3I                            I     *                5.80361
      4I                            I        *             8.44524
55    1I                            I      *               6.56178
      2I              *             I                      -5.77316
      3I                           I*                      .84313
      4I                         *  I                      -1.28301
56    1I                            I    *                 4.67463
      2I                     *      I                      -3.57085
      3I            *               I                      -7.84927
      4I                        *   I                      -2.75695
57    1I                      *     I                      -3.54497
      2I                      *     I                      -3.22793
      3I                    *       I                      -4.70163
      4I                    *       I                      -5.68577
58    1I             *              I                      -8.0039
      2I                            I          *           7.74935
      3I                            I           *          9.13354
      4I                          *-I                      -.412668
59    1I             *              I                      -7.52986
      2I                           I*  *                   2.75238
      3I                           I*                      1.50601
      4I                    *       I                      -4.54789
60    1I                    *       I                      -4.69571
      2I                        *   I                      -1.7446
      3I                      *     I                      -2.58725
      4I             *              I                      -6.06926
61    1I                        *   I                      -2.68722
      2I                           I*                      1.55935
      3I                           I*                      1.83576
      4I                   *        I                      -3.10748
62    1I                        *   I                      -2.35336
      2I                            I*                     3.15239
      3I                            I*                     2.60331
      4I                            .                      .750476E-01
63    1I                       *    I                      -2.45959
      2I                            I             *        10.2001
      3I                       *    I                      2.24239
      4I                            I*                     3.53091
64    1I                        *-I                        -1.92622
      2I                       *-I                         -1.29648
      3I                            I       *              7.34791
      4I                  *         I                      -4.30317
65    1I                          I*                       1.17636
      2I                          I*                       .466607
      3I                           I*                      3.3933
      4I                            I *                    4.53996
66    1I           *                I                      -6.97247
      2I     *                      I                      -14.2321
      3I          *                 I                      -10.9149
      4I             *              I                      -5.9918
67    1I                            I             *        13.0638
      2I                        *   I                      -2.48941
      3I                     *      I                      -4.35976
      4I                           I*                      .195246
      I+++++++++++++++++++++++++++++++++++++++++++++++++++
      @ THIS RESIDUAL FALLS OUTSIDE 2 STD DEV LIMIT
```

Figure C4.9 Residuals for model (C4.9).

Figure C4.8 shows that model (C4.9) is satisfactory. All estimated coefficients are statistically significant and they meet the required stationarity and invertibility conditions. (What are those conditions?) The RMSE is smaller than the one for model (C4.8). None of the absolute correlations among the estimated coefficients exceeds 0.9. Diagnostic checking using the residual acf indicates that we have an adequate model. None of the residual autocorrelations in Figure C4.8 has an absolute t-value larger than our practical warning values, and the chi-squared statistic is not significant at the 10% level. (What is the critical chi-squared value at the 10% level with 17 degrees of freedom?) Note especially that the residual autocorrelations at lags 1 and 3 are now satisfactorily small. We were wise to focus on the residual autocorrelation at lag 2 in Figure C4.7 since it was so much larger than its neighboring values.

Forecasts from model (C4.9) appear in Table C4.1, and the estimation residuals are plotted in Figure C4.9. See if you can calculate the forecasts using the difference-equation form of (C4.9), the estimated coefficients from Figure C4.8, the realization in Figure C4.1, and the residuals in Figure C4.9.

Alternative models.   Compare the results for (C4.9) with those for (C4.7). Estimation and diagnostic-checking results for the latter model appear in Figure C4.10. The coefficient $\hat{\phi}_2$ is so insignificant ($|t| = 0.62$) that we should drop it from the model. Yet the largest residual autocorrelation is at lag 2. Therefore, we replace the AR term at lag 2 with an MA term and estimate the following ARMA(3, 2) model with ϕ_2 and θ_1 excluded:

$$\left(1 - \phi_1 B - \phi_3 B^3\right)\tilde{z}_t = \left(1 - \theta_2 B^2\right)a_t \qquad (C4.10)$$

or

$$z_t = C + \phi_1 z_{t-1} + \phi_3 z_{t-3} - \theta_2 a_{t-2} + a_t$$

Table C4.1   Forecasts from model (C4.9)

Time		Forecast Values	80% Confidence Limits		Future Observed Values	Percent Forecast Errors
			Lower	Upper		
68	1	113.6505	104.5264	122.7745	n.a.[a]	n.a.
	2	112.2701	98.0002	126.5401	n.a.	n.a.
	3	110.5568	91.8589	129.2547	n.a.	n.a.
	4	109.2279	88.5988	129.8570	n.a.	n.a.

[a]n.a. = not available.

```
+ + + + + + + + +ECOSTAT UNIVARIATE B-J RESULTS+ + + + + + + + + +
+ FOR DATA SERIES:  HOUSING PERMITS                                    +
+ DIFFERENCING:        0                          DF    = 77           +
+ AVAILABLE:              DATA = 84    BACKCASTS = 0   TOTAL = 84       +
+ USED TO FIND SSR:  DATA = 81    BACKCASTS = 0   TOTAL = 81           +
+ (LOST DUE TO PRESENCE OF AUTOREGRESSIVE TERMS:              3)       +
```

COEFFICIENT		ESTIMATE	STD ERROR	T-VALUE
PHI	1	1.128	0.104	10.86
PHI	2	-0.099	0.161	-0.61
PHI	3	-0.337	0.103	-3.26
CONSTANT		33.4358	5.90394	5.6633
MEAN		108.414	2.55597	42.416

```
ADJUSTED RMSE =  7.0908    MEAN ABS % ERR =   5.12
     CORRELATIONS
     1      2      3      4
1   1.00
2  -0.84   1.00
3   0.56  -0.85   1.00
4  -0.03   0.02   0.00   1.00
```

```
++RESIDUAL ACF++
 COEF  T-VAL LAG                              0
 0.07   0.67   1      [              0>>>>>>>                    ]
 0.22   1.98   2      [              0>>>>>>>>>>>>>>>>>>>>>>>]
-0.18  -1.50   3   [      <<<<<<<<<<<<<<<<<<O                       ]
 0.05   0.39   4   [                    0>>>>>                      ]
-0.07  -0.55   5   [             <<<<<<<O                           ]
-0.11  -0.91   6   [          <<<<<<<<<<<O                          ]
-0.03  -0.26   7   [                  <<<O                          ]
-0.10  -0.86   8   [           <<<<<<<<<<O                          ]
 0.14   1.16   9   [                    0>>>>>>>>>>>>>>             ]
-0.09  -0.70  10              <<<<<<<<<O
 0.04   0.34  11                       0>>>>
-0.21  -1.70  12       <<<<<<<<<<<<<<<<<<<<<O
 0.01   0.08  13                       0>
-0.14  -1.05  14            <<<<<<<<<<<<<<O
 0.04   0.27  15                       0>>>>
-0.05  -0.40  16                   <<<<<O
 0.05   0.40  17                       0>>>>>
 0.13   0.97  18                       0>>>>>>>>>>>>>
 0.04   0.33  19                       0>>>>
 0.09   0.67  20                       0>>>>>>>>>
-0.04  -0.27  21                   <<<<O
     CHI-SQUARED* =  22.95 FOR DF =  17
```

Figure C4.10 Estimation and diagnostic-checking results for model (C4.7).

See Figure C4.11 for estimation results and the residual acf for (C4.10). The results are disappointing. Both $\hat{\phi}_1$ and $\hat{\phi}_3$ are still significant, but the absolute t-value attached to $\hat{\theta}_2$ is only 1.51. Furthermore, the residual acf has a spike at lag 1 with a t-value substantially larger than the relevant warning level of 1.25.

We try adding a θ_1 coefficient to model (C4.10). This seems reasonable in light of the residual acf spike at lag 1 in Figure C4.11. We also leave θ_2 in

```
+ + + + + + + + + +ECOSTAT UNIVARIATE B-J RESULTS+ + + + + + + + +
+ FOR DATA SERIES:  HOUSING PERMITS                                  +
+ DIFFERENCING:     0                         DF   = 77              +
+ AVAILABLE:        DATA = 84   BACKCASTS = 0  TOTAL = 84            +
+ USED TO FIND SSR: DATA = 81   BACKCASTS = 0  TOTAL = 81            +
+ (LOST DUE TO PRESENCE OF AUTOREGRESSIVE TERMS:            3)       +
```

```
COEFFICIENT    ESTIMATE      STD ERROR      T-VALUE
PHI     1       1.015         0.078         13.08
PHI     3      -0.360         0.065         -5.51
THETA   2      -0.225         0.149         -1.51
CONSTANT       37.3364        6.73429        5.54423

MEAN           108.233        2.73445       39.5813

ADJUSTED RMSE =  6.96192    MEAN ABS % ERR =   5.06
      CORRELATIONS
        1      2      3      4
1     1.00
2    -0.64   1.00
3     0.64  -0.48   1.00
4    -0.03   0.03   0.00   1.00
```

```
++RESIDUAL ACF++
  COEF   T-VAL  LAG _____0_____
  0.19    1.70   1   [                      0>>>>>>>>>>>>>>>>>>>>  ]
  0.02    0.14   2   [                      0>>                    ]
 -0.14   -1.23   3   [        <<<<<<<<<<<<<<0                      ]
  0.06    0.48   4   [                      0>>>>>>                ]
 -0.02   -0.15   5   [                     <<0                     ]
 -0.11   -0.96   6   [         <<<<<<<<<<<<0                       ]
 -0.08   -0.70   7   [           <<<<<<<<0                         ]
 -0.08   -0.69   8   [           <<<<<<<<0                         ]
  0.11    0.93   9   [                      0>>>>>>>>>>>           ]
 -0.04   -0.34  10   [                  <<<<0                      ]
 -0.02   -0.19  11   [                    <<0                      ]
 -0.18   -1.51  12   [        <<<<<<<<<<<<<<<<<0                   ]
 -0.01   -0.12  13                        <0
 -0.07   -0.59  14                  <<<<<<<0
  0.03    0.21  15                        0>>>
 -0.05   -0.36  16                   <<<<<0
  0.05    0.41  17                        0>>>>>
  0.14    1.08  18                        0>>>>>>>>>>>>>>
  0.06    0.47  19                        0>>>>>
  0.06    0.48  20                        0>>>>>>
 -0.03   -0.23  21                     <<<0
      CHI-SQUARED* =   16.13 FOR DF =   17
```

Figure C4.11 Estimation and diagnostic-checking results for model (C4.10).

the model even though its absolute t-value in Figure C4.11 is only 1.51. We leave it in because the residual acf spike at lag 2 in Figure C4.10 has a large t-value, which is why we added θ_2 in the first place. It would be unusual to find an insignificant $\hat{\theta}_2$ coefficient at the estimation stage (as we found in Figure C4.11) when there is a corresponding large spike in a preceding residual acf (as we found in Figure C4.10). Therefore, we include both θ_1 and θ_2 in our next model, an ARMA(3, 2) with $\phi_2 = 0$. If $\hat{\theta}_2$ continues to be insignificant we can drop it later. Our new model is

$$(1 - \phi_1 B - \phi_3 B_3)\tilde{z}_t = (1 - \theta_1 B - \theta_2 B^2)a_t \qquad (C4.11)$$

or

$$z_t = C + \phi_1 z_{t-1} + \phi_3 z_{t-3} - \theta_1 a_{t-1} - \theta_2 a_{t-2} + a_t$$

As seen from the estimation results and the residual acf in Figure C4.12, model (C4.11) is satisfactory. All estimated coefficients are highly significant. The residual acf supports the hypothesis that the shocks of (C4.11) are independent: there are no residual acf absolute t-values larger than the relevant practical warning levels and the chi-squared statistic is insignificant at the 10% level. According to the correlation matrix printed above the residual acf, the estimated coefficients are not too highly correlated since none of the absolute correlation coefficients exceeds our rule-of-thumb warning value of 0.9.

Model (C4.11) provides an adequate representation of the available data. It is slightly unusual to need four estimated parameters to adequately model a nonseasonal series, and it is also somewhat uncommon to find a nonseasonal model whose order is greater than two. (The order of the AR portion of the model is $p = 3$.) But neither of these results is so rare as to cause great concern. Forecasts from (C4.11) are shown in Table C4.2. They are roughly similar to the forecasts in Table C4.1, though model (C4.9) forecasts a decline in housing permits in the second quarter of 1968, while model (C4.11) forecasts a slight rise.

Cycles and AR models. The housing industry has historically moved in a strong cyclical pattern matching the overall business cycle fairly closely, though with a lead. Box and Jenkins [1, p. 59] point out that AR(2) processes where $\phi_1^2 + 4\phi_2 < 0$ produce quasiperiodic patterns. For model (C4.9), $\hat{\phi}_1^2 + 4\hat{\phi}_2 = 1.447 - 2.120 = -0.673$, so there is a quasiperiodic element in the model.

Interpreting the estimated acf in Figure C4.2 as a damped sine wave consistent with the AR portion of (C4.9), we see that wave repeating itself starting at about lag 18. This implies that the cycle in the housing-permit series has a period of about 18 quarters or $4\frac{1}{2}$ years. This result is consistent with the length of the typical post-World War II business cycle which has been about three to five years.

Jenkins [34, pp. 59–60] emphasizes that the quasiperiodic cycles generated by AR(2) models, where $\phi_1^2 + 4\phi_2 < 0$, are not deterministic. They show stochastic changes in period, phase, and amplitude depending on the

```
+ + + + + + + + + +ECOSTAT UNIVARIATE B-J RESULTS+ + + + + + + + + +
+ FOR DATA SERIES:   HOUSING PERMITS                                 +
+ DIFFERENCING:      0                            DF    = 76         +
+ AVAILABLE:          DATA = 84    BACKCASTS = 0   TOTAL = 84         +
+ USED TO FIND SSR:  DATA = 81    BACKCASTS = 0   TOTAL = 81         +
+ (LOST DUE TO PRESENCE OF AUTOREGRESSIVE TERMS:              3)     +

COEFFICIENT    ESTIMATE      STD ERROR     T-VALUE
   PHI    1      0. 859        0. 105        8. 20
   PHI    3     -0. 243        0. 091       -2. 68
   THETA  1     -0. 378        0. 127       -2. 98
   THETA  2     -0. 448        0. 136       -3. 29
   CONSTANT    41. 6334        9. 10872      4. 57072

   MEAN       108. 388         3. 50826     30. 8952

   ADJUSTED RMSE =  6. 72294    MEAN ABS % ERR =   4. 82
        CORRELATIONS
        1      2      3      4      5
   1    1. 00
   2   -0. 65   1. 00
   3    0. 56  -0. 47   1. 00
   4    0. 60  -0. 56   0. 52   1. 00
   5   -0. 06   0. 07  -0. 02  -0. 04   1. 00

++RESIDUAL ACF++
   COEF   T-VAL  LAG _____0_____
  -0. 02  -0. 15   1     [                      <<0                            ]
   0. 03   0. 31   2     [                      0>>>                           ]
  -0. 03  -0. 23   3     [                     <<<0                            ]
   0. 06   0. 58   4     [                      0>>>>>>                         ]
  -0. 10  -0. 90   5     [             <<<<<<<<<<<0                            ]
  -0. 12  -1. 10   6     [            <<<<<<<<<<<<0                            ]
  -0. 05  -0. 42   7     [                  <<<<<0                            ]
  -0. 12  -1. 06   8     [            <<<<<<<<<<<<0                            ]
   0. 10   0. 84   9   [                        0>>>>>>>>>>>                    ]
  -0. 07  -0. 56  10   [                 <<<<<<<0                              ]
   0. 02   0. 18  11   [                        0>>                            ]
  -0. 16  -1. 36  12   [        <<<<<<<<<<<<<<<<0                              ]
   0. 04   0. 34  13   [                        0>>>>                          ]
  -0. 08  -0. 67  14   [                 <<<<<<<0                              ]
   0. 07   0. 54  15   [                        0>>>>>>>                        ]
  -0. 05  -0. 41  16   [                   <<<<<0                              ]
   0. 05   0. 37  17   [                        0>>>>>                          ]
   0. 13   1. 05  18   [                        0>>>>>>>>>>>>>                  ]
   0. 03   0. 27  19   [                        0>>>                            ]
   0. 08   0. 66  20   [                        0>>>>>>>>                        ]
  -0. 04  -0. 36  21   [                    <<<<0                              ]
        CHI-SQUARED* =   12. 85 FOR DF =   16
```

Figure C4.12   Estimation and diagnostic-checking results for model (C4.11).

Table C4.2   Forecasts from model (C4.11)

Time		Forecast Values	80% Confidence Limits		Future Observed Values	Percent Forecast Errors
			Lower	Upper		
68	1	114.5840	105.9786	123.1894	n.a.[a]	n.a.
	2	115.0582	101.3689	128.7475	n.a.	n.a.
	3	112.8267	93.9488	131.7046	n.a.	n.a.
	4	110.6943	89.7483	131.6403	n.a.	n.a.

[a] n.a. = not available.

behavior of the random shock a_t. Models of this type can be important in describing cyclical economic data since the cycles in these series are not deterministic.

Final comments.   We can summarize as follows:

1. We have found two parsimonious mixed models (C4.9) and (C4.11) that fit the available data satisfactorily; (C4.9) is more parsimonious, but (C4.11) has a smaller RMSE and chi-squared statistic. It would be wise to monitor the forecast performance of both because one might regularly produce better forecasts.

2. Models (C4.9) and (C4.11) might be nearly the same mathematically. We encountered this phenomenon in Case 3. The reader is encouraged to review the steps followed in Case 3 for checking the mathematical equivalence of two models and to apply the same procedure to (C4.9) and (C4.11).

3. This case illustrates the practical value of starting with a pure AR model when it proves difficult to identify the order of a mixed model. Estimating just a few AR terms often provides a residual acf that clarifies the MA pattern considerably.

Group B

NONSTATIONARY, NONSEASONAL MODELS

CASE 5.   RAIL FREIGHT

In this case we model the quarterly freight volume carried by Class I railroads in the United States measured in billions of ton-miles. The data cover the period 1965–1978, a total of 56 observations. This is close to the minimum number of observations (50) for building a univariate ARIMA model. As in some earlier cases, these data have been seasonally adjusted for simplicity since this is one of the early case studies.*

The data are plotted in Figure C5.1. The series rises through time, so its mean may not be stationary. But deciding if the mean is stationary with visual inspection can be misleading. (This is especially true when the origin of the graph is not zero; the origin in Figure C5.1 is 166.8.) We must rely on the appearance of the estimated acf and on the values of AR coefficients at the estimation stage to decide if the mean of the rail-freight data is stationary.

Inspection of Figure C5.1 suggests that the variance of the series might be increasing along with its level. There is not much supportive evidence, however; there are just a few observations (late 1974 and late 1978) showing greater variability. For the time being, we will analyze the realization in Figure C5.1 without transforming it.

Identification. We begin the analysis by calculating $n/4$ autocorrelation coefficients using the undifferenced data, where n is the number of observa-

---

*The original, unadjusted data are taken from the U.S. Department of Commerce publication *Business Statistics*, 1977 ed., pp. 120 and 264, and various issues of the *Survey of Current Business*.

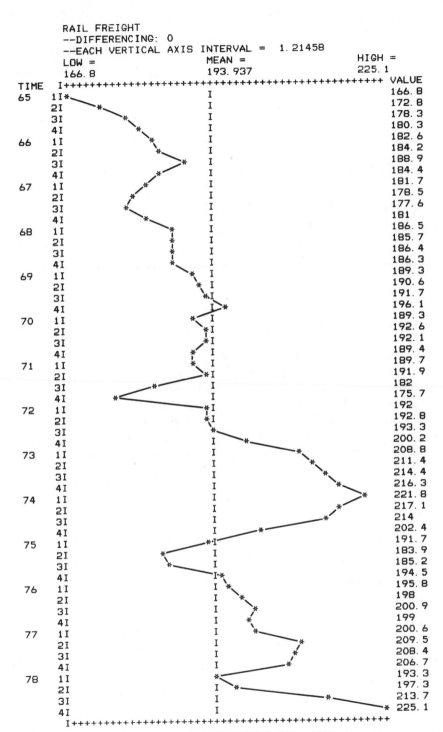

```
RAIL FREIGHT
--DIFFERENCING: 0
--EACH VERTICAL AXIS INTERVAL =    1.21458
   LOW =                    MEAN =              HIGH =
   166.8                    193.937             225.1
TIME  I+++++++++++++++++++++++++++++++++++++++++++++++ VALUE
65    1I*                         I                    166.8
      2I    *                     I                    172.8
      3I        *                 I                    178.3
      4I         *                I                    180.3
66    1I           *              I                    182.6
      2I            *             I                    184.2
      3I               *          I                    188.9
      4I             *            I                    184.4
67    1I           *              I                    181.7
      2I         *                I                    178.5
      3I        *                 I                    177.6
      4I          *               I                    181
68    1I              *           I                    186.5
      2I             *-           I                    185.7
      3I              *-          I                    186.4
      4I              *           I                    186.3
69    1I               *          I                    189.3
      2I                *         I                    190.6
      3I                 *I                            191.7
      4I                   I *                         196.1
70    1I                *  I                           189.3
      2I                 *I                            192.6
      3I                 *I                            192.1
      4I                *- I                           189.4
71    1I                *  I                           189.7
      2I                 *I                            191.9
      3I          *        I                           182
      4I      *             I                          175.7
72    1I                 *I                            192
      2I                 *I                            192.8
      3I                  I*                           193.3
      4I                  I     *                      200.2
73    1I                  I          *                 208.8
      2I                  I             *              211.4
      3I                  I              *             214.4
      4I                  I               *            216.3
74    1I                  I                   *        221.8
      2I                  I                 *          217.1
      3I                  I               *            214
      4I                  I         *                  202.4
75    1I                *·I                            191.7
      2I          *         I                          183.9
      3I           *        I                          185.2
      4I                 I·*                            194.5
76    1I                 I *                           195.8
      2I                 I  *                          198
      3I                 I    *                         200.9
      4I                 I  *                          199
77    1I                 I       *                     200.6
      2I                 I          *                  209.5
      3I                 I         *-                  208.4
      4I                 I        *                    206.7
78    1I                *         I                    193.3
      2I                 I *                           197.3
      3I                 I          *                  213.7
      4I                 I                  *          225.1
      I++++++++++++++++++++++++++++++++++++++++++++++++
```

Figure C5.1 Rail-freight data, 1965–1978.

tions. With $n = 56$ we calculate $56/4 = 14$ autocorrelations and partial autocorrelations shown in Figure C5.2.

Our first concern at the identification stage is whether the mean is stationary. To decide this we look at the estimated acf. If the acf fails to damp out quickly to statistical zeros at longer lags, we suspect nonstationarity. The estimated acf in Figure C5.2 suggests that the mean of the undifferenced realization is stationary: only the first three autocorrelations are statistically different from zero at about the 5% level, and only the first four have absolute t-values larger than the practical warning level discussed in Chapter 12. (The square brackets on the acf are placed at about the 5% significance level corresponding to an absolute t-value of about 2.0. With nonseasonal data, the practical warning level for all absolute t-values at the initial identification stage is around 1.6.)

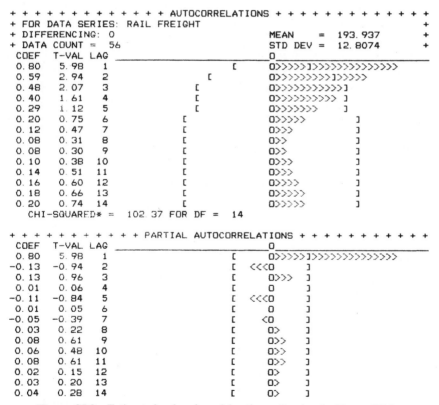

Figure C5.2 Estimated acf and pacf for the realization in Figure C5.1.

Although the mean of the realization in Figure C5.1 seems stationary so that differencing is not needed, we will examine the estimated acf of the first differences anyway. This should allow any seasonal pattern that has not been removed to stand out more clearly. Figure C5.3 shows the estimated acf for the first-differenced series. With quarterly data the seasonal lags are multiples of 4. The autocorrelation at lag 4 is insignificant, while the one at lag 8 has an absolute *t*-value exceeding the seasonal-lag practical warning level of 1.25. Although there might be some slight seasonality remaining in the data represented by the autocorrelation at lag 8, we will ignore it for the time being.

Our next concern at the identification stage is whether we should estimate an AR, an MA, or an ARMA model. According to the estimated acf in Figure C5.2 we should tentatively select an AR model. One piece of evidence is that the acf declines fairly smoothly toward the zero line rather than cutting off abruptly to statistical insignificance. As discussed in Chapter 6, a decaying acf suggests that an AR model may fit the data fairly well.

While an ARMA (mixed) model could also explain a decaying acf, such models are frequently difficult to identify initially. It is helpful to start with a pure AR model when the initial acf decays and move to a mixed model when the residual acf suggests it. There are exceptions to this strategy. For example, an initial acf decaying toward zero entirely from the negative side would be strong evidence for an ARMA(1, 1) model. The decay would rule out an MA model, and the series of entirely negative autocorrelations would rule out the common AR models—AR(1) or AR(2). [See Chapter 6 or 12 for examples of ARMA(1, 1) acf's of this type.]

```
+ + + + + + + + + + + + AUTOCORRELATIONS + + + + + + + + + + + + +
+ FOR DATA SERIES: RAIL FREIGHT                                    +
+ DIFFERENCING: 1                         MEAN    =  1.06          +
+ DATA COUNT =  55                        STD DEV =  5.91598       +
  COEF   T-VAL  LAG _____0_____
  0.27   2.01    1                         0>>>>>>>>>>>>>>1
 -0.11  -0.73    2            [       <<<<<0              ]
 -0.09  -0.62    3            [       <<<<<0              ]
 -0.05  -0.34    4            [         <<0               ]
 -0.11  -0.74    5            [       <<<<<0              ]
 -0.05  -0.33    6            [         <<0               ]
 -0.08  -0.56    7            [       <<<<0               ]
 -0.22  -1.48    8            [ <<<<<<<<<<<0              ]
 -0.21  -1.35    9          [   <<<<<<<<<<<0                ]
 -0.04  -0.23   10          [           <<0                ]
  0.12   0.72   11          [             0>>>>>>          ]
  0.11   0.70   12          [             0>>>>>>          ]
  0.12   0.73   13          [             0>>>>>>          ]
 -0.04  -0.25   14          [            <<0               ]
    CHI-SQUARED* =   16.37 FOR DF =  14
```

Figure C5.3 Estimated acf for the first differences of the rail-freight data.

In this case we tentatively choose an AR model. Next, we turn to the estimated pacf in Figure C5.2 to select the order of the AR model. The single spike at lag 1 in the pacf calls for an AR(1). Spikes in the pacf at both lags 1 and 2 would suggest an AR(2). In general, pacf spikes through lag p followed by a cutoff to zero suggest an AR(p) model. Here we have tentatively identified this model:

$$(1 - \phi_1 B)\tilde{z}_t = a_t \qquad (C5.1)$$

Estimation and diagnostic checking.   Figure C5.4 shows the results of estimating (C5.1). The large t-value (13.79) attached to $\hat{\phi}_1$ indicates it is significantly different from zero at better than the 5% level. With $\hat{\phi}_1 = 0.897$, the condition $|\hat{\phi}_1| < 1$ is met, so it seems that the model is stationary. But we should be cautious in reaching this conclusion because $\hat{\phi}_1$ is less than two

```
+ + + + + + + + + +ECOSTAT UNIVARIATE B-J RESULTS+ + + + + + + + + +
+ FOR DATA SERIES:    RAIL FREIGHT                                    +
+ DIFFERENCING:       0                        DF     = 53            +
+ AVAILABLE:          DATA = 56    BACKCASTS = 0    TOTAL = 56         +
+ USED TO FIND SSR:   DATA = 55    BACKCASTS = 0    TOTAL = 55         +
+ (LOST DUE TO PRESENCE OF AUTOREGRESSIVE TERMS:          1)          +

COEFFICIENT     ESTIMATE       STD ERROR      T-VALUE
  PHI    1       0.897          0.065          13.79
  CONSTANT      20.9471        12.6035          1.662

  MEAN          203.678        10.1054         20.1554

  ADJUSTED RMSE =   5.88928    MEAN ABS % ERR =    2.13
        CORRELATIONS
          1       2
  1    1.00
  2    0.65    1.00

++RESIDUAL ACF++
  COEF   T-VAL  LAG _____0_____
  0.30    2.19    1          [            0>>>>>>>>>>>>>>]>
 -0.09   -0.59    2          [         <<<<0            ]
 -0.08   -0.56    3          [         <<<<0            ]
 -0.02   -0.16    4          [            <0            ]
 -0.08   -0.54    5          [         <<<<0            ]
 -0.02   -0.14    6          [            <0            ]
 -0.08   -0.51    7          [         <<<<0            ]
 -0.22   -1.44    8          [ <<<<<<<<<<<<0            ]
 -0.21   -1.34    9        [   <<<<<<<<<<<0              ]
 -0.05   -0.31   10        [            <<0              ]
  0.10    0.60   11        [            0>>>>>           ]
  0.10    0.63   12        [            0>>>>>           ]
  0.09    0.57   13        [            0>>>>>           ]
 -0.06   -0.36   14        [          <<<0              ]
        CHI-SQUARED* =   15.27  FOR DF =  12
```

Figure C5.4   Estimation and diagnostic-checking results for model (C5.1).

standard errors away from 1.0. That is, testing the null hypothesis H_0: $\phi_1 = 1$ by calculating a t-statistic gives

$$t = \frac{\hat{\phi}_1 - \phi_1}{s(\hat{\phi}_1)}$$

$$= \frac{0.897 - 1}{0.065}$$

$$= -1.58$$

In these calculations we replace ϕ_1 with its hypothesized value (1). The estimated standard error of $\hat{\phi}_1$ is printed at the top of Figure C5.4 under the heading STD ERROR. From these calculations we see that $\hat{\phi}_1$ is 1.58 standard errors below 1.0. Therefore, $\hat{\phi}_1$ is not different from 1.0 at the 10% level. (The critical t-value at the 10% level is about -1.67.)

There is no fixed rule telling us how to proceed in a situation like this; we have to exercise some judgment. There is a practical rule which says to difference the data if we are in doubt about its stationarity. As pointed out in Chapter 7 the chief advantage of this rule is that forecasts from a differenced series are not tied to a fixed mean, so we gain some flexibility by differencing.

But unnecessary differencing typically increases the residual variance of the final model, so we do not want to difference without good reason. Our estimate of ϕ_1 comes close to being different from 1.0 at the 10% level. Furthermore, the residual acf in Figure C5.4 suggests that model (C5.1) is misspecified: the large spike at lag 1 in the residual acf is significant at about the 5% level, and its t-value exceeds the residual acf short-lag practical warning level (1.25) by a substantial margin. So we have evidence that the shocks associated with (C5.1) are not independent, meaning the model is not adequate. The point is this: $\hat{\phi}_1$ could change greatly if another coefficient is added to account for the residual acf spike at lag 1. It might move closer to 1.0 or further away. For now we specify a new model for the undifferenced data rather than differencing. We can always difference the data later if this choice gives unsatisfactory results.

Further identification.   For guidance in modifying (C5.1) we return to the initial acf and pacf calculated from the original data, and we also consider the residual acf in Figure C5.4.

Unfortunately, the original acf and pacf in Figure C5.2 offer no clues about how to modify (C5.1). They are consistent with an AR(1) but with little else.

However, the acf calculated from the residuals of model (C5.1) is more helpful. It says that an MA coefficient at lag 1 should improve our results. An MA term at lag 1 is called for because the only significant residual autocorrelation is at lag 1, but we already have an AR term at that lag. This makes adding an MA term at lag 1 the only sensible alternative. In addition, the residual acf cuts off to statistical insignificance after lag 1, suggesting that an MA term at lag 1 is appropriate. Therefore, we entertain an ARMA(1, 1) model:

$$(1 - \phi_1 B)\tilde{z}_t = (1 - \theta_1 B)a_t \qquad (C5.2)$$

Further estimation and diagnostic checking. Estimating ϕ_1 and θ_1 with a nonlinear least-squares method gives the results in Figure C5.5. The estimated AR coefficient ($\hat{\phi}_1 = 0.834$) is less than 1.0 in absolute value as is the

```
+ + + + + + + + + +ECOSTAT UNIVARIATE B-J RESULTS+ + + + + + + + + +
+ FOR DATA SERIES:  RAIL FREIGHT                                    +
+ DIFFERENCING:    0                            DF    = 52          +
+ AVAILABLE:            DATA = 56   BACKCASTS = 0   TOTAL = 56       +
+ USED TO FIND SSR: DATA = 55   BACKCASTS = 0   TOTAL = 55          +
+ (LOST DUE TO PRESENCE OF AUTOREGRESSIVE TERMS:            1)      +

COEFFICIENT    ESTIMATE        STD ERROR      T-VALUE
PHI     1       0.834          0.086           9.67
THETA   1      -0.403          0.139          -2.90
CONSTANT       33.2991         16.6956         1.99448

MEAN           200.1           7.18721        27.8412

ADJUSTED RMSE =  5.55445    MEAN ABS % ERR =   2.09
      CORRELATIONS
      1      2      3
1    1.00
2    0.37   1.00
3    0.49   0.17   1.00

++RESIDUAL ACF++
 COEF   T-VAL LAG _____0_____
 0.01    0.11   1                       0>
 0.00    0.03   2                       0
-0.05   -0.41   3                  <<<<<0
 0.03    0.25   4                       0>>>
-0.07   -0.51   5                <<<<<<<0
 0.04    0.26   6                       0>>>>
-0.01   -0.05   7                      <0
-0.15   -1.07   8         <<<<<<<<<<<<<<<0
-0.13   -0.97   9           <<<<<<<<<<<<<0
-0.03   -0.22  10                    <<<0
 0.09    0.66  11                       0>>>>>>>>>
 0.02    0.15  12                       0>>
 0.13    0.92  13                       0>>>>>>>>>>>>>
-0.13   -0.88  14          <<<<<<<<<<<<<0
      CHI-SQUARED* =    6.56 FOR DF =   11
```

Figure C5.5   Estimation and diagnostic-checking results for model (C5.2).

estimated MA coefficient ($\hat{\theta}_1 = -0.403$), so we might conclude that (C5.2) is both stationary and invertible. Since the absolute t-values attached to the estimated coefficients are both larger than 2.0, we conclude that ϕ_1 and θ_1 are nonzero and should be included in the model.

Next, we want to test whether the shocks (the a_t elements) of (C5.2) are independent. If they are not independent, we have violated an important assumption made in UBJ–ARIMA modeling and (C5.2) is not adequate. We would then try to identify another model.

The estimation stage residuals ($\hat{a}_t$) are used to construct an estimated acf to test the independence of the shocks. The residual acf for model (C5.2) is plotted in Figure C5.5 below the estimation results. All the absolute t-values of the residual autocorrelations are smaller than the practical warning levels stated in Chapter 12. (These are 1.25 at the first two or three lags in a residual acf, and 1.6 at other lags for nonseasonal data.) Note that the residual autocorrelation at lag 8 now has an absolute t-value less than the seasonal-lag warning level of 1.25. The chi-squared statistic is also insignificant. We conclude that the shocks of (C5.2) are independent and that this model is a statistically adequate representation of the available data. This does not mean we have identified the true process generating these observations. It does mean that we have found a model that fits the available data. It is also important that we have found a parsimonious model: (C5.2) contains only two estimated parameters ($\hat{\phi}_1$ and $\hat{\theta}_1$) in addition to the mean.

An alternative model.   Model (C5.2) has one troubling aspect: the estimated AR coefficient falls on the borderline if we test to see if it is significantly different from 1.0. Testing the null hypothesis $H_0: \phi_1 = 1.0$, we obtain a t-statistic of -1.93; the critical value for 52 degrees of freedom falls between -2.0 and -2.01 for a 5% two-sided test. Although (C5.2) is defensible, we could also defend differencing the realization since there is some evidence against the hypothesis $\phi_1 = 1.0$.

As pointed out in Chapter 6, differencing is preferred in ambiguous circumstances. Therefore, we entertain an ARIMA(0, 1, 1) model based on the estimated acf in Figure C5.3. (Do you agree with this analysis of that acf? Why?) In backshift form the ARIMA(0, 1, 1) is

$$(1 - B)\tilde{z}_t = (1 - \theta_1 B)a_t \qquad \text{(C5.3)}$$

The similarity between the estimated acf in Figure C5.3 and the residual acf in Figure C5.4 is no accident. Each represents the acf for the series z_t after it has been filtered through the operator $(1 - \phi_1 B)$. In one case (Figure C5.3), ϕ_1 has the value 1.0 since the filter is simply the differencing operator $(1 - B)$; in the other case (Figure C5.4), ϕ_1 has the value 0.897

estimated from the data. The numbers 0.897 and 1.0 are close in value, so the two filtering operations produce similar series with similar acf's.

Figure C5.6 shows the results of fitting model (C5.3) to the data and performing diagnostic checks on the residuals. The estimated MA coefficient ($\hat{\theta}_1 = -0.363$) has an absolute value substantially less than 1.0, thus satisfying the invertibility requirement $|\hat{\theta}_1| < 1$. It also has an absolute t-value greater than 2.0; we conclude that $\hat{\theta}_1$ is significantly different from zero. The residual acf shows no absolute t-values greater than the relevant warning levels, and the residual acf chi-squared is not significant. Our conclusion is that model (C5.3) provides a good representation of the rail-freight realization.

Final comments. We conclude:

1. Either model (C5.2) or (C5.3) could be used to forecast. However, our practical rule that differencing should be used in ambiguous cases leads to model (C5.3).

```
+ + + + + + + + +ECOSTAT UNIVARIATE B-J RESULTS+ + + + + + + + +
+ FOR DATA SERIES:   RAIL FREIGHT                                +
+ DIFFERENCING:      1                           DF    = 54      +
+ AVAILABLE:         DATA = 55    BACKCASTS = 0   TOTAL = 55      +
+ USED TO FIND SSR: DATA = 55    BACKCASTS = 0   TOTAL = 55      +
+ (LOST DUE TO PRESENCE OF AUTOREGRESSIVE TERMS:          0)     +

COEFFICIENT    ESTIMATE      STD ERROR      T-VALUE
  THETA  1      -0.363        0.128         -2.84

ADJUSTED RMSE =  5.72658    MEAN ABS % ERR =   2.21

++RESIDUAL ACF++
 COEF   T-VAL LAG _____0_____
-0.02  -0.18   1                        <<0
-0.06  -0.43   2                   <<<<<<0
-0.06  -0.46   3                   <<<<<<0
-0.01  -0.08   4                       <0
-0.10  -0.74   5              <<<<<<<<<<0
-0.01  -0.10   6                      <0
-0.02  -0.12   7                     <<0
-0.16  -1.19   8     <<<<<<<<<<<<<<<<<0
-0.15  -1.07   9       <<<<<<<<<<<<<<<0
-0.02  -0.15  10                    <<0
 0.11   0.79  11                       0>>>>>>>>>>>
 0.03   0.18  12                       0>>>
 0.15   1.06  13                       0>>>>>>>>>>>>>>>
-0.11  -0.74  14           <<<<<<<<<<<0
      CHI-SQUARED* =    8.16 FOR DF =  13
```

Figure C5.6 Estimation and diagnostic-checking results for model (C5.3).

Table C5.1   Forecasts from the ARMA(1, 1) model (C5.2)

Time		Forecast Values	80% Confidence Limits		Future Observed Values	Percent Forecast Errors
			Lower	Upper		
79	1	224.3772	217.2676	231.4869	n.a.[a]	n.a.
	2	220.3373	209.0321	231.6424	n.a.	n.a.
	3	216.9696	203.4977	230.4415	n.a.	n.a.
	4	214.1623	199.3705	228.9541	n.a.	n.a.

[a] n.a. = not available.

2. Tables C5.1 and C5.2 show the forecasts from models (C5.2) and (C5.3), respectively. These two forecast profiles are quite different. The forecasts from (C5.2) converge gradually toward the estimated mean (200.1) in Figure C5.5 because that model is stationary. But the forecasts from (C5.3) are not tied to a fixed mean because that model is nonstationary. This makes (C5.3) more flexible and thus preferable. (See Chapter 10 for examples of how forecasts are calculated.)

3. Suppose for the sake of argument that we do not difference the data and thus arrive at the ARMA(1, 1) model (C5.2). Then this case shows how difficult it can be to identify a mixed model at the initial identification stage. The original acf and pacf in Figure C5.2 do not provide evidence that a mixed model is appropriate. Sometimes with hindsight we can see acf or pacf patterns that are obscure at the first examination, but in this case even hindsight examination of Figure C5.2 is of little help.

Table C5.2   Forecasts from the nonstationary ARIMA(0, 1, 1) model (C5.3)

Time		Forecast Values	80% Confidence Limits		Future Observed Values	Percent Forecast Errors
			Lower	Upper		
79	1	227.4996	220.1696	234.8296	n.a.[a]	n.a.
	2	227.4996	215.1082	239.8910	n.a.	n.a.
	3	227.4996	211.5822	243.4170	n.a.	n.a.
	4	227.4996	208.7065	246.2927	n.a.	n.a.

[c] n.a. = not available.

However, the residual acf in Figure C5.5 is not difficult to analyze. It gives clear evidence that we should estimate an MA coefficient at lag 1, in addition to the previously estimated AR term. This illustrates a practical rule: unless the initial evidence for a mixed model is quite clear, start with a pure AR model and rely on the residual acf for guidance in modifying the original model. This rule is especially helpful in modeling the data in Case 4.

CASE 6.   AT & T STOCK PRICE

The data in Figure C6.1 are the weekly closing price of American Telephone and Telegraph (AT & T) common shares for the year 1979. The observations were taken from various issues of the *Wall Street Journal* (AT & T shares are traded on the New York Stock Exchange). With only 52 observations, we are close to the minimum (about 50) recommended for building an ARIMA model.

Although the variance of this series appears to be roughly constant through time, the mean seems to decline. We will be alert for an estimated acf that decays slowly to zero indicating a nonstationary mean and the need for differencing.

With weekly data the length of seasonality would be 52; but with only 52 observations we cannot calculate 52 autocorrelations, nor can we estimate an AR or MA coefficient with a lag length of 52. Therefore, we ignore the possibility of seasonality with length 52.

Identification. We start by examining the estimated acf and pacf for the undifferenced data in Figure C6.2. The estimated acf falls to zero slowly, indicating that the mean of the data is nonstationary and that nonseasonal differencing is required. We will see if estimation-stage results confirm the need for differencing.

Assuming for the moment that differencing is not needed, we start with an AR(1) model. This is consistent with the combination of a decaying pattern in the estimated acf and the cutoff to zero after lag 1 in the estimated pacf. For a pure MA model we would see just the opposite: spikes

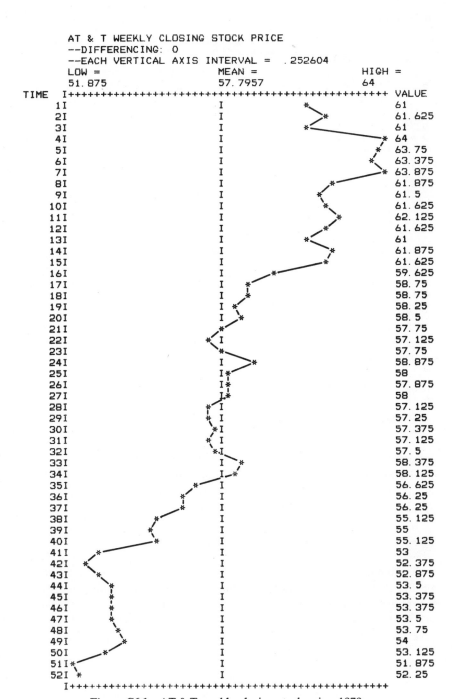

```
          AT & T WEEKLY CLOSING STOCK PRICE
          --DIFFERENCING: 0
          --EACH VERTICAL AXIS INTERVAL =  .252604
          LOW =                 MEAN =              HIGH =
          51.875                57.7957             64
TIME   I++++++++++++++++++++++++++++++++++++++++++++++++++ VALUE
     1I                          I                   *        61
     2I                          I                    *       61.625
     3I                          I                 *          61
     4I                          I                          * 64
     5I                          I                         *  63.75
     6I                          I                        *   63.375
     7I                          I                          * 63.875
     8I                          I                    *       61.875
     9I                          I               *            61.5
    10I                          I                *           61.625
    11I                          I                  *         62.125
    12I                          I                 *          61.625
    13I                          I              *             61
    14I                          I                *           61.875
    15I                          I                *           61.625
    16I                          I           *                59.625
    17I                          I      *                     58.75
    18I                          I      *                     58.75
    19I                          I    *                       58.25
    20I                          I    *                       58.5
    21I                        *  I                           57.75
    22I                        *  I                           57.125
    23I                        *  I                           57.75
    24I                          I  *                         58.875
    25I                        I *                            58
    26I                        I*                             57.875
    27I                        I *                            58
    28I                      * I                              57.125
    29I                      * I                              57.25
    30I                       *I                              57.375
    31I                      * I                              57.125
    32I                      *I                               57.5
    33I                          I *                          58.375
    34I                        I.*                            58.125
    35I                    *     I                            56.625
    36I                   *      I                            56.25
    37I                    *     I                            56.25
    38I                 *        I                            55.125
    39I               *          I                            55
    40I              *           I                            55.125
    41I          *               I                            53
    42I        *                 I                            52.375
    43I         *                I                            52.875
    44I           *              I                            53.5
    45I           *              I                            53.375
    46I           *              I                            53.375
    47I           *              I                            53.5
    48I            *             I                            53.75
    49I             *            I                            54
    50I          *               I                            53.125
    51I*                         I                            51.875
    52I *                        I                            52.25
       I++++++++++++++++++++++++++++++++++++++++++++++++++
```

Figure C6.1 AT & T weekly closing stock price, 1979.

```
+ + + + + + + + + + + AUTOCORRELATIONS + + + + + + + + + + + + +
+ FOR DATA SERIES: AT & T WEEKLY CLOSING STOCK PRICE              +
+ DIFFERENCING: 0                          MEAN    =  57.7957     +
+ DATA COUNT =  52                         STD DEV =  3.4136       +
  COEF   T-VAL  LAG _____0_____
  0.93   6.74    1                           [       0>>>>>]>>>>>>>>>>>>>>>>>>>>
  0.86   3.75    2                       [       0>>>>>>>>>>>]>>>>>>>>>>>
  0.81   2.85    3                   [        0>>>>>>>>>>>>>]>>>>>>>
  0.75   2.29    4              [        0>>>>>>>>>>>>>>>]>>>
  0.68   1.91    5           [         0>>>>>>>>>>>>>>>>]
  0.62   1.62    6       [          0>>>>>>>>>>>>>>>      ]
  0.55   1.38    7       [          0>>>>>>>>>>>>>        ]
  0.49   1.19    8       [          0>>>>>>>>>>>>         ]
  0.44   1.03    9    [             0>>>>>>>>>>>              ]
  0.38   0.87   10    [             0>>>>>>>>>               ]
  0.29   0.65   11    [             0>>>>>>>                 ]
  0.22   0.49   12    [             0>>>>>                   ]
  0.18   0.39   13    [             0>>>>                    ]
      CHI-SQUARED* =  280.32 FOR DF =   13

+ + + + + + + + + + PARTIAL AUTOCORRELATIONS + + + + + + + + + + +
  COEF   T-VAL  LAG _____0_____
  0.93   6.74    1                          [       0>>>>>]>>>>>>>>>>>>>>>>>>>>>>
 -0.09  -0.64    2                          [   <<0       ]
  0.16   1.15    3                          [    0>>>>     ]
 -0.20  -1.46    4                          [<<<<<0        ]
  0.05   0.35    5                          [    0>        ]
 -0.10  -0.75    6                          [   <<<0       ]
 -0.03  -0.22    7                          [    <0        ]
  0.00   0.00    8                          [    0         ]
  0.03   0.20    9                          [    0>        ]
 -0.14  -1.03   10                          [ <<<<0        ]
 -0.21  -1.54   11                          [<<<<<0        ]
  0.05   0.36   12                          [    0>        ]
  0.16   1.14   13                          [    0>>>>     ]
```

Figure C6.2   Estimated acf and pacf for the realization in Figure C6.1.

followed by a cutoff in the acf, and a tailing off of the pacf. For a mixed model we would see a damping out in both the acf and the pacf. An AR(1) is called for rather than a higher-order AR model because the pacf has a spike at lag 1 only. Therefore, the model is

$$(1 - \phi_1 B)\tilde{z}_t = a_t \qquad\qquad (C6.1)$$

Estimation and diagnostic checking.   Figure C6.3 shows the estimation results and the corresponding residual acf. The most striking fact in the estimation results (Figure C6.3) is that $\hat{\phi}_1$ is virtually 1.0. The stationarity condition for an AR(1) is $|\hat{\phi}_1| < 1$. Testing the null hypothesis that $\phi_1 = 1$ gives this t-statistic:

$$t = \frac{0.986 - 1}{0.036} = -0.39$$

```
+ + + + + + + + +ECOSTAT UNIVARIATE B-J RESULTS+ + + + + + + + +
+ FOR DATA SERIES:   AT&T WEEKLY CLOSING STOCK PRICE              +
+ DIFFERENCING:      0                         DF   = 49          +
+ AVAILABLE:         DATA = 52   BACKCASTS = 0  TOTAL = 52         +
+ USED TO FIND SSR.  DATA = 51   BACKCASTS = 0  TOTAL = 51         +
+ (LOST DUE TO PRESENCE OF AUTOREGRESSIVE TERMS:          1)      +
```

COEFFICIENT	ESTIMATE	STD ERROR	T-VALUE
PHI 1	0.986	0.036	27.33
CONSTANT	.629241	2.09346	.300575

MEAN	45.4988	33.5527	1.35604

```
ADJUSTED RMSE =   .865084    MEAN ABS % ERR =    1.04
      CORRELATIONS
        1      2
  1   1.00
  2  -0.97   1.00
```

```
++RESIDUAL ACF++
 COEF   T-VAL LAG _____0_____
-0.03  -0.23   1                            <<<0
-0.19  -1.36   2          <<<<<<<<<<<<<<<<<<<<<0
 0.13   0.89   3                             0>>>>>>>>>>>>>
-0.10  -0.68   4                  <<<<<<<<<<0
-0.07  -0.46   5                     <<<<<<<0
-0.02  -0.15   6                          <<0
 0.00   0.02   7                            0
-0.14  -0.91   8                <<<<<<<<<<<<<0
 0.08   0.54   9                             0>>>>>>>>
 0.21   1.37  10                             0>>>>>>>>>>>>>>>>>>>>>
-0.17  -1.05  11             <<<<<<<<<<<<<<<<<0
-0.17  -1.02  12            <<<<<<<<<<<<<<<<<<0
 0.12   0.75  13                             0>>>>>>>>>>>>
     CHI-SQUARED* =    13.22 FOR DF =   11
```

Figure C6.3 Estimation and diagnostic-checking results for model (C6.1).

$\hat{\phi}_1$ is only 0.39 standard errors below 1.0. This could easily happen just by chance in a sample even if $\phi_1 = 1$, so we accept the null hypothesis that $\phi_1 = 1$.

The estimated mean is highly correlated with $\hat{\phi}_1$ ($r = -0.97$). As indicated in Chapter 12 this is a common result when AR coefficients are estimated for a nonstationary series if the data are not properly differenced. The estimated mean in such cases tends to be unstable. This conclusion is reinforced by the estimated value of the mean ($\hat{\mu} = 45.4988$). The smallest value in the realization is 51.875. An estimated mean that falls entirely outside the range of the data makes little sense.

A practical rule is to difference when we have serious doubts about whether the mean of a series is stationary. We have powerful support for the hypothesis that the mean is nonstationary in this case. Therefore, we try modeling the differenced series.

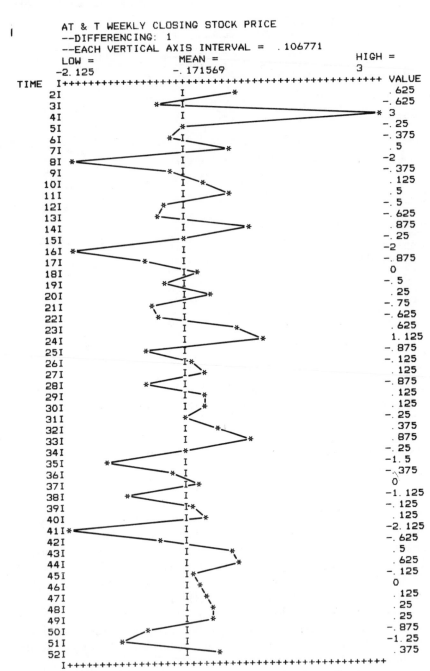

Figure C6.4 First differences of the stock-price realization.

Further identification.   Figure C6.4 is a plot of the first differences of the original data. This series does not show the gradual downward movement displayed by the data in Figure C6.1. Although the mean of the original data falls over time, the first differences seem to fluctuate around a constant mean. It is unlikely that the data require differencing more then once.

The estimated acf for the first differences in Figure C6.5 confirms that further differencing is unnecessary. None of the estimated autocorrelations has an absolute *t*-value larger than the practical warning level of 1.6. (There are no square brackets showing the 5% significance level because they fall beyond the bounds permitted by the scale of this graph. Thus none of the estimated autocorrelations is significant at the 5% level.) With patternless, statistically insignificant autocorrelations we get no further information from the estimated pacf.

The first-differenced series is apparently composed of statistically independent elements. This is the random-walk model discussed in Chapter 5.

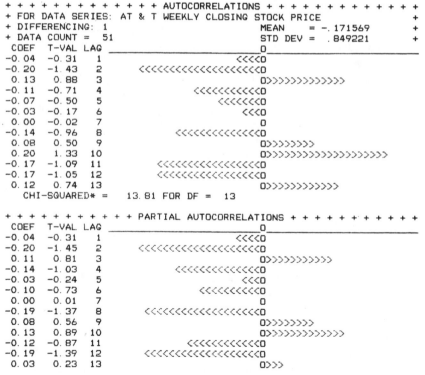

Figure C6.5   Estimated acf and pacf for the data in Figure C6.4.

The random walk is an ARIMA(0, 1, 0) model written as

$$(1 - B)\tilde{z}_t = a_t \tag{C6.2}$$

where $(1 - B)$ is the differencing operator as discussed in Chapter 5. In the backshift notation introduced in that same chapter, the differencing operator is always raised to the power d, where d is the number of times the data are differenced. In this case $d = 1$, so we write $(1 - B)^d$ simply as $(1 - B)$.

Forecasting. Model (C6.2) offers a statistically adequate explanation of the available data. The estimated acf of the first differences (Figure C6.5) does not indicate the need for any further AR or MA terms beyond the implicit AR term ($\phi_1 = 1$) imposed by the differencing procedure. Writing (C6.2) in difference-equation form we get

$$z_t = z_{t-1} + a_t \tag{C6.3}$$

From (C6.3) we see that the one-step-ahead ARIMA forecast ($\hat{z}_t$) for our data is simply the last observed value: $\hat{z}_t = z_{t-1}$. Since we do not know a_t or have an estimate of it when forecasting for time t, we set a_t equal to its expected value of zero. However, we do know the last observed z (z_{t-1}) and that becomes our forecast.

Model (C6.3) contains no constant term. This is because z_t is not stationary; it is not varying around a fixed mean. Furthermore, the first-differenced series ($w_t = z_t - z_{t-1}$) appears to have a fixed mean of about zero, so there is no deterministic (constant) trend element in z_t. If w_t had a nonzero mean, then (C6.2) and (C6.3) would have a constant term (C) equal to the mean of w_t (see Chapter 7). Then the model for w_t would be

$$(w_t - \mu_w) = a_t$$

or $\hspace{12cm}$ (C6.4)

$$w_t = C + a_t$$

where $C = \mu_w$. Substituting $z_t - z_{t-1}$ for w_t, (C6.4) shows that z_t is changing by a constant amount (C) each period, and our forecast then would be

$$\hat{z}_t = C + z_{t-1}$$

Further estimation. As an exercise let us return to the estimated acf in Figure C6.5 and try to construct a model for the first differences. The most significant autocorrelation is at lag 2. The absolute t-value (1.43) does not exceed the relevant warning level (1.6), but we will estimate an MA coefficient at lag 2 to see if it proves significant. (Why is the warning level not 1.25?) An MA coefficient is appropriate since the estimated autocorrelations cut off to zero after lag 2. Let us also suppose the series has a deterministic trend component as discussed in the last section and in Chapter 7. Let $\tilde{w}_t = w_t - \mu_w$. The model to be estimated is an ARIMA(0, 1, 2) with $\theta_1 = 0$:

$$\tilde{w}_t = (1 - \theta_2 B) a_t \tag{C6.5}$$

Estimation of (C6.5) produces this surprising result (absolute t-values are in parentheses):

$$w_t = -0.174 - 0.311 \hat{a}_{t-2} + \hat{a}_t$$
$$(2.12) \quad (2.28)$$

Both the estimated mean (-0.174) and $\hat{\theta}_2$ (0.311) are significant at the 5% level.

In Chapter 5 we said ARIMA modeling is based more on statistical appearances than on prior reasoning regarding what should appear in a model. Nevertheless, an ARIMA model can sometimes be rationalized. But the results of estimating (C6.5) do not make sense. The price of AT & T stock surely does not trend downward every week by the fixed amount 0.174; at that rate the price would reach zero in about 300 weeks (0.174 × 300 = 52.2). We reject (C6.5) as nonsensical, despite the statistical significance of the results, based on our knowledge of the nature of the data.

Estimating (C6.5) without a constant term produces this result:

$$w_t = -0.987 \hat{a}_{t-2} + \hat{a}_t$$
$$(1.32)$$

$\hat{\theta}_2$ is no longer significant. Thus we have further evidence that (C6.2) is an adequate model for the available data.

Final comments.   We make the following points:

1. The estimated acf in Figure C6.2 is fairly typical for a series with a nonstationary mean. Many such series produce an acf with an autocorrelation at lag 1 close to 1.0 (0.9 or larger) followed by a slow decay. But the key characteristic of a series with a nonstationary mean is the *slow decline* of the estimated autocorrelations, not their level. The first autocorrelation for some nonstationary series may be

substantially less than 1.0, but the remaining autocorrelations will move toward zero very slowly. It is the latter characteristic which is the critical indicator of a nonstationary mean. The next two case studies show examples of acf's whose slow decay indicates non-stationarity, but starting from relatively small coefficients.

2. The residual acf in Figure C6.3 is quite similar to the acf for the first differences in Figure C6.5. This should not be surprising. The residual acf in Figure C6.3 represents the behavior of the original data after they have been filtered through the operator $(1 - \hat\phi_1 B)$, with $\hat\phi_1 = 0.986$. The acf in Figure C6.5 represents the behavior of the original data after they have been filtered through the same operator $(1 - \phi_1 B)$, but with $\phi_1 = 1$. Since we have applied the same operator $(1 - \phi_1 B)$ to the same data, and since ϕ_1 in this operator is nearly the same in both cases (0.986 vs. 1.0), we should expect the two resulting series to behave similarly. Their estimated acf's should look much the same, and they do.

3. Model (C6.2) is the random walk. It is a good ARIMA model for many stock-price series. This empirical result is consistent with the weak form of the efficient-markets hypothesis summarized in Chapter 5.

4. Not every ARIMA model with significant estimates is a sensible one. Model (C6.5) is an example of a model which happens to fit a short realization, but which is not acceptable on other grounds. We strive for parsimonious models which fit the data adequately with significant coefficients, but we should also temper our statistical results with insight into the nature of the underlying data.

CASE 7. REAL-ESTATE LOANS

The series analyzed here is the monthly volume of commercial bank real-estate loans, in billions of dollars, from January 1973 to October 1978, a total of 70 observations. The data are derived from reports to the Federal Reserve System from large commercial banks. Figure C7.1 is a plot of the observations.*

From inspection of the data we might suspect that nonseasonal differencing is needed since the series trends upward over time, although it is quite stable in the middle-third of the data set. However, we must examine the estimated acf and possibly some estimated AR coefficients to determine whether differencing is needed. Although the variance of the series seems to be stationary, a logarithmic transformation could also be defended.

Identification. The estimated acf and pacf of the undifferenced data appear in Figure C7.2. There is no evidence of seasonal variation since the coefficient at lag 12 is statistically insignificant. With monthly data we expect observations at multiples of lag 12 (12, 24, 36, ...) to be correlated if a seasonal pattern is present. However, a weak seasonal element can be obscured in the acf for undifferenced data; the acf of the differenced series may yet reveal seasonality.

The estimated autocorrelations for the undifferenced data in Figure C7.2 decay slowly; they do not cross the zero line even by the eighteenth lag and the t-values exceed the 1.6 warning level out to the seventh lag. This

*These data are published by the U.S. Department of Commerce in *Business Statistics*, 1975, p. 90, and various issues of the *Survey of Current Business*. The data used here have been rounded to the nearest hundred million dollars.

411

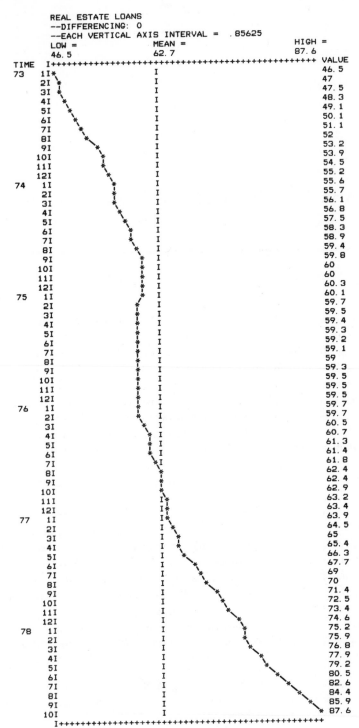

Figure C7.1 Real-estate-loans realization.

```
+ + + + + + + + + + + + AUTOCORRELATIONS + + + + + + + + + + + +
+ FOR DATA SERIES: REAL ESTATE LOANS                                    +
+ DIFFERENCING: 0                          MEAN    =  62.7              +
+ DATA COUNT =  70                         STD DEV =  9.42795           +
  COEF   T-VAL LAG                              0
  0.93   7.74   1                         [     0>>>>>]>>>>>>>>>>>>>>>>>>
  0.85   4.33   2                      [        0>>>>>>>>>>]>>>>>>>>>>>>
  0.78   3.20   3                    [          0>>>>>>>>>>]>>>>>>>>>
  0.71   2.57   4                  [            0>>>>>>>>>>>>>]>>>>
  0.65   2.16   5              [                0>>>>>>>>>>>>>>>]
  0.59   1.85   6             [                 0>>>>>>>>>>>>>>>]
  0.54   1.60   7             [                 0>>>>>>>>>>>>>> ]
  0.49   1.40   8         [                     0>>>>>>>>>>>>     ]
  0.44   1.23   9        [                      0>>>>>>>>>>>      ]
  0.39   1.07  10       [                       0>>>>>>>>>        ]
  0.35   0.93  11       [                       0>>>>>>>>         ]
  0.30   0.80  12       [                       0>>>>>>>          ]
  0.26   0.69  13      [                        0>>>>>>>            ]
  0.22   0.57  14      [                        0>>>>>>             ]
  0.18   0.48  15      [                        0>>>>>              ]
  0.15   0.39  16      [                        0>>>>               ]
  0.12   0.32  17      [                        0>>>                ]
  0.10   0.26  18      [                        0>>>                ]
        CHI-SQUARED* =   370.57 FOR DF =  18

+ + + + + + + + + + PARTIAL AUTOCORRELATIONS + + + + + + + + + +
  COEF   T-VAL LAG                              0
  0.93   7.74   1                         [     0>>>>>]>>>>>>>>>>>>>>>>>>
 -0.03  -0.21   2                         [    <0    ]
 -0.03  -0.24   3                         [    <0    ]
 -0.01  -0.11   4                         [     0    ]
  0.00   0.01   5                         [     0    ]
 -0.01  -0.10   6                         [     0    ]
 -0.01  -0.07   7                         [     0    ]
 -0.01  -0.11   8                         [     0    ]
 -0.01  -0.08   9                         [     0    ]
 -0.02  -0.19  10                         [    <0    ]
 -0.03  -0.24  11                         [    <0    ]
 -0.01  -0.08  12                         [     0    ]
 -0.02  -0.17  13                         [    <0    ]
 -0.02  -0.17  14                         [    <0    ]
  0.00  -0.03  15                         [     0    ]
 -0.01  -0.04  16                         [     0    ]
  0.01   0.06  17                         [     0    ]
  0.02   0.15  18                         [     0    ]
```

Figure C7.2 Estimated acf and pacf of the realization in Figure C7.1.

supports our earlier observation that the mean of the series is nonstationary. A check on this is to estimate an AR(1) model. This model is implied by the decaying acf and the single significant pacf spike at lag 1 in Figure C7.2. Estimation results (t-values in parentheses) are

$$(1 - 1.031B)(z_t - 42.951) = \hat{a}_t$$
$$\quad (169.79) \qquad (10.17)$$

With $|\hat{\phi}_1| > 1$ the nonstationary character of the data is confirmed. As with many models estimated for nonstationary series, the estimated mean is highly correlated with the estimated AR coefficient ($r = 0.91$); this tends to produce an unreliable estimate of μ. In this case we do not get a sensible result since $\hat{\mu} = 42.951$ is smaller than any of the observed values in Figure C7.1.

First differencing ($d = 1$) is clearly needed. Figure C7.3 is a plot of the first differences and Figure C7.4 is the estimated acf and pacf of the first differences. The first-differenced series does not show the same upward trend displayed by the original realization, but the middle-third of the differenced series lies below the overall mean and the last-third lies above it.

The acf for the first differences in Figure C7.4 moves toward zero more quickly than the acf for the original data (Figure C7.2), but the decline is still not rapid. For the time being we will suppose that the first-differenced series stationary, especially since differencing twice is rarely needed with business and economic data. The autocorrelation at lag 12 has an absolute t-value less than 1.25, so we still are not concerned about a seasonal pattern.

We are now in a position to formulate some tentative models. There are at least two plausible interpretations of the acf and pacf in Figure C7.4:

1. An ARIMA(2, 1, 0) is a good choice. Letting $I = 1$ indicates that we have differenced the data once, so we must subsequently integrate the differenced data once to regain the original series. We set $p = 2$ because the acf decays, suggesting an AR model; the pacf has two spikes followed by a cutoff to zero, implying an AR order of two. (The pacf spike at lag 4 is best ignored for now. An AR model of order four is unusual.) We therefore consider the following model:

$$\left(1 - \phi_1 B - \phi_2 B^2\right)\left(1 - B\right)\tilde{z}_t = a_t \qquad (C7.1)$$

2. An ARIMA(1, 1, 1) is also a possibility. A decaying acf like the one in Figure C7.4 is consistent with a mixed model as well as a pure AR model; the pacf could be described as decaying rapidly. When both the acf and pacf tail off rather than cut off, a mixed model is called for. Choosing an appropriate mixed model at this stage is difficult, but the ARMA(1, 1) is fairly common. [See Chapter 6 or 12 for theoretical acf's and pacf's for ARMA(1, 1) processes.] In this instance, the data are differenced once, so the model is an ARIMA(1, 1, 1):

$$(1 - \phi_1 B)(1 - B)\tilde{z}_t = (1 - \theta_1 B)a_t \qquad (C7.2)$$

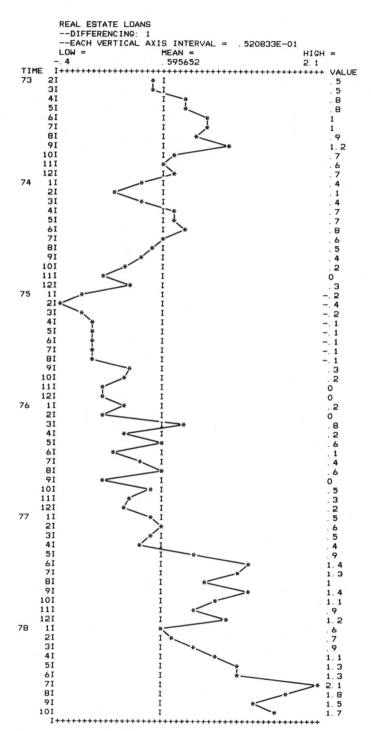

Figure C7.3 First differences of the real-estate-loans realization.

```
+ + + + + + + + + + + + AUTOCORRELATIONS + + + + + + + + + + + +
+ FOR DATA SERIES:  REAL ESTATE LOANS                              +
+ DIFFERENCING:  1                         MEAN    =  .595652      +
+ DATA COUNT =  69                         STD DEV =  .525973      +
   COEF   T-VAL LAG                             0
   0.80    6.65   1                     [       0>>>>>]>>>>>>>>>>>>>>>
   0.73    4.03   2                  [          0>>>>>>>>>]>>>>>>>>
   0.65    2.94   3               [             0>>>>>>>>>>>]>>>>>
   0.50    2.03   4               [             0>>>>>>>>>>>]>
   0.44    1.67   5            [                0>>>>>>>>>>>    ]
   0.35    1.30   6            [                0>>>>>>>>>     ]
   0.34    1.21   7            [                0>>>>>>>>     ]
   0.31    1.08   8            [                0>>>>>>>>     ]
   0.32    1.11   9            [                0>>>>>>>>     ]
   0.37    1.25  10            [                0>>>>>>>>>    ]
   0.33    1.11  11         [                   0>>>>>>>>       ]
   0.29    0.96  12         [                   0>>>>>>>        ]
   0.26    0.85  13         [                   0>>>>>>>        ]
   0.18    0.58  14         [                   0>>>>>          ]
   0.11    0.37  15         [                   0>>>            ]
   0.05    0.14  16         [                   0>             ]
  -0.05   -0.16  17         [                  <0             ]
  -0.12   -0.37  18         [                <<<0             ]
      CHI-SQUARED* =  224.76 FOR DF =   18
```

```
+ + + + + + + + + + PARTIAL AUTOCORRELATIONS + + + + + + + + + + +
   COEF   T-VAL LAG                             0
   0.80    6.65   1                     [       0>>>>>]>>>>>>>>>>>>>>>
   0.25    2.11   2                     [       0>>>>>]
   0.02    0.21   3                     [       0>     ]
  -0.23   -1.92   4                     [<<<<<0     ]
   0.04    0.31   5                     [       0>     ]
   0.01    0.09   6                     [       0      ]
   0.18    1.49   7                     [       0>>>> ]
   0.01    0.06   8                     [       0      ]
   0.11    0.95   9                     [       0>>>   ]
   0.14    1.14  10                     [       0>>>   ]
  -0.11   -0.92  11                     [  <<<0        ]
  -0.19   -1.59  12                     [<<<<<0        ]
  -0.02   -0.18  13                     [      <0       ]
  -0.06   -0.51  14                     [      <<0      ]
  -0.03   -0.24  15                     [      <0       ]
  -0.06   -0.48  16                     [      <0       ]
  -0.15   -1.25  17                     [  <<<<0        ]
  -0.10   -0.81  18                     [      <<0      ]
```

Figure C7.4   Estimated acf and pacf of the first differences in Figure C7.3.

Estimation.   Estimation of (C7.1) and (C7.2) gives these results (t-values in parentheses):

$$(1 - 0.637B - 0.353B^2)(1 - B)\tilde{z}_t = \hat{a}_t$$
$$\quad (5.43) \qquad (2.94)$$

$$(1 - 1.005B)(1 - B)\tilde{z}_t = (1 - 0.381B)\hat{a}_t$$
$$\quad (31.82) \qquad\qquad\qquad (3.14)$$

(Both models were also tried with a constant term but it was insignificant.)
The striking result is that both models are nonstationary. For (C7.2) the
condition $|\hat{\phi}_1| < 1$ is clearly violated; for (C7.1), $\hat{\phi}_1 + \hat{\phi}_2 = 0.99$. Allowing
for sampling error this result is easily consistent with the hypothesis
$\phi_1 + \phi_2 = 1$, thus violating one of the stationarity requirements for an AR
model of order 2. Estimation results show that both the original series and
its first differences have a nonstationary mean, so we should difference the
data again.

Further identification.   Figure C7.5 is the second differences of the
real-estate-loans data. Figure C7.6 is the estimated acf and pacf for the
second differences. The data now appear to fluctuate around a fixed mean,
and the estimated acf is consistent with the stationarity hypothesis. The acf
cuts off to zero sharply at lag 2 and then bounces closely around the zero
level thereafter. A significant spike at lag 1 followed by insignificant
autocorrelations signifies a stationary MA(1) process. Since the data are
differenced twice, the appropriate model is an ARIMA$(0, 2, 1)$:

$$(1 - B)^2 \tilde{z}_t = (1 - \theta_1 B) a_t \qquad (C7.3)$$

In the backshift notation presented in Chapter 5 the nonseasonal differencing operator is $(1 - B)^d$, where d is the number of times the data are
differenced. We have differenced twice so $d = 2$.

The pacf in Figure C7.6 is really not needed at this point, but you should
verify that its behavior is roughly consistent with an MA(1) model for the
twice-differenced data.

Before estimating (C7.3) consider whether we overlooked any early clues
about the need for differencing twice to achieve a stationary mean. In
general, differencing once is needed if the overall level of a series is shifting
through time. This certainly seems to characterize the original data series in
Figure C7.1. Differencing twice is needed if both the level and slope of a
series are changing. This, too, is true of the series in Figure C7.1. The slope
is positive for the first-third of the original data, about zero across the
middle-third, and positive again in the last-third. Thus there was a clue at
the beginning of the analysis that differencing twice is necessary.

Consider the first differences in Figure C7.3. We observed earlier that the
mean seems lower for the middle-third of the first differences than for the
remaining values. It is difficult to determine from inspection if these
apparent shifts in the mean are statistically significant. But estimation
results for models (C7.1) and (C7.2) violate stationarity conditions, thus

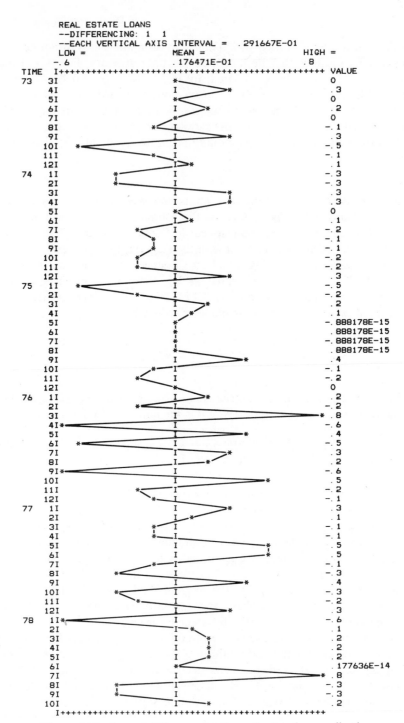

Figure C7.5 Second differences of the real-estate-loan realization.

418

```
+ + + + + + + + + + + + AUTOCORRELATIONS + + + + + + + + + + + +
+ FOR DATA SERIES:  REAL ESTATE LOANS                                  +
+ DIFFERENCING:  1   1                       MEAN    =  .176471E-01    +
+ DATA COUNT =  68                           STD DEV =  .305797        +
  COEF   T-VAL LAG _____0_____
 -0.36  -2.98   1          <<<<<<<[<<<<<<<<<<<<<0                      ]
 -0.01  -0.04   2                 [                 0                   ]
  0.14   1.02   3                 [                 0>>>>>>>            ]
 -0.11  -0.77   4                 [            <<<<<0                   ]
  0.05   0.34   5                 [                 0>>                 ]
 -0.16  -1.17   6                 [          <<<<<<<0                   ]
  0.08   0.54   7                 [                 0>>>>               ]
 -0.10  -0.71   8                 [            <<<<<0                   ]
 -0.09  -0.62   9                 [            <<<<<0                   ]
  0.15   1.02  10                 [                 0>>>>>>>            ]
  0.05   0.37  11                 [                 0>>>               ]
 -0.07  -0.48  12                 [             <<<<0                   ]
  0.11   0.74  13                 [                 0>>>>>              ]
  0.02   0.16  14                 [                 0>                 ]
 -0.05  -0.31  15                 [               <<0                   ]
  0.06   0.38  16                 [                 0>>>               ]
  0.00   0.03  17                 [                 0                   ]
       CHI-SQUARED* =    19.74 FOR DF =   17
```

```
+ + + + + + + + + + + PARTIAL AUTOCORRELATIONS + + + + + + + + + + +
  COEF   T-VAL LAG _____0_____
 -0.36  -2.98   1          <<<<<<<[<<<<<<<<<<<<<0                      ]
 -0.16  -1.30   2                 [        <<<<<<<<0                   ]
  0.09   0.78   3                 [                 0>>>>>             ]
 -0.02  -0.16   4                 [                <0                  ]
  0.02   0.15   5                 [                 0>                 ]
 -0.19  -1.60   6                 [ <<<<<<<<<<<<<0                     ]
 -0.04  -0.36   7                 [               <<0                  ]
 -0.13  -1.09   8                 [          <<<<<<<0                   ]
 -0.16  -1.30   9                 [        <<<<<<<<0                   ]
  0.02   0.20  10                 [                 0>                 ]
  0.17   1.44  11                 [                 0>>>>>>>>>>        ]
  0.02   0.14  12                 [                 0>                 ]
  0.07   0.61  13                 [                 0>>>>              ]
  0.04   0.31  14                 [                 0>>               ]
 -0.03  -0.23  15                 [                <0                  ]
  0.04   0.32  16                 [                 0>>               ]
  0.07   0.61  17                 [                 0>>>>             ]
```

Figure C7.6 Estimated acf and pacf of the second differences in Figure C7.5.

giving clear evidence that the mean of the first differences shifts by statistically significant amounts.

Does the estimated acf for the first differences (Figure C7.4) suggest a nonstationary mean? The first five autocorrelations have *t*-values larger than the relevant warning value (1.6). Although the autocorrelations from lags 6 through 12 are not very significant, they barely decline at all. This evidence for a nonstationary mean is easier to see by hindsight than it was earlier. Note that the first autocorrelation in Figure C7.4 is not very close to 1.0. The primary evidence of nonstationarity in the first differences is the *slow decay* of the acf, not the height from which the acf begins its descent.

Further estimation and diagnostic checking.   Estimation results for (C7.3) in Figure C7.7 show that the estimated MA coefficient ($\hat{\theta}_1 = 0.375$) is significant at better than the 5% level since its t-value is greater than 2.0. $\hat{\theta}_1$ also easily satisfies the invertibility condition $|\hat{\theta}_1| < 1$. The mean absolute percent error (0.36%) indicates that model (C7.3) is exceptionally accurate, at least insofar as it fits the past.

Next, we subject the residuals to autocorrelation analysis to test whether the shocks of (C7.3) are independent. The residual acf in Figure C7.7 reveals no significant residual autocorrelation coefficients since all absolute t-values are less than the relevant warning levels. Furthermore, the chi-squared statistic is small enough to allow acceptance of the null hypothesis that the shocks are independent as a set. The critical chi-squared statistic for 16 degrees of freedom at the 10% level is 23.5 compared with our calculated value of 12.55. The conclusion is that an ARIMA(0, 2, 1) provides an adequate representation of the observed real-estate-loans data.

Before using the model to forecast we should examine the residuals (Figure C7.8). One aspect of this plot should lead us to question model (C7.3): there is some tendency for the variance of the residuals to increase

```
+ + + + + + + + +ECOSTAT UNIVARIATE B-J RESULTS+ + + + + + + + +
+ FOR DATA SERIES:   REAL ESTATE LOANS                            +
+ DIFFERENCING:     1  1                      DF     = 67         +
+ AVAILABLE:     DATA = 68     BACKCASTS = 0   TOTAL = 68         +
+ USED TO FIND SSR: DATA = 68  BACKCASTS = 0   TOTAL = 68         +
+ (LOST DUE TO PRESENCE OF AUTOREGRESSIVE TERMS:           0)     +

COEFFICIENT    ESTIMATE      STD ERROR     T-VALUE
  THETA   1     0. 375        0. 113         3. 32

   ADJUSTED RMSE  =   . 286679    MEAN ABS % ERR =    0. 36

++RESIDUAL ACF++
  COEF   T-VAL LAG                           0
 -0. 03  -0. 28   1  [               <<<0                          ]
  0. 02   0. 20   2  [               0>>                           ]
  0. 12   1. 02   3  [               0>>>>>>>>>>>>                  ]
 -0. 08  -0. 69   4  [          <<<<<<<<0                          ]
 -0. 04  -0. 34   5  [               <<<0                          ]
 -0. 19  -1. 54   6  [  <<<<<<<<<<<<<<<<<<<0                       ]
 -0. 03  -0. 23   7            <<<0
 -0. 14  -1. 10   8     <<<<<<<<<<<<<<0
 -0. 09  -0. 66   9         <<<<<<<<<0
  0. 16   1. 23  10               0>>>>>>>>>>>>>>>>
  0. 11   0. 83  11               0>>>>>>>>>>>
  0. 01   0. 07  12               0>
  0. 14   1. 01  13               0>>>>>>>>>>>>>>
  0. 07   0. 48  14               0>>>>>>>
 -0. 01  -0. 04  15              <0
  0. 06   0. 42  16               0>>>>>>
  0. 00   0. 02  17               0
     CHI-SQUARED* =   12. 55 FOR DF =   16
```

Figure C7.7   Estimation and diagnostic-checking results for model (C7.3).

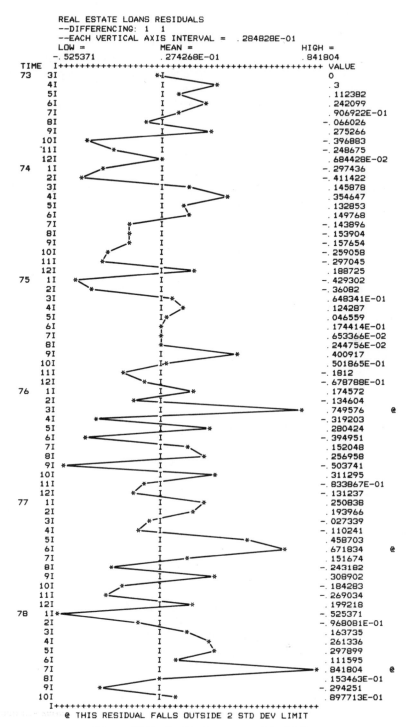

Figure C7.8 Residuals from model (C7.3).

over time. Because the level of the original data also increases over time, a logarithmic transformation might be useful. (As discussed in Chapter 7, the logarithmic transformation is appropriate when the variance is proportional to the mean.) Since the evidence favoring the logarithmic transformation is not overwhelming, we leave it as an exercise for the reader to model these data in log form.

Forecasting. Forecasts for up to six periods ahead are shown on the left of Table C7.1. As pointed out in Chapter 10, they are dominated by the trend imparted by the second-differencing since the effect of the MA term is lost after the first forecast. The difference-equation forecast form of (C7.3) is

$$\hat{z}_t = 2z_{t-1} - z_{t-2} - \hat{\theta}_1 \hat{a}_{t-1} \tag{C7.4}$$

An estimate of a_{t-1} is available when forecasting for period 71: it is the estimation residual ($\hat{a}_{t-1} = 0.0897713$) for period 70 (October 1978) shown in Figure C7.8. But with a forecast lead time $l = 2$ from origin $t = 70$ we do not even have an estimate of the corresponding a_{t-1}: the estimation residual for period 71 is unknown because our last available observation is for

Table C7.1 Forecasts from model (C7.3)

Time		Forecast Values	80% Confidence Limits		Future Observed Values	Percent Forecast Errors
			Lower	Upper		
78	11	89.2664	88.8994	89.6333	n.a.[a]	n.a.
	12	90.9327	90.2325	91.6330	n.a.	n.a.
79	1	92.5991	91.5163	93.6820	n.a.	n.a.
	2	94.2655	92.7534	95.7776	n.a.	n.a.
	3	95.9319	93.9476	97.9161	n.a.	n.a.
	4	97.5982	95.1021	100.0944	n.a.	n.a.
	5	99.2646	96.2196	102.3096	n.a.	n.a.
	6	100.9310	97.3025	104.5595	n.a.	n.a.
	7	102.5973	98.3525	106.8422	n.a.	n.a.
	8	104.2637	99.3714	109.1560	n.a.	n.a.
	9	105.9301	100.3605	111.4997	n.a.	n.a.
	10	107.5965	101.3211	113.8718	n.a.	n.a.

[a] n.a. = not available.

period 70. To forecast beyond one period ahead from origin $t = 70$, we simply use the expected value of a_{t-1} which is zero. Thus the forecasts for period 72 and further ahead are dominated by the first two terms, $2z_{t-1} - z_{t-2}$, in equation (C7.4).

Of course, the estimation residual for period 70 continues to affect these forecasts because it affects $\hat{z}_{71}$, which in turn affects $\hat{z}_{72}$, which affects $\hat{z}_{73}$, and so forth. Starting with forecasts for lead time $l = 3$ (January 1979), we are generating strictly bootstrap forecasts: the z_{t-2} used in producing the period 73 forecast must be the period 71 forecast ($\hat{z}_{t-2}$). At forecast origin $t = 70$ we have not yet observed any data for period 71. All we have is the forecast for that period. Once again we make the point that ARIMA models should generally be reestimated and new forecasts produced as new data become available.

Additional checks. As an added check on the stability of the model we delete the last seven observations and reestimate. The outcome (Figure C7.9) gives no cause for alarm. The reestimated $\hat{\theta}_1$ remains statistically

```
+ + + + + + + + +ECOSTAT UNIVARIATE B-J RESULTS+ + + + + + + + +
+ FOR DATA SERIES:   REAL ESTATE LOANS                          +
+ DIFFERENCING:      1  1                      DF    = 60       +
+ AVAILABLE:         DATA = 61   BACKCASTS = 0  TOTAL = 61       +
+ USED TO FIND SSR:  DATA = 61   BACKCASTS = 0  TOTAL = 61       +
+ (LOST DUE TO PRESENCE OF AUTOREGRESSIVE TERMS:          0)    +

 COEFFICIENT    ESTIMATE      STD ERROR      T-VALUE
   THETA  1      0.418         0.117          3.58

   ADJUSTED RMSE =  .274524    MEAN ABS % ERR =    0.36

 ++RESIDUAL ACF++
   COEF   T-VAL  LAG _____0_____
  -0.04  -0.28    1                      <<<<0
   0.05   0.35    2                      0>>>>>
   0.13   1.05    3                      0>>>>>>>>>>>>>
  -0.08  -0.63    4                <<<<<<<<0
  -0.01  -0.09    5                      <0
  -0.11  -0.84    6              <<<<<<<<<<<0
  -0.09  -0.70    7               <<<<<<<<<0
  -0.16  -1.21    8          <<<<<<<<<<<<<<<<0
  -0.05  -0.35    9                      <<<<<0
   0.08   0.57   10                      0>>>>>>>>
   0.11   0.76   11                      0>>>>>>>>>>>
  -0.02  -0.11   12                      <<0
   0.03   0.21   13                      0>>>
   0.01   0.09   14                      0>
   0.04   0.28   15                      0>>>>
     CHI-SQUARED* =    7.00 FOR DF =   14
```

Figure C7.9 Estimation and diagnostic-checking results for model (C7.3) using the first 63 observations.

Table C7.2   Forecasts from model (C7.3) using the first 63 observations

Time		Forecast Values	80% Confidence Limits		Future Observed Values	Percent Forecast Errors
			Lower	Upper		
78	4	77.6371	77.2857	77.9884	77.9000	0.34
	5	78.4741	77.8165	79.1317	79.2000	0.92
	6	79.31	78.3059	80.3164	80.5000	1.48
	7	80.1482	78.7549	81.5416	82.6000	2.97
	8	80.9853	79.1663	82.8042	84.4000	4.05
	9	81.8223	79.5429	84.1018	85.9000	4.75
	10	82.6594	79.8869	85.4319	87.6000	5.64

significant and is within 0.1 of the original $\hat{\theta}_1$ estimate in Figure C7.7. The residual acf shows no absolute t-values worthy of attention and the chi-squared statistic is insignificant.

Using this model with the shortened data set to forecast history, we find that it performs quite satisfactorily for the first few periods ahead as shown in Table C7.2. UBJ–ARIMA models usually forecast better over the very near term than over a long forecast horizon and this one is no exception. The percent forecast errors grow fairly steadily across the first half-dozen forecasts. This is not unusual and it suggests once again that an ARIMA model should be reestimated and new forecasts generated as new data become available.

The reader is encouraged to forecast history (deleting the last 7 observations, as above) with the data modeled in logarithmic form.

CASE 8.   PARTS AVAILABILITY

The data for this case are adapted from a series provided by a large U.S. corporation. There are 90 weekly observations showing the percent of the time that parts for an industrial product are available when needed. UBJ–ARIMA models are especially useful for forecasting data of this type where the time interval between observations is short.

Look at the data in Figure C8.1. Does this series appear to be stationary? The variance seems constant, though the mean may rise and fall through time. We cannot tell from inspection of the data if these apparent changes in the mean are statistically significant. We must rely on autocorrelation analysis and perhaps estimated AR coefficients to decide if the mean is stationary.

Identification.   The estimated acf and pacf for these data are shown in Figure C8.2. With 90 observations we may safely examine up to about 23 autocorrelations. Our first concern is to decide if the mean is stationary without differencing; for this we focus on the estimated acf. If the estimated acf fails to die out rapidly toward zero, we conclude the mean is nonstationary and the series must be differenced.

The estimated acf in Figure C8.2 indicates that the mean is nonstationary. Rather than dying down along the first several lags, the estimated autocorrelations actually increase up to lag 3. After lag 3 they decline very slowly. The *t*-values either exceed or come close to the warning level of 1.6 out to lag 7. This is the same phenomenon we observed in the last two cases, where we found nonstationary series. There we saw initial estimated acf's

425

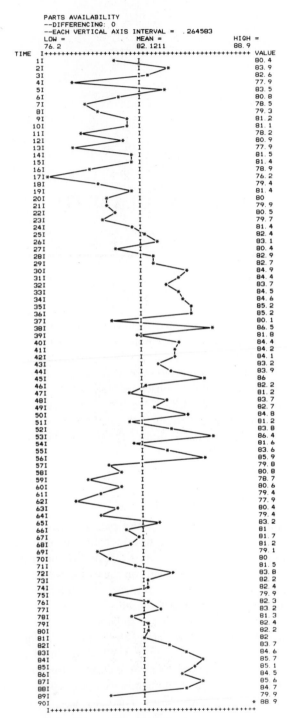

Figure C8.1 Parts-availability realization.

```
+ + + + + + + + + + + + AUTOCORRELATIONS + + + + + + + + + + + + +
+ FOR DATA SERIES: PARTS AVAILABILITY                             +
+ DIFFERENCING: 0                        MEAN    =   82.1211       +
+ DATA COUNT =  90                       STD DEV =   2.36796       +
  COEF   T-VAL LAG _____0_____
  0.37   3.51   1                         [         0>>>>>>>>>>]>>>>>>>>>>
  0.42   3.54   2                      [            0>>>>>>>>>>>>]>>>>>>>>>>
  0.47   3.46   3                   [               0>>>>>>>>>>>>>>]>>>>>>>>>>
  0.37   2.47   4             [                     0>>>>>>>>>>>>>>>]>>>
  0.36   2.23   5             [                     0>>>>>>>>>>>>>>>]>>
  0.26   1.52   6             [                     0>>>>>>>>>>>>>   ]
  0.26   1.52   7          [                        0>>>>>>>>>>>>>   ]
  0.15   0.85   8          [                        0>>>>>>>>       ]
  0.12   0.65   9          [                        0>>>>>>         ]
  0.20   1.11  10          [                        0>>>>>>>>>>     ]
  0.08   0.42  11          [                        0>>>>           ]
  0.07   0.40  12          [                        0>>>>           ]
 -0.04  -0.22  13          [                      <<0              ]
 -0.05  -0.27  14          [                      <<0              ]
 -0.11  -0.59  15          [                   <<<<<0              ]
 -0.10  -0.54  16          [                   <<<<<0              ]
 -0.17  -0.90  17          [                <<<<<<<<0              ]
 -0.18  -0.94  18          [                <<<<<<<<<0             ]
 -0.21  -1.11  19          [             <<<<<<<<<<<0              ]
 -0.14  -0.75  20       [                   <<<<<<<0                  ]
 -0.28  -1.46  21       [            <<<<<<<<<<<<<<0                  ]
 -0.26  -1.34  22       [             <<<<<<<<<<<<<0                  ]
 -0.27  -1.35  23       [            <<<<<<<<<<<<<<0                  ]
     CHI-SQUARED* =   143.16 FOR DF =   23

+ + + + + + + + + + + PARTIAL AUTOCORRELATIONS + + + + + + + + + + + +
  COEF   T-VAL LAG _____0_____
  0.37   3.51   1                          [         0>>>>>>>>>>]>>>>>>>>>>
  0.33   3.13   2                          [         0>>>>>>>>>>]>>>>>>
  0.31   2.95   3                          [         0>>>>>>>>>>]>>>>>>
  0.12   1.15   4                          [         0>>>>>>>    ]
  0.07   0.65   5                          [         0>>>        ]
 -0.08  -0.78   6                          [     <<<<0           ]
 -0.02  -0.19   7                          [        <0           ]
 -0.13  -1.23   8                          [   <<<<<<0           ]
 -0.09  -0.82   9                          [     <<<<0           ]
  0.11   1.07  10                          [         0>>>>>>     ]
  0.00   0.02  11                          [         0           ]
  0.00   0.01  12                          [         0           ]
 -0.17  -1.61  13                        [ <<<<<<<<<0            ]
 -0.13  -1.19  14                          [   <<<<<0            ]
 -0.15  -1.43  15                         [. <<<<<<<<0           ]
  0.01   0.07  16                          [         0           ]
 -0.05  -0.50  17                          [     <<<0            ]
  0.04   0.38  18                          [         0>>          ]
 -0.03  -0.24  19                          [        <0           ]
  0.09   0.90  20                          [         0>>>>>       ]
 -0.18  -1.66  21                        [<<<<<<<<<<0            ]
 -0.14  -1.35  22                          [   <<<<<<0           ]
 -0.13  -1.19  23                          [     <<<<<0          ]
```

Figure C8.2 Estimated acf and pacf for the realization in Figure C8.1.

(see Figures C6.2 and C7.2) with t-values of 1.6 or larger out to lags 6 or 7. We have good reason for concluding that the mean of the data in the present case is also nonstationary.

Estimated autocorrelations need not start out close to 1.0 to suggest a nonstationary mean. The key indicator is when the estimated acf fails to die out quickly at longer lags. The estimated autocorrelations in Figure C8.2 do not start out close to 1.0, but that does not matter. What matters is their slow decline toward zero.

With a nonstationary mean we should calculate the first differences and find the estimated acf and pacf for this new series. Figure C8.3 is a plot of the first differences of the parts-availability data; Figure C8.4 is the estimated acf and pacf for the first differences. Inspection of Figure C8.3 suggests the mean is now stationary. The estimated acf clearly suggests the first differences are stationary and that they can be represented by an MA(1) model: the spike followed by a cutoff to zero in the estimated acf says an MA model is appropriate, and since the spike occurs at lag 1 we choose an MA model of order $q = 1$. The estimated pacf is consistent with an MA(1) model: pure MA models of order one are typically associated with pacf's that tail off toward zero starting at lag 1; the estimated pacf in Figure C8.4 displays this behavior.

From the preceding analysis we tentatively select an ARIMA(0, 1, 1) model. In backshift form the model is

$$(1 - B)\tilde{z}_t = (1 - \theta_1 B)a_t \qquad (C8.1)$$

Estimation and diagnostic checking.   Estimation results and the residual acf appear in Figure C8.5. All indications are that (C8.1) is satisfactory. The estimated MA coefficient is significant judging by its large absolute t-value (9.84), and the invertibility condition $|\hat{\theta}_1| < 1$ is satisfied.

According to the residual acf in Figure C8.5, we may accept the hypothesis that the shocks of model (C8.1) are independent. There are no absolute t-values in the residual acf exceeding any of the relevant practical warning levels, and the chi-squared statistic is insignificant at the 10% level. We conclude that model (C8.1) is a statistically adequate representation of the available data.

Alternative models.   We have argued that the estimated acf in Figure C8.2 indicates that the original realization is nonstationary. The important clue is that the estimated acf does not die out to zero rapidly enough. As a learning exercise suppose we fail to see that important clue. How would we try to model the undifferenced realization?

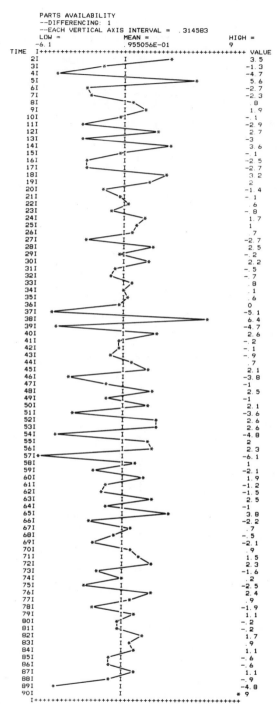

PARTS AVAILABILITY
--DIFFERENCING: 1
--EACH VERTICAL AXIS INTERVAL = .314583
LOW =   MEAN =   HIGH =
-6.1   .955056E-01   9

Figure C8.3   First differences of the parts-availability series.

```
+ + + + + + + + + + + AUTOCORRELATIONS + + + + + + + + + + + +
+ FOR DATA SERIES: PARTS AVAILABILITY                                    +
+ DIFFERENCING: 1                          MEAN     =   .955056E-01      +
+ DATA COUNT =   89                        STD DEV =  2.56563            +
  COEF   T-VAL LAG                                  0
 -0.47   -4.45   1  <<<<<<<<<<<<<<<<[<<<<<<<<<<0              ]
 -0.05   -0.42   2               [        <<<0              ]
  0.09    0.72   3               [         0>>>>>           ]
 -0.04   -0.30   4               [        <<0               ]
  0.06    0.46   5               [         0>>>             ]
 -0.11   -0.84   6               [       <<<<<0             ]
  0.12    0.91   7               [         0>>>>>>          ]
 -0.05   -0.40   8           [        <<<0                  ]
 -0.08   -0.62   9           [       <<<<0                  ]
  0.17    1.25  10           [         0>>>>>>>>>           ]
 -0.10   -0.72  11           [       <<<<<0                 ]
  0.10    0.72  12           [         0>>>>>               ]
 -0.10   -0.75  13           [       <<<<<0                 ]
  0.05    0.39  14           [         0>>>                 ]
 -0.04   -0.29  15           [        <<0                   ]
  0.03    0.21  16           [         0>                   ]
 -0.04   -0.28  17           [        <<0                   ]
  0.01    0.06  18           [         0                    ]
 -0.07   -0.48  19           [        <<<0                  ]
  0.19    1.42  20           [         0>>>>>>>>>>          ]
 -0.12   -0.88  21           [       <<<<<<0                ]
 -0.01   -0.04  22           [         0                    ]
  0.02    0.16  23           [         0>                   ]
       CHI-SQUARED* =    38.93 FOR DF =   23

+ + + + + + + + + + + PARTIAL AUTOCORRELATIONS + + + + + + + + + + +
  COEF   T-VAL LAG                                  0
 -0.47   -4.45   1  <<<<<<<<<<<<<<<<[<<<<<<<<<<0              ]
 -0.35   -3.35   2       <<<<<<<<<<[<<<<<<<<<<0              ]
 -0.16   -1.52   3               [ <<<<<<<<0                ]
 -0.11   -1.07   4               [   <<<<<<0                ]
  0.02    0.19   5               [         0>               ]
 -0.09   -0.87   6               [     <<<<<0               ]
  0.05    0.43   7               [         0>>              ]
  0.01    0.06   8               [         0                ]
 -0.10   -0.98   9               [     <<<<<0               ]
  0.07    0.63  10               [         0>>>             ]
  0.02    0.18  11               [         0>               ]
  0.15    1.43  12               [         0>>>>>>>>        ]
  0.03    0.28  13               [         0>>              ]
  0.06    0.56  14               [         0>>>             ]
 -0.05   -0.51  15               [        <<<0              ]
  0.02    0.21  16               [         0>               ]
 -0.09   -0.85  17               [     <<<<<0               ]
 -0.04   -0.40  18               [        <<0               ]
 -0.17   -1.64  19              [<<<<<<<<<<0                ]
  0.13    1.24  20               [         0>>>>>>>         ]
  0.05    0.43  21               [         0>>              ]
  0.03    0.27  22               [         0>               ]
 -0.01   -0.13  23               [        <0                ]
```
Figure C8.4 Estimated acf and pacf of the first differences in Figure C8.3.

```
+ + + + + + + + +ECOSTAT UNIVARIATE B-J RESULTS+ + + + + + + + +
+ FOR DATA SERIES:  PARTS AVAILABILITY                           +
+ DIFFERENCING:      1                          DF    = 88       +
+ AVAILABLE:         DATA = 89   BACKCASTS = 0   TOTAL = 89       +
+ USED TO FIND SSR: DATA = 89   BACKCASTS = 0   TOTAL = 89       +
+ (LOST DUE TO PRESENCE OF AUTOREGRESSIVE TERMS:           0)    +

COEFFICIENT    ESTIMATE      STD ERROR     T-VALUE
   THETA  1     0.725         0.074          9.84

   ADJUSTED RMSE =  2.02987   MEAN ABS % ERR =   1.97

++RESIDUAL ACF++
  COEF  T-VAL LAG _____0_____
 -0.10  -0.94  1    [            <<<<<<<<<<<0                    ]
 -0.01  -0.14  2    [                    <0                      ]
  0.11   1.07  3    [                     0>>>>>>>>>>>           ]
  0.04   0.34  4    [                     0>>>>                  ]
  0.06   0.59  5    [                     0>>>>>>               ]
 -0.05  -0.43  6    [                <<<<<0                      ]
  0.07   0.64  7    [                     0>>>>>>>              ]
 -0.05  -0.49  8    [                <<<<<0                      ]
 -0.05  -0.47  9    [                <<<<<0                      ]
  0.15   1.39 10    [                     0>>>>>>>>>>>>>>>       ]
  0.00  -0.01 11    [                     0                      ]
  0.06   0.51 12    [                     0>>>>>>               ]
 -0.08  -0.70 13    [             <<<<<<<<0                      ]
 -0.02  -0.22 14    [                   <<0                      ]
 -0.08  -0.72 15    [             <<<<<<<<0                      ]
 -0.02  -0.20 16    [                   <<0                      ]
 -0.07  -0.59 17    [              <<<<<<<0                      ]
 -0.03  -0.26 18    [                  <<<0                      ]
 -0.04  -0.33 19    [                 <<<<0                      ]
  0.13   1.15 20   [                      0>>>>>>>>>>>>>        ]
 -0.10  -0.86 21   [             <<<<<<<<<<0                     ]
 -0.05  -0.45 22   [                 <<<<<0                      ]
 -0.03  -0.27 23   [                  <<<0                       ]
   CHI-SQUARED* =   12.71 FOR DF =   22
```

Figure C8.5 Estimation and diagnostic-checking results for model (C8.1).

It should be clear that a pure MA model is not appropriate for the undifferenced data. The estimated acf in Figure C8.2 does not display spikes followed by a sharp cutoff to zero. The decaying pattern calls for either a pure AR model or an ARMA model.

The estimated acf and pacf in Figure C8.2 together suggest that a pure AR model of order three might fit the undifferenced data. The acf decays and the estimated pacf has three spikes (at lags 1, 2, and 3) followed by a cutoff to zero (at lag 4 and beyond). The number of spikes in the estimated pacf indicates the order of a pure AR model. We tentatively select an AR(3) model, written in backshift form as:

$$\left(1 - \phi_1 B - \phi_2 B^2 - \phi_3 B^3\right)\tilde{z}_t = a_t \qquad (C8.2)$$

Estimation results for this model (not shown) produced a t-value for $\hat{\phi}_1$ of only 1.45. With $\hat{\phi}_1$ not different from zero at the 5% level, we remove that coefficient from our model. This gives us an AR(3) with ϕ_1 constrained to zero:

$$\left(1 - \phi_2 B^2 - \phi_3 B^3\right)\tilde{z}_t = a_t \qquad (C8.3)$$

```
+ + + + + + + + + +ECOSTAT UNIVARIATE B-J RESULTS+ + + + + + + + + +
+ FOR DATA SERIES:   PARTS AVAILABILITY                            +
+ DIFFERENCING:      0                          DF     = 84        +
+ AVAILABLE:            DATA = 90    BACKCASTS = 0    TOTAL = 90    +
+ USED TO FIND SSR:  DATA = 87    BACKCASTS = 0    TOTAL = 87       +
+ (LOST DUE TO PRESENCE OF AUTOREGRESSIVE TERMS:            3)      +

COEFFICIENT    ESTIMATE      STD ERROR     T-VALUE
PHI    2       0.294         0.107         2.74
PHI    3       0.389         0.108         3.61
CONSTANT       25.9915       9.17251       2.83362

MEAN           82.233        .688507       119.437

ADJUSTED RMSE =  2.02339    MEAN ABS % ERR =   1.91
         CORRELATIONS
         1      2      3
1     1.00
2    -0.46   1.00
3     0.02   0.06   1.00

++RESIDUAL ACF++
COEF   T-VAL LAG _____0_____
0.11    1.00   1    [                   0>>>>>>>>>>>          ]
-0.07  -0.66   2    [           <<<<<<<<0                     ]
-0.05  -0.50   3    [           <<<<<0                        ]
0.14    1.31   4    [                   0>>>>>>>>>>>>>>        ]
0.14    1.30   5    [                   0>>>>>>>>>>>>>>        ]
0.03    0.23   6    [.                  0>>>                  ]
0.01    0.05   7    [                   0>                    ]
-0.08  -0.70   8    [           <<<<<<<<0                     ]
-0.01  -0.09   9    [                  <0                     ]
0.19    1.68  10    [                   0>>>>>>>>>>>>>>>>>>>   ]
0.11    0.96  11  [                     0>>>>>>>>>>>             ]
0.05    0.40  12  [                     0>>>>>                  ]
-0.11  -0.91  13  [          <<<<<<<<<<<0                       ]
-0.02  -0.20  14  [                   <<0                       ]
-0.05  -0.38  15  [                <<<<<0                       ]
0.00   -0.04  16  [                     0                       ]
-0.13  -1.10  17  [         <<<<<<<<<<<<<0                       ]
-0.07  -0.56  18  [             <<<<<<<<0                       ]
-0.02  -0.13  19  [                  <<0                        ]
0.12    0.97  20  [                     0>>>>>>>>>>>>           ]
-0.07  -0.57  21  [             <<<<<<<<0                       ]
-0.09  -0.74  22  [           <<<<<<<<<<0                       ]
-0.10  -0.76  23         <<<<<<<<<<<0
        CHI-SQUARED* =   20.00 FOR DF =   20
```

Figure C8.6 Estimation and diagnostic-checking results for model (C8.3).

Figure C8.6 shows the estimation and diagnostic-checking results for this model. The full set of stationarity conditions for an AR(3) model are difficult to determine and we will not do so here. In any case, a necessary (but not sufficient) condition is that the AR coefficients must sum to less than 1.0, and $\hat{\phi}_2 + \hat{\phi}_3 = 0.683$. Model (C8.3) is not as satisfactory as (C8.1). The residual acf in Figure C8.6 has larger t-values at lags 4, 5, and 10. The residual autocorrelations in Figure C8.5 are less significant judging by their smaller t-values and by the smaller chi-squared statistic.

Now consider if a mixed model is consistent with the estimated acf and pacf in Figure C8.2. The estimated acf is consistent with a mixed model because it decays. We could argue that the estimated pacf also decays, though it does so very slowly along the first three lags. When both the acf and pacf decay toward zero, a mixed model is called for. In practice it is often difficult to determine from an initial estimated acf and pacf which mixed model is best. Perhaps the best we can do in this case is to begin with the common ARMA(1, 1):

$$(1 - \phi_1 B)\tilde{z}_t = (1 - \theta_1 B)a_t \qquad (C8.4)$$

The estimation and diagnostic-checking results for (C8.4) in Figure C8.7 show that it is superior to model (C8.3). The RMSE is about the same while the residual acf is cleaner. The most important thing to note is that $\hat{\phi}_1$ is very close to 1.0: it falls only about one standard error below 1.0. (The estimated standard error of $\hat{\phi}_1$ is 0.059. One *minus* this value is 0.941, which is approximately equal to ϕ_1.) This result suggests that the data are nonstationary and should be differenced at least once. This brings us back to model (C8.1), which we have already discussed.

The ARIMA(0, 1, 1) as an EWMA.   Forecasts and confidence intervals from (C8.1) are plotted in Figure C8.8. You should be able to show from the material presented in Chapter 5, Section 5.5 that a one-step-ahead forecast from (C8.1) is an exponentially weighted moving average (EWMA) with these weights applied to past observations:

Past Observation	Weight
z_{t-1}	0.275
z_{t-2}	0.199
z_{t-3}	0.145
z_{t-4}	0.105
z_{t-5}	0.076
z_{t-6}	0.055
$\vdots$	$\vdots$

```
+ + + + + + + + +ECOSTAT UNIVARIATE B-J RESULTS+ + + + + + + + +
+ FOR DATA SERIES:   PARTS AVAILABILITY                          +
+ DIFFERENCING:      0                         DF    = 86        +
+ AVAILABLE:             DATA = 90   BACKCASTS = 0  TOTAL = 90   +
+ USED TO FIND SSR: DATA = 89   BACKCASTS = 0  TOTAL = 89        +
+ (LOST DUE TO PRESENCE OF AUTOREGRESSIVE TERMS:           1)    +
```

```
COEFFICIENT   ESTIMATE      STD ERROR     T-VALUE
PHI      1     0.944          0.059        16.09
THETA    1     0.699          0.112         6.23
CONSTANT       4.63683        4.81488      .963021

MEAN          82.8587         1.43932      57.568
```

```
ADJUSTED RMSE =  2.0271   MEAN ABS % ERR =   1.93
        CORRELATIONS
        1       2      3
1     1.00
2     0.72   1.00
3     0.57   0.38   1.00
```

```
++RESIDUAL ACF++
 COEF  T-VAL LAG _____0_____
-0.10  -0.90   1   [            <<<<<<<<<<<0                        ]
 0.00  -0.03   2   [                       0                        ]
 0.13   1.19   3   [                       0>>>>>>>>>>>>>>          ]
 0.05   0.45   4   [                       0>>>>>                   ]
 0.07   0.69   5   [                       0>>>>>>>                 ]
-0.04  -0.33   6   [                   <<<<0                        ]
 0.08   0.70   7   [                       0>>>>>>>>               ]
-0.05  -0.45   8   [                  <<<<<0                        ]
-0.05  -0.47   9   [                  <<<<<0                        ]
 0.15   1.35  10   [                       0>>>>>>>>>>>>>>>>        ]
 0.00  -0.04  11   [                       0                        ]
 0.05   0.47  12   [                       0>>>>>                   ]
-0.08  -0.72  13   [                <<<<<<<<0                        ]
-0.03  -0.27  14   [                    <<<0                        ]
-0.09  -0.79  15   [               <<<<<<<<<0                        ]
-0.03  -0.26  16   [                    <<<0                        ]
-0.08  -0.66  17   [                <<<<<<<<0                        ]
-0.04  -0.34  18  [                    <<<<0                       ]
-0.05  -0.46  19  [                   <<<<<0                       ]
 0.11   0.98  20  [                       0>>>>>>>>>>>>            ]
-0.12  -0.99  21  [            <<<<<<<<<<<<0                       ]
-0.06  -0.55  22  [                  <<<<<<0                       ]
-0.04  -0.36  23  [                   <<<<0                       ]
        CHI-SQUARED* =   13.71 FOR DF =   20
```

Figure C8.7   Estimation and diagnostic-checking results for model (C8.4).

We did not choose this EWMA arbitrarily. The identification stage led us to difference the data and choose an MA(1) for the differenced series. The estimation and diagnostic-checking stages confirmed this choice and provided an optimal estimate of the weighting factor. A great advantage of the UBJ method is that we are led to appropriate models from the behavior of the data, rather than having to choose a model ahead of time.

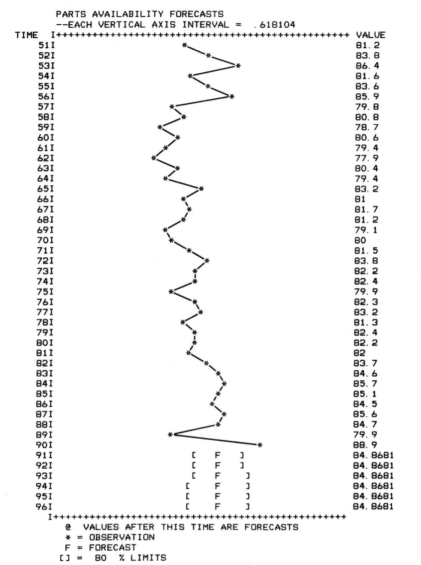

```
        PARTS AVAILABILITY FORECASTS
        --EACH VERTICAL AXIS INTERVAL =   .618104
TIME   I+++++++++++++++++++++++++++++++++++++++++++++++++ VALUE
    51I                      *                            81. 2
    52I                         *                         83. 8
    53I                            *                      86. 4
    54I                      *                             81. 6
    55I                     *                             83. 6
    56I                          *                        85. 9
    57I                  *                                79. 8
    58I                   *                               80. 8
    59I                *                                  78. 7
    60I                 *                                 80. 6
    61I                *                                  79. 4
    62I              *                                    77. 9
    63I               *                                   80. 4
    64I               *                                   79. 4
    65I                        *                          83. 2
    66I               *                                   81
    67I                *                                  81. 7
    68I                  *                                81. 2
    69I               *                                   79. 1
    70I               *                                   80
    71I                 *                                 81. 5
    72I                    *                              83. 8
    73I                  *                                82. 2
    74I                  *                                82. 4
    75I               *                                   79. 9
    76I                *                                  82. 3
    77I                  *                                83. 2
    78I                *                                  81. 3
    79I                 *                                 82. 4
    80I                 *                                 82. 2
    81I                 *                                 82
    82I                   *                               83. 7
    83I                    *                              84. 6
    84I                     *                             85. 7
    85I                    *                              85. 1
    86I                  *                                84. 5
    87I                    *                              85. 6
    88I                   *                               84. 7
    89I           *                                       79. 9
    90I                            *                      88. 9        @
    91I                   [    F    ]                     84. 8681
    92I                   [    F    ]                     84. 8681
    93I                   [    F    ]                     84. 8681
    94I                  [    F    ]                      84. 8681
    95I                  [    F    ]                      84. 8681
    96I                  [    F    ]                      84. 8681
       I++++++++++++++++++++++++++++++++++++++++++++++++++
        @   VALUES AFTER THIS TIME ARE FORECASTS
        * = OBSERVATION
        F = FORECAST
        [] =   80  % LIMITS
```

Figure C8.8   Forecasts from model (C8.1).

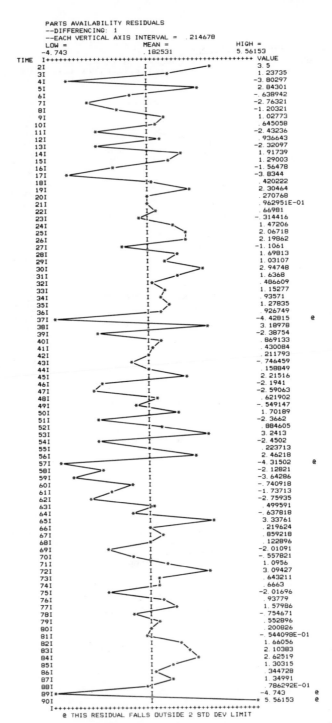

```
PARTS AVAILABILITY RESIDUALS
--DIFFERENCING: 1
--EACH VERTICAL AXIS INTERVAL =  .214678
  LOW =                    MEAN =              HIGH =
  -4.743                   .182531             5.56153
TIME  I++++++++++++++++++++++++++++++++++++++++++++++++++ VALUE
  2I                       I                *                3.5
  3I                       I    *                            1.23735
  4I       *               I                                -3.80297
  5I                       I                    *            2.84301
  6I                   *   I                                - .638942
  7I          *          I                                  -2.76321
  8I              *     I                                    -1.20321
  9I                       I        *                        1.02773
 10I                     I-*                                  .645058
 11I             *         I                                -2.43236
 12I                     I->*                                 .936643
 13I              *        I                                -2.32097
 14I                       I             *                    1.91739
 15I                       I      *                           1.29003
 16I                  *    I                                 -1.56478
 17I       *               I                                -3.8344
 18I                     I-*                                   .420222
 19I                       I                *                 2.30464
 20I                       *                                   .270768
 21I                      I<                                   .962951E-01
 22I                     I>                                    .66981
 23I                    *I                                   - .314416
 24I                       I          *                       1.47206
 25I                       I              *                   2.06718
 26I                       I              *                   2.19862
 27I              *        I                                 -1.1061
 28I                       I           *                      1.69813
 29I                       I       *                          1.03107
 30I                       I                    *             2.94748
 31I                       I          *                       1.6368
 32I                      I*                                   .486609
 33I                       I    *                             1.15277
 34I                       I   *->                            .93571
 35I                       I     *                            1.27835
 36I                     I-*                                   .926749
 37I  *                    I                                 -4.42815    @
 38I                       I                        *         3.18978
 39I          *            I   *                             -2.38754
 40I                       I>*                                 .869133
 41I                      I*                                   .430084
 42I                     I*                                    .211793
 43I              *        I                                 - .746459
 44I                     I*                                    .158849
 45I                       I                   *              2.21516
 46I             *         I                                 -2.1941
 47I          *            I                                 -2.59063
 48I                     I=>*                                  .621902
 49I               *       I                                 - .549147
 50I                       I           *                      1.70189
 51I             *         I                                 -2.3662
 52I                       I   *                               .884605
 53I                       I                    *             3.2413
 54I             *         I                                 -2.4502
 55I                      *I                                   .223713
 56I                       I                  *               2.46218
 57I  *                    I                                 -4.31502    @
 58I              *        I                                 -2.12821
 59I         *             I                                 -3.64286
 60I                   *   I                                 - .740918
 61I                 *     I                                 -1.73713
 62I           *           I                                 -2.75935
 63I                      I*                                   .499591
 64I                    *  I                                 - .637818
 65I                       I                      *           3.33761
 66I                       I*                                  .219624
 67I                       I  *                                .859218
 68I                       I*                                  .122896
 69I             *         I                                 -2.01091
 70I                  *    I                                 - .557821
 71I                       I     *                            1.0956
 72I                       I                   *              3.09427
 73I                       I *->                              .643211
 74I                     I->*                                  .6663
 75I             *         I                                 -2.01696
 76I                     I->*                                  .93779
 77I                       I     *                            1.57986
 78I             *         I                                 - .754671
 79I                     I*                                    .552896
 80I                      I*                                   .200826
 81I                     I*                                   - .544098E-01
 82I                       I    *->                           1.66056
 83I                       I          *                       2.10383
 84I                       I             *                    2.62519
 85I                       I     *                            1.30315
 86I                      I*                                   .344728
 87I                       I    *                             1.34991
 88I                     *-I                                   .786292E-01
 89I*                      I                                  -4.743      @
 90I                       I                              *   5.56153    @
    I++++++++++++++++++++++++++++++++++++++++++++++++++
       @ THIS RESIDUAL FALLS OUTSIDE 2 STD DEV LIMIT
```

Figure C8.9 Residuals from model (C8.1).

436

Residuals.   The residuals obtained by estimating model (C8.1) are plotted in Figure C8.9. The 37th, the 57th, and the last two are especially large, falling more than two standard deviations away from the mean of the residuals. Sometimes this happens when data are recorded incorrectly; residual analysis can help detect these errors. A check of the data source in this case reveals that these four observations were recorded correctly. Further study of those weeks might reveal events that caused these deviant observations (e.g., the breakdown of a critical machine, severe weather that curtailed parts shipments, etc.). If specific, potentially repeatable events can be identified, forecasts might be improved by expanding (C8.1) into a multivariate intervention model. (See Case 2 for references about this method.) Of course, large residuals can occur just by chance and we must be prepared to accept that possibility.

Group C

NONSTATIONARY, SEASONAL MODELS

CASE 9. AIR-CARRIER FREIGHT

This is the first case study in this text where the data have seasonal variation. All subsequent cases involve data with a seasonal pattern. Experience has shown that ARIMA models can often forecast seasonal series rather well; but because most seasonal series also have nonseasonal patterns, finding an appropriate model is often challenging. The seasonal and nonseasonal elements are mixed together in the estimated acf's and pacf's at the identification stage; separating these two patterns in one's eye and mind may not be easy. Nevertheless, adherence to the UBJ model–building cycle of identification, estimation, and diagnostic checking, along with the practical modeling rules in Chapter 12, usually guides the analyst to an adequate model. You may want to review seasonal models in Chapter 11 and the practical rules pertaining to them in Chapter 12 before reading the remaining cases.

In this case we study the volume of freight, measured in ton-miles, hauled by air carriers in the United States. There are 120 monthly observations covering the years 1969–1978.*

Identification. The first step in identification is to look at a plot of the data (see Figure C9.1). Three features of this series stand out: (i) the overall level rises through time, suggesting that the mean is not stationary and that

---

*The data are taken from various issues of the *Survey of Current Business* published by the U.S. Department of Commerce.

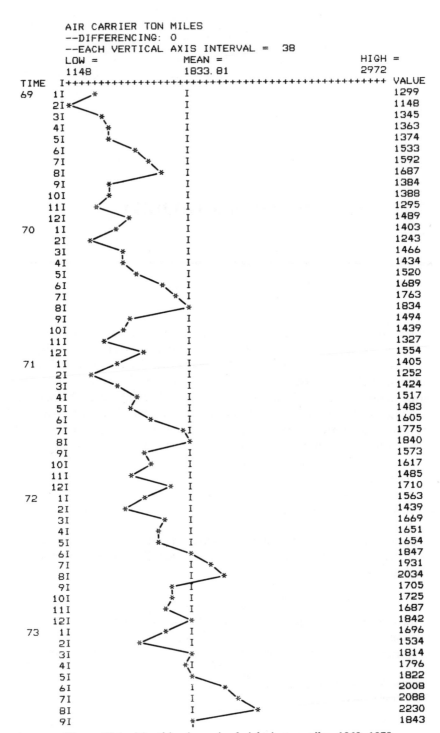

Figure C9.1 Monthly air-carrier freight in ton-miles, 1969–1978.

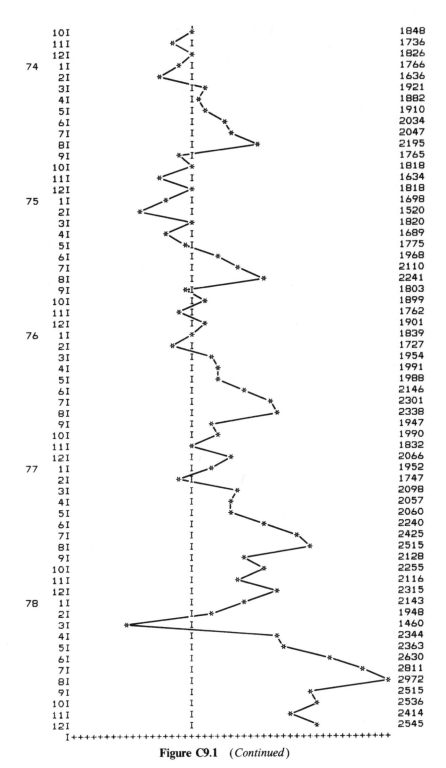

	10I	1848
	11I	1736
	12I	1826
74	1I	1766
	2I	1636
	3I	1921
	4I	1882
	5I	1910
	6I	2034
	7I	2047
	8I	2195
	9I	1765
	10I	1818
	11I	1634
	12I	1818
75	1I	1698
	2I	1520
	3I	1820
	4I	1689
	5I	1775
	6I	1968
	7I	2110
	8I	2241
	9I	1803
	10I	1899
	11I	1762
	12I	1901
76	1I	1839
	2I	1727
	3I	1954
	4I	1991
	5I	1988
	6I	2146
	7I	2301
	8I	2338
	9I	1947
	10I	1990
	11I	1832
	12I	2066
77	1I	1952
	2I	1747
	3I	2098
	4I	2057
	5I	2060
	6I	2240
	7I	2425
	8I	2515
	9I	2128
	10I	2255
	11I	2116
	12I	2315
78	1I	2143
	2I	1948
	3I	1460
	4I	2344
	5I	2363
	6I	2630
	7I	2811
	8I	2972
	9I	2515
	10I	2536
	11I	2414
	12I	2545

Figure C9.1 (*Continued*)

443

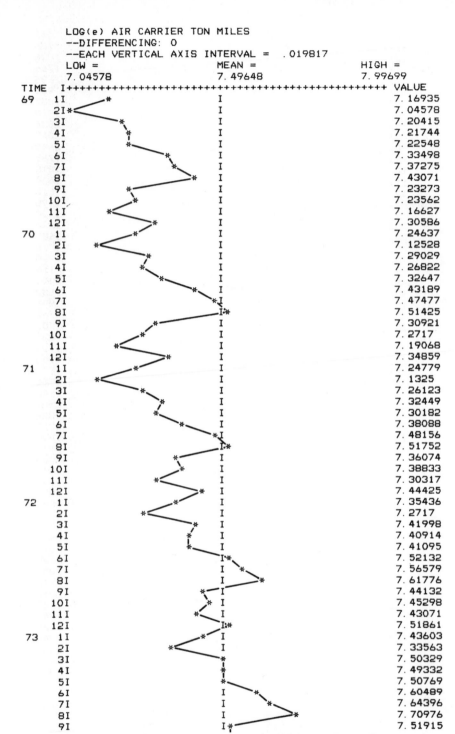

```
       LOG(e) AIR CARRIER TON MILES
       --DIFFERENCING: 0
       --EACH VERTICAL AXIS INTERVAL =  .019817
       LOW =                MEAN =                 HIGH =
       7.04578              7.49648                7.99699
 TIME  I++++++++++++++++++++++++++++++++++++++++++++++++++++ VALUE
 69  1I        *                   I                         7.16935
     2I*                           I                         7.04578
     3I          *                 I                         7.20415
     4I           *                I                         7.21744
     5I          *                 I                         7.22548
     6I             *              I                         7.33498
     7I              *             I                         7.37275
     8I                *           I                         7.43071
     9I          *                 I                         7.23273
    10I           *                I                         7.23562
    11I       *                    I                         7.16627
    12I            *               I                         7.30586
 70  1I          *                 I                         7.24637
     2I      *                     I                         7.12528
     3I           *                I                         7.29029
     4I          *                 I                         7.26822
     5I            *               I                         7.32647
     6I              *             I                         7.43189
     7I                *           I                         7.47477
     8I                          I.*                         7.51425
     9I          *                 I                         7.30921
    10I          *                 I                         7.2717
    11I       *                    I                         7.19068
    12I            *               I                         7.34859
 71  1I          *                 I                         7.24779
     2I     *                      I                         7.1325
     3I          *                 I                         7.26123
     4I            *               I                         7.32449
     5I           *               I                          7.30182
     6I             *              I                         7.38088
     7I                *           I                         7.48156
     8I                          I.*                         7.51752
     9I             *             I                          7.36074
    10I             *             I                          7.38833
    11I         *                 I                          7.30317
    12I              *            I                          7.44425
 72  1I            *              I                          7.35436
     2I         *                 I                          7.2717
     3I              *            I                          7.41998
     4I            *              I                          7.40914
     5I            *              I                          7.41095
     6I                         I*                           7.52132
     7I                         I  *                         7.56579
     8I                         I      *                     7.61776
     9I            *          I                              7.44132
    10I            *          I                              7.45298
    11I          *           I                              7.43071
    12I                        I.*                           7.51861
 73  1I            *          I                              7.43603
     2I        *              I                              7.33563
     3I                      *                               7.50329
     4I                     *                                7.49332
     5I                      *                               7.50769
     6I                       I    *                         7.60489
     7I                       I      *                       7.64396
     8I                       I          *                   7.70976
     9I                      I*                              7.51915
```

Figure C9.2 Natural logarithms of the realization in Figure C9.1.

444

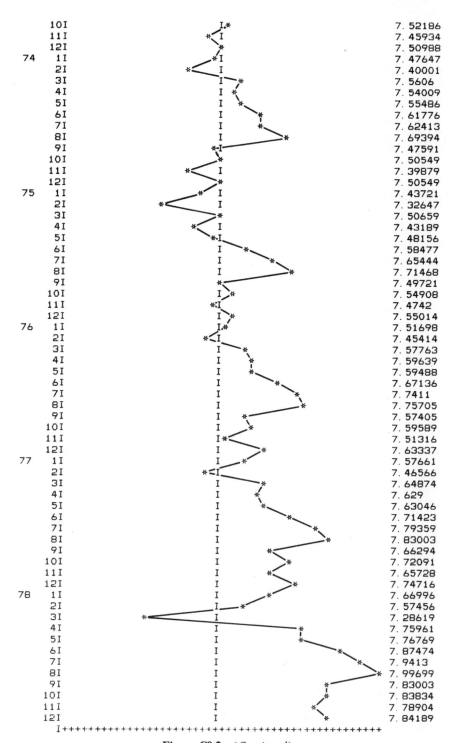

10I	I.*	7.52186
11I	*.I	7.45934
12I	*.I	7.50988
74 1I	*I	7.47647
2I	* I	7.40001
3I	I *	7.5606
4I	I *	7.54009
5I	I *	7.55486
6I	I *	7.61776
7I	I *	7.62413
8I	I *	7.69394
9I	*-I	7.47591
10I	I *	7.50549
11I	* I	7.39879
12I	*I	7.50549
75 1I	*I	7.43721
2I	* I	7.32647
3I	I*	7.50659
4I	* I	7.43189
5I	*I	7.48156
6I	I *	7.58477
7I	I *	7.65444
8I	I *	7.71468
9I	*	7.49721
10I	I *	7.54908
11I	*·I	7.4742
12I	I *	7.55014
76 1I	I*	7.51698
2I	*·I	7.45414
3I	I *	7.57763
4I	I *	7.59639
5I	I *	7.59488
6I	I *	7.67136
7I	I *	7.7411
8I	I *	7.75705
9I	I *	7.57405
10I	I *	7.59589
11I	I*	7.51316
12I	I *	7.63337
77 1I	I *	7.57661
2I	*<I	7.46566
3I	I *	7.64874
4I	I *	7.629
5I	I *	7.63046
6I	I *	7.71423
7I	I *	7.79359
8I	I *	7.83003
9I	I *	7.66294
10I	I *	7.72091
11I	I *	7.65728
12I	I *	7.74716
78 1I	I *	7.66996
2I	I *	7.57456
3I	* I	7.28619
4I	I *	7.75961
5I	I *	7.76769
6I	I *	7.87474
7I	I *	7.9413
8I	I *	7.99699
9I	I *	7.83003
10I	I *	7.83834
11I	I *	7.78904
12I	I *	7.84189
I+++		

Figure C9.2   (*Continued*)

445

nonseasonal differencing is required; (ii) there is an obvious seasonal pattern within each year, with high values occurring in August and low values in February; (iii) the variance appears to rise over time, so that a transformation is needed to induce a stationary variance.

We consider item (iii) first. Often a logarithmic transformation is adequate when the variance of a series is nonstationary; however, this transformation should not be used arbitrarily. It is appropriate only when the variance is at least roughly proportional to the mean. The data in Figure C9.1 may qualify on this count: the variability of the data is greater for the later observations when the mean also is larger.

There are some sophisticated statistical tests to check for constancy of variance, but visual inspection of the data is as effective as any other procedure. Figure C9.2 is a plot of the natural logarithms of the original data. Inspection indicates that the log transformation is acceptable since the log series has a roughly constant variance.

Throughout the rest of this case we analyze the log series. But our purpose is to construct a model to forecast the original series, not the log series. After modeling the log data, we must be careful to correctly translate the log forecasts into forecasts of the original series. (See Chapters 7 and 10 for discussions of the use of the log transformation.)

Figure C9.3 is the estimated acf and pacf of the log data. The slow decay of the acf along the first six lags shows that the mean of the data is nonstationary; calculating nonseasonal first differences is appropriate. We must be careful when making this decision. Sometimes strong seasonality can make the nonseasonal pattern in the estimated acf appear nonstationary when the nonseasonal element is, in fact, stationary. Large autocorrelations at the seasonal lags can be highly correlated with other autocorrelations, pulling them up and preventing them from dropping rapidly to zero. In this case the slow decay of the acf confirms our tentative conclusion based on visual analysis of the data in Figures C9.1 and C9.2: we see there that the level of the series is rising through time. If the mean of the data is nonstationary, the estimated pacf provides no additional useful information.

The acf in Figure C9.3 has peaks at lags 12 and 24; with monthly data these are the seasonal lags. As discussed in Chapter 11, the frequency with which data are recorded determines the length of periodicity. Monthly data have seasonality of length 12, quarterly data have seasonality of length 4, and so forth.

Even if the mean of the original series appears stationary, it is still wise to examine the estimated acf and pacf of the first differences as an aid in identifying the seasonal pattern. Often the acf and pacf of the first differences reveal more clearly the nature of the seasonal element. In this case we have two reasons for considering the acf and pacf of the first differences:

```
+ + + + + + + + + + + + + AUTOCORRELATIONS + + + + + + + + + + + + +
+ FOR DATA SERIES: LOG(e) AIR CARRIER TON MILES                    +
+ DIFFERENCING:  0                      MEAN    =  7.49648          +
+ DATA COUNT =  120                     STD DEV =  .187338          +
  COEF   T-VAL  LAG _____0_____
  0.81   8.85    1                          [    0>>>]>>>>>>>>>>>>>>>>>>>
  0.71   5.10    2                         [    0>>>>>]>>>>>>>>>>>>>
  0.61   3.67    3                    [      0>>>>>>>]>>>>>>>
  0.54   2.97    4                [    0>>>>>>>>]>>>>
  0.46   2.34    5              [    0>>>>>>>>>]>
  0.39   1.89    6              [    0>>>>>>>>>]
  0.42   1.96    7              [    0>>>>>>>>>]
  0.44   2.02    8              [    0>>>>>>>>]>
  0.45   2.01    9           [     0>>>>>>>>>>>]
  0.51   2.20   10          [     0>>>>>>>>>>>]>
  0.55   2.26   11          [     0>>>>>>>>>>>]>>
  0.64   2.54   12         [     0>>>>>>>>>>>>]>>>>
  0.52   1.98   13        [     0>>>>>>>>>>>>>]
  0.44   1.60   14       [     0>>>>>>>>>>>  ]
  0.37   1.31   15       [     0>>>>>>>>>    ]
  0.33   1.16   16       [     0>>>>>>>>     ]
  0.26   0.91   17       [     0>>>>>>       ]
  0.19   0.64   18       [     0>>>>>        ]
  0.22   0.75   19       [     0>>>>>        ]
  0.24   0.83   20       [     0>>>>>>       ]
  0.26   0.90   21       [     0>>>>>>       ]
  0.29   0.97   22       [     0>>>>>>>      ]
  0.32   1.08   23       [     0>>>>>>>      ]
  0.41   1.37   24       [     0>>>>>>>>>>   ]
  0.31   1.01   25    [        0>>>>>>>         ]
  0.24   0.78   26    [        0>>>>>>          ]
  0.18   0.58   27    [        0>>>>>           ]
  0.14   0.47   28    [        0>>>>            ]
  0.08   0.25   29    [        0>>              ]
  0.01   0.03   30    [        0               ]
     CHI-SQUARED* =  702.85 FOR DF =  30

+ + + + + + + + + + + + PARTIAL AUTOCORRELATIONS + + + + + + + + + + + +
  COEF   T-VAL  LAG _____0_____
  0.81   8.85    1                    [    0>>>]>>>>>>>>>>>>>>>>>>>>
  0.15   1.69    2                    [   0>>>]
  0.00   0.05    3                    [   0    ]
  0.06   0.61    4            /       [   0>   ]
 -0.06  -0.66    5                    [   <0   ]
 -0.03  -0.30    6                    [   <0   ]
  0.26   2.88    7                    [   0>>>]>>>
  0.13   1.44    8                    [   0>>>]
  0.04   0.42    9                    [   0>   ]
  0.22   2.45   10                    [   0>>>]>>
  0.05   0.59   11                    [   0>   ]
  0.28   3.05   12                    [   0>>>]>>>
 -0.40  -4.33   13            <<<<<<<[<<<0   ]
 -0.16  -1.78   14                 [<<<0   ]
 -0.02  -0.17   15                    [   0    ]
  0.07   0.76   16                    [   0>>  ]
 -0.04  -0.47   17                    [   <0   ]
 -0.07  -0.76   18                    [  <<0   ]
  0.08   0.89   19                    [   0>>  ]
  0.00  -0.03   20                    [   0    ]
  0.08   0.91   21                    [   0>>  ]
 -0.12  -1.28   22                    [<<<0   ]
  0.06   0.64   23                    [   0>  ]
  0.17   1.85   24                    [   0>>>]
 -0.23  -2.51   25                 <<[<<<0   ]
 -0.02  -0.26   26                    [  <0   ]
 -0.02  -0.17   27                    [   0   ]
 -0.04  -0.40   28                    [   <0  ]
 -0.03  -0.36   29                    [   <0  ]
  0.02   0.19   30                    [   0   ]
```

Figure C9.3   Estimated acf and pacf for the log data in Figure C9.2.

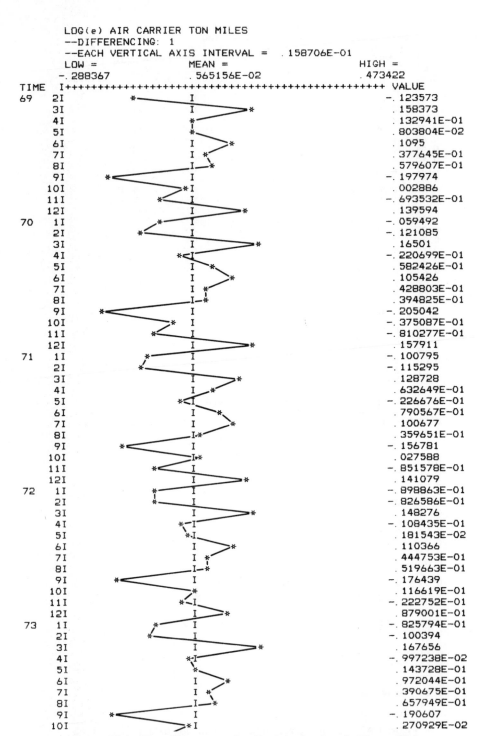

LOG(e) AIR CARRIER TON MILES
--DIFFERENCING: 1
--EACH VERTICAL AXIS INTERVAL = . 158706E-01
LOW =   MEAN =   HIGH =
-. 288367   . 565156E-02   . 473422

TIME		VALUE
69	2I	-. 123573
	3I	. 158373
	4I	. 132941E-01
	5I	. 803804E-02
	6I	. 1095
	7I	. 377645E-01
	8I	. 579607E-01
	9I	-. 197974
	10I	. 002886
	11I	-. 693532E-01
	12I	. 139594
70	1I	-. 059492
	2I	-. 121085
	3I	. 16501
	4I	-. 220699E-01
	5I	. 582426E-01
	6I	. 105426
	7I	. 428803E-01
	8I	. 394825E-01
	9I	-. 205042
	10I	-. 375087E-01
	11I	-. 810277E-01
	12I	. 157911
71	1I	-. 100795
	2I	-. 115295
	3I	. 128728
	4I	. 632649E-01
	5I	-. 226676E-01
	6I	. 790567E-01
	7I	. 100677
	8I	. 359651E-01
	9I	-. 156781
	10I	. 027588
	11I	-. 851578E-01
	12I	. 141079
72	1I	-. 898863E-01
	2I	-. 826586E-01
	3I	. 148276
	4I	-. 108435E-01
	5I	. 181543E-02
	6I	. 110366
	7I	. 444753E-01
	8I	. 519663E-01
	9I	-. 176439
	10I	. 116619E-01
	11I	-. 222752E-01
	12I	. 879001E-01
73	1I	-. 825794E-01
	2I	-. 100394
	3I	. 167656
	4I	-. 997238E-02
	5I	. 143728E-01
	6I	. 972044E-01
	7I	. 390675E-01
	8I	. 657949E-01
	9I	-. 190607
	10I	. 270929E-02

Figure C9.4 First differences ($d = 1$) of the log data in Figure C9.3.

448

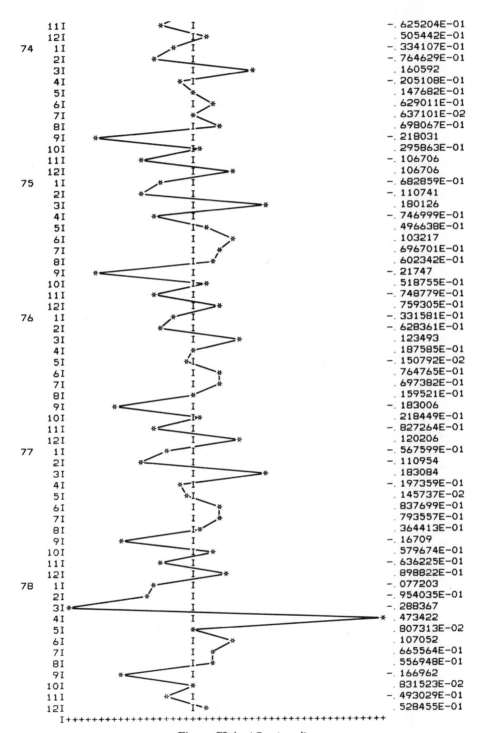

	11I	-.625204E-01
	12I	.505442E-01
74	1I	-.334107E-01
	2I	-.764629E-01
	3I	.160592
	4I	-.205108E-01
	5I	.147682E-01
	6I	.629011E-01
	7I	.637101E-02
	8I	.698067E-01
	9I	-.218031
	10I	.295863E-01
	11I	-.106706
	12I	.106706
75	1I	-.682859E-01
	2I	-.110741
	3I	.180126
	4I	-.746999E-01
	5I	.496638E-01
	6I	.103217
	7I	.696701E-01
	8I	.602342E-01
	9I	-.21747
	10I	.518755E-01
	11I	-.748779E-01
	12I	.759305E-01
76	1I	-.331581E-01
	2I	-.628361E-01
	3I	.123493
	4I	.187585E-01
	5I	-.150792E-02
	6I	.764765E-01
	7I	.697382E-01
	8I	.159521E-01
	9I	-.183006
	10I	.218449E-01
	11I	-.827264E-01
	12I	.120206
77	1I	-.567599E-01
	2I	-.110954
	3I	.183084
	4I	-.197359E-01
	5I	.145737E-02
	6I	.837699E-01
	7I	.793557E-01
	8I	.364413E-01
	9I	-.16709
	10I	.579674E-01
	11I	-.636225E-01
	12I	.898822E-01
78	1I	-.077203
	2I	-.954035E-01
	3I	-.288367
	4I	.473422
	5I	.807313E-02
	6I	.107052
	7I	.665564E-01
	8I	.556948E-01
	9I	-.166962
	10I	.831523E-02
	11I	-.493029E-01
	12I	.528455E-01

Figure C9.4 (*Continued*)

449

we think first differencing is needed to induce a stationary mean and we want to get a better picture of the seasonal pattern.

The first differences are plotted in Figure C9.4. This series no longer has an upward trend, suggesting the nonseasonal portion is now stationary. The estimated acf and pacf of the first differences of the log data are shown in Figure C9.5. The acf confirms that the nonseasonal part of the series is now stationary since the acf drops quickly to zero: the autocorrelations at lag 2 through 5 are not significantly different from zero. But the seasonal part of the data could be nonstationary and seasonal differencing might be required.

The slow decay of the autocorrelations at lags 12 and 24 suggests that seasonal differencing is needed. If they decayed exponentially the value at lag 24 would be $(0.61)^2 = 0.37$, instead of 0.55. Visual examination of the first differences of the log values (Figure C9.4) confirms that March and December observations, for example, regularly lie above September observations.

It is possible that the acf spikes at lags 12 and 24 in Figure C9.5 represent a seasonal MA(2) time structure. Then the seasonal pattern would automatically be stationary since pure MA models are always stationary, and seasonal differencing would not be needed. Ideally, we would have more observations allowing us to calculate enough autocorrelations to decide if the estimated acf decays (indicating an AR model) or cuts off (indicating an MA model) at further multiples of lag 12, but with 120 observations we may safely use only about $120/4 = 30$ estimated autocorrelations. In this case inspection of Figure C9.4 suggests that the mean of the realization varies regularly by months; in addition, the pacf in Figure C9.5 offers mild confirmation that the seasonal element in the first differences is autoregressive because the partial autocorrelation at lag 24 has a t-value less than 2.0. All things considered, seasonal differencing seems appropriate.

The noticeable spikes at the half-seasonal lags (6, 18, 30) in Figure C9.5 are best ignored for now. As pointed out in Chapter 11, strong seasonality can produce reflections in the estimated acf at fractional-seasonal lags. These reflections frequently disappear in residual acf's after the seasonal element has been modeled with seasonal differencing or appropriate seasonal AR and MA terms.

Figure C9.6 is the estimated acf and pacf after both seasonal and nonseasonal first differencing. It is now fairly easy to identify an appropriate model. Note that the autocorrelations are insignificant at lags 6 and 18. The spike at lag 1 suggests a nonseasonal MA term, and the spike at lag 12 calls for a seasonal MA coefficient. Although the autocorrelation at lag 12 has an absolute t-value of only 1.54, this exceeds the approximate

```
+ + + + + + + + + + + + AUTOCORRELATIONS + + + + + + + + + + + + +
+ FOR DATA SERIES: LOG(e) AIR CARRIER TON MILES                    +
+ DIFFERENCING: 1                        MEAN     = .565156E-02     +
+ DATA COUNT =  119                      STD DEV = .107985          +
  COEF   T-VAL LAG _____0_____
-0.29   -3.20   1                 <<<<[<<<<<0        ]
 0.01    0.10   2                      [    0         ]
-0.10   -1.00   3                      [ <<<0         ]
 0.02    0.21   4                      [    0>        ]
 0.00   -0.01   5                      [    0         ]
-0.28   -2.81   6                 <<<[<<<<<0         ]
 0.05    0.44   7                      [    0>>         ]
-0.01   -0.09   8                      [    0          ]
-0.03   -0.25   9                      [   <0          ]
-0.01   -0.10  10                      [    0          ]
-0.18   -1.64  11                      [ <<<<<<0          ]
 0.61    5.56  12                      [        0>>>>>>>]>>>>>>>>>>>>
-0.11   -0.79  13                      [  <<<<0          ]
-0.02   -0.15  14                   [       <0         ]
-0.09   -0.70  15                   [   <<<0          ]
 0.07    0.49  16                   [       0>>         ]
 0.05    0.34  17                  ·[       0>>         ]
-0.29   -2.16  18                   [<<<<<<<<<<0         ]
 0.05    0.39  19                   [       0>>         ]
-0.05   -0.33  20                   [     <<0          ]
-0.02   -0.12  21                   [      <0          ]
-0.01   -0.07  22                   [       0          ]
-0.16   -1.14  23                   [  <<<<<0          ]
 0.55    3.85  24                   [       0>>>>>>>>>>]>>>>>>>>
-0.11   -0.70  25                   [   <<<<0          ]
-0.02   -0.11  26                   [      <0          ]
-0.05   -0.32  27                   [     <<0          ]
 0.04    0.26  28                   [       0>         ]
 0.02    0.14  29                   [       0>         ]
-0.23   -1.44  30                   [ <<<<<<<<0         ]
       CHI-SQUARED* =  154.46 FOR DF =   30

+ + + + + + + + + + + + PARTIAL AUTOCORRELATIONS + + + + + + + + + + + +
  COEF   T-VAL LAG _____0_____
-0.29   -3.20   1                 <<<<[<<<<<0        ]
-0.08   -0.91   2                      [  <<<0        ]
-0.13   -1.45   3                      [ <<<<0        ]
-0.06   -0.60   4                      [   <<0        ]
-0.02   -0.27   5                      [    <0        ]
-0.34   -3.69   6                 <<<<<[<<<<<0        ]
-0.20   -2.15   7                   <[<<<<<<0        ]
-0.14   -1.58   8                     [<<<<<0        ]
-0.23   -2.48   9                   <<[<<<<<0        ]
-0.22   -2.43  10                   <[<<<<<<0        ]
-0.52   -5.68  11          <<<<<<<<<<<<[<<<<<0        ]
 0.23    2.52  12                      [    0>>>>>]>>
 0.14    1.52  13                      [    0>>>>>]
-0.04   -0.41  14                      [   <0        ]
-0.06   -0.67  15                      [  <<0        ]
 0.03    0.28  16                      [    0>        ]
 0.14    1.57  17                      [    0>>>>>]
 0.02    0.21  18                      [    0>        ]
 0.03    0.34  19                      [    0>        ]
-0.02   -0.24  20                      [   <0        ]
-0.08   -0.92  21                      [  <<<0        ]
-0.03   -0.31  22                      [   <0        ]
-0.17   -1.88  23                      [<<<<<0        ]
 0.17    1.85  24                      [    0>>>>>]
 0.01    0.15  25                      [    0        ]
-0.10   -1.11  26                      [  <<<0        ]
 0.06    0.64  27                      [    0>>        ]
 0.00   -0.01  28                      [    0        ]
-0.08   -0.92  29                      [  <<<0        ]
 0.01    0.11  30                      [    0        ]
```

Figure C9.5 Estimated acf and pacf of the first differences of the log data ($d = 1$) in Figure C9.4.

451

```
+ + + + + + + + + + + AUTOCORRELATIONS + + + + + + + + + + + +
+ FOR DATA SERIES: LOG(e) AIR CARRIER TON MILES                      +
+ DIFFERENCING: 12  1                        MEAN     = .165448E-03  +
+ DATA COUNT =   107                         STD DEV  = .744335E-01   +
  COEF   T-VAL LAG                            0
-0.47   -4.86   1   <<<<<<<<<<<<<<<[<<<<<<<<<0              ]
 0.05    0.39   2            [           0>>                ]
 0.02    0.13   3            [           0>                 ]
-0.05   -0.46   4            [        <<<0                   ]
-0.01   -0.06   5            [           0                   ]
-0.04   -0.36   6            [         <<0                   ]
 0.06    0.48   7            [           0>>>                ]
-0.02   -0.21   8            [          <0                   ]
 0.02    0.16   9            [           0>                  ]
 0.00    0.00  10            [           0                   ]
 0.08    0.67  11            [           0>>>>               ]
-0.18   -1.54  12            [    <<<<<<<<<0                 ]
 0.10    0.80  13            [           0>>>>>              ]
 0.00    0.02  14            [           0                   ]
-0.10   -0.82  15            [      <<<<<0                   ]
 0.08    0.62  16            [           0>>>>               ]
 0.00    0.04  17            [           0                   ]
-0.04   -0.30  18            [         <<0                   ]
 0.06    0.47  19            [           0>>>                ]
-0.04   -0.37  20            [         <<0                   ]
 0.04    0.34  21            [           0>>                 ]
-0.01   -0.11  22            [          <0                   ]
-0.10   -0.83  23            [      <<<<<0                   ]
 0.14    1.10  24            [           0>>>>>>>            ]
-0.09   -0.70  25            [       <<<<0                   ]
 0.01    0.05  26            [           0                   ]
 0.06    0.49  27            [           0>>>                ]
     CHI-SQUARED* =    40.21 FOR DF =   27

+ + + + + + + + + + + PARTIAL AUTOCORRELATIONS + + + + + + + + + + +
  COEF   T-VAL LAG                            0
-0.47   -4.86   1   <<<<<<<<<<<<<<<[<<<<<<<<<0              ]
-0.22   -2.33   2            <[<<<<<<<<<0                    ]
-0.09   -0.90   3            [       <<<<0                   ]
-0.10   -1.03   4            [      <<<<<0                   ]
-0.10   -1.06   5            [      <<<<<0                   ]
-0.14   -1.46   6            [    <<<<<<<0                   ]
-0.05   -0.56   7            [        <<<0                   ]
-0.04   -0.46   8            [         <<0                   ]
-0.02   -0.20   9            [          <0                   ]
-0.02   -0.17  10            [          <0                   ]
 0.10    1.05  11            [           0>>>>>              ]
-0.12   -1.27  12            [      <<<<<0                   ]
-0.06   -0.67  13            [        <<<0                   ]
-0.01   -0.07  14            [           0                   ]
-0.12   -1.20  15            [      <<<<<0                   ]
-0.07   -0.70  16            [        <<<0                   ]
-0.02   -0.19  17            [          <0                   ]
-0.08   -0.81  18            [       <<<<0                   ]
 0.00    0.03  19            [           0                   ]
-0.04   -0.44  20            [         <<0                   ]
 0.00    0.02  21            [           0                   ]
 0.02    0.18  22            [           0>                  ]
-0.12   -1.26  23            [      <<<<<0                   ]
-0.01   -0.09  24            [           0                   ]
-0.04   -0.37  25            [         <<0                   ]
-0.06   -0.65  26            [        <<<0                   ]
-0.02   -0.19  27            [          <0                   ]
```

Figure C9.6 Estimated acf and pacf for the differences of the log data ($d = 1$, $D = 1$).

practical warning level (1.25) for seasonal lags suggested in Chapters 11 and 12.

A multiplicative model as presented in Chapter 11 is a good starting place for data with both seasonal and nonseasonal variation. In this case we have selected an ARMA$(0, 1)(0, 1)_{12}$ model for the stationary (logged and differenced) series w_t:

$$w_t = (1 - \Theta_{12}B^{12})(1 - \theta_1 B)a_t \qquad \text{(C9.1)}$$

```
+ + + + + + + + +ECOSTAT UNIVARIATE B-J RESULTS+ + + + + + + + +
+ FOR DATA SERIES:   LOG(e) AIR CARRIER TON MILES                    +
+ DIFFERENCING:      12  1                        DF    = 105        +
+ AVAILABLE:         DATA = 107   BACKCASTS = 13  TOTAL = 120        +
+ USED TO FIND SSR:  DATA = 107   BACKCASTS = 13  TOTAL = 120        +
+ (LOST DUE TO PRESENCE OF AUTOREGRESSIVE TERMS:               0)    +

COEFFICIENT    ESTIMATE      STD ERROR     T-VALUE
   THETA  1     0.766         0.058         13.23
   THETA* 12    0.857         0.057         14.96

   ADJUSTED RMSE =  .584345E-01    MEAN ABS % ERR =    0.47
        CORRELATIONS
            1     2
   1     1.00
   2     0.00   1.00

++RESIDUAL ACF++
   COEF   T-VAL  LAG _____0_____
   0.08    0.83    1     [              0>>>>>>>>>               ]
   0.08    0.79    2     [              0>>>>>>>>>               ]
   0.02    0.21    3     [              0>>                      ]
  -0.04   -0.42    4     [           <<<<0                       ]
  -0.04   -0.38    5     [           <<<<0                       ]
  -0.04   -0.40    6     [           <<<<0                       ]
   0.04    0.39    7     [              0>>>>                     ]
  -0.04   -0.45    8     [           <<<<0                       ]
   0.06    0.61    9     [              0>>>>>>                   ]
   0.02    0.19   10     [              0>>                      ]
   0.01    0.10   11     [              0>                       ]
   0.01    0.10   12     [              0>                       ]
   0.00   -0.04   13     [              0                        ]
  -0.04   -0.40   14     [           <<<<0                       ]
  -0.06   -0.59   15     [          <<<<<<0                      ]
   0.01    0.11   16     [              0>                       ]
  -0.01   -0.14   17     [             <0                        ]
   0.00    0.00   18     [              0                        ]
   0.04    0.40   19     [              0>>>>                     ]
  -0.03   -0.29   20     [            <<<0                       ]
  -0.02   -0.15   21     [             <<0                       ]
  -0.05   -0.54   22     [           <<<<<0                      ]
  -0.04   -0.42   23     [            <<<<0                      ]
   0.03    0.30   24     [              0>>>                     ]
  -0.08   -0.78   25     [        <<<<<<<<0                      ]
  -0.02   -0.22   26     [             <<0                       ]
   0.03    0.26   27     [              0>>>                     ]
      CHI-SQUARED* =    5.64 FOR DF =    25
```

Figure C9.7 Estimation and diagnostic-checking results for model (C9.3).

Model (C9.1) for w_t implies a model for the original series z_t. We know that

$$w_t = (1 - B)(1 - B^{12})\ln z_t \qquad (C9.2)$$

Substituting (C9.2) into (C9.1) gives the following model for z_t, an ARIMA$(0, 1, 1)(0, 1, 1)_{12}$, where z_t' stands for the log value of z_t:

$$(1 - B)(1 - B^{12})\tilde{z}_t' = (1 - \Theta_{12}B^{12})(1 - \theta_1 B)a_t \qquad (C9.3)$$

Estimation.   Estimation results for model (C9.3) are shown in Figure C9.7. Both estimated coefficients have absolute t-values exceeding 2.0 and each satisfies its invertibility condition: $|\hat{\theta}_1| < 1$ and $|\hat{\Theta}_{12}| < 1$. Recall from Chapter 11 that in a multiplicative model the invertibility requirements apply separately to the seasonal and nonseasonal coefficients.

Note that $\hat{\Theta}_{12}$ is quite large and highly significant despite the fact that the t-value at lag 12 in Figure C9.6 is only -1.54. This common result illustrates why the practical warning value (about 1.25) for seasonal-lag t-values is substantially less than 2.0.

As shown at the top of Figure C9.7 we employed backcasting in estimating this model. (Backcasting is discussed in Appendix 8B of Chapter 8.) Backcasting is especially valuable in estimating seasonal models. We employ the backcasting technique in all remaining cases in this text since they all involve data with seasonal variation.

Diagnostic checking.   The diagnostic checking of seasonal models is essentially the same as the checking of nonseasonal models, except we pay special attention to residual autocorrelations at the seasonal lags, as well as those at the short lags.

The residual acf for (C9.3) appears below the estimation results in Figure C9.7. This model appears to explain the available data exceptionally well: none of the residual autocorrelations has an absolute t-value even approaching the practical warning values summarized in Chapter 12, and the chi-squared statistic is quite insignificant. There is no evidence that (C9.3) must be reformulated.

Residual plot.   The residuals plotted in Figure C9.8 reveal one troublesome aspect of our results. The residual for March 1978 is quite large. A check of the data source confirms that the corresponding observation was copied correctly. A single aberrant value like this one can sometimes affect estimation results substantially.

As a rough check on the effect of this observation we drop the last 12 observations and reestimate. The result (t-values in parentheses) is

$$w_t = (1 - 0.873B^{12})(1 - 0.361B)a_t$$
$$(10.87)(3.85)$$

While $\hat{\Theta}_{12}$ is virtually unchanged, $\hat{\theta}_1$ is less than half the value obtained when estimating the full data set. This is in sharp contrast to the results obtained in Case 2 where an outlying residual (and observation) occurred. There the estimation results were quite stable when the data segment starting with the aberrant observation was removed.

Outlying residuals can be dealt with in a variety of ways.

1. They can be effectively removed if investigation reveals data transcription errors.

2. They may be ignored on the grounds that they reflect the chance element inherent in the underlying process. Further aberrant observations are likely to arise as more data are gathered; the analyst can choose to let current deviant observations reflect the stochastic possibilities of future values.

3. They may be accounted for with a more sophisticated technique such as a multivariate intervention model, referred to at the end of Case 2. Careful investigation of the time periods involved sometimes reveals specific, identifiable events that are logically responsible for large deviations in the data. When this occurs this method of intervention analysis can be helpful.

4. They can be removed through adjustment of the data. If thorough investigation reveals no data errors or identifiable causal events, the analyst might choose to alter the offending observation. This must be done very cautiously. Arbitrary manipulation of data to produce a better-fitting model can be a dangerously misleading practice. On the other hand, the analyst may be convinced that an inexplicably deviant observation reveals little about the true stochastic properties of the underlying process, and that this value is likely to have seriously negative effects on estimation and forecasting results. Then adjustment of that observation can be considered. A reasonable adjustment procedure, if enough data are available, is to estimate an ARIMA model using all available data *preceding* the deviant observation, and to produced a one-step-ahead forecast from this model. Then replace the aberrant observation with this forecast value and construct an ARIMA model using the full data set.

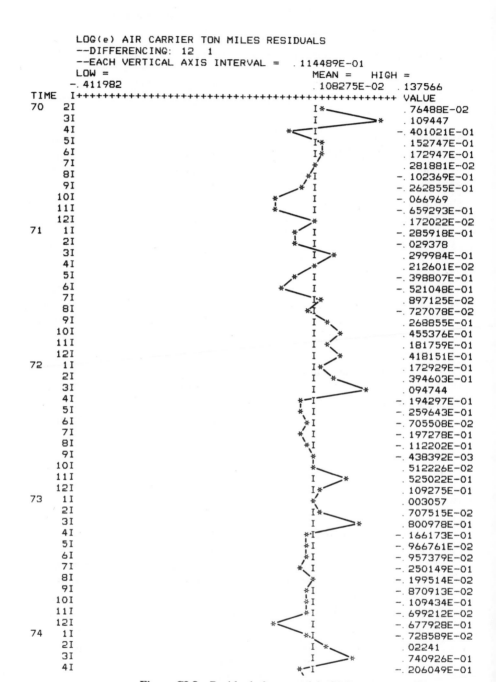

```
LOG(e) AIR CARRIER TON MILES RESIDUALS
--DIFFERENCING: 12  1
--EACH VERTICAL AXIS INTERVAL =  .114489E-01
   LOW =                              MEAN =    HIGH =
  -.411982                          .108275E-02  .137566
TIME   I+++++++++++++++++++++++++++++++++++++++++++++++ VALUE
70   2I                                        I*        .76488E-02
     3I                                        I         .109447
     4I                                   *    I        -.401021E-01
     5I                                        I*        .152747E-01
     6I                                        I*        .172947E-01
     7I                                        I         .281881E-02
     8I                                       *I        -.102369E-01
     9I                                     *  I        -.262855E-01
    10I                               *        I        -.066969
    11I                               *        I        -.659293E-01
    12I                                        *         .172022E-02
71   1I                                    *   I        -.285918E-01
     2I                                    *   I        -.029378
     3I                                        I *       .299984E-01
     4I                                        *         .212601E-02
     5I                                    *   I        -.398807E-01
     6I                               *        I        -.521048E-01
     7I                                        I*        .897125E-02
     8I                                       *I        -.727078E-02
     9I                                        I  *      .268855E-01
    10I                                        I   *     .455376E-01
    11I                                        I *       .181759E-01
    12I                                        I  *      .418151E-01
72   1I                                        I*        .172929E-01
     2I                                        I  *      .394603E-01
     3I                                        I      *  .094744
     4I                                       *I        -.194297E-01
     5I                                      * I        -.259643E-01
     6I                                       *I        -.705508E-02
     7I                                       *I        -.197278E-01
     8I                                       *I        -.112202E-01
     9I                                       *I        -.438392E-03
    10I                                        *         .512226E-02
    11I                                        I   *     .525022E-01
    12I                                        I*        .109275E-01
73   1I                                        *         .003057
     2I                                        I*        .707515E-02
     3I                                        I    *    .800978E-01
     4I                                       *I        -.166173E-01
     5I                                       *I        -.966761E-02
     6I                                       *I        -.957379E-02
     7I                                      * I        -.250149E-01
     8I                                       *I        -.199514E-02
     9I                                       *I        -.870913E-02
    10I                                       *I        -.109434E-01
    11I                                       *I        -.699212E-02
    12I                                  *     I        -.677928E-01
74   1I                                       *I        -.728589E-02
     2I                                        I *       .02241
     3I                                        I   *     .740926E-01
     4I                                       *I        -.206049E-01
```

Figure C9.8 Residuals from model (C9.3).

456

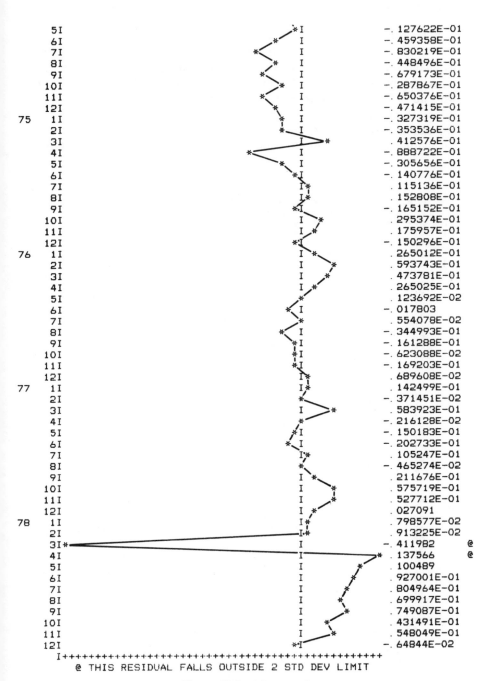

5I		−.127622E−01
6I		−.459358E−01
7I		−.830219E−01
8I		−.448496E−01
9I		−.679173E−01
10I		−.287867E−01
11I		−.650376E−01
12I		−.471415E−01
75 1I		−.327319E−01
2I		−.353536E−01
3I		.412576E−01
4I		−.888722E−01
5I		−.305656E−01
6I		−.140776E−01
7I		.115136E−01
8I		.152808E−01
9I		−.165152E−01
10I		.295374E−01
11I		.175957E−01
12I		−.150296E−01
76 1I		.265012E−01
2I		.593743E−01
3I		.473781E−01
4I		.265025E−01
5I		.123692E−02
6I		−.017803
7I		.554078E−02
8I		−.344993E−01
9I		−.161288E−01
10I		−.623088E−02
11I		−.169203E−01
12I		.689608E−02
77 1I		.142499E−01
2I		−.371451E−02
3I		.583923E−01
4I		−.216128E−02
5I		−.150183E−01
6I		−.202733E−01
7I		.105247E−01
8I		−.465274E−02
9I		.211676E−01
10I		.575719E−01
11I		.527712E−01
12I		.027091
78 1I		.798577E−02
2I		.913225E−02
3I*		−.411982 @
4I		.137566 @
5I		.100489
6I		.927001E−01
7I		.804964E−01
8I		.699917E−01
9I		.749087E−01
10I		.431491E−01
11I		.548049E−01
12I		−.64844E−02

    I++
         @ THIS RESIDUAL FALLS OUTSIDE 2 STD DEV LIMIT

Figure C9.8   (*Continued*)

457

Forecasting. We will forecast using (C9.3) on the assumption that the observation (and corresponding residual) for March 1978 could have arisen by chance due to the stochastic nature of the underlying process. Forecasts for lead times $l = 1, 2, \ldots, 24$ appear in Table C9.1 and Figure C9.9 along with an 80% confidence interval for each forecast and the corresponding future observed values. These forecasts are for the original series (i.e., they are not in log form). As discussed in Chapter 10 they are found by calculating the antilog of the sum of the log forecast *plus* one-half the

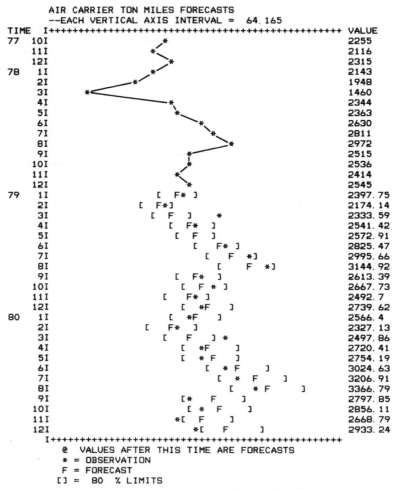

Figure C9.9   Forecasts from model (C9.3).

estimated forecast-error variance of the log forecast. The confidence intervals for the original-data forecasts, however, are found by calculating the antilog of the upper and lower bounds of the corresponding log forecast confidence interval.

Model (C9.3) forecasts rather well. The forecasts faithfully track the strong seasonal pattern. Only four of the 24 confidence intervals fail to contain the observed values—the two March forecasts and the last two forecasts. Since the data are seasonally differenced, the starting place for each forecast is the observed (or forecast) value 12 months earlier. Thus the March 1979 forecast starts from the observed value in March 1978; the

Table C9.1 Forecasts^a from model (C9.3)

Time		Forecast Values	80% Confidence Limits		Future Observed Values	Percent Forecast Errors
			Lower	Upper		
79	1	2397.7509	2221.1560	2579.5630	2445.0000	1.93
	2	2174.1441	2009.7517	2343.5264	2275.0000	4.43
	3	2333.5893	2152.6838	2520.1290	2857.0000	18.32
	4	2541.4241	2339.6756	2749.6116	2601.0000	2.29
	5	2572.9062	2363.9847	2788.6532	2593.0000	0.77
	6	2825.4678	2591.0242	3067.7437	2939.0000	3.86
	7	2995.6592	2741.8957	3258.0836	3149.0000	4.87
	8	3144.9185	2873.1702	3426.1340	3333.0000	5.64
	9	2613.3908	2383.2235	2851.7364	2650.0000	1.38
	10	2667.7343	2428.4303	2915.7043	2764.0000	3.48
	11	2492.6981	2265.1076	2728.6828	2608.0000	4.42
	12	2739.6237	2485.1858	3003.6135	2668.0000	−2.68
80	1	2566.3956	2817.9193	2824.6117	2536.0000	−1.20
	2	2327.1284	2097.3398	2566.1104	2415.0000	3.64
	3	2497.8648	2246.5091	2759.4766	2883.0000	13.36
	4	2720.4086	2441.6324	3010.7770	2635.0000	−3.24
	5	2754.1870	2466.9577	3053.5803	2665.0000	−3.35
	6	3024.6305	2703.8161	3359.2728	2914.0000	−3.80
	7	3206.9105	2861.1584	3567.8215	3050.0000	−5.14
	8	3366.7923	2998.0209	3752.0007	3236.0000	−4.04
	9	2797.8457	2486.6678	3123.1170	2540.0000	−10.15
	10	2856.1069	2533.7045	3193.3391	2629.0000	−8.64
	11	2668.7879	2363.1668	2988.6799	2379.0000	−12.18
	12	2933.2412	2592.6131	3290.0089	2590.0000	−13.25

^aForecasts are in original metric.

March 1980 forecast starts from the March 1979 forecast. Both of the March forecasts are substantially lower than the corresponding observed values. Apparently, the unusually low March 1978 value is truly aberrant rather than indicating some new, lower overall level for March observations.

We may, of course, forecast the log series instead of the original series. See Chapter 10 for a discussion of how log forecasts are interpreted as percent change forecasts of the original series. Log forecasts are calculated as shown in Chapter 10, using the difference-equation form of (C9.3):

$$\hat{z}'_t = z'_{t-1} + z'_{t-12} - z'_{t-13} - \hat{\Theta}_{12}\hat{a}_{t-12} - \hat{\theta}_1\hat{a}_{t-1} + \hat{\theta}_1\hat{\Theta}_{12}\hat{a}_{t-13} \quad (C9.4)$$

where z' represents a log value.

Use the observed log values in Figure C9.2 and the residuals (the $\hat{a}$ values) in Figure C9.8. Substituting forecast values for unobserved z's and the expected value of zero for unobserved $\hat{a}$ terms where necessary, the first two log forecasts from (C9.4) are:

$$\hat{z}'_{120}(1) = z'_{120} + z'_{109} - z'_{108} - \hat{\Theta}_{12}\hat{a}_{109} - \hat{\theta}_1\hat{a}_{120} + \hat{\theta}_1\hat{\Theta}_{12}\hat{a}_{108}$$

$$= 7.84189 + 7.66996 - 7.74716 - 0.857(0.00799)$$

$$- 0.766(-0.00648) + (0.857)(0.766)(0.02709)$$

$$= 7.78057$$

$$\hat{z}'_{120}(2) = \hat{z}'_{121} + z'_{110} - z'_{109} - \hat{\Theta}_{12}\hat{a}_{110} - \hat{\theta}_1\hat{a}_{121} + \hat{\theta}_1\hat{\Theta}_{12}\hat{a}_{109}$$

$$= 7.78057 + 7.57456 - 7.66996 - 0.857(0.00913)$$

$$- 0.766(0) + (0.857)(0.766)(0.00799)$$

$$= 7.68260$$

Alternative models.   As an exercise consider some alternatives to (C9.3). Suppose we are not sure that differencing is needed to induce a stationary mean. The estimated acf and pacf for the undifferenced data in Figure C9.3 suggest an ARIMA$(1, 0, 0)(1, 0, 0)_{12}$ model. The acf decays after lag 1, while the estimated pacf has a spike at lag 1 followed by a cutoff to zero; this calls for a nonseasonal AR(1) term. The acf decay at lags 12 and 24 suggests a seasonal AR coefficient. Therefore, we estimate this model:

$$(1 - \phi_1 B)(1 - \Phi_{12}B^{12})\tilde{z}'_t = a_t \quad (C9.5)$$

```
+ + + + + + + + +ECOSTAT UNIVARIATE B-J RESULTS+ + + + + + + + +
+ FOR DATA SERIES:  LOG(e) AIR CARRIER TON MILES                         +
+ DIFFERENCING:      0                          DF    = 117              +
+ AVAILABLE:          DATA = 120   BACKCASTS = 380  TOTAL = 500          +
+ USED TO FIND SSR:  DATA = 120   BACKCASTS = 367  TOTAL = 487          +
+ (LOST DUE TO PRESENCE OF AUTOREGRESSIVE TERMS:              13)        +

COEFFICIENT    ESTIMATE       STD ERROR     T-VALUE
  PHI     1     0.666          0.071          9.37
  PHI*   12     0.996          0.033         30.56
  CONSTANT      .104031E-01   .795126E-01   .130836

  MEAN          8.46256       10.2684        .824135

ADJUSTED RMSE =  .648528E-01    MEAN ABS % ERR =    0.47
        CORRELATIONS
         1      2      3
  1    1.00
  2   -0.18   1.00
  3   -0.20   0.97   1.00
```

```
++RESIDUAL ACF++
 COEF   T-VAL LAG _____0_____
-0.35  -3.60   1           <<<<<<<[<<<<<<<<<<0            ]
 0.09   0.87   2           [         0>>>>>              ]
 0.06   0.53   3           [         0>>>                ]
-0.02  -0.15   4           [        <0                   ]
 0.02   0.15   5           [         0>                  ]
-0.02  -0.16   6           [        <0                   ]
 0.07   0.61   7           [         0>>>                ]
-0.01  -0.09   8           [         0                   ]
 0.02   0.14   9           [         0>                  ]
 0.01   0.06  10           [         0                   ]
 0.07   0.62  11           [         0>>>                ]
-0.17  -1.57  12          [<<<<<<<<<0                    ]
 0.08   0.70  13        [           0>>>>                ]
-0.01  -0.07  14        [           0                    ]
-0.10  -0.91  15        [        <<<<<0                  ]
 0.06   0.52  16        [           0>>>                 ]
 0.00  -0.04  17        [           0                    ]
-0.04  -0.37  18        [          <<0                   ]
 0.04   0.37  19        [           0>>                  ]
-0.05  -0.45  20        [         <<<0                   ]
 0.02   0.22  21        [           0>                   ]
-0.03  -0.26  22        [          <0                    ]
-0.11  -0.93  23        [        <<<<<0                  ]
 0.11   0.96  24        [           0>>>>>>              ]
-0.09  -0.78  25        [        <<<<<0                  ]
-0.01  -0.04  26        [           0                    ]
 0.05   0.40  27        [           0>>                  ]
     CHI-SQUARED* =    27.91 FOR DF =   24
```

Figure C9.10 Estimation and diagnostic-checking results for model (C9.5).

461

Estimation results in Figure C9.10 show that the seasonal AR coefficient $\hat{\Phi}_{12}$ is close enough to 1.0 to warrant seasonal differencing. As commonly happens when estimating a nonstationary series, the mean is highly correlated with the offending AR coefficient ($r = 0.97$ in this case). The estimated mean makes little sense since it is larger than any of the observed log values in Figure C9.2. The residual acf in Figure C9.9 has spikes at lags 1 and 12 suggesting the addition of MA terms at those lags. This leads us to an ARIMA$(1, 0, 1)(0, 1, 1)_{12}$:

$$(1 - \phi_1 B)(1 - B^{12})\tilde{z}_t' = (1 - \Theta_{12}B^{12})(1 - \theta_1 B)a_t \qquad (C9.6)$$

The key result of estimating this model (not shown) is that $\hat{\phi}_1 = 0.999$, which is virtually 1.0. This indicates that nonseasonal differencing is required. With both seasonal and nonseasonal first differencing we are led back to the estimated acf and pacf in Figure C9.6 and to model (C9.3).

As demonstrated in some of the preceding case studies, the need for differencing can be confirmed with estimation results. Although excessive differencing must be avoided, in borderline cases it is better to difference than not. Differencing frees the forecasts from a fixed mean; this added flexibility often produces more accurate forecasts.

Final comments. In Chapter 5 we demonstrated that forecasts from an ARIMA$(0, 1, 1)$ model are exponentially weighted moving averages (EWMA's) of the past values of a series. The same is true for an ARIMA$(0, 1, 1)_s$ model, but in the seasonal case the exponential weighting applies to observations that are $s, 2s, 3s, \ldots$ periods in the past rather than $1, 2, 3, \ldots$ periods in the past. Thus the forecasts from model (C9.3) may be interpreted roughly as the combination of two EWMA's, one (the nonseasonal part) which is a weighting of all past observations after the seasonal element has been accounted for, and the other (the seasonal part) a weighting of all past observations of the month being forecast.

CASE 10.   PROFIT MARGIN

The data in this case are after-tax profits, measured in cents per dollar of sales, for all U.S. manufacturing corporations.* As shown in Figure C10.1 these quarterly data cover the period 1953–1972, a total of 80 observations.

Identification. Inspection of the data in Figure C10.1 is not greatly revealing. The variance appears to be stationary. The series does not have an obvious rising or falling trend, but this does not guarantee that the mean is stationary. It is always possible to calculate a single mean for a data series (the calculated mean for the present series is the line through the center of the graph), but this single mean may not be appropriate for the entire data set. In fact, the data seem to spend rather long periods of time entirely above or entirely below the calculated mean, suggesting that the mean may be shifting. We rely on autocorrelation analysis and estimation results to determine whether the mean of this series is fixed.

Unlike the air-freight data in the preceding case, the profit-margin data do not display an obvious seasonal variation. Any clues about seasonality will have to come from estimated autocorrelation coefficients at the seasonal lags. With quarterly data, the seasonal lags are multiples of 4 (4, 8, 12,...).

Figure C10.2 is the estimated acf and pacf for the original data. The acf drops to zero fairly quickly so the mean of the data seems to be stationary. The decay toward zero (rather than a cutoff to zero) implies that an AR model is a good starting point. The estimated pacf has a single spike at lag 1

*The data are taken from *Business Conditions Digest*, January 1978, p. 103.

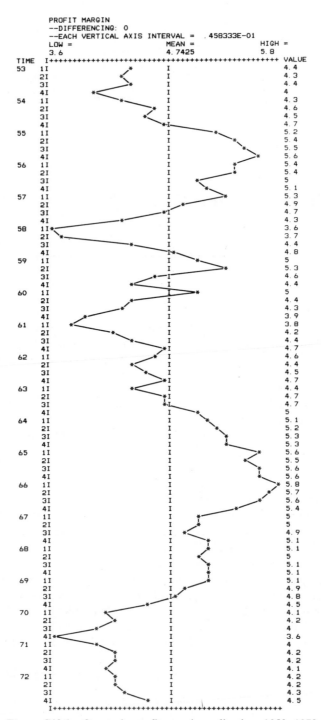

Figure C10.1 Quarterly-profit-margin realization, 1953–1972.

followed by a cutoff to zero, suggesting an AR(1) with $\hat{\phi}_1$ approximately equal to 0.87 (the value of $\hat{\phi}_{11}$).

Although the mean of the data may be stationary, we examine the estimated acf of the nonseasonal first differences. This should be done routinely whenever the data might have seasonal or other periodic variation, even if nonseasonal differencing is not needed to induce a fixed mean. Often

```
+ + + + + + + + + + + + AUTOCORRELATIONS + + + + + + + + + + + + +
+ FOR DATA SERIES: PROFIT MARGIN                                   +
+ DIFFERENCING: 0                      MEAN     =   4. 7425        +
+ DATA COUNT =  80                     STD DEV =  . 541474         +
  COEF   T-VAL LAG _____0_____
  0. 87   7. 78   1                          [       0>>>>>]>>>>>>>>>>>>>>>>>
  0. 71   4. 01   2                        [       0>>>>>>>]>>>>>>>>>>>
  0. 56   2. 69   3                     [       0>>>>>>>>>>>]>>>>
  0. 40   1. 74   4            [        0>>>>>>>>>> ]
  0. 29   1. 24   5            [        0>>>>>>>     ]
  0. 21   0. 88   6            [        0>>>>>       ]
  0. 11   0. 46   7            [        0>>>         ]
  0. 02   0. 09   8            [        0>           ]
 -0. 02  -0. 07   9            [        0            ]
 -0. 04  -0. 15  10            [       <0            ]
 -0. 05  -0. 19  11            [       <0            ]
 -0. 08  -0. 34  12            [      <<0            ]
 -0. 12  -0. 48  13            [      <<<0           ]
 -0. 16  -0. 63  14            [     <<<<0           ]
 -0. 21  -0. 84  15            [    <<<<<0           ]
 -0. 24  -0. 98  16            [    <<<<<<0          ]
 -0. 26  -1. 05  17            [   <<<<<<<0          ]
 -0. 31  -1. 23  18            [   <<<<<<<0          ]
 -0. 36  -1. 40  19           [  <<<<<<<<<0          ]
 -0. 40  -1. 50  20         [   <<<<<<<<<<0                ]
     CHI-SQUARED* =  223. 02 FOR DF =  20

+ + + + + + + + + + PARTIAL AUTOCORRELATIONS + + + + + + + + + + +
  COEF   T-VAL LAG _____0_____
  0. 87   7. 78   1              [       0>>>>>]>>>>>>>>>>>>>>>>>>
 -0. 18  -1. 63   2              [<<<<<0       ]
 -0. 04  -0. 32   3              [    <0       ]
 -0. 20  -1. 75   4              [<<<<<0       ]
  0. 19   1. 72   5              [       0>>>>>]
 -0. 07  -0. 66   6              [    <<0      ]
 -0. 13  -1. 18   7              [    <<<0     ]
 -0. 09  -0. 83   8              [    <<0      ]
  0. 20   1. 77   9              [       0>>>>>]
  0. 00   0. 00  10              [       0     ]
 -0. 05  -0. 45  11              [    <0       ]
 -0. 25  -2. 28  12              [<<<<<0       ]
  0. 12   1. 09  13              [       0>>>  ]
 -0. 05  -0. 44  14              [    <0       ]
 -0. 11  -1. 03  15              [    <<<0     ]
 -0. 14  -1. 29  16              [ <<<<0       ]
  0. 11   0. 97  17              [       0>>>  ]
 -0. 17  -1. 54  18              [ <<<<0       ]
 -0. 09  -0. 79  19              [    <<0      ]
 -0. 18  -1. 64  20              [<<<<<0       ]
```

Figure C10.2   Estimated acf and pacf for the realization in Figure C10.1.

```
+ + + + + + + + + + + + AUTOCORRELATIONS + + + + + + + + + + + + +
+ FOR DATA SERIES: PROFIT MARGIN                                     +
+ DIFFERENCING: 1                        MEAN     =  .126582E-02    +
+ DATA COUNT =  79                       STD DEV =  .274436          +
  COEF   T-VAL LAG _____0_____
  0.09   0.84   1           [                  0>>>>>>>>>               ]
 -0.04  -0.37   2           [                <<<<0                      ]
  0.06   0.55   3           [                  0>>>>>>                  ]
 -0.25  -2.18   4  <<<[<<<<<<<<<<<<<<<<<<<<<<<<0                        ]
 -0.06  -0.50   5  [                 <<<<<<0                            ]
  0.07   0.58   6  [                       0>>>>>>>                     ]
 -0.04  -0.29   7  [                    <<<<0                           ]
 -0.19  -1.58   8  [      <<<<<<<<<<<<<<<<<<<0                          ]
 -0.05  -0.39   9                     <<<<<0
 -0.03  -0.21  10                       <<<0
  0.11   0.88  11                          0>>>>>>>>>>>
  0.00   0.01  12                       .  0
  0.02   0.13  13                          0>>
  0.04   0.33  14                          0>>>>
 -0.05  -0.42  15                     <<<<<0
 -0.05  -0.38  16                     <<<<<0
  0.10   0.75  17                          0>>>>>>>>>>
 -0.02  -0.14  18                        <<0
 -0.07  -0.56  19                   <<<<<<<0
  0.08   0.65  20                          0>>>>>>>>
        CHI-SQUARED* =   15.17 FOR DF =  20

+ + + + + + + + + + + + PARTIAL AUTOCORRELATIONS + + + + + + + + + + +
  COEF   T-VAL LAG _____0_____
  0.09   0.84   1                 [        0>>>>>          ]
 -0.05  -0.46   2                 [      <<<0              ]
  0.07   0.64   3                 [        0>>>>           ]
 -0.27  -2.39   4  <[<<<<<<<<<<<<<0                        ]
  0.00   0.04   5                 [        0               ]
  0.05   0.41   6                 [        0>>             ]
 -0.01  -0.11   7                 [       <0               ]
 -0.26  -2.35   8  <[<<<<<<<<<<<<<0                        ]
 -0.03  -0.26   9                 [       <0               ]
  0.00  -0.04  10                 [        0               ]
  0.16   1.46  11                 [        0>>>>>>>>        ]
 -0.18  -1.64  12                 [  <<<<<<<<<0            ]
  0.02   0.17  13                 [        0>              ]
  0.03   0.22  14                 [        0>              ]
  0.05   0.47  15                 [        0>>>            ]
 -0.18  -1.61  16                 [  <<<<<<<<<0            ]
  0.11   0.98  17                 [        0>>>>>          ]
 -0.04  -0.36  18                 [       <<0              ]
  0.04   0.32  19                 [        0>>             ]
 -0.05  -0.47  20                 [      <<<0              ]
```

Figure C10.3 Estimated acf and pacf of the first differences of the profit-margin realization.

the nature of a seasonal pattern emerges more clearly in the acf of the differenced series.

Figure C10.3 is the estimated acf and pacf of the first differences. The striking characteristic of the acf is the spikes at lags 4 and 8. Since these are seasonal lags the relevant practical warning level for their absolute t-values is 1.25. Both of these spikes have t-values exceeding this level, so they deserve special attention in our modeling plans.

To identify the seasonal part of a model, we focus on the estimated autocorrelations and partial autocorrelations at the seasonal lags while suppressing the nonseasonal acf and pacf patterns in our minds. Figure C10.3 has a spike at lag 4, a spike at lag 8, and a cutoff to zero at lags 12, 16, and 20. This pattern suggests an $MA(2)_4$ model for the seasonal part of the data. This tentative conclusion is confirmed by the estimated pacf; it decays (roughly) at the seasonal lags (4, 8, 12, 16, 20) rather than cutting off to zero. When an acf cuts off while the pacf decays, an MA model is called for. Consult Chapter 6 or 12 for theoretical acf's and pacf's of MA(2) processes.

The preceding analysis suggests that we should estimate an $ARIMA(1, 0, 0)(0, 0, 2)_4$ model. However, we will proceed more slowly for pedagogical purposes. Suppose we fail to examine the acf and pacf of the first differences and thus miss the seasonal pattern. The acf and pacf for the undifferenced data (Figure C10.2) call for an AR(1):

$$(1 - \phi_1 B)\tilde{z}_t = a_t \qquad (C10.1)$$

Estimation and diagnostic checking. Estimation results for (C10.1) appear in Figure C10.4. The estimated AR coefficient $\hat{\phi}_1$ is significantly different from zero and satisfies the stationarity condition $|\hat{\phi}_1| < 1$: it is almost 16 estimated standard errors above zero and more than two standard errors below 1.0.

Diagnostic-checking reveals that (C10.1) is inadequate. The spikes in the residual acf at the seasonal lags 4 and 8 have absolute t-values larger than the seasonal-lag warning level of 1.25.

This diagnostic check illustrates very well how the UBJ method guides us to a proper model: even if we start with an inappropriate model because of incomplete analysis at the identification stage, the diagnostic-checking stage gives warning signals that the model must be reformulated. See Chapter 9 for guidelines on reformulating a model from the residual acf.

There is a strong similarity between the estimated acf for the first differences w_t in Figure C10.3 and the acf for the residuals $\hat{a}_t$ in Figure C10.4. This is no accident. The first differences w_t are the result of filtering

```
+ + + + + + + + + +ECOSTAT UNIVARIATE B-J RESULTS+ + + + + + + + + +
+ FOR DATA SERIES:   PROFIT MARGIN                               +
+ DIFFERENCING:       0                           DF    = 78     +
+ AVAILABLE:          DATA = 80    BACKCASTS = 39  TOTAL = 119    +
+ USED TO FIND SSR: DATA = 80    BACKCASTS = 38  TOTAL = 118    +
+ (LOST DUE TO PRESENCE OF AUTOREGRESSIVE TERMS:            1)   +

COEFFICIENT      ESTIMATE      STD ERROR      T-VALUE
  PHI    1        0. 876        0. 055         15. 99
  CONSTANT       . 583765      . 258963        2. 25424

  MEAN           4. 70189      . 197549       23. 8011

  ADJUSTED RMSE =  . 267297    MEAN ABS % ERR =   4. 58
        CORRELATIONS
         1      2
   1    1. 00
   2   -0. 01   1. 00

++RESIDUAL ACF++
  COEF   T-VAL LAG _____0_____
  0. 14   1. 27   1   [                    0>>>>>>>>>>>>>>         ]
  0. 01   0. 06   2   [                    0>                     ]
  0. 09   0. 80   3   [                    0>>>>>>>>>             ]
 -0. 21  -1. 79   4   [   <<<<<<<<<<<<<<<<<<<<<<0                 ]
 -0. 04  -0. 30   5   [                <<<<0                     ]
  0. 08   0. 67   6   [                    0>>>>>>>>             ]
 -0. 02  -0. 19   7   [                  <<0                     ]
 -0. 17  -1. 41   8   [   <<<<<<<<<<<<<<<<<<0                     ]
 -0. 04  -0. 31   9   [                <<<<0                     ]
 -0. 02  -0. 15  10   [                  <<0                     ]
  0. 11   0. 89  11   [                    0>>>>>>>>>>           ]
  0. 00   0. 02  12                        0
  0. 01   0. 10  13                        0>
  0. 03   0. 24  14                        0>>>
 -0. 06  -0. 50  15                   <<<<<<0
 -0. 06  -0. 47  16                   <<<<<<0
  0. 07   0. 57  17                        0>>>>>>>
 -0. 04  -0. 31  18                    <<<<0
 -0. 10  -0. 76  19               <<<<<<<<<<0
  0. 04   0. 33  20                        0>>>>
       CHI-SQUARED* =    13. 48 FOR DF =  18
```

Figure C10.4 Estimation and diagnostic-checking results for model (C10.1).

the original series through the differencing operator $(1 - B)$:

$$(1 - B)\tilde{z}_t = w_t \qquad (C10.2)$$

From (C10.1) we see that the residuals $\hat{a}_t$ used to construct the acf in Figure C10.4 are the result of filtering z_t through the AR operator $(1 - \phi_1 B)$, where the estimate $\hat{\phi}_1 = 0.876$ is substituted for the unknown ϕ_1:

$$(1 - 0.876B)\tilde{z}_t = \hat{a}_t \qquad (C10.3)$$

Comparison of (C10.2) and (C10.3) shows that w_t and $\hat{a}_t$ are quite similar. The implicit coefficient of B in (C10.2) is 1.0, while in (C10.3) the explicit coefficient of B is 0.876. These two coefficients are close in value, so the two filtering operations produce similar series with similar estimated acf's.

Further estimation and diagnostic checking.   Based on our initial identification analysis and on the estimation and diagnostic-checking results in the last section, we now tentatively consider an ARIMA$(1, 0, 0)(0, 0, 2)_4$ model:

$$(1 - \phi_1 B)\tilde{z}_t = \left(1 - \Theta_4 B^4 - \Theta_8 B^8\right)a_t \qquad \text{(C10.4)}$$

```
+ + + + + + + + +ECOSTAT UNIVARIATE B-J RESULTS+ + + + + + + + +
+ FOR DATA SERIES:  PROFIT MARGIN                                   +
+ DIFFERENCING:     1                        DF    = 77             +
+ AVAILABLE:           DATA = 79   BACKCASTS = 8    TOTAL = 87       +
+ USED TO FIND SSR: DATA = 79   BACKCASTS = 8    TOTAL = 87         +
+ (LOST DUE TO PRESENCE OF AUTOREGRESSIVE TERMS:            0)      +

COEFFICIENT    ESTIMATE      STD ERROR    T-VALUE
  THETA   4     0.529         0.104        5.08
  THETA   8     0.396         0.105        3.76

  ADJUSTED RMSE =  .223258    MEAN ABS % ERR =   3.86
          CORRELATIONS
          1     2
   1    1.00
   2   -0.66   1.00

++RESIDUAL ACF++
 COEF   T-VAL LAG _____0_____
 0.13   1.16   1   [                    0>>>>>>>>>>>>>            ]
 0.02   0.19   2   [                    0>>                       ]
 0.11   0.99   3   [                    0>>>>>>>>>>               ]
-0.10  -0.86   4   [          <<<<<<<<<<0                         ]
 0.02   0.15   5   [                    0>>                       ]
 0.09   0.80   6   [                    0>>>>>>>>>                ]
 0.02   0.14   7   [                    0>>                       ]
 0.00   0.04   8   [                    0                         ]
 0.02   0.13   9   [                    0>>                       ]
 0.01   0.12  10   [                    0>                        ]
 0.11   0.90  11   [                    0>>>>>>>>>>               ]
-0.05  -0.45  12   [               <<<<<0                         ]
 0.00   0.00  13   [                    0                         ]
-0.02  -0.14  14   [                  <<0                         ]
-0.08  -0.68  15   [            <<<<<<<<0                         ]
-0.07  -0.60  16   [             <<<<<<<0                         ]
 0.00  -0.01  17   [                    0                         ]
-0.09  -0.74  18   [          <<<<<<<<<<0                         ]
-0.10  -0.84  19   [          <<<<<<<<<<0                         ]
 0.00  -0.04  20   [                    0                         ]
   CHI-SQUARED* =    8.71 FOR DF =   18
```

Figure C10.5 Estimation and diagnostic-checking results for model (C10.6).

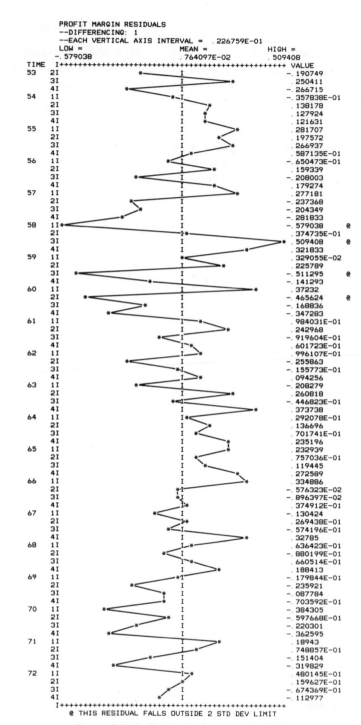

Figure C10.6 Residuals from model (C10.6).

470

Estimation of this model produces these results (t-values in parentheses):

$$(1 - 0.945B)\tilde{z}_t = (1 - 0.515B^4 - 0.420B^8)\hat{a}_t$$
$$\quad\;\; (22.08) \qquad\qquad (4.54) \qquad (3.87) \qquad\qquad\qquad \text{(C10.5)}$$

The critical aspect of these results is the value of $\hat{\phi}_1$: it is close enough to 1.0 to warrant nonseasonal first differencing. Although $\hat{\phi}_1$ is less than 1.0 it is not significantly different from 1.0. Its estimated standard error is 0.043, so $\hat{\phi}_1$ is less than 1.3 standard errors below 1.0. This is good evidence that the data do not have a stationary mean and should be differenced. A rule of thumb is that we should difference when in doubt; we do not want to tie our forecasts to a fixed mean if the mean does not seem to be stationary.

This analysis leads us back to the estimated acf and pacf of the first differences in Figure C10.3. An MA(2)$_4$ model is appropriate for the differenced data:

$$(1 - B)\tilde{z}_t = \left(1 - \Theta_4 B^4 - \Theta_8 B^8\right)a_t \qquad\qquad \text{(C10.6)}$$

The estimation and diagnostic-checking results for this model (Figure C10.5) indicate that it provides an adequate representation of the data. Both coefficients have t-values greater than 2.0 and together they satisfy the invertibility requirements for an MA(2)$_4$:

$$|\hat{\Theta}_8| = 0.396 < 1$$

$$\Theta_8 + \Theta_4 = 0.396 + 0.529 = 0.925 < 1$$

$$\Theta_8 - \Theta_4 = 0.396 - 0.529 = -0.133 < 1$$

The coefficients are not too highly correlated since the absolute correlation between them is less than 0.9. The residual acf has no absolute t-value in excess of the practical warning levels, and the chi-squared statistic is insignificant, so we accept the hypothesis that the random shocks in (C10.6) are independent. The residual plot in Figure C10.6 shows that neither the very early nor the very late portions of the data are poorly explained by the model.

Final comments. The seasonal coefficient $\hat{\Theta}_8$ in Figure C10.5 is highly significant: its t-value (3.76) is well in excess of 2.0. Yet the estimated autocorrelation at lag 8 in Figure C10.3 has an absolute t-value of only 1.58. This illustrates the importance of using the practical warning value of 1.25 for seasonal autocorrelations at the identification and diagnostic-checking stages.

CASE 11.   BOSTON ARMED ROBBERIES

In 1977, S. J. Deutsch and F. B. Alt published a journal article [35] in which they present an ARIMA model for the number of armed robberies reported each month in Boston, Massachusetts. Their work was subsequently criticized by R. A. Hay, Jr. and R. McCleary [36], who present an ARIMA model different from the Deutsch and Alt model. (See also McCleary and Hay [32].)

In this case study we examine the same data set analyzed by these authors and contrast our identification procedures with theirs. We will emphasize the importance of the practical warning levels for acf *t*-values summarized in Chapter 12.

Identification. Inspection of the data in Figure C11.1 leads to the tentative conclusion that the data are nonstationary in two ways: both the mean and the variance rise over time. As pointed out in Chapter 7, the natural log transformation will induce a constant variance when the variance is proportional to the mean. Fortunately, many series with nonstationary variances come close to displaying this property. The armed-robbery data might satisfy this criterion since the variance rises along with the mean.

The log values of the armed-robbery data are shown in Figure C11.2. The variance is now roughly constant throughout the series, so we construct an ARIMA model using the log values. After building this model, we translate the log forecasts back into the units of the original data.

The estimated acf for the log values is shown in Figure C11.3 and confirms that the data have a nonstationary mean: the acf approaches zero very slowly. Therefore, nonseasonal first differencing ($d = 1$) is required. The first differences in Figure C11.4 no longer trend upward; this series also confirms that the log transformation is acceptable since it has an approximately constant variance. The estimated acf and pacf of the differenced series appear in Figure C11.5; the differenced data apparently have a stationary mean since the estimated acf falls quickly to zero.

With monthly data a seasonal pattern produces significant autocorrelations at multiples of lag 12. The autocorrelation at lag 12 in Figure C11.5 has a relatively small t-value, but it exceeds the practical seasonal-lag warning value of 1.25 suggested in Chapters 11 and 12. The autocorrelation at lag 24 is about zero, so the seasonal acf pattern matches an MA(1)$_{12}$ process: there is a spike at lag 12 followed by a cutoff to zero at lag 24.

The nonseasonal pattern suggests an MA(2): there are two short-lag spikes (at lags 1 and 2) with absolute t-values larger than 1.6, followed by a cutoff to zero. This interpretation is reinforced by the appearance of the estimated pacf: it decays (irregularly) toward zero. As discussed in Chapters 3 and 6, MA processes have a decaying pattern in the pacf. But as emphasized in Chapter 3, estimated acf's and pacf's do not match their theoretical counterparts exactly because of sampling error. This may explain why the estimated pacf in Figure C11.4 does not decay smoothly like a theoretical MA(2) pacf.

We have tentatively identified an ARIMA$(0, 1, 2)(0, 0, 1)_{12}$ model for the log series. Letting z'_t represent the log values, we have

$$(1 - B)\tilde{z}'_t = \left(1 - \Theta_{12}B^{12}\right)\left(1 - \theta_1 B - \theta_2 B^2\right)a_t \qquad (C11.1)$$

Estimation and diagnostic checking.   The results of estimating this model (Figure C11.6) show that it is largely satisfactory. One troubling aspect is that the t-value associated with $\hat{\theta}_2$ is only 1.91, slightly less than the rule-of-thumb minimum value of 2.0. But as pointed out in Chapter 8, estimated standard errors and t-values are only approximate whenever MA terms are present in an ARIMA model. An MA(2) is sufficiently common and the estimated t-value is sufficiently close to 2.0 that we may legitimately leave θ_2 in the model. We return to this question shortly.

This model satisfies all the relevant invertibility conditions:

$$|\hat{\Theta}_{12}| = 0.206 < 1$$

$$|\hat{\theta}_2| = 0.177 < 1$$

$$\hat{\theta}_2 + \hat{\theta}_1 = 0.177 + 0.319 = 0.496 < 1$$

$$\hat{\theta}_2 - \hat{\theta}_1 = 0.177 - 0.319 = -0.142 < 1$$

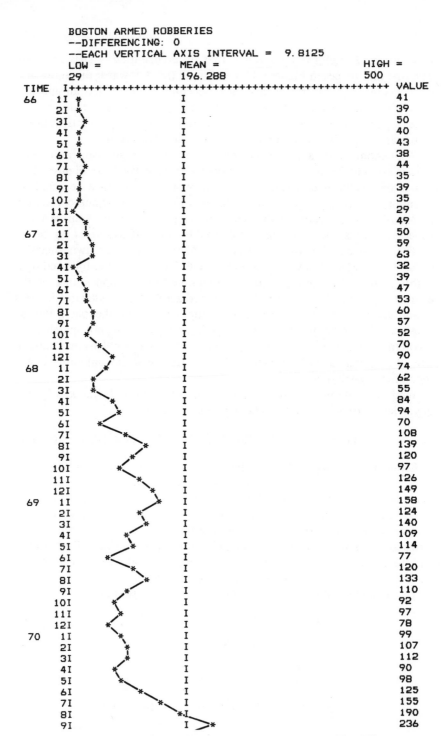

Figure C11.1 Monthly Boston armed robberies, 1966–1975.

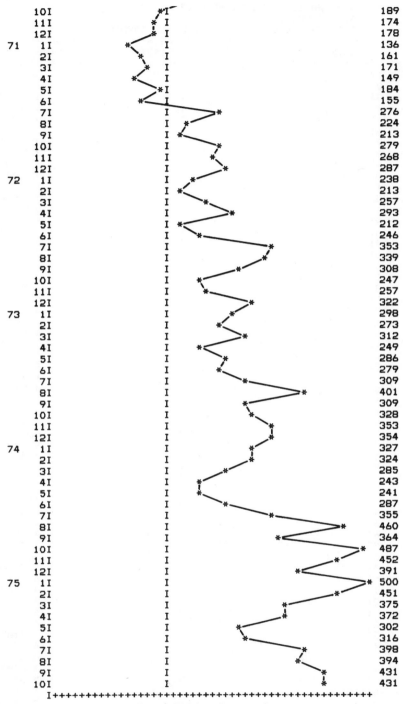

	10I	189
	11I	174
	12I	178
71	1I	136
	2I	161
	3I	171
	4I	149
	5I	184
	6I	155
	7I	276
	8I	224
	9I	213
	10I	279
	11I	268
	12I	287
72	1I	238
	2I	213
	3I	257
	4I	293
	5I	212
	6I	246
	7I	353
	8I	339
	9I	308
	10I	247
	11I	257
	12I	322
73	1I	298
	2I	273
	3I	312
	4I	249
	5I	286
	6I	279
	7I	309
	8I	401
	9I	309
	10I	328
	11I	353
	12I	354
74	1I	327
	2I	324
	3I	285
	4I	243
	5I	241
	6I	287
	7I	355
	8I	460
	9I	364
	10I	487
	11I	452
	12I	391
75	1I	500
	2I	451
	3I	375
	4I	372
	5I	302
	6I	316
	7I	398
	8I	394
	9I	431
	10I	431

Figure C11.1 (*Continued*)

475

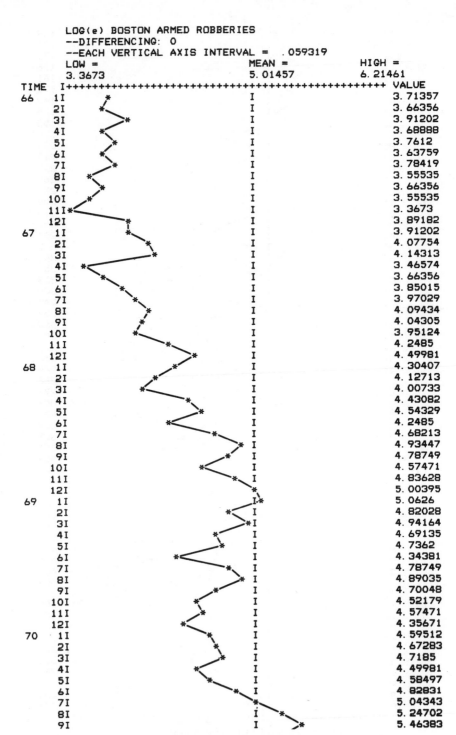

Figure C11.2 Natural logarithms of the realization in Figure C11.1.

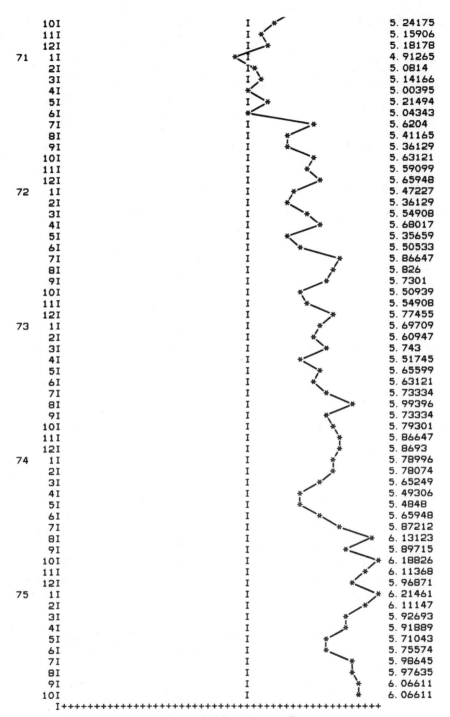

10I		5. 24175
11I		5. 15906
12I		5. 18178
71 1I		4. 91265
2I		5. 0814
3I		5. 14166
4I		5. 00395
5I		5. 21494
6I		5. 04343
7I		5. 6204
8I		5. 41165
9I		5. 36129
10I		5. 63121
11I		5. 59099
12I		5. 65948
72 1I		5. 47227
2I		5. 36129
3I		5. 54908
4I		5. 68017
5I		5. 35659
6I		5. 50533
7I		5. 86647
8I		5. 826
9I		5. 7301
10I		5. 50939
11I		5. 54908
12I		5. 77455
73 1I		5. 69709
2I		5. 60947
3I		5. 743
4I		5. 51745
5I		5. 65599
6I		5. 63121
7I		5. 73334
8I		5. 99396
9I		5. 73334
10I		5. 79301
11I		5. 86647
12I		5. 8693
74 1I		5. 78996
2I		5. 78074
3I		5. 65249
4I		5. 49306
5I		5. 4848
6I		5. 65948
7I		5. 87212
8I		6. 13123
9I		5. 89715
10I		6. 18826
11I		6. 11368
12I		5. 96871
75 1I		6. 21461
2I		6. 11147
3I		5. 92693
4I		5. 91889
5I		5. 71043
6I		5. 75574
7I		5. 98645
8I		5. 97635
9I		6. 06611
10I		6. 06611

Figure C11.2 (*Continued*)

477

```
+ + + + + + + + + + + + AUTOCORRELATIONS + + + + + + + + + + + +
+ FOR DATA SERIES: LOG(e) BOSTON ARMED ROBBERIES                +
+ DIFFERENCING: 0                         MEAN    =  5.01457     +
+ DATA COUNT =  118                       STD DEV =  .783586     +
  COEF   T-VAL LAG _____0_____
  0.95  10.28   1                            [    0>>>]>>>>>>>>>>>>>>>>>>>>>>
  0.91   5.89   2                       [         0>>>>>>>]>>>>>>>>>>>>>>>>>
  0.88   4.55   3                     [           0>>>>>>>>>]>>>>>>>>>>>>>>>
  0.85   3.79   4                 [               0>>>>>>>>>>>]>>>>>>>>>>>
  0.84   3.33   5                 [               0>>>>>>>>>>>]>>>>>>>>>>>
  0.81   2.95   6             [                   0>>>>>>>>>>>>>]>>>>>>
  0.78   2.67   7             [                   0>>>>>>>>>>>>>]>>>>>>
  0.76   2.45   8           [                     0>>>>>>>>>>>>>]>>>
  0.73   2.23   9           [                     0>>>>>>>>>>>>>]>>
  0.70   2.05  10           [                     0>>>>>>>>>>>>>]>
  0.67   1.91  11       [                         0>>>>>>>>>>>>>]
  0.65   1.79  12       [                         0>>>>>>>>>>>>  ]
  0.62   1.66  13       [                         0>>>>>>>>>>>   ]
  0.59   1.55  14     [                           0>>>>>>>>>>>     ]
  0.56   1.44  15     [                           0>>>>>>>>>>      ]
  0.53   1.34  16     [                           0>>>>>>>>>       ]
  0.50   1.25  17     [                           0>>>>>>>>>       ]
  0.48   1.18  18     [                           0>>>>>>>>        ]
  0.46   1.13  19     [                           0>>>>>>>>        ]
  0.44   1.07  20     [                           0>>>>>>>         ]
  0.42   0.99  21     [                           0>>>>>>>         ]
  0.40   0.94  22   [                             0>>>>>>>           ]
  0.38   0.89  23   [                             0>>>>>>>           ]
  0.36   0.85  24   [                             0>>>>>>            ]
  0.34   0.79  25   [                             0>>>>>>            ]
  0.32   0.73  26   [                             0>>>>>>            ]
  0.28   0.64  27   [                             0>>>>>>            ]
  0.26   0.59  28   [                             0>>>>>             ]
  0.24   0.56  29   [                             0>>>>>             ]
  0.22   0.51  30   [                             0>>>>>             ]
        CHI-SQUARED* = 1436.13 FOR DF =   30
```

Figure C11.3   Estimated acf for the log data in Figure C11.2.

The residual acf indicates that (C11.1) is statistically adequate. The *t*-value at lag 7 exceeds the relevant practical warning value (1.6), but it is not unusual for several residual autocorrelations out of 30 to be moderately significant just by chance. Furthermore, lag 7 is not one that warrants special attention, unlike the seasonal lags (12, 24) or the short lags (1, 2, perhaps 3). Finally, the chi-squared statistic is not significant. We have constructed a common, parsimonious, and statistically adequate model to represent the available data.

Forecasting.   Model (C11.1) was built using the logs of the original data, but our interest is in forecasting the original data. Therefore, we must translate forecasts of the log values into appropriate antilog units.

Recall from Chapter 10 that if the random shocks of the log form of a model are Normally distributed, then the shocks of the corresponding model for the original data are log-Normally distributed. It can then be shown that each forecast for the original data is the antilog of the sum of the corresponding log forecast and one-half the log forecast-error variance. The confidence limits for each forecast for the original data, however, are simply the antilogs of the limits for the corresponding log forecast. It follows that the confidence interval around each forecast for the original data is asymmetric, as seen in Table C11.1

The residuals in Figure C11.7 are satisfactory. The few large ones that appear could occur just by chance. The model fits the data fairly uniformly throughout the series.

Comparison with earlier models. Deutsch and Alt propose an ARIMA$(0, 1, 1)(0, 1, 1)_{12}$ model for the original data. This can be criticized for two reasons. First, modeling the original data ignores the nonstationary variance. Second, there is no evidence that seasonal differencing ($D = 1$) is required.

McCleary and Hay propose a more defensible ARIMA$(0, 1, 1)(0, 0, 1)_{12}$ model for the log data. The only difference between their model and (C11.1) is that they exclude θ_2.

McCleary and Hay [32] admit to some difficulty justifying the presence of a seasonal term in the model. They indicate that the autocorrelation at lag 12 is insignificant (in Figure C11.5), but they claim to see seasonal variation in the plotted data. They also suggest that following the overfitting strategy would probably have led to the inclusion of a seasonal term. But as discussed in Chapter 12, experience suggests that seasonal autocorrelations with absolute t-values of about 1.25 or greater call for further attention since the corresponding coefficients are often much more significant at the estimation stage. This practical rule leads immediately to consideration of a model with a seasonal component.

McCleary and Hay do not discuss including an MA term at lag 2. We are led to consider such a term by using the practical nonseasonal-lag warning level of about 1.6 for absolute acf t-values. It is instructive to compare our results for (C11.1) with the ARIMA$(0, 1, 1)(0, 0, 1)_{12}$ model proposed by McCleary and Hay:

$$(1 - B)\tilde{z}_t' = \left(1 - \Theta_{12}B^{12}\right)\left(1 - \theta_1 B\right)a_t \qquad (C11.2)$$

Results in Figure C11.8 show that both $\hat{\theta}_1$ and $\hat{\Theta}_{12}$ are significantly different from zero since their absolute t-values exceed 2.0. Both estimated coefficients also satisfy their respective invertibility conditions since their absolute values are less than 1.0. (See Chapter 11 for a discussion of the invertibility conditions for multiplicative models.)

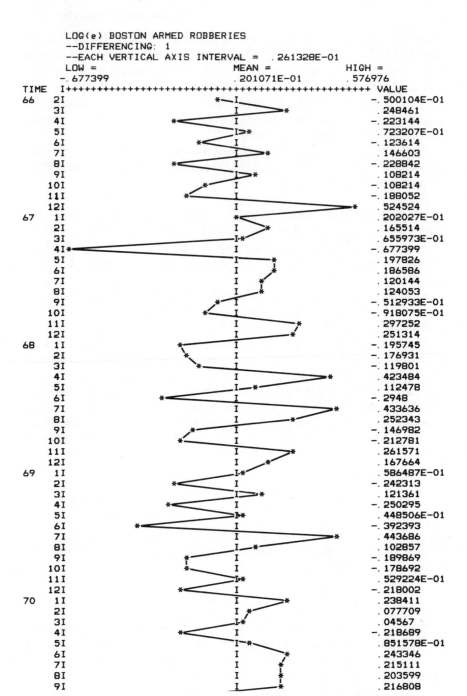

```
     LOG(e) BOSTON ARMED ROBBERIES
     --DIFFERENCING: 1
     --EACH VERTICAL AXIS INTERVAL =  .261328E-01
        LOW =                      MEAN =              HIGH =
        -.677399                   .201071E-01         .576976
TIME  I+++++++++++++++++++++++++++++++++++++++++++++++++++ VALUE
 66    2I                        *  I                          -.500104E-01
       3I                           I            *              .248461
       4I                    *      I                          -.223144
       5I                           I>*                         .723207E-01
       6I                  *        I                          -.123614
       7I                           I        *                  .146603
       8I                  *        I                          -.228842
       9I                           I>*                         .108214
      10I                *  *       I                          -.108214
      11I               *           I                          -.188052
      12I                           I                        *  .524524
 67    1I                           *                           .202027E-01
       2I                           I    *                      .165514
       3I                          I*                           .655973E-01
       4I*                          I                          -.677399
       5I                           I      *                    .197826
       6I                           I    *                      .186586
       7I                           I   *                       .120144
       8I                           I   *                       .124053
       9I                        *  I                          -.512933E-01
      10I                   *       I                          -.918075E-01
      11I                           I           *               .297252
      12I                           I          *                .251314
 68    1I                      *    I                          -.195745
       2I                     *     I                          -.176931
       3I                      *    I                          -.119801
       4I                           I            *              .423484
       5I                           I_*                         .112478
       6I                 *         I                          -.2948
       7I                           I             *             .433636
       8I                           I       *                   .252343
       9I                     *     I                          -.146982
      10I                 *         I                          -.212781
      11I                           I          *                .261571
      12I                           I    *                      .167664
 69    1I                          I.*                          .586487E-01
       2I                  *        I                          -.242313
       3I                          I.*                          .121361
       4I                  *        I                          -.250295
       5I                          I**                          .448506E-01
       6I                *          I                          -.392393
       7I                           I             *             .443686
       8I                           I_*                         .102857
       9I                 *         I                          -.189869
      10I                *          I                          -.178692
      11I                         I*                            .529224E-01
      12I                 *         I                          -.218002
 70    1I                           I          *                .238411
       2I                           I *                         .077709
       3I                         I.*                           .04567
       4I                 *        I                           -.218689
       5I                         I_*                           .851578E-01
       6I                           I       *                   .243346
       7I                           I      *                    .215111
       8I                           I     *                     .203599
       9I                           I      *                    .216808
```

Figure C11.4   First differences of the log data in Figure C11.2.

480

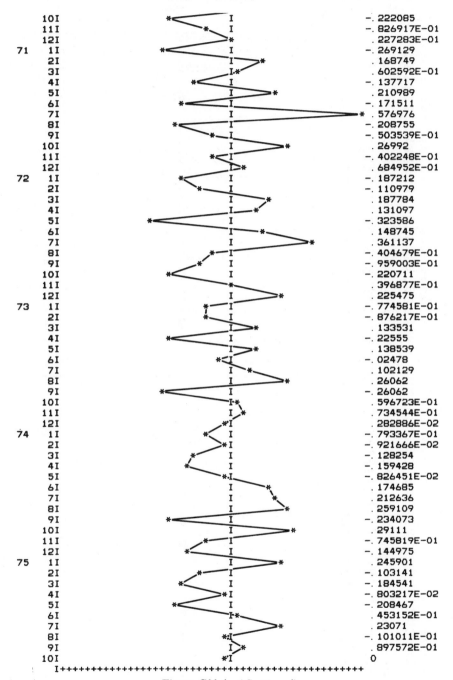

10I		-.222085
11I		-.826917E-01
12I		.227283E-01
71 1I		-.269129
2I		.168749
3I		.602592E-01
4I		-.137717
5I		.210989
6I		-.171511
7I		.576976
8I		-.208755
9I		-.503539E-01
10I		.26992
11I		-.402248E-01
12I		.684952E-01
72 1I		-.187212
2I		-.110979
3I		.187784
4I		.131097
5I		-.323586
6I		.148745
7I		.361137
8I		-.404679E-01
9I		-.959003E-01
10I		-.220711
11I		.396877E-01
12I		.225475
73 1I		-.774581E-01
2I		-.876217E-01
3I		.133531
4I		-.22555
5I		.138539
6I		-.02478
7I		.102129
8I		.26062
9I		-.26062
10I		.596723E-01
11I		.734544E-01
12I		.282886E-02
74 1I		-.793367E-01
2I		-.921666E-02
3I		-.128254
4I		-.159428
5I		-.826451E-02
6I		.174685
7I		.212636
8I		.259109
9I		-.234073
10I		.29111
11I		-.745819E-01
12I		-.144975
75 1I		.245901
2I		-.103141
3I		-.184541
4I		-.803217E-02
5I		-.208467
6I		.453152E-01
7I		.23071
8I		-.101011E-01
9I		.897572E-01
10I		0

Figure C11.4 (*Continued*)

481

```
+ + + + + + + + + + + AUTOCORRELATIONS + + + + + + + + + + + + + +
+ FOR DATA SERIES: LOG(e) BOSTON ARMED ROBBERIES                  +
+ DIFFERENCING: 1                        MEAN    = .201071E-01    +
+ DATA COUNT = 117                       STD DEV = .206181        +
  COEF   T-VAL LAG                            0
-0.23  -2.48   1   <<<<<[<<<<<<<<<<<<<<<<<<<<0                     ]
-0.16  -1.67   2      [   <<<<<<<<<<<<<<<<<<<0                     ]
 0.00  -0.04   3      [                       0                    ]
-0.08  -0.79   4      [            <<<<<<<<0                       ]
 0.10   1.02   5      [                      0>>>>>>>>>>           ]
-0.04  -0.39   6      [                  <<<<0                     ]
-0.13  -1.33   7      [        <<<<<<<<<<<<<0                      ]
 0.16   1.58   8      [                      0>>>>>>>>>>>>>>>>     ]
-0.11  -1.07   9      [          <<<<<<<<<<<0                      ]
-0.12  -1.13  10    [         <<<<<<<<<<<<0                        ]
 0.08   0.77  11    [                    0>>>>>>>>                 ]
 0.14   1.35  12    [                    0>>>>>>>>>>>>>>           ]
-0.03  -0.27  13    [              <<<0                            ]
 0.05   0.49  14    [                    0>>>>>                    ]
-0.14  -1.24  15    [         <<<<<<<<<<<<<<0                      ]
 0.02   0.21  16    [                    0>>                       ]
 0.01   0.07  17    [                    0>                        ]
-0.06  -0.57  18    [              <<<<<<0                         ]
 0.05   0.46  19    [                    0>>>>>                    ]
 0.02   0.22  20    [                    0>>                       ]
-0.12  -1.04  21    [         <<<<<<<<<<<<0                        ]
 0.06   0.49  22    [                    0>>>>>>                   ]
-0.03  -0.24  23    [              <<<0                            ]
 0.06   0.53  24    [                    0>>>>>>                   ]
 0.03   0.29  25    [                    0>>>                      ]
 0.06   0.53  26    [                    0>>>>>>                   ]
-0.10  -0.88  27    [           <<<<<<<<<<0                        ]
-0.03  -0.24  28    [              <<<0                            ]
-0.02  -0.21  29    [               <<0                            ]
-0.03  -0.29  30    [              <<<0                            ]
     CHI-SQUARED* =    34.30 FOR DF = 30

+ + + + + + + + + + + PARTIAL AUTOCORRELATIONS + + + + + + + + + + +
  COEF   T-VAL LAG                            0
-0.23  -2.48   1   <<<<<[<<<<<<<<<<<<<<<<<<<<0                     ]
-0.23  -2.45   2   <<<<<[<<<<<<<<<<<<<<<<<<<<0                     ]
-0.11  -1.23   3      [      <<<<<<<<<<<<0                         ]
-0.17  -1.80   4   [<<<<<<<<<<<<<<<<<<<<0                          ]
 0.01   0.13   5      [                    0>                      ]
-0.06  -0.68   6      [          <<<<<<<<0                         ]
-0.17  -1.83   7   [<<<<<<<<<<<<<<<<<<<<0                          ]
 0.06   0.67   8      [                    0>>>>>                  ]
-0.12  -1.32   9      [       <<<<<<<<<<<<0                        ]
-0.20  -2.19  10   <<[<<<<<<<<<<<<<<<<<<<0                         ]
-0.08  -0.90  11      [         <<<<<<<<0                          ]
 0.11   1.23  12      [                   0>>>>>>>>>>>             ]
-0.02  -0.25  13      [                 <<0                        ]
 0.10   1.08  14      [                   0>>>>>>>>>>              ]
-0.04  -0.47  15      [             <<<<0                          ]
-0.04  -0.40  16      [             <<<<0                          ]
-0.07  -0.76  17      [           <<<<<<0                          ]
-0.06  -0.66  18      [            <<<<<0                          ]
-0.03  -0.30  19      [               <<<0                         ]
-0.02  -0.23  20      [                <<0                         ]
-0.09  -0.93  21      [          <<<<<<<<<0                        ]
-0.02  -0.18  22      [                <<0                         ]
-0.05  -0.57  23      [             <<<<<0                         ]
-0.02  -0.19  24      [                <<0                         ]
-0.03  -0.30  25      [               <<<0                         ]
 0.10   1.07  26      [                   0>>>>>>>>>>              ]
-0.06  -0.68  27      [            <<<<<<0                         ]
-0.07  -0.75  28      [           <<<<<<0                          ]
-0.04  -0.41  29      [              <<<<0                         ]
-0.12  -1.34  30      [        <<<<<<<<<<<<<0                      ]
```

Figure C11.5 Estimated acf and pacf for the first differences in Figure C11.4.

482

```
+ + + + + + + + +ECOSTAT UNIVARIATE B-J RESULTS+ + + + + + + + +
+ FOR DATA SERIES:  LOG(e) BOSTON ARMED ROBBERIES                    +
+ DIFFERENCING:      1                        DF    = 114            +
+ AVAILABLE:          DATA = 117   BACKCASTS = 14   TOTAL = 131       +
+ USED TO FIND SSR: DATA = 117   BACKCASTS = 14   TOTAL = 131        +
+ (LOST DUE TO PRESENCE OF AUTOREGRESSIVE TERMS:              0)     +

COEFFICIENT   ESTIMATE      STD ERROR     T-VALUE
   THETA   1     0.319        0.093         3.45
   THETA   2     0.177        0.093         1.91
   THETA* 12    -0.206        0.094        -2.19

  ADJUSTED RMSE =  .194059    MEAN ABS % ERR =   3.20
       CORRELATIONS
         1     2     3
  1    1.00
  2   -0.41  1.00
  3   -0.11  0.10  1.00

++RESIDUAL ACF++
  COEF  T-VAL LAG _____0_____
 -0.04  -0.39   1         [          <<<<0            ]
 -0.03  -0.29   2         [          <<<0             ]
  0.01   0.14   3         [             0>             ]
 -0.11  -1.22   4         [     <<<<<<<<<<<0           ]
  0.08   0.80   5         [             0>>>>>>>>      ]
 -0.05  -0.50   6         [          <<<<0            ]
 -0.17  -1.78   7         [<<<<<<<<<<<<<<<<<0         ]
  0.10   1.04   8          [          0>>>>>>>>>>     ]
 -0.12  -1.19   9          [   <<<<<<<<<<<<0          ]
 -0.10  -1.00  10          [    <<<<<<<<<<0           ]
  0.09   0.94  11          [          0>>>>>>>>>      ]
 -0.04  -0.40  12          [          <<<<0           ]
  0.01   0.12  13          [          0>               ]
  0.04   0.40  14          [          0>>>>           ]
 -0.11  -1.11  15          [   <<<<<<<<<<<0           ]
  0.01   0.06  16          [          0>               ]
 -0.04  -0.39  17          [          <<<0            ]
 -0.05  -0.48  18          [          <<<<0           ]
  0.03   0.33  19          [          0>>>            ]
 -0.03  -0.30  20          [          <<<0            ]
 -0.06  -0.63  21          [         <<<<<<0          ]
  0.05   0.47  22          [          0>>>>>          ]
 -0.03  -0.30  23          [          <<<0            ]
  0.06   0.60  24          [          0>>>>>>         ]
  0.04   0.40  25          [          0>>>>           ]
  0.06   0.57  26          [          0>>>>>>         ]
 -0.10  -0.97  27          [   <<<<<<<<<<0            ]
 -0.07  -0.63  28          [      <<<<<<<0            ]
 -0.04  -0.37  29         [          <<<<0            ]
 -0.05  -0.44  30         [          <<<<0            ]
     CHI-SQUARED* =   20.28 FOR DF =   27
```

Figure C11.6 Estimation and diagnostic-checking results for model (C11.1).

Table C11.1   Forecasts[a] from model (C11.1)

Time		Forecast Values	80% Confidence Limits		Future Observed Values	Percent Forecast Errors
			Lower	Upper		
75	11	422.5603	323.4702	531.6038	n.a.[b]	n.a.
	12	415.4990	299.2967	545.8881	n.a.	n.a.
76	1	440.4777	307.9883	590.5194	n.a.	n.a.
	2	433.2207	294.5597	591.5899	n.a.	n.a.
	3	421.6933	279.2184	585.7065	n.a.	n.a.
	4	424.9268	274.3275	599.5765	n.a.	n.a.
	5	409.9089	258.2832	586.9745	n.a.	n.a.
	6	412.4238	253.8581	598.8143	n.a.	n.a.
	7	430.8649	259.2769	633.8249	n.a.	n.a.
	8	428.2626	252.1193	637.8522	n.a.	n.a.
	9	441.7800	254.5900	665.7815	n.a.	n.a.
	10	438.8901	247.7250	668.8947	n.a.	n.a.

[a] Forecasts are in original metric.
[b] n.a. = not available.

Is (C11.2) better than (C11.1)? (C11.1) has a smaller RMSE, a smaller chi-squared statistic associated with its residual acf, and a more satisfactory residual acf across the very important short lags. However, these advantages for (C11.1) are slight. The advantage of (C11.2) is that it is more parsimonious. Nevertheless, (C11.1) is not profligate in its number of estimated parameters.

It is not clear which of these two models should be used for forecasting. One way of deciding is to use the one which forecasts history better. We drop the last few observations, refit each model, forecast, and compare the forecasts with the last few observed values. The results of this exercise are shown in Figures C11.9 and C11.10.

Model (C11.2) is the better choice for these reasons: (i) Its estimated coefficients are more statistically significant. Two of the t-values for (C11.1) have dropped below 1.9. (ii) It forecasts the immediate future somewhat better as shown in Tables C11.2 and C11.3. It has smaller percent forecast errors for three of the first four forecasts. These near-term forecasts are the most important since ARIMA models are best suited to short-term forecasting. [Considering all 12 forecasts, (C11.1) performs somewhat better: the mean of its percent forecast errors for all 12 forecasts is somewhat smaller.] (iii) Model (C11.2) is more parsimonious. When two models are quite

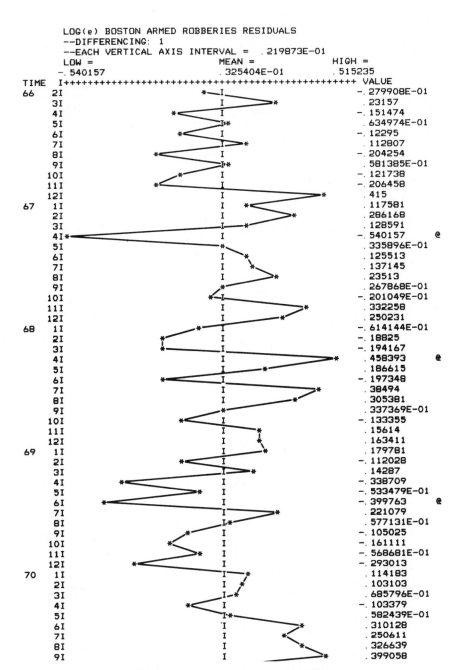

Figure C11.7 Residuals for model (C11.1).

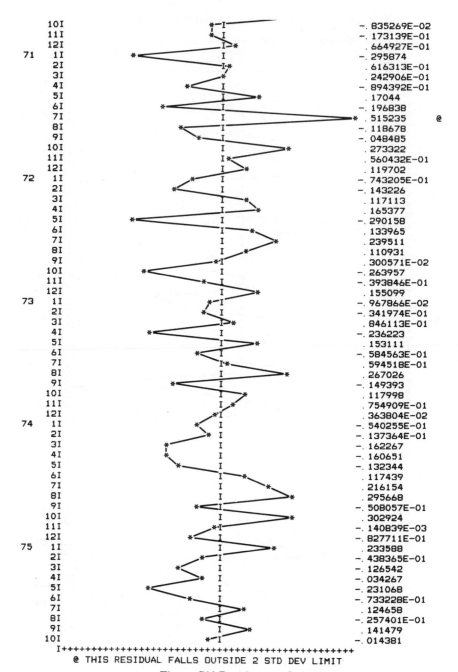

10I		-. 835269E-02
11I		-. 173139E-01
12I		. 664927E-01
71 1I		-. 295874
2I		. 616313E-01
3I		. 242906E-01
4I		-. 894392E-01
5I		. 17044
6I		-. 196838
7I		. 515235 @
8I		-. 118678
9I		-. 048485
10I		. 273322
11I		. 560432E-01
12I		. 119702
72 1I		-. 743205E-01
2I		-. 143226
3I		. 117113
4I		. 165377
5I		-. 290158
6I		. 133965
7I		. 239511
8I		. 110931
9I		. 300571E-02
10I		-. 263957
11I		-. 393846E-01
12I		. 155099
73 1I		-. 967866E-02
2I		-. 341974E-01
3I		. 846113E-01
4I		-. 236223
5I		. 153111
6I		-. 584563E-01
7I		. 594518E-01
8I		. 267026
9I		-. 149393
10I		. 117998
11I		. 754909E-01
12I		. 363804E-02
74 1I		-. 540255E-01
2I		-. 137364E-01
3I		-. 162267
4I		-. 160651
5I		-. 132344
6I		. 117439
7I		. 216154
8I		. 295668
9I		-. 508057E-01
10I		. 302924
11I		-. 140839E-03
12I		-. 827711E-01
75 1I		. 233588
2I		-. 438365E-01
3I		-. 126542
4I		-. 034267
5I		-. 231068
6I		-. 733228E-01
7I		. 124658
8I		-. 257401E-01
9I		. 141479
10I		-. 014381

I++
@ THIS RESIDUAL FALLS OUTSIDE 2 STD DEV LIMIT

Figure C11.7   (*Continued*)

486

```
+ + + + + + + + +ECOSTAT UNIVARIATE B-J RESULTS+ + + + + + + + +
+ FOR DATA SERIES:   LOG(e) BOSTON ARMED ROBBERIES                      +
+ DIFFERENCING:      1                            DF    = 115           +
+ AVAILABLE:         DATA = 117   BACKCASTS = 13   TOTAL = 130          +
+ USED TO FIND SSR:  DATA = 117   BACKCASTS = 13   TOTAL = 130          +
+ (LOST DUE TO PRESENCE OF AUTOREGRESSIVE TERMS:              0)        +

COEFFICIENT    ESTIMATE      STD ERROR     T-VALUE
  THETA   1     0. 394        0. 085        4. 61
  THETA* 12    -0. 227        0. 093       -2. 45

  ADJUSTED RMSE =  . 195704    MEAN ABS % ERR =   3. 29
           CORRELATIONS
            1       2
   1     1. 00
   2    -0. 07   1. 00

++RESIDUAL ACF++
  COEF   T-VAL  LAG _____O_____
  0. 04   0. 45   1           [            O>>>>                ]
 -0. 15  -1. 64   2           [ <<<<<<<<<<<<<<<O                ]
 -0. 05  -0. 50   3           [       <<<<<O                    ]
 -0. 11  -1. 14   4           [ <<<<<<<<<<<O                    ]
  0. 08   0. 83   5         [              O>>>>>>>>              ]
 -0. 05  -0. 49   6         [          <<<<<O                     ]
 -0. 15  -1. 57   7         [ <<<<<<<<<<<<<<<O                    ]
  0. 10   0. 97   8         [              O>>>>>>>>>>            ]
 -0. 10  -1. 00   9         [      <<<<<<<<<<O                    ]
 -0. 10  -0. 95  10         [      <<<<<<<<<<O                    ]
  0. 10   0. 95  11         [              O>>>>>>>>>>            ]
 -0. 03  -0. 31  12         [           <<<O                     ]
  0. 02   0. 22  13         [              O>>                    ]
  0. 04   0. 38  14         [              O>>>>                  ]
 -0. 10  -0. 95  15         [      <<<<<<<<<<O                    ]
  0. 00   0. 00  16         [              O                     ]
 -0. 03  -0. 31  17         [           <<<O                     ]
 -0. 04  -0. 35  18         [          <<<<O                     ]
  0. 04   0. 41  19         [              O>>>>                  ]
 -0. 03  -0. 32  20         [           <<<O                     ]
 -0. 06  -0. 58  21         [        <<<<<<O                     ]
  0. 03   0. 33  22         [              O>>>                   ]
 -0. 02  -0. 21  23         [            <<O                     ]
  0. 06   0. 59  24         [              O>>>>>>                ]
  0. 07   0. 70  25         [              O>>>>>>>               ]
  0. 06   0. 58  26         [              O>>>>>>                ]
 -0. 10  -0. 96  27         [      <<<<<<<<<<O                    ]
 -0. 08  -0. 78  28        [        <<<<<<<<O                      ]
 -0. 04  -0. 36  29        [           <<<<O                       ]
 -0. 04  -0. 36  30        [           <<<<O                       ]
  CHI-SQUARED* =    21. 75 FOR DF =    28
```

Figure C11.8 Estimation and diagnostic-checking results for model (C11.2).

```
+ + + + + + + + + +ECOSTAT UNIVARIATE B-J RESULTS+ + + + + + + + + +
+ FOR DATA SERIES:   LOG(e) BOSTON ARMED ROBBERIES                    +
+ DIFFERENCING:      1                          DF    = 102          +
+ AVAILABLE:         DATA = 105   BACKCASTS = 14  TOTAL = 119         +
+ USED TO FIND SSR:  DATA = 105   BACKCASTS = 14  TOTAL = 119         +
+ (LOST DUE TO PRESENCE OF AUTOREGRESSIVE TERMS:              0)      +

COEFFICIENT    ESTIMATE       STD ERROR      T-VALUE
  THETA   1      0.318          0.099          3.19
  THETA   2      0.169          0.099          1.70
  THETA* 12     -0.191          0.101         -1.89

  ADJUSTED RMSE =  .200939    MEAN ABS % ERR =    3.39
       CORRELATIONS
          1      2      3
  1    1.00
  2   -0.39   1.00
  3   -0.10   0.07   1.00

++RESIDUAL ACF++
  COEF   T-VAL  LAG _____0_____
 -0.04  -0.38    1       [              <<<<0                       ]
 -0.03  -0.35    2       [              <<<0                        ]
 -0.02  -0.16    3       [              <<0                         ]
 -0.08  -0.84    4       [          <<<<<<<<0                       ]
  0.07   0.76    5       [                  0>>>>>>>                 ]
 -0.06  -0.60    6       [             <<<<<<0                      ]
 -0.15  -1.51    7       [   <<<<<<<<<<<<<<<0                       ]
  0.11   1.12    8       [                  0>>>>>>>>>>>            ]
 -0.10  -0.99    9       [        <<<<<<<<<<0                       ]
 -0.08  -0.79   10       [          <<<<<<<<0                       ]
  0.08   0.75   11       [                  0>>>>>>>>               ]
 -0.04  -0.38   12       [              <<<<0                       ]
 -0.02  -0.15   13     [                <<0                          ]
  0.02   0.22   14     [                  0>>                        ]
 -0.14  -1.29   15     [     <<<<<<<<<<<<<<0                         ]
  0.02   0.22   16     [                  0>>                        ]
 -0.05  -0.51   17     [              <<<<<0                         ]
 -0.05  -0.49   18     [              <<<<<0                         ]
  0.05   0.49   19     [                  0>>>>>                     ]
 -0.05  -0.46   20     [              <<<<<0                         ]
 -0.05  -0.46   21     [              <<<<<0                         ]
  0.06   0.51   22     [                  0>>>>>>                    ]
 -0.05  -0.46   23     [              <<<<<0                         ]
  0.08   0.73   24     [                  0>>>>>>>>>                 ]
  0.01   0.10   25     [                  0>                         ]
  0.05   0.44   26     [                  0>>>>>                     ]
     CHI-SQUARED* =    15.25 FOR DF =   23
```

Figure C11.9 Estimation and diagnostic-checking results for model (C11.1) using the first 106 observations.

```
+ + + + + + + + +ECOSTAT UNIVARIATE B-J RESULTS+ + + + + + + + +
+ FOR DATA SERIES:   LOG(e) BOSTON ARMED ROBBERIES                +
+ DIFFERENCING:      1                              DF   = 103    +
+ AVAILABLE:          DATA = 105    BACKCASTS = 13   TOTAL = 118   +
+ USED TO FIND SSR: DATA = 105    BACKCASTS = 13   TOTAL = 118    +
+ (LOST DUE TO PRESENCE OF AUTOREGRESSIVE TERMS:            0)     +

COEFFICIENT   ESTIMATE      STD ERROR     T-VALUE
  THETA  1      0.387         0.091         4.24
  THETA* 12    -0.206         0.100        -2.06

  ADJUSTED RMSE =    202498    MEAN ABS % ERR =    3.48
        CORRELATIONS
          1      2
  1    1.00
  2   -0.07   1.00


++RESIDUAL ACF++
  COEF   T-VAL  LAG _____0_____
  0.04    0.39   1        [              0>>>>                ]
 -0.16   -1.62   2        [   <<<<<<<<<<<<<<<<0               ]
 -0.07   -0.72   3        [           <<<<<<<0               ]
 -0.08   -0.76   4        [          <<<<<<<<0               ]
  0.08    0.82   5        [              0>>>>>>>>           ]
 -0.06   -0.59   6        [            <<<<<<0               ]
 -0.14   -1.34   7        [   <<<<<<<<<<<<<<<0               ]
  0.11    1.05   8       [               0>>>>>>>>>>>        ]
 -0.09   -0.81   9      [          <<<<<<<<<0               ]
 -0.08   -0.77  10      [           <<<<<<<<0               ]
  0.08    0.79  11      [               0>>>>>>>>           ]
 -0.03   -0.28  12      [              <<<0               ]
  0.00   -0.03  13      [                 0               ]
  0.02    0.17  14      [                 0>>             ]
 -0.12   -1.11  15      [      <<<<<<<<<<<<0               ]
  0.02    0.17  16      [                 0>>             ]
 -0.04   -0.38  17      [             <<<<0               ]
 -0.04   -0.35  18      [             <<<<0               ]
  0.06    0.56  19      [                 0>>>>>>         ]
 -0.05   -0.46  20      [            <<<<<0               ]
 -0.05   -0.42  21      [            <<<<<0               ]
  0.04    0.38  22      [                 0>>>>           ]
 -0.04   -0.34  23      [             <<<<0               ]
  0.07    0.68  24      [                 0>>>>>>>        ]
  0.04    0.38  25      [                 0>>>>           ]
  0.05    0.43  26      [                 0>>>>>          ]
      CHI-SQUARED* =    16.54 FOR DF =    24
```

Figure C11.10 Estimation and diagnostic-checking results for model (C11.2) using the first 106 observations.

similar in most respects, we invoke the principle of parsimony in making the final selection for forecasting.

Regardless of which model is used we should produce forecasts with both models and monitor their relative performance over time, even if only one set of forecasts is used for making decisions.

Final comments.   This case illustrates the usefulness of the practical warning levels for absolute acf t-values as discussed in Chapter 12. Adherence to these warning values in this case study leads to the appropriate inclusion of a seasonal MA coefficient. It also leads to the possible inclusion of a θ_2 coefficient. Although we might not include θ_2 in the final model, estimation results indicate that its inclusion is worthy of consideration. Furthermore, we should continue to be alert to the possible addition of this term as more data become available.

The practical warning levels for t-values proposed in Chapter 12 are not foolproof. Sometimes a coefficient has a corresponding autocorrelation t-value exceeding the warning level, but the coefficient proves insignificant at the estimation stage.

Furthermore, the warning values provide only practical guidelines, not fixed rules. Sometimes we should accept a model with some significant residual autocorrelations since these can occur by chance. This is especially

Table C11.2   Forecasts^a from model (C11.1) using the first 106 observations

Time		Forecast Values	80% Confidence Limits		Future Observed Values	Percent Forecast Errors
			Lower	Upper		
74	11	461.3408	349.5850	584.7303	452.0000	− 2.07
	12	439.3036	312.3853	582.2915	391.0000	− 12.35
75	1	435.8429	300.1089	590.2795	500.0000	12.83
	2	438.3756	292.8728	605.4325	451.0000	2.80
	3	428.4718	278.1843	602.4750	375.0000	− 14.26
	4	421.7325	266.4408	602.9425	372.0000	− 13.37
	5	420.1122	258.5648	610.0131	302.0000	− 39.11
	6	437.6896	262.6817	644.8456	316.0000	− 38.51
	7	457.3997	267.9076	683.1800	398.0000	− 14.92
	8	478.6405	273.8085	724.2283	394.0000	− 21.48
	9	464.2856	259.5727	711.1977	431.0000	− 7.72
	10	491.4510	268.6884	761.6681	431.0000	− 14.03

^{a}Forecasts are in original metric.

Table C11.3   Forecasts[a] from model (C11.2) using the first 106 observations

Time		Forecast Values	80% Confidence Limits		Future Observed Values	Percent Forecast Errors
			Lower	Upper		
74	11	458.7373	346.8100	582.4088	452.0000	− 1.49
	12	458.3587	328.8100	603.9116	391.0000	− 17.23
75	1	455.4194	311.8031	619.1070	500.0000	8.92
	2	459.2084	301.2624	641.5250	451.0000	− 1.82
	3	448.8316	282.9934	642.4585	375.0000	− 19.69
	4	442.4875	268.7578	647.4576	372.0000	− 18.95
	5	441.1321	258.5906	658.5831	302.0000	− 46.07
	6	462.1112	261.8506	702.8154	316.0000	− 46.24
	7	485.4834	266.2687	751.1837	398.0000	− 21.98
	8	510.4304	271.2798	802.5810	394.0000	− 29.55
	9	495.2846	255.3317	790.5947	431.0000	− 14.92
	10	527.3068	263.9159	853.7406	431.0000	− 22.34

[a] Forecasts are in original metric.

so when the troublesome autocorrelation occurs somewhere other than at the short lags (1, 2, 3) or the first few seasonal lags. The point is that autocorrelation coefficients with t-values exceeding the warning levels should receive serious attention whether or not their corresponding coefficients are included in a final model.

CASE 12. MACHINE-TOOL SHIPMENTS

The realization for this case study (machine-tool shipments) is shown in Figure C12.1*. The variance appears to be stationary. However, the series appears to change level, suggesting that nonseasonal differencing may be needed to achieve a stationary mean. Since it also seems to change slope, with a negative slope from 1968 through 1971 and a positive slope through the remainder of the series, differencing twice may be required before the mean is stationary. But we must examine estimated acf's and perhaps obtain some estimates of AR coefficients to choose the proper degree of differencing.

This series seems to have some seasonal variation. For example, December observations are typically higher than November figures, and June higher than May. Since the data are recorded monthly we look closely at estimated acf and pacf coefficients at multiples of lag 12.

Identification. Figure C12.2 shows the estimated acf for the original data. With 84 observations our practical rule is that we may safely examine about $84/4 = 21$ autocorrelations.

The slow decay of the acf from lags 1 through 9 suggests that the mean of the original series is not stationary, so we difference the data once ($d = 1$). This may also help to expose more fully the nature of any seasonal variation.

The nonseasonal first differences in Figure C12.3 appear to fluctuate around a fixed mean. The estimated acf is shown in Figure C12.4. We have

---

*These data are taken from various issues of the *Survey of Current Business*, published by the U.S. Department of Commerce.

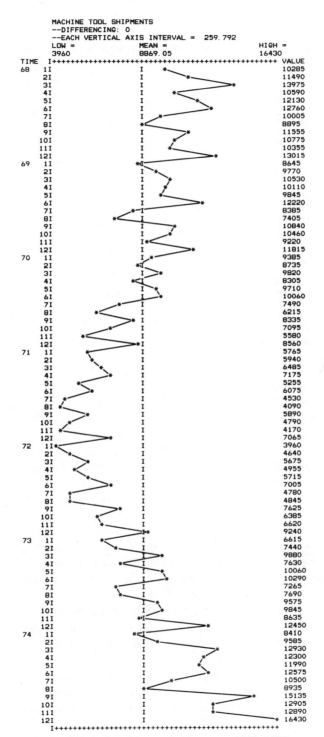

Figure C12.1 Machine-tool shipments, 1968–1974.

493

```
+ + + + + + + + + + + + AUTOCORRELATIONS + + + + + + + + + + + +
+ FOR DATA SERIES: MACHINE TOOL SHIPMENTS                        +
+ DIFFERENCING: 0                        MEAN    =  8869. 05     +
+ DATA COUNT =   84                      STD DEV =  2758. 98     +
  COEF   T-VAL  LAG _____0_____
  0. 68   6. 19   1                            [    0>>>>>]>>>>>>>>>>>>>
  0. 63   4. 17   2                        [        0>>>>>>>]>>>>>>>>>
  0. 74   4. 11   3                      [          0>>>>>>>]>>>>>>>>>>>
  0. 53   2. 49   4                    [            0>>>>>>>>>]>>>
  0. 53   2. 34   5              [                  0>>>>>>>>>>>>]>
  0. 68   2. 81   6              [                  0>>>>>>>>>>>>]>>>>>
  0. 49   1. 87   7        [                        0>>>>>>>>>>>>  ]
  0. 40   1. 45   8        [                        0>>>>>>>>>>  ]
  0. 51   1. 81   9        [                        0>>>>>>>>>>>>>]
  0. 30   1. 04  10        [                        0>>>>>>>>  ]
  0. 28   0. 95  11        [                        0>>>>>>>  ]
  0. 42   1. 41  12        [                        0>>>>>>>>>>>  ]
  0. 20   0. 64  13              [                  0>>>>>        ]
  0. 13   0. 42  14              [                  0>>>          ]
  0. 22   0. 70  15              [                  0>>>>>        ]
  0. 06   0. 18  16              [                  0>           ]
  0. 01   0. 04  17              [                  0            ]
  0. 13   0. 41  18              [                  0>>>          ]
 -0. 03  -0. 10  19              [                 <0            ]
 -0. 12  -0. 40  20              [              <<<0            ]
 -0. 04  -0. 14  21              [                 <0            ]
     CHI-SQUARED* =   329. 42 FOR DF =   21
```

Figure C12.2 Estimated acf for the realization in Figure C12.1.

pushed the number of autocorrelations to 24 to get a better picture of the seasonal pattern. This acf has an unusual pattern for differenced data. The most striking feature is the nondecaying positive autocorrelations at multiples of lag 3, suggesting a nonstationary periodic pattern of length three. It also resembles an AR(2) acf with both ϕ_1 and ϕ_2 negative. The critical fact, however, is that acf is not falling rapidly to zero, so the mean of the data is still nonstationary and further differencing is required.

We have three choices for further differencing. First, we could difference again by length one, setting $d = 2$. Second, we could perform seasonal differencing ($D = 1$). With monthly data we would difference by length 12. This is supported by the acf spikes at lags 12 and 24 that are decaying quite slowly (Figure C12.4). Third, we might difference by length three on the assumption that the nondecaying spikes at multiples of lag 3 reflect a nonstationary periodicity of length three.

A periodicity of length three in monthly data is unusual, though not so peculiar that we should rule it out. But the practical rule is to start with the common alternatives. Therefore, we consider the first and third possibilities, setting $d = 2$ or $D = 1$.

Nonseasonal second differencing is not often required, but we saw in Figure C12.1 that the data appear to have two different slopes. Seasonal differencing is commonly needed. We pursue both of these possibilities.

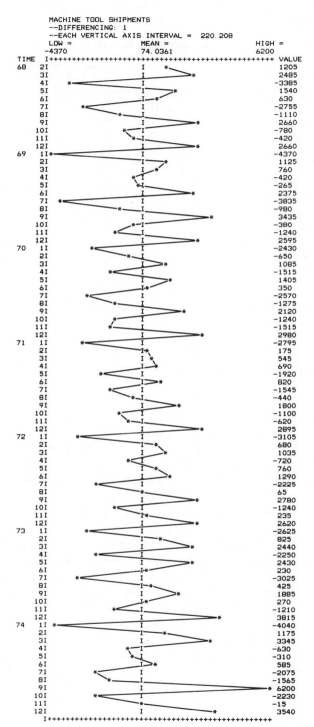

Figure C12.3 First differences of the realization in Figure C12.1.

```
+ + + + + + + + + + + + AUTOCORRELATIONS + + + + + + + + + + + +
+ FOR DATA SERIES: MACHINE TOOL SHIPMENTS                        +
+ DIFFERENCING: 1                      MEAN     =  74. 0361       +
+ DATA COUNT =   83                    STD DEV  = 2068. 1         +
  COEF   T-VAL LAG _____0_____
-0. 43   -3. 93  1            <<<<<<[<<<<<<<0           ]
-0. 29   -2. 22  2             <<[<<<<<<<<0             ]
 0. 53    3. 93  3            [              0>>>>>>>>>>>]>>>>>>>>
-0. 25   -1. 60  4            [ <<<<<<<<<0              ]
-0. 30   -1. 80  5            [<<<<<<<<<<0              ]
 0. 57    3. 33  6          [                0>>>>>>>>>>>>>]>>>>>>>
-0. 15   -0. 77  7          [        <<<<<0            ]
-0. 39   -2. 00  8         <[<<<<<<<<<<<<0             ]
 0. 55    2. 74  9        [                0>>>>>>>>>>>>>>]>>>>
-0. 25   -1. 14 10        [        <<<<<<<<0                 ]
-0. 28   -1. 27 11        [       <<<<<<<<<0                ]
 0. 59    2. 60 12      [                 0>>>>>>>>>>>>>>>>>]>>>>
-0. 20   -0. 83 13      [          <<<<<<<0                  ]
-0. 31   -1. 26 14      [         <<<<<<<<<0                 ]
 0. 45    1. 80 15      [                0>>>>>>>>>>>>>>>>>]
-0. 17   -0. 65 16    [            <<<<<0                      ]
-0. 28   -1. 07 17    [           <<<<<<<<<0                   ]
 0. 44    1. 65 18    [                 0>>>>>>>>>>>>>>>      ]
-0. 11   -0. 40 19    [             <<<<0                      ]
-0. 27   -0. 97 20    [           <<<<<<<<<0                   ]
 0. 36    1. 30 21    [                0>>>>>>>>>>>>           ]
-0. 13   -0. 47 22    [             <<<<0                      ]
-0. 26   -0. 91 23  [             <<<<<<<<<0                     ]
 0. 49    1. 70 24  [                  0>>>>>>>>>>>>>>>>        ]
   CHI-SQUARED* =  316. 95 FOR DF =   24
```

Figure C12.4   Estimated acf of the first differences in Figure C12.3.

Figure C12.5 is the estimated acf of the second differences of the data. It is barely different from the acf of the first differences in Figure C12.4. It does not fall rapidly to zero, so the mean of the data is still not stationary.

But seasonal first differencing (along with nonseasonal first differencing) induces a stationary mean. The estimated acf in Figure C12.6 moves rapidly to zero at the short lags, and the spike at lag 12 is followed by an insignificant autocorrelation at lag 24.

This result is an excellent illustration of two practical points. First, a strong seasonal pattern can sometimes make identification difficult. As pointed out in Chapter 11, strong seasonality can produce large autocorrelations at fractional-seasonal lags. In this example the seasonal element induced large autocorrelations at multiples of lag 3 in the acf in Figure C12.4. This pattern disappears completely after seasonal differencing. Second, we should always start with the common alternatives. The common step of seasonal first differencing (along with nonseasonal first differencing) has eliminated the unusual alternative of a nonstationary periodicity of length three. Indeed, experience show that common procedures and models

```
+ + + + + + + + + + + + AUTOCORRELATIONS + + + + + + + + + + + + +
+ FOR DATA SERIES: MACHINE TOOL SHIPMENTS                         +
+ DIFFERENCING: 1  1                      MEAN     =  28.4756     +
+ DATA COUNT =   82                       STD DEV =  3497.17      +
  COEF   T-VAL LAG                            0
-0.55   -4.96   1     <<<<<<<<<<<[<<<<<<<<0          ]
-0.22   -1.61   2             [  <<<<<<<0        ]
 0.53    3.71   3             [        0>>>>>>>>>>]>>>>>>>>
-0.24   -1.42   4           [    <<<<<<<<0          ]
-0.31   -1.84   5           [ <<<<<<<<<<0          ]
 0.55    3.08   6             [        0>>>>>>>>>>>>]>>>>>>
-0.17   -0.86   7       [          <<<<<0          ]
-0.41   -2.08   8       [<<<<<<<<<<<<<0          ]
 0.60    2.89   9       [        0>>>>>>>>>>>>>>>]>>>>>>
-0.27   -1.17  10         [      <<<<<<<<<0           ]
-0.30   -1.27  11         [      <<<<<<<<<<0          ]
 0.56    2.35  12         [      0>>>>>>>>>>>>>>>>]>>>
-0.23   -0.89  13         [       <<<<<<<<0          ]
-0.31   -1.23  14       [     <<<<<<<<<<0          ]
 0.49    1.87  15       [         0>>>>>>>>>>>>>>> ]
-0.18   -0.67  16       [          <<<<<0         ]
-0.28   -1.01  17       [       <<<<<<<<0         ]
 0.44    1.58  18       [         0>>>>>>>>>>>>>> ]
-0.15   -0.54  19       [          <<<<0         ]
-0.26   -0.91  20     [        <<<<<<<<0           ]
 0.38    1.33  21     [          0>>>>>>>>>>>>           ]
-0.13   -0.44  22     [            <<<<0             ]
-0.29   -0.99  23     [        <<<<<<<<<0            ]
 0.50    1.68  24     [          0>>>>>>>>>>>>>>>> ]
      CHI-SQUARED* =   334.16 FOR DF =  24
```

Figure C12.5 Estimated acf of the second differences ($d = 2$).

are nearly always adequate; the analyst must have the mental discipline to focus on these common alternatives.

Our task now is to tentatively choose one or more models based on the estimated acf and pacf in Figure C12.6. The spike at lag 12 followed by a cutoff to zero at lag 24 clearly calls for a seasonal MA coefficient. The absolute t-value at lag 12 (1.92) is well in excess of the practical warning level for seasonal lags (1.25).

The short-lag autocorrelations could be interpreted as one or two spikes followed by a cutoff to zero. The spike at lag 2 is only borderline significant; its t-value falls slightly short of the practical warning level of 1.6. The pacf is consistent with an MA model for the nonseasonal part of the data: it decays starting from lag 1.

We will entertain two models: an ARIMA$(0, 1, 1)(0, 1, 1)_{12}$ and an ARIMA$(0, 1, 2)(0, 1, 1)_{12}$:

$$(1 - B)(1 - B^{12})\tilde{z}_t = (1 - \Theta_{12}B^{12})(1 - \theta_1 B)a_t \qquad (C12.1)$$

$$(1 - B)(1 - B^{12})\tilde{z}_t = (1 - \Theta_{12}B^{12})(1 - \theta_1 B - \theta_2 B^2)a_t \quad (C12.2)$$

```
+ + + + + + + + + + + + AUTOCORRELATIONS + + + + + + + + + + + +
+ FOR DATA SERIES: MACHINE TOOL SHIPMENTS                                    +
+ DIFFERENCING: 12  1                        MEAN    =  79.1549              +
+ DATA COUNT =  71                           STD DEV =  1365.74              +
  COEF  T-VAL LAG _____O_____
 -0.62  -5.21   1   <<<<<<<<<<<<<<<[<<<<<<<O           ]
  0.24   1.54   2                       [       O>>>>>>>>> ]
 -0.10  -0.59   3                       [    <<<O           ]
  0.06   0.39   4                       [       O>>         ]
 -0.02  -0.10   5                       [      <O           ]
 -0.01  -0.05   6                       [       O           ]
  0.12   0.74   7                       [       O>>>         ]
 -0.17  -1.02   8                     [     <<<<<O               ]
  0.23   1.35   9                     [         O>>>>>>>>>      ]
 -0.30  -1.72  10                   [ <<<<<<<<<<O               ]
  0.35   1.94  11                     [         O>>>>>>>>>>>>]
 -0.36  -1.92  12                   [<<<<<<<<<<<<O               ]
  0.26   1.30  13                 [              O>>>>>>>>>          ]
 -0.19  -0.93  14                 [        <<<<<<O                  ]
  0.10   0.51  15                 [              O>>>               ]
 -0.01  -0.06  16                 [              O                  ]
 -0.11  -0.52  17                 [          <<<<O                  ]
  0.08   0.37  18                 [              O>>>               ]
  0.04   0.18  19                 [              O>                 ]
 -0.08  -0.38  20                 [           <<<O                  ]
 -0.09  -0.45  21                 [           <<<O                  ]
  0.20   0.95  22                 [              O>>>>>>>            ]
 -0.13  -0.60  23                 [          <<<<O                  ]
 -0.02  -0.11  24                 [             <O                  ]
    CHI-SQUARED* =    90.23 FOR DF =   24

+ + + + + + + + + + + + PARTIAL AUTOCORRELATIONS + + + + + + + + + + + +
  COEF  T-VAL LAG _____O_____
 -0.62  -5.21   1   <<<<<<<<<<<<<<<[<<<<<<<O           ]
 -0.23  -1.91   2                 [<<<<<<<O           ]
 -0.09  -0.77   3                 [    <<<O           ]
  0.01   0.05   4                 [       O           ]
  0.05   0.41   5                 [       O>>         ]
  0.02   0.15   6                 [       O>          ]
  0.20   1.68   7                 [       O>>>>>>>]
  0.02   0.18   8                 [       O>          ]
  0.18   1.55   9                 [       O>>>>>>     ]
 -0.13  -1.12  10                 [    <<<<O           ]
  0.13   1.12  11                 [       O>>>>       ]
 -0.17  -1.41  12               [ <<<<<<O           ]
 -0.08  -0.66  13                 [    <<<O           ]
 -0.18  -1.50  14               [ <<<<<<O           ]
 -0.10  -0.81  15                 [    <<<O           ]
 -0.01  -0.10  16                 [       O           ]
 -0.10  -0.84  17                 [    <<<O           ]
 -0.15  -1.24  18                 [  <<<<<O           ]
  0.23   1.91  19                 [       O>>>>>>>]
  0.00  -0.03  20                 [       O           ]
 -0.08  -0.68  21                 [    <<<O           ]
 -0.04  -0.31  22                 [      <O           ]
  0.20   1.64  23                 [       O>>>>>>>]
 -0.16  -1.33  24                 [   <<<<<O           ]
```

Figure C12.6 Estimated acf and pacf for the differenced series ($d = 1$, $D = 1$).

Estimation, diagnostic checking, reformulation. Results shown in Figures C12.7 and C12.8 indicate that (C12.2) is superior since its adjusted RMSE and chi-squared statistic are both smaller. However, both models have somewhat troubling residual autocorrelations at lags 6 and 7, and the residual autocorrelation at lag 9 in Figure C12.7 also has a t-value greater than 1.6. Although we must sometimes accept moderately large residual autocorrelations, especially at uncommon lags, both chi-squared statistics suggest our models are inadequate. They are both significant at the 10% level.

In Chapter 12 we emphasize that autocorrelations at certain lags require special attention: the short lags (1, 2, perhaps 3), the seasonal lags (multiples

```
+ + + + + + + + + +ECOSTAT UNIVARIATE B-J RESULTS+ + + + + + + + +
+ FOR DATA SERIES:   MACHINE TOOL SHIPMENTS                        +
+ DIFFERENCING:        12  1                       DF   = 69       +
+ AVAILABLE:           DATA = 71   BACKCASTS = 13   TOTAL = 84      +
+ USED TO FIND SSR:  DATA = 71   BACKCASTS = 13   TOTAL = 84       +
+ (LOST DUE TO PRESENCE OF AUTOREGRESSIVE TERMS:            0)     +

COEFFICIENT    ESTIMATE      STD ERROR     T-VALUE
   THETA   1     0.430         0.104         4.12
   THETA* 12     0.799         0.109         7.36

ADJUSTED RMSE =  926.592    MEAN ABS % ERR =    9.15
       CORRELATIONS
        1      2
  1    1.00
  2   -0.10   1.00

++RESIDUAL ACF++
  COEF   T-VAL  LAG _____0_____
 -0.10  -0.84    1  [          <<<<<<<<<<O                         ]
  0.10   0.87    2  [                    O>>>>>>>>>>              ]
  0.12   1.03    3  [                    O>>>>>>>>>>>>            ]
  0.11   0.89    4  [                    O>>>>>>>>>>>             ]
  0.08   0.64    5  [                    O>>>>>>>>               ]
  0.23   1.82    6  [                    O>>>>>>>>>>>>>>>>>>>>>>>]
  0.24   1.86    7                       O>>>>>>>>>>>>>>>>>>>>>>>>
 -0.03  -0.19    8                   <<<O
  0.24   1.76    9                       O>>>>>>>>>>>>>>>>>>>>>>>>
 -0.05  -0.33   10                  <<<<<O
  0.17   1.22   11                       O>>>>>>>>>>>>>>>>>
  0.03   0.23   12                       O>>>
  0.17   1.19   13                       O>>>>>>>>>>>>>>>>>
 -0.06  -0.37   14                 <<<<<<O
  0.07   0.46   15                       O>>>>>>>
  0.14   0.90   16                       O>>>>>>>>>>>>>>
 -0.18  -1.19   17    <<<<<<<<<<<<<<<<<<O
  0.13   0.84   18                       O>>>>>>>>>>>>>>
    CHI-SQUARED* =   30.86 FOR DF =   16
```

Figure C12.7 Estimation and diagnostic-checking results for model (C12.1).

```
+ + + + + + + + + +ECOSTAT UNIVARIATE B-J RESULTS+ + + + + + + + + +
+ FOR DATA SERIES:   MACHINE TOOL SHIPMENTS                         +
+ DIFFERENCING:       12  1                         DF    = 68      +
+ AVAILABLE:          DATA = 71    BACKCASTS = 14   TOTAL = 85      +
+ USED TO FIND SSR: DATA = 71    BACKCASTS = 14   TOTAL = 85      +
+ (LOST DUE TO PRESENCE OF AUTOREGRESSIVE TERMS:              0)   +
```

```
COEFFICIENT   ESTIMATE      STD ERROR      T-VALUE
  THETA  1      0. 620         0. 111         5. 59
  THETA  2     -0. 232         0. 113        -2. 07
  THETA* 12     0. 835         0. 087         9. 63
```

```
ADJUSTED RMSE =   913. 592    MEAN ABS % ERR =    9. 27
        CORRELATIONS
          1      2      3
  1     1. 00
  2    -0. 54   1. 00
  3     0. 02   0. 01   1. 00
```

```
++RESIDUAL ACF++
  COEF   T-VAL LAG _____0_____
  0. 08   0. 70   1            [              0>>>>>          ]
 -0. 03  -0. 27   2            [             <<0             ]
  0. 03   0. 21   3            [              0>             ]
  0. 04   0. 37   4            [              0>>            ]
  0. 10   0. 86   5            [              0>>>>>          ]
  0. 29   2. 42   6            [              0>>>>>>>>>>>>]>>>
  0. 26   1. 99   7          [                0>>>>>>>>>>>>>>]
  0. 00  -0. 04   8          [                0             ]
  0. 15   1. 10   9          [                0>>>>>>>>      ]
 -0. 03  -0. 19  10          [               <0             ]
  0. 13   0. 90  11          [                0>>>>>>>       ]
  0. 11   0. 77  12          [                0>>>>>         ]
  0. 16   1. 12  13          [                0>>>>>>>>      ]
 -0. 02  -0. 16  14          [               <0             ]
  0. 05   0. 35  15          [                0>>>          ]
  0. 06   0. 44  16          [                0>>>          ]
 -0. 17  -1. 15  17          [         <<<<<<<<0            ]
  0. 13   0. 90  18          [                0>>>>>>>       ]
        CHI-SQUARED* =   25. 82 FOR DF =   15
```

Figure C12.8 Estimation and diagnostic-checking results for model (C12.2).

of 12 in the present case), and fractional or near-seasonal lags. In this case lag 6 deserves primary consideration since it is the half-seasonal lag. Machine-tool shipments could have a six-month seasonal pattern in addition to a twelve-month pattern.

Adding an MA term at lag 6 to each previously estimated model gives these new models:

$$(1 - B)(1 - B^{12})\tilde{z}_t = (1 - \Theta_{12}B^{12})(1 - \theta_1 B - \theta_6 B^6)a_t \qquad (C12.3)$$

$$(1 - B)(1 - B^{12})\tilde{z}_t = (1 - \Theta_{12}B^{12})(1 - \theta_1 B - \theta_2 B^2 - \theta_6 B^6)a_t$$

$$(C12.4)$$

Estimation and diagnostic-checking results are displayed in Figures C12.9 and C12.10. In both cases $\hat{\theta}_6$ is highly significant with an absolute t-value well in excess of 2.0; all other coefficients remain significantly different from zero. Model (C12.4) has a substantially smaller chi-squared statistic, but the chi-squared for (C12.3) is also satisfactory. (C12.3) has a slightly larger RMSE.

As a final check on these two models, drop the last 12 observations, reestimate the parameters, forecast 12 periods ahead, and compare these forecasts with the last 12 available observations. The percent forecast errors for (C12.3) and (C12.4) are shown in the last column in Tables C12.1 and C12.2, respectively. (C12.4) forecasts with slightly more accuracy, except for the first forecast.

```
+ + + + + + + + +ECOSTAT UNIVARIATE B-J RESULTS+ + + + + + + + +
+ FOR DATA SERIES:   MACHINE TOOL SHIPMENTS                       +
+ DIFFERENCING:      12   1                      DF    = 68       +
+ AVAILABLE:            DATA = 71    BACKCASTS = 18   TOTAL = 89   +
+ USED TO FIND SSR: DATA = 71    BACKCASTS = 18   TOTAL = 89      +
+ (LOST DUE TO PRESENCE OF AUTOREGRESSIVE TERMS:           0)     +

COEFFICIENT    ESTIMATE     STD ERROR     T-VALUE
  THETA  1      0.579        0.089          6.49
  THETA  6     -0.473        0.095         -5.00
  THETA* 12     0.818        0.103          7.98

  ADJUSTED RMSE =   809.93   MEAN ABS % ERR =   8.10
        CORRELATIONS
        1      2      3
  1   1.00
  2   0.07   1.00
  3   0.02   0.20   1.00

++RESIDUAL ACF++
  COEF   T-VAL LAG _____0_____
 -0.14  -1.22   1  [           <<<<<<<<<<<<<<<0                      ]
  0.08   0.65   2  [                         0>>>>>>>>              ]
  0.08   0.63   3  [                         0>>>>>>>>              ]
  0.09   0.76   4  [                         0>>>>>>>>>             ]
  0.08   0.63   5  [                         0>>>>>>>>              ]
 -0.02  -0.14   6  [                       <<0                      ]
  0.12   0.94   7  [                         0>>>>>>>>>>>>          ]
 -0.09  -0.75   8              <<<<<<<<<0
  0.20   1.56   9                         0>>>>>>>>>>>>>>>>>>>>
 -0.12  -0.89  10          <<<<<<<<<<<<0
  0.13   1.02  11                         0>>>>>>>>>>>>>
 -0.05  -0.37  12                     <<<<<0
  0.08   0.59  13                         0>>>>>>>>
  0.00   0.00  14                         0
  0.02   0.14  15                         0>>
  0.11   0.80  16                         0>>>>>>>>>>>
 -0.17  -1.25  17          <<<<<<<<<<<<<<<<<0
  0.18   1.26  18                         0>>>>>>>>>>>>>>>>>>
       CHI-SQUARED* =   19.24 FOR DF =   15
```

Figure C12.9   Estimation and diagnostic-checking results for model (C12.3).

```
+ + + + + + + + +ECOSTAT UNIVARIATE B-J RESULTS+ + + + + + + + + +
+ FOR DATA SERIES:   MACHINE TOOL SHIPMENTS                       +
+ DIFFERENCING:     12  1                         DF    = 67      +
+ AVAILABLE:         DATA = 71    BACKCASTS = 18   TOTAL = 89      +
+ USED TO FIND SSR:  DATA = 71    BACKCASTS = 18   TOTAL = 89      +
+ (LOST DUE TO PRESENCE OF AUTOREGRESSIVE TERMS:             0)   +

COEFFICIENT     ESTIMATE       STD ERROR       T-VALUE
   THETA  1       0.812          0.104           7.84
   THETA  2      -0.224          0.103          -2.16
   THETA  6      -0.401          0.075          -5.35
   THETA* 12      0.808          0.107           7.58

ADJUSTED RMSE =  808.14    MEAN ABS % ERR =    7.84
      CORRELATIONS
          1      2      3      4
   1    1.00
   2   -0.76   1.00
   3    0.34  -0.26   1.00
   4   -0.05   0.10   0.09   1.00

++RESIDUAL ACF++
  COEF   T-VAL  LAG _____0_____
  0.06    0.54    1  [                       0>>>>>>                  ]
 -0.01   -0.10    2  [                      <0                        ]
  0.04    0.34    3  [                       0>>>>                     ]
  0.08    0.65    4  [                       0>>>>>>>>                ]
  0.11    0.93    5  [                       0>>>>>>>>>>>             ]
  0.05    0.44    6  [                       0>>>>>                   ]
  0.05    0.40    7  [                       0>>>>>                   ]
 -0.05   -0.42    8  [                  <<<<<0                        ]
  0.18    1.43    9  [                       0>>>>>>>>>>>>>>>>>>       ]
 -0.04   -0.34   10                     <<<<0
  0.08    0.64   11                         0>>>>>>>>
 -0.02   -0.16   12                        <0
  0.10    0.82   13                         0>>>>>>>>>>
  0.01    0.11   14                         0>
  0.02    0.17   15                         0>>
  0.06    0.47   16                         0>>>>>>
 -0.12   -0.95   17                 <<<<<<<<<<<<0
  0.18    1.34   18                         0>>>>>>>>>>>>>>>>>>
  CHI-SQUARED* =    11.67 FOR DF =   14
```

Figure C12.10   Estimation and diagnostic-checking results for model (C12.4).

The forecast accuracy of both models deteriorates sharply after the first two forecasts. Accuracy would presumably be improved if each model were reestimated and forecasts recalculated each period as a new observation becomes available. Much of the improved accuracy would stem from the reduction of bootstrapping. Because the data have been differenced, forecasts are based in part on past values of the machine-tool series; however, to forecast the longer lead times we must use forecast values in place of observed values. As indicated in Chapter 10 this substitution is known as bootstrapping. (See Case 3 for a numerical illustration of bootstrapping.)

Table C12.1   Forecasts from model (C12.3) using the first 72 observations

Time		Forecast Values	80% Confidence Limits		Future Observed Values	Percent Forecast Errors
			Lower	Upper		
74	1	8701.4441	7717.3491	9685.5391	8410.0000	− 3.47
	2	9498.4514	8430.0041	10566.8986	9585.0000	0.90
	3	10771.7091	9625.0984	11918.3198	12930.0000	16.69
	4	9939.0788	8719.3032	11158.8544	12300.0000	19.19
	5	10669.4823	9380.6887	11958.2760	11990.0000	11.01
	6	11820.8572	10466.5582	13175.1561	12575.0000	6.00
	7	9179.3928	7608.4636	10750.3221	10500.0000	12.58
	8	8868.8662	7107.7553	10629.9771	8935.0000	0.74
	9	11217.7873	9285.1195	13150.4550	15135.0000	25.88
	10	10535.2013	8445.0105	12625.3921	12905.0000	18.36
	11	9769.4898	7532.8426	12006.1370	12890.0000	24.21
	12	12834.3903	10460.3044	15208.4761	16430.0000	21.88

Table C12.2   Forecasts from model (C12.4) using the first 72 observations

Time		Forecast Values	80% Confidence Limits		Future Observed Values	Percent Forecast Errors
			Lower	Upper		
74	1	8784.5177	7817.3999	9751.6356	8410.0000	− 4.45
	2	9658.1291	8685.9142	10630.3441	9585.0000	− 0.76
	3	11043.9062	9971.9591	12115.8532	12930.0000	14.59
	4	10045.8285	8882.6694	11208.9876	12300.0000	18.33
	5	10929.8991	9682.1781	12177.6200	11990.0000	8.84
	6	11925.1353	10598.2305	13252.0400	12575.0000	5.17
	7	9293.1801	7770.4033	10815.9570	10500.0000	11.49
	8	8987.8747	7291.6959	10684.0535	8935.0000	− 0.59
	9	11316.3784	9462.9503	13169.8065	15135.0000	25.23
	10	10658.4266	8660.0851	12656.7682	12905.0000	17.41
	11	9905.2428	7771.8085	12038.6772	12890.0000	23.16
	12	13001.9596	10741.4916	15262.4275	16430.0000	20.86

Alternative models.   In this section we consider two other models based on different interpretations of the acf and pacf in Figure C12.6. Stretching our imaginations, we could interpret these graphs as representing an ARIMA$(1, 1, 0)(0, 1, 1)_{12}$. The acf seems to decay (*very* rapidly) while alternating in sign, and the pacf has one spike with an absolute t-value greater than 2.0.

Estimation (results not shown) of an ARIMA$(1, 1, 0)(0, 1, 1)_{12}$ gives satisfactory coefficients and t-values, but the residual acf reveals a significant spike at lag 6 and the chi-squared statistic is significant. Adding an MA coefficient at lag 6 gives the results shown in Figure C12.11. This model is inferior to both (C12.3) and (C12.4): it has a noticeably larger RMSE and chi-squared statistic.

```
+ + + + + + + + + +ECOSTAT UNIVARIATE B-J RESULTS+ + + + + + + + + +
+ FOR DATA SERIES:   MACHINE TOOL SHIPMENTS                            +
+ DIFFERENCING:      12  1                          DF    = 68        +
+ AVAILABLE:         DATA = 71    BACKCASTS = 429  TOTAL = 500        +
+ USED TO FIND SSR:  DATA = 71    BACKCASTS = 428  TOTAL = 499        +
+ (LOST DUE TO PRESENCE OF AUTOREGRESSIVE TERMS:               1)     +

COEFFICIENT    ESTIMATE      STD ERROR      T-VALUE
  PHI    1       -0.505        0.099         -5.09
  THETA  6       -0.284        0.119         -2.39
  THETA* 12       0.806        0.107          7.53

ADJUSTED RMSE =   877.901    MEAN ABS % ERR =    8.93
        CORRELATIONS
        1     2     3
  1   1.00
  2  -0.10  1.00
  3   0.09  0.18  1.00

++RESIDUAL ACF++
  COEF   T-VAL LAG _____0_____
 -0.11   -0.90  1  [             <<<<<<<<<<<<0                        ]
 -0.16   -1.33  2  [         <<<<<<<<<<<<<<<<<0                       ]
  0.13    1.04  3  [                          0>>>>>>>>>>>>>          ]
  0.11    0.85  4                             0>>>>>>>>>>>
 -0.02   -0.12  5                           <<0
  0.04    0.31  6                             0>>>>
  0.22    1.70  7                             0>>>>>>>>>>>>>>>>>>>>>>>
 -0.01   -0.07  8                            <0
  0.12    0.92  9                             0>>>>>>>>>>>
 -0.06   -0.46 10                       <<<<<<0
  0.17    1.24 11                             0>>>>>>>>>>>>>>>>>
  0.03    0.19 12                             0>>>
  0.01    0.09 13                             0>
 -0.04   -0.31 14                        <<<<<0
  0.08    0.57 15                             0>>>>>>>>
  0.08    0.55 16                             0>>>>>>>>
 -0.24   -1.75 17  <<<<<<<<<<<<<<<<<<<<<<<<<<<0
  0.15    1.06 18                             0>>>>>>>>>>>>>>>
    CHI-SQUARED* =    21.94 FOR DF =   15
```

Figure C12.11   Estimation and diagnostic-checking results for an alternative model.

The estimated acf and pacf in Figure C12.6 could also be interpreted as decaying, thus suggesting a mixed model for the nonseasonal part of the data. For example, we could try an ARIMA$(1, 1, 1)(0, 1, 1)_{12}$. Estimation produces unacceptably small t-values for $\hat{\phi}_1$ and $\hat{\theta}_1$ (both less than 1.6 in absolute value) so this alternative model is not worth pursuing further.

Final comments. This case illustrates several important points:

1. Focus on the very short lags (1, 2, 3) and the first few seasonal lags in estimated acf's and pacf's. At the identification stage, do not be distracted by significant values at near-seasonal lags, fractional-seasonal lags, or at lags which imply uncommon models (e.g., lag 7 with monthly data). For example, consider the moderately significant coefficients in the acf in Figure C12.6 at lags 9–11 and 13. These are substantially reduced in the residual acf's in Figure C12.7 and C12.8 without having to include coefficients in those lags. Likewise, the common procedure of seasonal differencing clears up the peculiar wave pattern in the acf in Figure C12.4.

2. After a common model has been fitted to a series, a seasonal element may still appear at residual acf fractional-seasonal lags or at lags very near the seasonal lags. Such residual autocorrelations deserve attention, but only after an otherwise satisfactory model has been found. The warning value for absolute t-values at these lags in the residual acf is the same as for seasonal lags (1.25). In the present case the residual acf t-value at lag 6 in Figure C12.7 is only 1.82; but $\hat{\theta}_6$ in Figure C12.9 has an absolute t-value of 5.0.

3. The chi-squared test for independence of the random shocks is a useful aid in diagnostic checking. Use of t-tests alone might lead us to incorrectly accept a model with a large number of residual autocorrelations that are only moderately significant. In the present case, the chi-squared statistic in Figure C12.7 forces us to identify another model, although none of the residual autocorrelations is significant at the 5% level.

CASE 13.   CIGAR CONSUMPTION

A practical rule in Chapter 12 states that modeling the seasonal element first sometimes makes identification of the nonseasonal pattern easier. As seen in earlier cases this is not always so; sometimes we see both the seasonal and nonseasonal patterns quite clearly in the estimated acf and pacf. In this case we find that removing the seasonal pattern first is quite helpful.

The data to be analyzed are plotted in Figure C13.1. They represent monthly cigar consumption (withdrawals from stock) for the years 1969–1976.* The mean of this series seems to fall over time, so nonseasonal differencing may be needed.

There is some evidence that the variance of the series falls along with the mean. In particular, the variability of the data during 1969 is greater than the variability during 1975 and 1976. This contrast aside, the variance over the rest of the series is fairly uniform. Though we could try a logarithmic transformation we conclude (quite subjectively) that the evidence is not strong enough to warrant this step. More data may provide a better picture of any changes in variance.

The series shows an obvious seasonal pattern with peak values in October and low values in December. With monthly data the length of seasonality is 12; we pay special attention to estimated autocorrelations and partial autocorrelations at lags that are multiples of 12.

---

*The data are taken from various issues of *Business Statistics* published by the U.S. Department of Commerce.

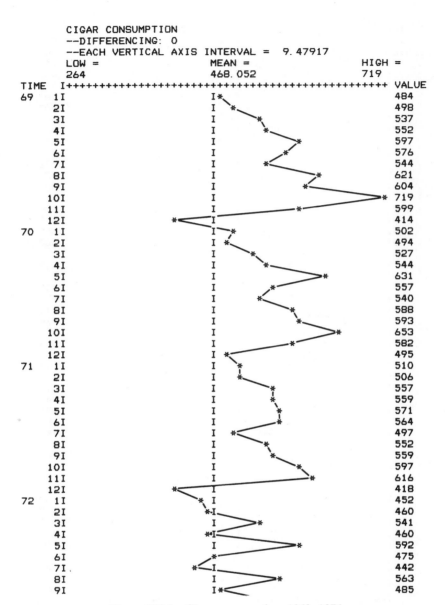

```
        CIGAR CONSUMPTION
        --DIFFERENCING: 0
        --EACH VERTICAL AXIS INTERVAL =  9.47917
        LOW =                    MEAN =              HIGH =
        264                      468.052             719
TIME    I++++++++++++++++++++++++++++++++++++++++++++++++++ VALUE
69   1I                          I*                          484
     2I                          I  *                        498
     3I                          I       *                   537
     4I                          I          *                552
     5I                          I              *            597
     6I                          I            *              576
     7I                          I        *                  544
     8I                          I                 *         621
     9I                          I              *            604
    10I                          I                         * 719
    11I                          I            *              599
    12I              *           I                           414
70   1I                          I *                         502
     2I                          I *                         494
     3I                          I   *                       527
     4I                          I      *                    544
     5I                          I              *            631
     6I                          I        *                  557
     7I                          I      *                    540
     8I                          I         *                 588
     9I                          I          *                593
    10I                          I              *            653
    11I                          I           *               582
    12I                        *  I                          495
71   1I                          I *                         510
     2I                          I *                         506
     3I                          I      *                    557
     4I                          I      *                    559
     5I                          I       *                   571
     6I                          I       *                   564
     7I                          I  *                        497
     8I                          I      *                    552
     9I                          I       *                   559
    10I                          I           *               597
    11I                          I              *            616
    12I                      *    I                          418
72   1I                         * I                          452
     2I                        *I                            460
     3I                          I      *                    541
     4I                        *I                            460
     5I                          I             *             592
     6I                         *                            475
     7I                    *    I                            442
     8I                          I            *              563
     9I                          I*                          485
```

Figure C13.1   Cigar consumption, 1969–1976.

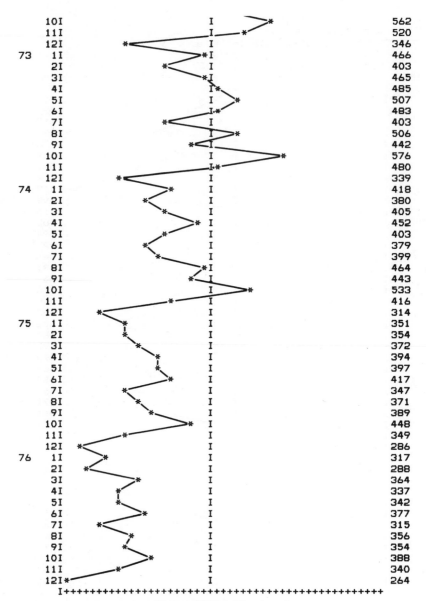

	10I	562
	11I	520
	12I	346
73	1I	466
	2I	403
	3I	465
	4I	485
	5I	507
	6I	483
	7I	403
	8I	506
	9I	442
	10I	576
	11I	480
	12I	339
74	1I	418
	2I	380
	3I	405
	4I	452
	5I	403
	6I	379
	7I	399
	8I	464
	9I	443
	10I	533
	11I	416
	12I	314
75	1I	351
	2I	354
	3I	372
	4I	394
	5I	397
	6I	417
	7I	347
	8I	371
	9I	389
	10I	448
	11I	349
	12I	286
76	1I	317
	2I	288
	3I	364
	4I	337
	5I	342
	6I	377
	7I	315
	8I	356
	9I	354
	10I	388
	11I	340
	12I	264

Figure C13.1 (*Continued*)

508

```
+ + + + + + + + + + + + AUTOCORRELATIONS + + + + + + + + + + + + +
+ FOR DATA SERIES: CIGAR CONSUMPTION                                +
+ DIFFERENCING: 0                        MEAN    =   468.052        +
+ DATA COUNT =  96                       STD DEV =   97.1571        +
  COEF   T-VAL  LAG _____0_____
  0.72   7.06    1                           [     0>>>>>]>>>>>>>>>>>>>>
  0.63   4.31    2                       [         0>>>>>>>]>>>>>>>>>
  0.57   3.35    3                       [         0>>>>>>>]>>>>>>
  0.52   2.74    4                    [            0>>>>>>>>>]>>>
  0.63   3.05    5                    [            0>>>>>>>>>]>>>>>>
  0.59   2.62    6                  [              0>>>>>>>>>>>]>>>
  0.57   2.37    7                  [              0>>>>>>>>>>>]>>
  0.46   1.82    8                  [              0>>>>>>>>>>]
  0.45   1.71    9               [                 0>>>>>>>>>>>   ]
  0.48   1.77   10               [                 0>>>>>>>>>>>>  ]
  0.54   1.93   11               [                 0>>>>>>>>>>>>>]
  0.67   2.32   12               [                 0>>>>>>>>>>>>>>]>>>
  0.47   1.54   13             [                   0>>>>>>>>>>>>    ]
  0.39   1.25   14             [                   0>>>>>>>>>>     ]
  0.33   1.04   15             [                   0>>>>>>>>>      ]
  0.29   0.90   16             [                   0>>>>>>>>       ]
  0.39   1.22   17             [                   0>>>>>>>>>>     ]
  0.33   1.00   18             [                   0>>>>>>>>>      ]
  0.32   0.95   19             [                   0>>>>>>>>       ]
  0.22   0.67   20             [                   0>>>>>>         ]
  0.19   0.57   21             [                   0>>>>>          ]
  0.22   0.64   22             [                   0>>>>>          ]
  0.27   0.80   23             [                   0>>>>>>>        ]
  0.37   1.09   24          [                      0>>>>>>>>>>         ]
     CHI-SQUARED* =   567.38 FOR DF =   24

+ + + + + + + + + + + PARTIAL AUTOCORRELATIONS + + + + + + + + + + +
  COEF   T-VAL  LAG _____0_____
  0.72   7.06    1                           [     0>>>>>]>>>>>>>>>>>>>>>
  0.23   2.22    2                           [     0>>>>>>]
  0.13   1.31    3                           [     0>>>  ]
  0.06   0.59    4                           [     0>>   ]
  0.37   3.62    5                           [     0>>>>>>]>>>
  0.02   0.24    6                           [     0>    ]
  0.06   0.60    7                           [     0>>   ]
 -0.19  -1.85    8                      [<<<<<0        ]
  0.12   1.19    9                           [     0>>>  ]
  0.04   0.41   10                           [     0>    ]
  0.23   2.27   11                           [     0>>>>>>]
  0.32   3.14   12                           [     0>>>>>]>>
 -0.40  -3.91   13                  <<<<[<<<<<0        ]
 -0.17  -1.67   14                       [ <<<<0        ]
 -0.12  -1.14   15                       [   <<<0        ]
 -0.03  -0.25   16                       [    <0        ]
  0.02   0.23   17                       [     0>        ]
 -0.08  -0.80   18                       [    <<0        ]
  0.05   0.54   19                       [     0>        ]
 -0.03  -0.26   20                       [     <0        ]
  0.02   0.17   21                       [     0         ]
 -0.07  -0.73   22                       [    <<0        ]
  0.05   0.52   23                       [     0>        ]
  0.11   1.12   24                       [     0>>>      ]
```

Figure C13.2 Estimated acf and pacf for the realization in Figure C13.1.

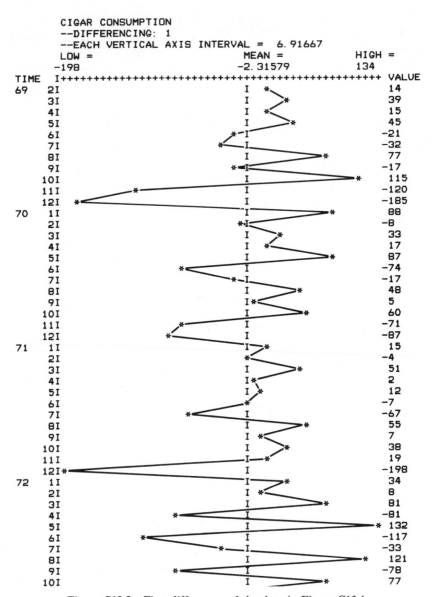

Figure C13.3   First differences of the data in Figure C13.1.

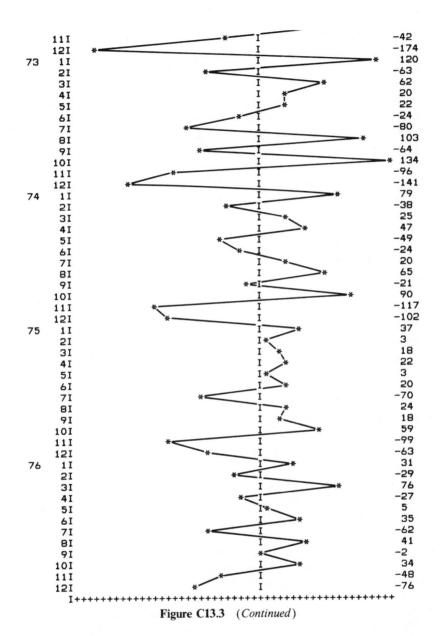

Figure C13.3 (*Continued*)

Identification.   The estimated acf in Figure C13.2 indicates that nonseasonal differencing is required to induce a stationary mean. This acf falls to zero very slowly; not until lag 13 do we find the *t*-values falling below the relevant practical warning levels.

After nonseasonal first differencing the data no longer trend downward (Figure C13.3); the relevant acf and pacf are displayed in Figure C13.4. The acf now cuts off to zero very sharply at lags 2 and 3, so further nonseasonal differencing is not needed. As often occurs, nonseasonal differencing clarifies the seasonal pattern. The autocorrelations at lags 12 and 24 decay very slowly, so seasonal differencing is called for. [An MA(2)$_{12}$ is inappropriate for the seasonal part of the data since the pacf cuts off at lag 24. A seasonal MA process would decay at multiples of the seasonal lag in the estimated pacf.]

Figure C13.5 shows the estimated acf and pacf with $d = 1$ and $D = 1$. We now have a fully stationary series with easily identifiable patterns. The spike at lag 1 in the acf followed by a cutoff to zero clearly suggests a nonseasonal MA(1). Likewise, the acf spike at lag 12 with a cutoff to zero at lag 24 calls for a seasonal MA(1)$_{12}$. (We ignore the significant spike at lag 11 at this stage.) Combining these two terms in a multiplicative model, we entertain an ARIMA(0, 1, 1)(0, 1, 1)$_{12}$:

$$(1 - B)(1 - B^{12})\tilde{z}_t = (1 - \Theta_{12}B^{12})(1 - \theta_1 B)a_t \qquad (C13.1)$$

Estimation and diagnostic checking.   The estimation results at the top of Figure C13.6 are satisfactory: both coefficients are statistically significant (with absolute *t*-values greater than 2.0) and both satisfy their respective invertibility requirements.

Unfortunately, the residual acf is a disaster. There are large spikes at lags 1, 3, 4, 6, 7 and 13 and the chi-squared statistic is quite significant. Model (C13.1) is clearly not acceptable.

Further identification.   When a model is as unsatisfactory as (C13.1) it is wise to reconsider the estimated acf and pacf used to identify that model. In this instance we go back to Figure C13.5, which shows the acf and pacf that indicated the data were stationary after both nonseasonal and seasonal first differencing. The residual acf and pacf for model (C13.1) may also be helpful. Between these pairs of acf's and pacf's, we should be able to find clues to a better model.

First, reconsider Figure C13.5. The acf is not very helpful in suggesting an alternative to (C13.1). The spike at lag 1 is quite significant, but none of the subsequent autocorrelations before lag 11 have *t*-statistics close to a

```
+ + + + + + + + + + + + AUTOCORRELATIONS + + + + + + + + + + + + +
+ FOR DATA SERIES: CIGAR CONSUMPTION                                          +
+ DIFFERENCING: 1                          MEAN     = -2.31579                +
+ DATA COUNT =   95                        STD DEV =  69.8459                  +
  COEF   T-VAL  LAG _____0_____
 -0.33  -3.25   1            <<<<<[<<<<<0      ]
 -0.06  -0.50   2            [    <<0          ]
 -0.02  -0.16   3            [    <0           ]
 -0.31  -2.70   4            <<[<<<<<<<0       ]
  0.26   2.11   5            [     0>>>>>>>]>
 -0.01  -0.08   6            [     0           ]
  0.15   1.19   7            [     0>>>>>       ]
 -0.19  -1.45   8            [ <<<<<0           ]
 -0.07  -0.56   9            [    <<0           ]
 -0.10  -0.72  10            [    <<<0          ]
 -0.12  -0.93  11            [   <<<<0          ]
  0.64   4.74  12            [     0>>>>>>>]>>>>>>>>>>>>>
 -0.21  -1.30  13          [ <<<<<<<0           ]
  0.01   0.04  14          [        0           ]
 -0.07  -0.40  15          [      <<0           ]
 -0.29  -1.76  16          [ <<<<<<<<<0         ]
  0.32   1.85  17          [        0>>>>>>>>>>>]
 -0.07  -0.39  18          [      <<0           ]
  0.14   0.80  19          [        0>>>>>       ]
 -0.12  -0.68  20          [     <<<<0          ]
 -0.12  -0.67  21          [     <<<<0          ]
 -0.06  -0.33  22          [      <<0           ]
 -0.10  -0.55  23          [     <<<0           ]
  0.52   2.87  24          [        0>>>>>>>>>>>>]>>>>>
       CHI-SQUARED* =  153.96 FOR DF =   24

+ + + + + + + + + + + PARTIAL AUTOCORRELATIONS + + + + + + + + + + +
  COEF   T-VAL  LAG _____0_____
 -0.33  -3.25   1          <<<<<<<[<<<<<<<<<<0           ]
 -0.19  -1.83   2          [<<<<<<<<<0           ]
 -0.12  -1.18   3          [   <<<<<<0           ]
 -0.44  -4.25   4      <<<<<<<<<<<<<[<<<<<<<<<0           ]
 -0.08  -0.78   5          [    <<<<0           ]
 -0.07  -0.72   6          [    <<<<0           ]
  0.15   1.46   7          [        0>>>>>>>     ]
 -0.20  -1.94   8          [<<<<<<<<<0           ]
 -0.11  -1.05   9          [    <<<<<0           ]
 -0.32  -3.10  10          <<<<<<<[<<<<<<<<<<0           ]
 -0.46  -4.48  11      <<<<<<<<<<<<<<<[<<<<<<<<<0           ]
  0.32   3.12  12          [        0>>>>>>>>>>]>>>>>>
  0.17   1.64  13          [        0>>>>>>>>> ]
  0.13   1.29  14          [        0>>>>>>>  ]
  0.04   0.37  15          [        0>>        ]
 -0.03  -0.32  16          [      <<0           ]
  0.09   0.84  17          [        0>>>>       ]
 -0.04  -0.36  18          [      <<0           ]
  0.02   0.22  19          [        0>          ]
 -0.03  -0.27  20          [       <0           ]
  0.03   0.34  21          [        0>>         ]
  0.01   0.06  22          [        0           ]
 -0.12  -1.19  23          [    <<<<<<0           ]
  0.11   1.04  24          [        0>>>>>       ]
```

Figure C13.4 Estimated acf and pacf for the first differences in Figure C13.3.

```
+ + + + + + + + + + + + AUTOCORRELATIONS + + + + + + + + + + + + +
+ FOR DATA SERIES: CIGAR CONSUMPTION                                 +
+ DIFFERENCING: 12  1                    MEAN     = -.481928         +
+ DATA COUNT =   83                      STD DEV =   52.5161         +
  COEF   T-VAL  LAG                              0
 -0.57  -5.20    1    <<<<<<<<<<<<<<[<<<<<<<0             ]
  0.03   0.20    2                     [         0>        ]
  0.16   1.16    3                     [         0>>>>>    ]
 -0.17  -1.16    4                     [   <<<<<<0         ]
  0.00   0.00    5                     [         0         ]
  0.13   0.92    6                     [         0>>>>     ]
 -0.19  -1.27    7                     [   <<<<<<0         ]
  0.11   0.74    8                     [         0>>>>     ]
  0.05   0.31    9                     [         0>>       ]
 -0.15  -1.00   10                     [    <<<<<0         ]
  0.31   2.01   11                     [         0>>>>>>>>>>]
 -0.31  -1.95   12                     [<<<<<<<<<0         ]
  0.01   0.06   13             [                 0                 ]
  0.12   0.71   14             [                 0>>>>             ]
  0.03   0.16   15             [                 0>               ]
 -0.16  -0.96   16             [           <<<<<0               ]
  0.15   0.90   17             [                 0>>>>>           ]
 -0.02  -0.10   18             [                <0               ]
 -0.07  -0.41   19             [               <<0               ]
  0.14   0.81   20             [                 0>>>>>           ]
 -0.10  -0.56   21             [              <<<0               ]
 -0.09  -0.52   22             [              <<<0               ]
  0.14   0.81   23             [                 0>>>>>           ]
 -0.10  -0.59   24             [              <<<0               ]
     CHI-SQUARED* =    75.55 FOR  DF =   24

+ + + + + + + + + + + PARTIAL AUTOCORRELATIONS + + + + + + + + + + +
  COEF   T-VAL  LAG                              0
 -0.57  -5.20    1    <<<<<<<<<<<<<[<<<<<<<0             ]
 -0.44  -4.02    2    <<<<<<<[<<<<<<<0             ]
 -0.12  -1.09    3              [   <<<0             ]
 -0.14  -1.29    4              [   <<<<0             ]
 -0.23  -2.10    5              [<<<<<<<0             ]
 -0.07  -0.66    6              [    <<0             ]
 -0.20  -1.79    7              [<<<<<<<0             ]
 -0.17  -1.51    8              [  <<<<<<0             ]
 -0.05  -0.48    9              [     <<0             ]
 -0.16  -1.43   10              [   <<<<<0             ]
  0.27   2.48   11              [         0>>>>>>>]>
  0.05   0.42   12              [         0>>        ]
 -0.18  -1.65   13              [  <<<<<0             ]
 -0.26  -2.39   14              <[<<<<<<<0             ]
  0.07   0.67   15              [         0>>        ]
 -0.02  -0.23   16              [        <0             ]
 -0.14  -1.31   17              [   <<<<0             ]
  0.03   0.24   18              [         0>        ]
 -0.01  -0.06   19              [         0         ]
  0.10   0.91   20              [         0>>>       ]
  0.14   1.26   21              [         0>>>>>     ]
 -0.12  -1.06   22              [   <<<<0             ]
  0.07   0.65   23              [         0>>        ]
  0.01   0.10   24              [         0         ]
```

Figure C13.5 Estimated acf and pacf for the differenced series ($d = 1$ and $D = 1$).

```
+ + + + + + + + + +ECOSTAT UNIVARIATE B-J RESULTS+ + + + + + + + + +
+ FOR DATA SERIES:  CIGAR CONSUMPTION                                    +
+ DIFFERENCING:      12  1                          DF    = 81           +
+ AVAILABLE:         DATA = 83    BACKCASTS = 13    TOTAL = 96           +
+ USED TO FIND SSR:  DATA = 83    BACKCASTS = 13    TOTAL = 96           +
+ (LOST DUE TO PRESENCE OF AUTOREGRESSIVE TERMS:                0)       +
```

COEFFICIENT	ESTIMATE	STD ERROR	T-VALUE
THETA 1	0.824	0.058	14.18
THETA* 12	0.849	0.090	9.43

```
ADJUSTED RMSE =  29.2016    MEAN ABS % ERR =   5.34
        CORRELATIONS
        1       2
1    1.00
2   -0.21    1.00
```

```
++RESIDUAL ACF++
 COEF   T-VAL LAG _____O_____
-0.26  -2.40   1              <<<[<<<<<<<<<<O          ]
 0.01   0.12   2              [            O>           ]
 0.22   1.89   3              [            O>>>>>>>>>>>>]
-0.22  -1.84   4              [<<<<<<<<<<<O            ]
 0.00   0.03   5              [            O            ]
 0.25   1.98   6              [            O>>>>>>>>>>>>]>
-0.28  -2.07   7             [<<<<<<<<<<<<<O           ]
 0.08   0.57   8              [            O>>>>        ]
 0.14   0.99   9              [            O>>>>>>>     ]
-0.13  -0.89  10              [       <<<<<<O           ]
 0.12   0.84  11              [            O>>>>>>      ]
 0.05   0.37  12              [            O>>>         ]
-0.25  -1.74  13              [<<<<<<<<<<<<O            ]
 0.14   0.95  14              [            O>>>>>>>     ]
 0.03   0.23  15            [              O>>            ]
-0.20  -1.35  16            [        <<<<<<<<<O          ]
 0.20   1.28  17            [              O>>>>>>>>>>    ]
 0.06   0.40  18            [              O>>>           ]
-0.18  -1.14  19            [       <<<<<<<<<O           ]
 0.13   0.84  20            [              O>>>>>>>       ]
        CHI-SQUARED* =   56.44 FOR DF =   18
+RESIDUAL  PACF+
 COEF   T-VAL LAG _____O_____
-0.26  -2.40   1              <<<[<<<<<<<<<<O          ]
-0.06  -0.55   2              [       <<<O              ]
 0.23   2.05   3              [            O>>>>>>>>>>]>
-0.12  -1.10   4              [      <<<<<<O            ]
-0.10  -0.89   5              [      <<<<<O             ]
 0.22   2.02   6              [            O>>>>>>>>>>]>
-0.11  -1.03   7              [      <<<<<<O            ]
-0.05  -0.49   8              [       <<<O              ]
 0.10   0.87   9              [            O>>>>>       ]
 0.07   0.67  10              [            O>>>         ]
 0.04   0.35  11              [            O>>          ]
 0.00   0.03  12              [            O            ]
```

Figure C13.6 Estimation and diagnostic-checking results for model (C13.1).

warning value. An MA(1) seems to be the only reasonable alternative for the nonseasonal part of the model based on this estimated acf.

The pacf in Figure C13.5 could be interpreted as decaying, thus supporting the idea of an MA model (or perhaps a mixed model) for the nonseasonal element. But it could also be interpreted as displaying two spikes followed by a cutoff to zero which suggests an AR(2). Is the acf consistent with an AR(2)? We would have to stretch our imaginations to see a decaying pattern in the acf. If it decays, it certainly decays quite rapidly—so rapidly that it cuts off to zero at lag 2.

Now consider the residual acf and pacf in Figure C13.6. We could assume that (C13.1) is correct as far as it goes, but that it is incomplete. Then the residual acf and pacf for that model might tell us how it could be reformulated. In fact, this residual acf and pacf are confusing. The acf has a series of nondecaying residual autocorrelations at lags 1, 3, 4, 6, and 7, suggesting that additional differencing might be needed to induce a stationary mean. However, the quick decline to zero in the acf in Figure C13.5 is strong evidence that differencing beyond $d = 1$ and $D = 1$ is not required.

Because the residual autocorrelation at lag 1 in Figure C13.6 is significant despite the presence of a θ_1 coefficient in the model, we can consider adding a ϕ_1 coefficient. This brings us to an ARIMA$(1, 1, 1)(0, 1, 1)_{12}$ model. This possibility is reinforced by the spike at lag 1 followed by a cutoff to zero at lag 2 in the residual pacf. Unfortunately, it is difficult to see how adding a ϕ_1 coefficient to (C13.1) could account for the residual autocorrelations at lags 3, 4, 6, and 7. Furthermore, the estimated acf in Figure C13.5 does not decay from lag 1 as it should if an ARMA(1, 1) were appropriate for the differenced series; likewise, the residual acf in Figure C13.6 does not decay as it should if the residuals of model (C13.1) follow an AR(1) pattern with $\phi_1 < 0$.

We have been stymied in our effort to find a reasonable alternative to (C13.1). Perhaps the best we can do is adopt a strategy of estimating the seasonal element first, then use the residual acf and pacf to help identify the nonseasonal part of the model. We estimate an ARIMA$(0, 1, 0)(0, 1, 1)_{12}$ which has only a seasonal MA term (after differencing):

$$(1 - B)(1 - B^{12})\tilde{z}_t = (1 - \Theta_{12}B^{12})a_t \qquad (C13.2)$$

We expect this model to be incomplete since it contains no terms to account for the nonseasonal pattern.

Further estimation and diagnostic checking. Figure C13.7 shows the results of estimating and checking (C13.2). Remember our strategy is to remove the seasonal pattern so the residual acf and pacf give a better picture of the nonseasonal pattern.

The residual acf in Figure C13.7 decays in a wavelike pattern. The decay is rather irregular, but the contrast with the acf in Figure C13.5 is marked. The earlier acf cuts off sharply to zero after lag 1, suggesting an MA(1) for the nonseasonal part of the model. Now it appears that an AR or mixed

```
+ + + + + + + + + +ECOSTAT UNIVARIATE B-J RESULTS+ + + + + + + + + +
+ FOR DATA SERIES:  CIGAR CONSUMPTION                                 +
+ DIFFERENCING:     12  1                    DF    = 82               +
+ AVAILABLE:        DATA = 83   BACKCASTS = 12  TOTAL = 95            +
+ USED TO FIND SSR: DATA = 83   BACKCASTS = 12  TOTAL = 95            +
+ (LOST DUE TO PRESENCE OF AUTOREGRESSIVE TERMS:              0)       +

COEFFICIENT   ESTIMATE      STD ERROR     T-VALUE
   THETA* 12    0.879         0.078         11.23

   ADJUSTED RMSE =  41.2423    MEAN ABS % ERR =   7.66

++RESIDUAL ACF++
  COEF  T-VAL LAG _____0_____
 -0.60  -5.46   1      <<<<<<<<<<<<<<[<<<<<<<<O          ]
  0.03   0.21   2                 [          O>           ]
  0.25   1.77   3                 [          O>>>>>>>>     ]
 -0.26  -1.76   4                 [<<<<<<<<<O              ]
 -0.01  -0.10   5                 [          O             ]
  0.31   2.02   6                 [          O>>>>>>>>>>>1
 -0.34  -2.08   7              <[<<<<<<<<<O                ]
  0.11   0.64   8                 [          O>>>>               ]
  0.12   0.73   9                 [          O>>>>               ]
 -0.20  -1.14  10                 [    <<<<<<<O                ]
  0.12   0.66  11                 [          O>>>>               ]
  0.10   0.55  12                 [          O>>>                ]
 -0.26  -1.49  13                 [   <<<<<<<<<O               ]
  0.18   1.02  14                 [          O>>>>>>              ]
  0.06   0.31  15                 [          O>>                 ]
 -0.25  -1.35  16                 [    <<<<<<<O                ]
  0.20   1.06  17                 [          O>>>>>>>>           ]
  0.05   0.26  18                 [          O>>                 ]
 -0.22  -1.14  19                 [     <<<<<<O                ]
  0.21   1.11  20                 [          O>>>>>>>           ]
      CHI-SQUARED* =  103.00 FOR DF =  19
+RESIDUAL  PACF+
  COEF  T-VAL LAG _____0_____
 -0.60  -5.46   1      <<<<<<<<<<<<<<<[<<<<<<<<O          ]
 -0.51  -4.67   2        <<<<<<<<<<<[<<<<<<<<O          ]
 -0.05  -0.49   3                 [     <<O                ]
 -0.09  -0.79   4                 [    <<<O                ]
 -0.33  -3.00   5              <<<[<<<<<<<<O               ]
  0.06   0.56   6                 [          O>>                 ]
  0.00   0.03   7                 [          O                  ]
 -0.10  -0.94   8                 [     <<<O                ]
 -0.04  -0.41   9                 [      <O                 ]
  0.00   0.02  10                 [          O                  ]
  0.03   0.31  11                 [          O>                 ]
  0.13   1.16  12                 [          O>>>>               ]
```

Figure C13.7 Estimation and diagnostic-checking results for model (C13.2).

structure is more appropriate. The residual pacf in Figure C13.7 is consistent with an AR(2): it has two spikes before cutting off to zero at lags 3 and 4. The pacf spike at lag 5 is best ignored for now. Analysis of the residuals of (C13.2) leads to an ARIMA(2, 1, 0)(0, 1, 1)$_{12}$:

$$\left(1 - \phi_1 B - \phi_2 B^2\right)\left(1 - B\right)\left(1 - B^{12}\right)\tilde{z}_t = \left(1 - \Theta_{12} B^{12}\right)a_t \quad (C13.3)$$

The results for model (C13.3) appear in Figure C13.8. All estimated coefficients are statistically significant with absolute t-values well in excess

```
+ + + + + + + + + +ECOSTAT UNIVARIATE B-J RESULTS+ + + + + + + + + +
+ FOR DATA SERIES:   CIGAR CONSUMPTION                              +
+ DIFFERENCING:      12  1                          DF    = 80      +
+ AVAILABLE:         DATA = 83    BACKCASTS = 417  TOTAL = 500      +
+ USED TO FIND SSR:  DATA = 83    BACKCASTS = 415  TOTAL = 498      +
+ (LOST DUE TO PRESENCE OF AUTOREGRESSIVE TERMS:            2)      +

COEFFICIENT     ESTIMATE      STD ERROR     T-VALUE
PHI    1        -0.843         0.088         -9.56
PHI    2        -0.495         0.088         -5.63
THETA* 12        0.901         0.064         14.14

ADJUSTED RMSE =  28.2222    MEAN ABS % ERR =    4.91
       CORRELATIONS
        1      2     3
1     1.00
2     0.58   1.00
3     0.12   0.03   1.00

++RESIDUAL ACF++
COEF   T-VAL  LAG _____O_____
-0.16  -1.40   1      [     <<<<<<<<<<<<<<<<<<<O                        ]
-0.03  -0.29   2      [              <<<O                               ]
-0.19  -1.68   3      [    <<<<<<<<<<<<<<<<<<<<<O                       ]
-0.07  -0.57   4    [              <<<<<<<O                             ]
-0.02  -0.20   5    [                 <<O                               ]
 0.19   1.61   6    [                   O>>>>>>>>>>>>>>>>>>>>            ]
-0.19  -1.57   7    [   <<<<<<<<<<<<<<<<<<<<<O                          ]
 0.02   0.16   8                       O>>
 0.08   0.62   9                       O>>>>>>>>
 0.03   0.26  10                       O>>>
 0.06   0.46  11                       O>>>>>>
 0.00   0.00  12                       O
-0.21  -1.65  13          <<<<<<O<<<<<<<<<<<<<<<<<<O
 0.05   0.40  14                       O>>>>>
 0.01   0.09  15                       O>
-0.09  -0.68  16              <<<<<<<<<O
 0.15   1.17  17                       O>>>>>>>>>>>>>>>
 0.15   1.09  18                       O>>>>>>>>>>>>>>>
-0.10  -0.76  19             <<<<<<<<<<O
 0.00  -0.02  20                       O
       CHI-SQUARED* =   24.73 FOR DF =   17
```

Figure C13.8   Estimation and diagnostic-checking results for model (C13.3).

of 2.0. The model is invertible since $\hat{\Theta}_{12}$ satisfies the condition $|\hat{\Theta}_{12}| < 1$. It is also stationary because $\hat{\phi}_1$ and $\hat{\phi}_2$ meet the necessary conditions:

$$|\hat{\phi}_2| = 0.495 < 1$$

$$\hat{\phi}_2 + \hat{\phi}_1 = -0.495 - 0.843 = -1.338 < 1$$

$$\hat{\phi}_2 - \hat{\phi}_1 = -0.495 + 0.843 = 0.348 < 1$$

The residual acf indicates that the residuals of this model might satisfy the independence assumption. The t-values of the autocorrelations at lags 1, 3, 6, and 7 are around the warning level of 1.6, and the t-value at 13 exceeds the near-seasonal warning level of 1.25. Nevertheless, this residual acf is far better than the ones in Figures C13.6 and C13.7. The chi-squared statistic is insignificant at the 10% level, but just barely. We will experiment with some other models, but we might end up using (C13.3).

How could we improve on the preceding results? A sensible strategy is to try adding an MA coefficient at each lag where the residual autocorrelation t-value approaches or exceeds the relevant warning level. By all means we should adhere to the principle of parsimony by trying one additional MA coefficient at a time. We should also favor those lags suggested by experience and common sense. Therefore, lag 1, with a residual acf absolute t-value greater than the practical warning level of 1.25, is a preferred choice. Second would come lags 3, 6, and 13 because they are either short lags or possibly related to the seasonal pattern. Finally, we might consider lag 7.

To make a long story short, $\hat{\theta}_1$ is significant but it reduces $\hat{\phi}_2$ to insignificance and the residual acf is unacceptable. $\hat{\theta}_6$ is insignificant ($t = -1.15$). $\hat{\theta}_{13}$ is significant ($t = 2.19$), but the residual acf is improved only moderately. However, as shown in Figure C13.9, adding an MA term at lag 3 produces excellent results. Our model is

$$\left(1 - \phi_1 B - \phi_2 B^2\right)\left(1 - B\right)\left(1 - B^{12}\right)\tilde{z}_t = \left(1 - \Theta_{12} B^{12}\right)\left(1 - \theta_3 B^3\right)a_t$$

$$(C13.4)$$

Not only is $\hat{\theta}_3$ significant, the residual acf clears up beautifully. None of its t-values exceeds the relevant warning levels and the chi-squared statistic is quite insignificant. The residual plot in Figure C13.10 causes no problems.

The reader should check to see that all stationarity and invertibility conditions for this model are met. In doing so it may help to remember that, with a multiplicative model, the invertibility condition applies to $\hat{\Theta}_{12}$ and $\hat{\theta}_3$ separately.

```
+ + + + + + + + +ECOSTAT UNIVARIATE B-J RESULTS+ + + + + + + + +
+ FOR DATA SERIES:   CIGAR CONSUMPTION                            +
+ DIFFERENCING:      12  1                    DF    = 79          +
+ AVAILABLE:          DATA = 83   BACKCASTS = 417  TOTAL = 500    +
+ USED TO FIND SSR:  DATA = 83   BACKCASTS = 415  TOTAL = 498    +
+ (LOST DUE TO PRESENCE OF AUTOREGRESSIVE TERMS:              2)  +
```

```
COEFFICIENT     ESTIMATE      STD ERROR     T-VALUE
PHI    1         -1.017         0.059        -17.20
PHI    2         -0.837         0.074        -11.36
THETA  3          0.514         0.112          4.59
THETA* 12         0.890         0.073         12.24
```

```
ADJUSTED RMSE =  26.4654    MEAN ABS % ERR =   4.64
     CORRELATIONS
        1      2     3      4
 1    1.00
 2    0.57   1.00
 3   -0.24  -0.63  1.00
 4    0.06  -0.04 -0.01   1.00
```

```
++RESIDUAL ACF++
 COEF   T-VAL LAG _____0_____
-0.05  -0.46   1   [              <<<<<0                          ]
 0.06   0.55   2   [                  0>>>>>>                      ]
-0.04  -0.39   3   [              <<<<0                            ]
 0.02   0.20   4   [                  0>>                          ]
-0.03  -0.25   5   [               <<<0                            ]
 0.10   0.89   6   [                  0>>>>>>>>>>                  ]
-0.07  -0.63   7   [            <<<<<<<0                           ]
 0.00   0.01   8   [                  0                            ]
 0.03   0.25   9   [                  0>>>                         ]
 0.11   0.93  10   [                  0>>>>>>>>>>>                 ]
 0.03   0.23  11   [                  0>>>                         ]
-0.04  -0.39  12   [               <<<<0                           ]
-0.13  -1.10  13   [       <<<<<<<<<<<<<0                          ]
 0.03   0.26  14   [                  0>>>                         ]
 0.01   0.11  15   [                  0>                           ]
-0.06  -0.53  16   [              <<<<<0                           ]
 0.11   0.95  17   [                  0>>>>>>>>>>>                 ]
 0.13   1.08  18   [                  0>>>>>>>>>>>>>               ]
-0.09  -0.77  19   [          <<<<<<<<<0                           ]
-0.03  -0.26  20   [               <<<0                            ]
   CHI-SQUARED* =     9.80 FOR DF =   16
```

Figure C13.9 Estimation and diagnostic-checking results for model (C13.4).

Final comments.   These cigar-consumption data are unusually challenging to model. It is instructive to repeat several practical principles that
helped us find a satisfactory model:

1. When an initial model is grossly inadequate, return to the original acf
   and pacf with a fresh eye.
2. Model the seasonal element first to get a clearer picture of the
   nonseasonal pattern in the residual acf and pacf than is available in
   the initial acf and pacf.

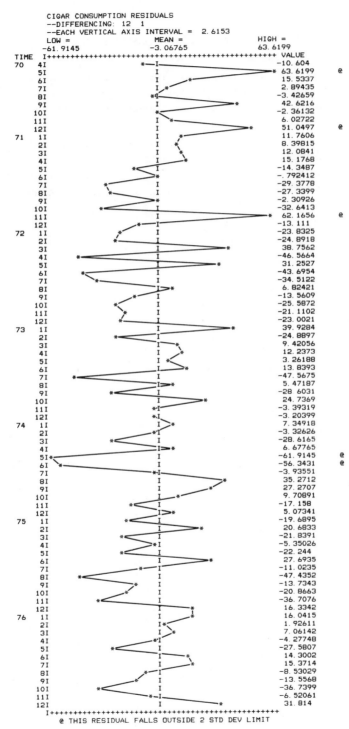

Figure C13.10 Residuals from model (C13.4).

521

3. Add coefficients to an existing model one at a time. That is, be guided throughout by the principle of parsimony.

4. Give preference to commonly occurring or common-sense models. Coefficients at lags 1, 2, and 3, seasonal lags, and near-seasonal lags (13 in this case) or fractional-seasonal lags (3 and 6 in this case) should receive the closest attention.

5. Ignore all but the short lags and the first few seasonal lags early in the identification stage. For example, in this case the significant autocorrelation at lag 11 in Figure C13.5 effectively disappears in all subsequent residual acf's without any coefficients being estimated at that lag.

CASE 14.   COLLEGE ENROLLMENT

College enrollment is typically larger in the fall semester than in the spring semester. New students usually start in the fall but some withdraw between semesters because of academic or social problems. Therefore, we expect that new-student enrollment will have a seasonal pattern with $s = 2$.

The data in Figure C14.1 confirm this expectation. They show new-student enrollment at a college starting with the fall semester, 1954. Visual analysis confirms that the second half of each year (the fall semester) invariably has higher enrollment than the subsequent spring semester. We might think of these observations as having been drawn from two different probability distributions, one (for fall semesters) with a larger mean and the other (for spring semesters) with a smaller mean. If the mean of this series changes each period, but such that observations separated by two time periods have a similar overall level, we expect that differencing by length two is required to induce a stationary mean.

The series appears to have a roughly constant variance. Aside from seasonal variation, its overall level does not change much. Therefore, transformations other than seasonal differencing do not seem necessary.

Identification.   Figure C14.2 provides evidence that the original series has a nonstationary mean: the estimated acf drops toward zero quite slowly. The t-values do not fall well below the 1.6 warning level until lag 9. Furthermore, the warning level of 1.25 is relevant at all lags that are multiples of 2 because these are seasonal lags; the t-values do not pierce this

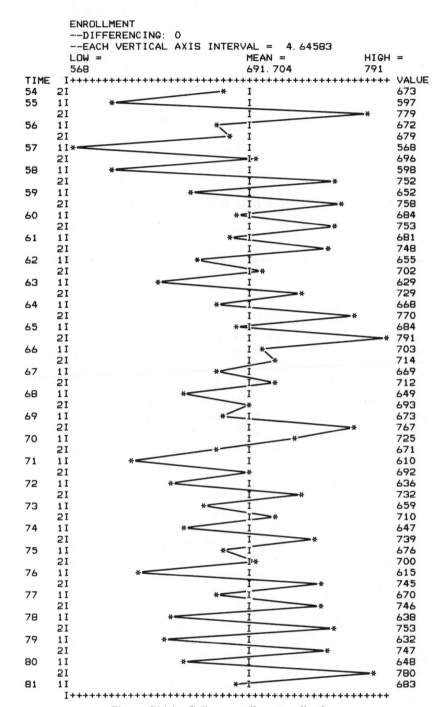

Figure C14.1 College-enrollment realization.

level until lag 12. With a nonstationary series the estimated pacf provides no additional useful information.

The seasonal autocorrelations (at lags $s = 2$, $2s = 4$, $3s = 6, \dots$) do not start their slow decay from a high level. As pointed out in Chapter 7, estimated autocorrelations need not be large to indicate a nonstationary mean. It is their slow decay that indicates nonstationarity.

If we were not alert to the strong seasonal pattern in these data, we might conclude that nonseasonal differencing is required because of the slowly decaying acf. However, visual analysis of the data and our knowledge of how seasonality of length two can affect an estimated acf lead us to conclude that seasonal differencing is needed.

As a point of information, Figure C14.3 shows the estimated acf after nonseasonal first differencing ($d = 1$). Its pattern is roughly the same as

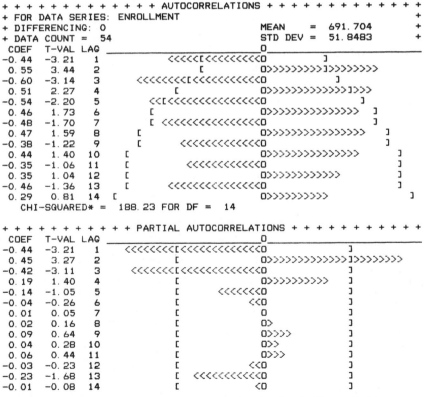

Figure C14.2 Estimated acf and pacf for the data in Figure C14.1.

that of the original acf. Most importantly, it does not drop rapidly to zero. A novice analyst, unfamiliar with the effects of nonstationary seasonal variation, might proceed with even further nonseasonal differencing by setting $d = 2$; this will not solve the fundamental problem. Differencing once ($D = 1$) by the seasonal length ($s = 2$) is needed: we must calculate the values $w_t = z_t - z_{t-2}$.

The acf after differencing once by length two (Figure C14.4) suggests the data now have a stationary mean: all the t-values after lag 3 are quite small. The highly significant spike at lag 2 indicates there is still a seasonal pattern in the data. This pattern appears to be of the MA variety since the acf cuts off to zero at lags 4, 6, 8, and 12. (Remember that lags s, $2s$, $3s$,... are the important ones when we want to identify the seasonal pattern. With $s = 2$, we must focus on lags 2, 4, 6,... .) The decay (rather than a cutoff) to zero, on the negative side, at lags 2, 4, and 6 in the pacf is consistent with an $MA(1)_2$ model for the seasonal part of the data.

The strong spike at lag 2 in the acf could influence the values of the adjacent autocorrelations at lags 1 and 3. Recall from Chapter 3 that estimated autocorrelation coefficients can be correlated (positively or negatively) with each other. Thus it is difficult in this case to identify the nonseasonal pattern. The seasonal pattern, in the form of a spike at lag 2, intrudes itself into the middle of the nonseasonal (short-lag) pattern. It is wise to estimate a purely seasonal model first, letting the residual acf and pacf guide us in identifying the nonseasonal pattern. We tentatively entertain an $ARIMA(0, 0, 0)(0, 1, 1)_2$ realizing that it could be incomplete since it

```
+ + + + + + + + + + + + AUTOCORRELATIONS + + + + + + + + + + + + +
+ FOR DATA SERIES: ENROLLMENT                                       +
+ DIFFERENCING: 1                          MEAN     =  .188679      +
+ DATA COUNT =   53                        STD DEV =  88.6911       +
  COEF   T-VAL LAG _____0_____
 -0.85  -6.16   1    <<<<<<<<<<<<<<<<<[<<<<<0      ]
  0.75   3.51   2                      [     0>>>>>>>>>]>>>>>>>>>>
 -0.79  -3.05   3    <<<<<<<<<[<<<<<<<<<<<<<0           ]
  0.75   2.48   4            [         0>>>>>>>>>>>>>>>>>>]>>>
 -0.71  -2.13   5      <<[<<<<<<<<<<<<<<<<<<<0              ]
  0.68   1.87   6            [         0>>>>>>>>>>>>>>>>>]
 -0.66  -1.70   7     [   <<<<<<<<<<<<<<<<<0                ]
  0.63   1.55   8     [         0>>>>>>>>>>>>>>>>>      ]
 -0.58  -1.37   9   [     <<<<<<<<<<<<<<<<0                    ]
  0.56   1.28  10   [         0>>>>>>>>>>>>>>>           ]
 -0.52  -1.15  11   [     <<<<<<<<<<<<<<0                      ]
  0.53   1.14  12   [         0>>>>>>>>>>>>>>>               ]
 -0.55  -1.15  13   [     <<<<<<<<<<<<<<<0                    ]
  0.50   1.03  14   [         0>>>>>>>>>>>>>                   ]
  CHI-SQUARED* =   377.53 FOR DF =   14
```

Figure C14.3 Estimated acf for the first differences ($d = 1$) of the data in Figure C14.1.

```
+ + + + + + + + + + + + + AUTOCORRELATIONS + + + + + + + + + + + + +
+ FOR DATA SERIES: ENROLLMENT                                       +
+ DIFFERENCING: 2                    MEAN    =  3.71154             +
+ DATA COUNT =  52                   STD DEV =  46.49               +
  COEF   T-VAL LAG _____0_____
  0.32    2.30   1              [              0>>>>>>>>>>>>>>]>>
 -0.42   -2.75   2    <<<<<[<<<<<<<<<<<<<<<<<<0              ]
 -0.36   -2.08   3        [<<<<<<<<<<<<<<<<<<<0              ]
 -0.11   -0.57   4        [          <<<<<0                 ]
 -0.01   -0.05   5        [              0                  ]
 -0.04   -0.24   6        [             <<0                 ]
 -0.03   -0.14   7        [             <0                  ]
  0.10    0.52   8        [              0>>>>>             ]
  0.15    0.81   9        [              0>>>>>>>>          ]
  0.15    0.81  10      [                0>>>>>>>>            ]
  0.14    0.70  11      [                0>>>>>>>            ]
 -0.07   -0.37  12      [             <<<<0                  ]
 -0.21   -1.08  13      [        <<<<<<<<<<<0                ]
     CHI-SQUARED* =   32.29 FOR DF =  13

+ + + + + + + + + + + PARTIAL AUTOCORRELATIONS + + + + + + + + + + +
  COEF   T-VAL LAG _____0_____
  0.32    2.30   1              [              0>>>>>>>>>>]>
 -0.58   -4.18   2    <<<<<<<<<<[<<<<<<<<<<<0              ]
  0.06    0.44   3              [          0>>                ]
 -0.32   -2.27   4        <[<<<<<<<<<<0                   ]
 -0.08   -0.54   5              [    <<<0                   ]
 -0.29   -2.08   6        [<<<<<<<<<<0                    ]
 -0.09   -0.62   7              [    <<<0                   ]
 -0.10   -0.75   8              [    <<<0                   ]
 -0.03   -0.21   9              [     <0                    ]
  0.13    0.96  10              [      0>>>>                 ]
  0.17    1.26  11              [      0>>>>>>               ]
 -0.01   -0.10  12              [      0                     ]
  0.16    1.14  13              [      0>>>>>                ]
```

Figure C14.4 Estimated acf and pacf for the first seasonal differences ($D = 1$) of the college-enrollment data.

has no nonseasonal element:

$$(1 - B^2)\tilde{z}_t = (1 - \Theta_2 B^2)a_t \qquad (C14.1)$$

Estimation and diagnostic checking. In Figure C14.5 we see that $\hat{\Theta}_2$ is statistically significant so we keep it in the model. Our major concern is to identify the nonseasonal pattern using the residual acf and pacf.

The most striking feature of the residual acf is the significant spike at lag 1 with a cutoff to zero at lag 2. This calls for the addition of a θ_1 coefficient to (C14.1). The acf spike at lag 3 also has an absolute t-value greater than the residual acf short-lag warning level of 1.25, but the principle of parsimony dictates adding one coefficient at a time.

Adding an MA term at lag 1 could reduce the residual autocorrelation at lag 3 to insignificance. In a multiplicative model, there are implicit coefficients at lags which are sums of the lags of the multiplied coefficients. In the

```
+ + + + + + + + + +ECOSTAT UNIVARIATE B-J RESULTS+ + + + + + + + + +
+ FOR DATA SERIES:   ENROLLMENT                                        +
+ DIFFERENCING:      2                              DF    = 51         +
+ AVAILABLE:         DATA = 52    BACKCASTS = 2     TOTAL = 54         +
+ USED TO FIND SSR:  DATA = 52    BACKCASTS = 2     TOTAL = 54         +
+ (LOST DUE TO PRESENCE OF AUTOREGRESSIVE TERMS:               0)     +

COEFFICIENT    ESTIMATE        STD ERROR      T-VALUE
  THETA* 2      0. 945          0. 060         15. 73

  ADJUSTED RMSE =  33. 9611    MEAN ABS % ERR =   4. 11

++RESIDUAL ACF++
 COEF   T-VAL LAG _____0_____
 0. 36   2. 63   1                   [              O>>>>>>>>>>>>>>>>]>>>>
 0. 09   0. 58   2            [        O>>>>>               ]
-0. 25  -1. 60   3            [  <<<<<<<<<<<<<<O            ]
-0. 14  -0. 87   4            [      <<<<<<<<O              ]
-0. 21  -1. 28   5            [  <<<<<<<<<<<<O              ]
-0. 12  -0. 69   6         [         <<<<<<O                  ]
-0. 09  -0. 49   7         [          <<<<O                   ]
 0. 11   0. 63   8         [              O>>>>>              ]
 0. 13   0. 76   9         [              O>>>>>>>            ]
 0. 16   0. 90  10         [              O>>>>>>>>           ]
 0. 12   0. 68  11         [              O>>>>>>             ]
-0. 04  -0. 21  12         [             <<O                  ]
-0. 20  -1. 07  13         [      <<<<<<<<<<O                 ]
       CHI-SQUARED* =   24. 18 FOR DF =   12
+RESIDUAL  PACF+
 COEF   T-VAL LAG _____0_____
 0. 36   2. 63   1                   [              O>>>>>>>>>>>>>>>>]>>>>
-0. 05  -0. 35   2                   [            <<O    ]
-0. 31  -2. 23   3            <[<<<<<<<<<<<<<<<O          ]
 0. 07   0. 51   4                   [         O>>>>      ]
-0. 17  -1. 26   5                   [  <<<<<<<<<O        ]
-0. 08  -0. 60   6                   [      <<<<O         ]
-0. 01  -0. 08   7                   [         <O         ]
 0. 09   0. 63   8                   [         O>>>>      ]
 0. 02   0. 16   9                   [         O>         ]
 0. 04   0. 31  10                   [         O>>        ]
 0. 10   0. 74  11                   [         O>>>>>     ]
-0. 14  -1. 02  12                   [   <<<<<<<O         ]
-0. 12  -0. 89  13                   [    <<<<<<O         ]
```

Figure C14.5 Estimation and diagnostic-checking results for model (C14.1).

present case, including θ_1 and Θ_2 in a multiplicative model creates an implicit coefficient at lag 3. This may be seen as follows. In backshift form the model we are entertaining is

$$(1 - B^2)\tilde{z}_t = (1 - \Theta_2 B^2)(1 - \theta_1 B)a_t \qquad (C14.2)$$

Expanding the RHS and writing it in common algebraic form (using the definitions of operator B) gives

$$(1 - B^2)\tilde{z}_t = a_t - \Theta_2 a_{t-2} - \theta_1 a_{t-1} + \theta_1 \Theta_2 a_{t-3}$$

This shows that (C14.2) contains an implicit term at lag 3 with the coefficient constrained to be $\theta_1\Theta_2$.

Aside from the implicit terms in multiplicative models, adding one appropriate coefficient can improve the overall fit sufficiently to reduce the sizes of residual autocorrelations at many different lags. (This is illustrated in a striking way in Case 13 where adding θ_3 clears up the residual acf

```
+ + + + + + + + + +ECOSTAT UNIVARIATE B-J RESULTS+ + + + + + + + +
+ FOR DATA SERIES:   ENROLLMENT                                    +
+ DIFFERENCING:      2                           DF   = 50         +
+ AVAILABLE:         DATA = 52   BACKCASTS = 3   TOTAL = 55         +
+ USED TO FIND SSR:  DATA = 52   BACKCASTS = 3   TOTAL = 55         +
+ (LOST DUE TO PRESENCE OF AUTOREGRESSIVE TERMS:            0)      +

COEFFICIENT    ESTIMATE        STD ERROR    T-VALUE
  THETA  1      -0.454          0.144        -3.15
  THETA* 2       0.677          0.126         5.39

ADJUSTED RMSE =  33.5676    MEAN ABS % ERR =   3.89
        CORRELATIONS
         1      2
1      1.00
2      0.50   1.00

++RESIDUAL ACF++
  COEF   T-VAL LAG _____0_____
  0.03    0.20   1              [                O>          ]
  0.01    0.11   2              [                O>          ]
 -0.26   -1.88   3              [<<<<<<<<<<<<<<O             ]
 -0.07   -0.45   4              [            <<<O            ]
 -0.13   -0.85   5              [         <<<<<<O            ]
 -0.02   -0.14   6            [              <O             ]
 -0.06   -0.39   7            [            <<<O             ]
  0.15    1.03   8            [              O>>>>>>>>       ]
  0.12    0.80   9            [              O>>>>>>         ]
  0.14    0.91  10            [              O>>>>>>>        ]
  0.12    0.74  11            [              O>>>>>>         ]
  0.00    0.00  12            [              O              ]
 -0.18   -1.14  13            [         <<<<<<<<<O          ]
        CHI-SQUARED* =  12.58 FOR DF =  11
+RESIDUAL   PACF+
  COEF   T-VAL LAG _____0_____
  0.03    0.20   1              [                O>          ]
  0.01    0.10   2              [                O>          ]
 -0.26   -1.89   3              [<<<<<<<<<<<<<<O             ]
 -0.06   -0.40   4            [            <<<O             ]
 -0.12   -0.88   5            [          <<<<<<O            ]
 -0.09   -0.66   6            [          <<<<<O             ]
 -0.10   -0.71   7            [          <<<<<O             ]
  0.09    0.68   8            [              O>>>>>         ]
  0.09    0.62   9            [              O>>>>          ]
  0.09    0.68  10            [              O>>>>          ]
  0.19    1.37  11            [              O>>>>>>>>>      ]
  0.08    0.56  12            [              O>>>>          ]
 -0.09   -0.65  13            [          <<<<<O             ]
```

Figure C14.6 Estimation and diagnostic-checking results for model (C14.2).

dramatically.) Thus we modify (C14.1) by adding one MA coefficient at lag 1.

Further results.   Model (C14.2) gives good results (see Figure C14.6). Each estimated coefficient is significant and each satisfies its respective invertibility condition. The former residual acf spike at lag 1 (Figure C14.5) has effectively disappeared, and the chi-squared statistic has been cut in half.

There remains a residual acf spike at lag 3 with an absolute t-value greater than 1.25. Adding an MA term at lag 3 is appropriate. We choose an MA term over an AR term since the residual acf cuts off to zero after lag 3 rather than decaying.

Using the standard multiplicative form, our ARIMA$(0, 0, 3)(0, 1, 1)_2$ model appears in backshift form as:

$$(1 - B^2)\tilde{z}_t = (1 - \Theta_2 B^2)(1 - \theta_1 B - \theta_3 B^3)a_t \qquad (C14.3)$$

```
+ + + + + + + + + +ECOSTAT UNIVARIATE B-J RESULTS+ + + + + + + + + +
+ FOR DATA SERIES:   ENROLLMENT                                    +
+ DIFFERENCING:      2                          DF    = 49         +
+ AVAILABLE:         DATA = 52    BACKCASTS = 5  TOTAL = 57         +
+ USED TO FIND SSR:  DATA = 52    BACKCASTS = 5  TOTAL = 57         +
+ (LOST DUE TO PRESENCE OF AUTOREGRESSIVE TERMS:            0)      +

COEFFICIENT    ESTIMATE       STD ERROR      T-VALUE
  THETA   1    -0. 434          0. 126         -3. 45
  THETA   3     0. 255          0. 125          2. 05
  THETA*  2     0. 799          0. 092          8. 64

  ADJUSTED RMSE =   32. 2218    MEAN ABS % ERR =    3. 74
        CORRELATIONS
          1      2      3
  1     1. 00
  2    -0. 06  1. 00
  3     0. 19  0. 13  1. 00

++RESIDUAL ACF++
  COEF    T-VAL LAG _____O_____
  0. 00    0. 03   1                         O
  0. 09    0. 67   2                         O>>>>>>>>>
 -0. 05   -0. 39   3                   <<<<<O
 -0. 14   -0. 97   4           <<<<<<<<<<<<<<O
 -0. 10   -0. 73   5              <<<<<<<<<<O
 -0. 01   -0. 06   6                        <O
 -0. 12   -0. 84   7            <<<<<<<<<<<<O
  0. 17    1. 17   8                         O>>>>>>>>>>>>>>>>>
  0. 11    0. 71   9                         O>>>>>>>>>>>
  0. 07    0. 45  10                         O>>>>>>>
  0. 15    0. 97  11                         O>>>>>>>>>>>>>>>
 -0. 01   -0. 06  12                        <O
 -0. 18   -1. 17  13            <<<<<<<<<<<<<<<<<<O
      CHI-SQUARED* =     10. 10 FOR DF =   10
```

Figure C14.7   Estimation and diagnostic-checking results for model (C14.3).

This form gives a complicated coefficient at lag 3: it is a combination of the explicit coefficient θ_3 and the implicit coefficient $\theta_1 \Theta_2$. This can be seen by expanding the RHS of (C14.3) and expressing it in common algebraic form:

$$(1 - B^2)\tilde{z}_t = a_t - \theta_1 a_{t-1} - \Theta_2 a_{t-2} + (\theta_1 \Theta_2 - \theta_3) a_{t-3}$$

It may be preferable to estimate θ_3 freely by considering this additive model:

$$(1 - B^2)\tilde{z}_t = (1 - \theta_1 B - \Theta_2 B^2 - \theta_3 B^3) a_t \qquad \text{(C14.4)}$$

See Chapter 11 for a discussion of additive models. There is no *a priori* reason to choose (C14.3) or (C14.4). If both produce acceptable results we may simply choose the one giving the best fit.

A comparison of the estimation and diagnostic-checking results in Figures C14.7 and C14.8 shows that model (C14.4) is slightly superior: its

```
+ + + + + + + + + +ECOSTAT UNIVARIATE B-J RESULTS+ + + + + + + + + +
+ FOR DATA SERIES:   ENROLLMENT                                      +
+ DIFFERENCING:      2                            DF     = 49        +
+ AVAILABLE:             DATA = 52    BACKCASTS = 3   TOTAL = 55     +
+ USED TO FIND SSR: DATA = 52    BACKCASTS = 3   TOTAL = 55         +
+ (LOST DUE TO PRESENCE OF AUTOREGRESSIVE TERMS:              0)     +

COEFFICIENT    ESTIMATE      STD ERROR     T-VALUE
  THETA   1     -0.412         0.121        -3.40
  THETA   2      0.812         0.086         9.41
  THETA   3      0.520         0.124         4.21

  ADJUSTED RMSE  =  31.4406    MEAN ABS % ERR =   3.69
       CORRELATIONS
        1      2      3
  1   1.00
  2   0.13   1.00
  3  -0.70   0.14   1.00

++RESIDUAL ACF++
  COEF   T-VAL LAG _____0_____
  0.01    0.11   1                                  0>
  0.06    0.44   2                                  0>>>>>>
 -0.15   -1.07   3                 <<<<<<<<<<<<<<<<<0
 -0.15   -1.06   4                 <<<<<<<<<<<<<<<<<0
 -0.04   -0.27   5                             <<<<0
 -0.14   -0.96   6                 <<<<<<<<<<<<<<<<<0
 -0.02   -0.16   7                             <<0
  0.06    0.41   8                                  0>>>>>>
  0.12    0.78   9                                  0>>>>>>>>>>>>
  0.08    0.51  10                                  0>>>>>>>>
  0.08    0.53  11                                  0>>>>>>>>
  0.02    0.10  12                                  0>>
 -0.21   -1.40  13            <<<<<<<<<<<<<<<<<<<<<<0
  CHI-SQUARED* =   9.36 FOR DF =   10
```

Figure C14.8 Estimation and diagnostic-checking results for model (C14.4).

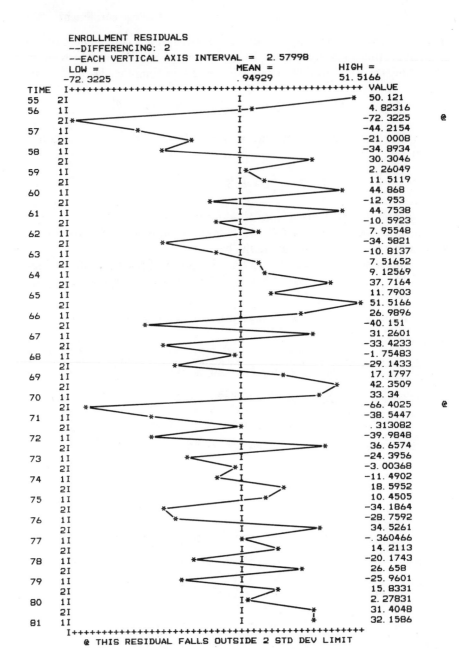

Figure C14.9 Residuals from model (C14.4).

Table C14.1   Forecasts from model C14.4

Time		Forecast Values	80% Confidence Limits		Future Observed Values	Percent Forecast Errors
			Lower	Upper		
81	2	766.5591	726.3151	806.8031	n.a.[a]	n.a.
82	1	640.5447	597.0222	684.0672	n.a.	n.a.
	2	749.8210	705.6451	793.9969	n.a.	n.a.
83	1	640.5447	596.1527	684.9367	n.a.	n.a.

[a] n.a. = not available.

RMSE and chi-squared statistic are both smaller. Nevertheless, both models are acceptable and either one could be used.

The residual plot for (C14.4) in Figure C14.9 does not point to any problems. The forecasts from (C14.4) in Table C14.1 show that it mimics the strong seasonal pattern in the data quite well.

CASE 15.   EXPORTS

Box and Jenkins [1] suggest that about 50 observations are required to build an ARIMA model. However, Jenkins [34, p. 64] presents an example of a (multivariate) ARIMA model based on only 14 observations. The key is not necessarily the absolute number of observations but rather the amount of "statistical noise" in the data. If the noise factor (the variance of the random shocks) is small, it may be possible to extract enough information from relatively few observations to construct a useful ARIMA model.

In this case we build an ARIMA model using only 42 observations. Our purpose is not to contradict Box and Jenkins' guideline regarding the minimum number of observations. In fact, it should be emphasized that an analyst should actively consider alternatives to an ARIMA model when less than 50 observations are available. Nevertheless, we show in this case that it is possible to develop an acceptable ARIMA model even when the number of observations seems barely adequate.

In Chapter 7 we discuss the idea of a deterministic trend. In this case study we illustrate how a model with a deterministic trend element can sometimes provide a useful representation of a data series.

The data, shown in Figure C15.1,* represent quarterly U.S. exports to the European Economic Community (EEC) from the first quarter of 1958 through the second quarter of 1968. The upward trend of the data suggests that the mean is not stationary; nonseasonal differencing may therefore be required.

---

*These data are found in Table 6, p. 89, of *U.S. Exports and Imports*, published by the U.S. Census Bureau.

534

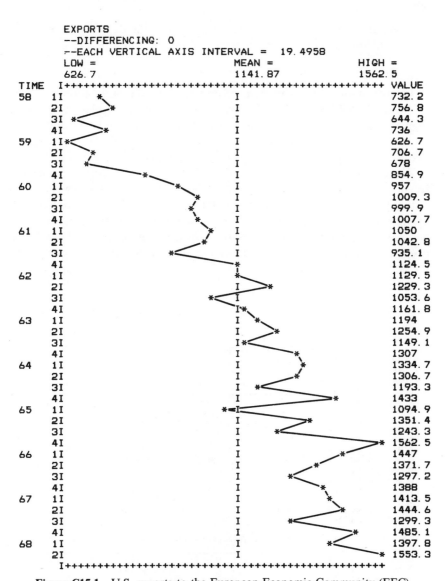

```
      EXPORTS
      --DIFFERENCING: 0
      --EACH VERTICAL AXIS INTERVAL =   19.4958
      LOW =                        MEAN =              HIGH =
      626.7                        1141.87             1562.5
TIME  I+++++++++++++++++++++++++++++++++++++++++++++++++  VALUE
58   1I        *              I                               732.2
     2I      *-*              I                               756.8
     3I   *-*                 I                               644.3
     4I      *-*              I                               736
59   1I  *                    I                               626.7
     2I    *-*                I                               706.7
     3I    *                  I                               678
     4I            *-*        I                               854.9
60   1I                *-*    I                               957
     2I                  *-*  I                               1009.3
     3I                 *-*   I                               999.9
     4I                 *-*   I                               1007.7
61   1I                  *-*  I                               1050
     2I                 *-*   I                               1042.8
     3I           *-*         I                               935.1
     4I              *        I                               1124.5
62   1I                       *                               1129.5
     2I                  I    *                               1229.3
     3I          *-*          I                               1053.6
     4I              I*       I                               1161.8
63   1I              I  *-*   I                               1194
     2I              I    *   I                               1254.9
     3I              I*       I                               1149.1
     4I              I      *-* I                             1307
64   1I              I        *-* I                           1334.7
     2I              I         *  I                           1306.7
     3I              I  *-*    I                               1193.3
     4I              I              *                          1433
65   1I          *-=I-*       I                               1094.9
     2I              I           *-*                          1351.4
     3I              I   *-*    I                              1243.3
     4I              I                    *                   1562.5
66   1I              I                *-* I                    1447
     2I              I          *-*      I                    1371.7
     3I              I     *-*           I                    1297.2
     4I              I         *-*       I                    1388
67   1I              I          *-*      I                    1413.5
     2I              I            *-*    I                    1444.6
     3I              I     *-*           I                    1299.3
     4I              I               *-* I                    1485.1
68   1I              I        *-*        I                    1397.8
     2I+++++++++++++++++++++++++++++++++*+++++++++++++++       1553.3
      I+++++++++++++++++++++++++++++++++++++++++++++++++
```

Figure C15.1 U.S. exports to the European Economic Community (EEC).

There also appears to be a seasonal pattern in the data so we expect to see significant autocorrelations at the seasonal lags (multiples of 4).

Identification.   The estimated acf and pacf for the original data appear in Figure C15.2. Although Box and Jenkins suggest calculating no more than $n/4$ autocorrelations, we stretch that guideline slightly by finding 12 autocorrelations. This should provide a clearer picture of the seasonal pattern by giving us three seasonal autocorrelations (at lags 4, 8, and 12).

The estimated acf decays to zero rather slowly suggesting that nonseasonal first differencing may be needed to induce a stationary mean. This is not surprising since inspection of the data shows that the data are trending upward. It is possible, however, that the seasonal pattern is obscuring the nonseasonal pattern: a large, positive autocorrelation at lag 4 could be positively correlated with the surrounding autocorrelations, thus preventing the acf from decaying rapidly.

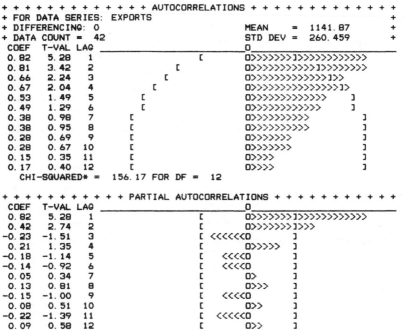

Figure C15.2   Estimated acf and pacf for the realization in Figure C15.1.

We should get a better picture of the seasonal pattern by examining the acf and pacf of the first differences. The acf in Figure C15.3 provides evidence that there is a nonstationary seasonal pattern in the data: the autocorrelations at lags 4, 8, and 12 have t-values greater than 1.25 and they do not decay at all. The autocorrelations at lags 4 and 8 are equal in value and the coefficient at lag 12 is larger than those at lags 4 and 8. Apparently seasonal first differencing (length four) is needed. This is supported by inspection of the first differences plotted in Figure C15.4. For example, all the third-quarter observations lie below the calculated mean, while all but one of the fourth-quarter observations lie above the mean.

The acf and pacf after seasonal differencing but without nonseasonal differencing (Figure C15.5) indicate that a seasonal element remains in the data: the negative acf spike at lag 4 has a large absolute t-value. The acf cuts off to zero at lags 8 and 12, suggesting that the remaining seasonal pattern is $MA(1)_4$. This is confirmed by the tendency of the estimated pacf to decay on the negative side at lags 4, 8, and 12.

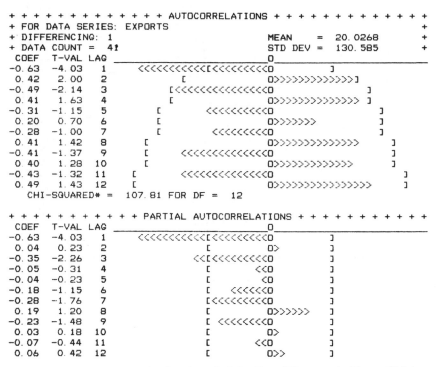

Figure C15.3 Estimated acf and pacf of the first differences in Figure C15.4.

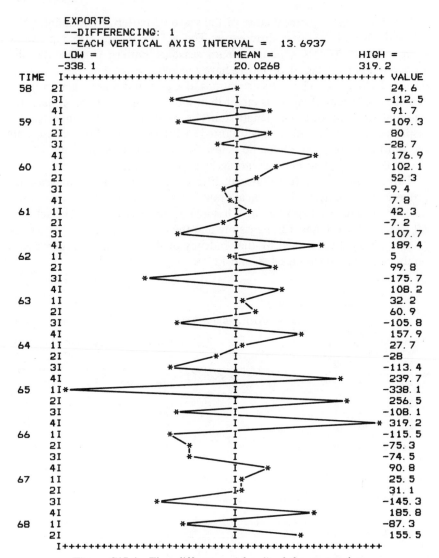

Figure C15.4 First differences ($d = 1$) of the export data.

```
+ + + + + + + + + + + + AUTOCORRELATIONS + + + + + + + + + + + + + +
+ FOR DATA SERIES: EXPORTS                                              +
+ DIFFERENCING: 4                            MEAN    = 75.4263          +
+ DATA COUNT =   38                          STD DEV = 121.859          +
   COEF  T-VAL LAG                              0
   0.35   2.17   1               [              0>>>>>>>>>>>>>>>]>>
   0.18   1.01   2              [               0>>>>>>>>>             ]
  -0.22  -1.19   3              [     <<<<<<<<<<<0                     ]
  -0.48  -2.51   4     <<<<[<<<<<<<<<<<<<<<<<<<<<0                         ]
  -0.12  -0.53   5             [          <<<<<<0                         ]
  -0.17  -0.76   6             [       <<<<<<<<<0                         ]
   0.24   1.06   7             [              0>>>>>>>>>>>>            ]
   0.12   0.50   8           [                 0>>>>>>                    ]
   0.14   0.61   9           [                 0>>>>>>>                   ]
   0.06   0.26  10           [                 0>>>                       ]
  -0.19  -0.80  11           [         <<<<<<<<<0                         ]
  -0.03  -0.11  12           [                <0                         ]
      CHI-SQUARED* =    28.02 FOR DF =   12

+ + + + + + + + + + PARTIAL AUTOCORRELATIONS + + + + + + + + + + + +
   COEF  T-VAL LAG                              0
   0.35   2.17   1               [              0>>>>>>>>>>>>>>>]>>
   0.07   0.42   2              [               0>>>                  ]
  -0.35  -2.15   3            <[<<<<<<<<<<<<<<<<<0                     ]
  -0.41  -2.55   4     <<<<<[<<<<<<<<<<<<<<<<<<<<0                     ]
   0.35   2.16   5              [               0>>>>>>>>>>>>>>>]>
  -0.12  -0.75   6             [          <<<<<<0                     ]
   0.10   0.59   7             [                0>>>>>                 ]
  -0.18  -1.12   8             [        <<<<<<<<0                     ]
   0.21   1.28   9             [                0>>>>>>>>>            ]
  -0.11  -0.65  10             [           <<<<<0                     ]
  -0.07  -0.43  11             [              <<<0                    ]
   0.00  -0.02  12             [                0                     ]
```

Figure C15.5 Estimated acf and pacf of the seasonal differences ($D = 1$) of the export data.

This acf has a surprising characteristic: it decays rapidly to zero, suggesting that nonseasonal first differencing is not required. This is unexpected considering the apparent upward trend of the data in Figure C15.1. Figure C15.6 is a plot of the seasonally differenced data. These numbers seem to fluctuate around a fixed mean, reinforcing the evidence from the acf that only seasonal differencing is needed to induce a stationary mean.

It is difficult at this point to clearly identify an appropriate model for the nonseasonal part of the data. The acf in Figure C15.5 suggests an MA(1) because of the spike at lag 1 with a cutoff to zero at lag 2. However, the large autocorrelation at lag 4 could be correlated with the surrounding autocorrelations, distorting our picture of the nonseasonal pattern. Therefore, we attempt to remove the seasonal pattern first and let the residual acf suggest an appropriate form for the nonseasonal part of the model. We choose an ARIMA$(0, 1, 1)_4$ model for the seasonal part of the data:

$$(1 - B^4)\tilde{z}_t = (1 - \Theta_4 B^4)a_t \tag{C15.1}$$

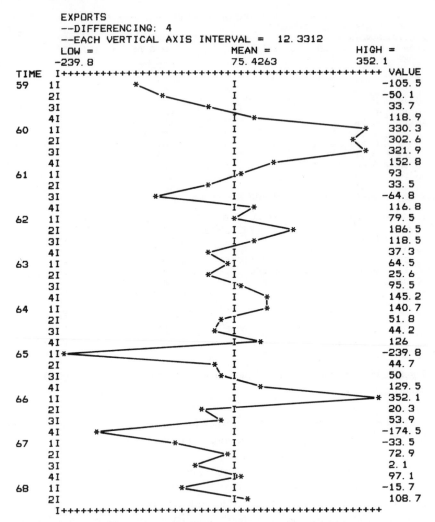

```
EXPORTS
--DIFFERENCING: 4
--EACH VERTICAL AXIS INTERVAL =   12. 3312
     LOW =                    MEAN =              HIGH =
     -239. 8                  75. 4263            352. 1
TIME  I+++++++++++++++++++++++++++++++++++++++++++++++++++  VALUE
59   1I              *            I                          -105. 5
     2I                 *         I                          -50. 1
     3I                   *       I                           33. 7
     4I                           I  *                        118. 9
60   1I                           I                    *      330. 3
     2I                           I                   *       302. 6
     3I                           I                    *      321. 9
     4I                           I    *                      152. 8
61   1I                         I.*                            93
     2I                       *   I                            33. 5
     3I           *               I                           -64. 8
     4I                        I-*                            116. 8
62   1I                         *I                             79. 5
     2I                           I        *                  186. 5
     3I                         I.*                           118. 5
     4I                      *    I                            37. 3
63   1I                         *I                             64. 5
     2I                       *   I                             25. 6
     3I                         I*                             95. 5
     4I                           I    *                      145. 2
64   1I                           I     *                     140. 7
     2I                       *-I                               51. 8
     3I                       *-I                               44. 2
     4I                        I-*                             126
65   1I*                         I                            -239. 8
     2I                       *   I                            44. 7
     3I                       *-I                               50
     4I                           I   *                        129. 5
66   1I                           I                        * 352. 1
     2I                      *    I                            20. 3
     3I                        *  I                            53. 9
     4I            *              I                           -174. 5
67   1I                     *     I                           -33. 5
     2I                         *I                             72. 9
     3I                      *    I                             2. 1
     4I                          I*                            97. 1
68   1I                      *    I                           -15. 7
     2I                          I-*                          108. 7
     I+++++++++++++++++++++++++++++++++++++++++++++++++++++
```

Figure C15.6 Seasonal differences ($D = 1$) of the export data.

Estimation and diagnostic checking.   Estimation and diagnostic-checking results for (C15.1) in Figure C15.7 are discouraging. The estimated coefficient is insignificant and the residual acf has a large spike remaining at lag 4. This is puzzling since the acf and pacf in Figure C15.5 provide strong evidence that a Θ_4 coefficient is needed.

Inspection of the seasonally differenced data in Figure C15.6 provides a clue. The mean of the series is about 75; this could be significantly different

```
+ + + + + + + + + +ECOSTAT UNIVARIATE B-J RESULTS+ + + + + + + + + +
+ FOR DATA SERIES:   EXPORTS                                        +
+ DIFFERENCING:      4                            DF    = 37        +
+ AVAILABLE:             DATA = 38   BACKCASTS = 4  TOTAL = 42        +
+ USED TO FIND SSR:  DATA = 38   BACKCASTS = 4  TOTAL = 42        +
+ (LOST DUE TO PRESENCE OF AUTOREGRESSIVE TERMS:            0)       +

COEFFICIENT    ESTIMATE        STD ERROR      T-VALUE
   THETA* 4      0.038           0.166         0.23

   ADJUSTED RMSE  =   145.054     MEAN ABS % ERR  =   10.00

++RESIDUAL ACF++
   COEF  T-VAL LAG _____0_____
   0.35   2.17   1              [              0>>>>>>>>>>>>>>>>>]>>
   0.19   1.04   2            [              0>>>>>>>>>          ]
  -0.21  -1.12   3            [     <<<<<<<<<<0                ]
  -0.46  -2.40   4   <<<[<<<<<<<<<<<<<<<<<<<<<0                    ]
  -0.11  -0.48   5   [            <<<<<0                          ]
  -0.17  -0.76   6   [          <<<<<<<<0                          ]
   0.23   1.04   7   [              0>>>>>>>>>>>>              ]
   0.10   0.44   8 [              0>>>>>                          ]
   0.14   0.60   9 [              0>>>>>>>>                        ]
   0.07   0.28  10 [              0>>>                            ]
   CHI-SQUARED*  =    24.35 FOR DF  =    9
```

Figure C15.7   Estimation and diagnostic-checking results for model (C15.1).

from zero when the data range from only -239.8 to 352.1. When the mean of a differenced series is nonzero, it implies that the data contain a deterministic trend. (This topic is introduced in Chapter 7.) This conclusion is reinforced by the behavior of the original data in Figure C15.1. We have not accounted for the upward trend in that series with nonseasonal differencing; the acf in Figure C15.5 drops quickly to zero, implying that nonseasonal differencing is not needed. Yet something must appear in the model to explain the upward drift of the original data. Apparently, a deterministic trend is called for. As discussed in Chapter 7 a deterministic trend may be included in a model by expressing the differenced variable w_t in deviations from its mean μ_w rather than assuming (as we usually do with data outside the physical sciences) that $\mu_w = 0$. Thus an alternative to (C15.1) is

$$\tilde{w}_t = \left(1 - \Theta_4 B^4\right) a_t \qquad (C15.2)$$

where $\tilde{w}_t$ is the seasonal differences of z_t expressed in deviations from the mean μ_w.

(C15.2) gives good results. Figure C15.8 shows that both $\hat{\Theta}_4$ and the constant are significant. (The estimated constant $\hat{C}$ is equal to the estimated mean $\hat{\mu}_w$ since there are no AR terms present.) The residual acf shows no evidence of a remaining seasonal pattern.

```
+ + + + + + + + + +ECOSTAT UNIVARIATE B-J RESULTS+ + + + + + + + + +
+ FOR DATA SERIES:   EXPORTS                                      +
+ DIFFERENCING:      4                          DF    = 36        +
+ AVAILABLE:              DATA = 38   BACKCASTS = 4   TOTAL = 42   +
+ USED TO FIND SSR:  DATA = 38   BACKCASTS = 4   TOTAL = 42        +
+ (LOST DUE TO PRESENCE OF AUTOREGRESSIVE TERMS:              0)   +
```

COEFFICIENT	ESTIMATE	STD ERROR	T-VALUE
THETA* 4	0.895	0.151	5.91
CONSTANT	82.1545	5.84458	14.0565
MEAN	82.1545	5.84458	14.0565

```
ADJUSTED RMSE =  100.583   MEAN ABS % ERR =   7.58
     CORRELATIONS
         1     2
  1    1.00
  2    0.81   1.00
```

```
++RESIDUAL ACF++
 COEF   T-VAL LAG _____0_____
 0.48   2.95   1                    [     0>>>>>>>>>>>>>>>>>>>]>>>>>>>>>
 0.38   1.93   2          [               0>>>>>>>>>>>>>>>>>>>>>]
 0.12   0.55   3   [                       0>>>>>>                      ]
 0.05   0.22   4   [                       0>>                         ]
 0.15   0.68   5   [                       0>>>>>>>                     ]
 0.05   0.22   6   [                       0>>                         ]
 0.19   0.88   7   [                       0>>>>>>>>>>                  ]
 0.04   0.19   8   [                       0>>                         ]
 0.06   0.26   9   [                       0>>>                        ]
 0.06   0.28  10   [                       0>>>                        ]
       CHI-SQUARED* =    19.53 FOR DF =  8
+RESIDUAL  PACF+
 COEF   T-VAL LAG _____0_____
 0.48   2.95   1               [             0>>>>>>>>>>>>>>>>>>>]>>>>>>>>>
 0.19   1.18   2               [             0>>>>>>>>>>          ]
-0.16  -0.99   3               [     <<<<<<<<0                    ]
-0.03  -0.17   4               [            <0                    ]
 0.23   1.41   5               [             0>>>>>>>>>>>          ]
-0.10  -0.59   6               [       <<<<<0                     ]
 0.14   0.87   7               [             0>>>>>>>             ]
-0.09  -0.54   8               [        <<<<0                     ]
-0.01  -0.07   9               [            <0                    ]
 0.09   0.55  10               [             0>>>>                ]
```

Figure C15.8 Estimation and diagnostic-checking results for model (C15.2).

Further identification. The residual acf in Figure C15.8 provides information for identifying the nonseasonal pattern. As pointed out in Chapter 12, filtering out the seasonal pattern sometimes permits the residual acf to give a better indication of the remaining nonseasonal pattern. In this case the residual acf in Figure C15.8 calls for a different nonseasonal structure than is suggested in the acf in Figure C15.5. The latter has a single spike at lag 1, implying that an MA(1) describes the nonseasonal pattern. But the residual autocorrelations at lags 1 and 2 in Figure C15.8 both have *t*-values larger than the practical warning level of 1.25; the acf then cuts off to zero

```
+ + + + + + + + +ECOSTAT UNIVARIATE B-J RESULTS+ + + + + + + + +
+ FOR DATA SERIES:   EXPORTS                                      +
+ DIFFERENCING:      4                          DF    = 35        +
+ AVAILABLE:         DATA = 38   BACKCASTS = 5   TOTAL = 43        +
+ USED TO FIND SSR:  DATA = 38   BACKCASTS = 5   TOTAL = 43        +
+ (LOST DUE TO PRESENCE OF AUTOREGRESSIVE TERMS:           0)     +


COEFFICIENT    ESTIMATE       STD ERROR      T-VALUE
  THETA   1    -0.331          0.159          -2.08
  THETA* 4      0.871          0.150           5.81
  CONSTANT     81.6018         5.61943        14.5214


  MEAN         81.6018         5.61943        14.5214


  ADJUSTED RMSE = 92.2867     MEAN ABS % ERR =    6.18
        CORRELATIONS
           1      2      3
  1     1.00
  2     0.02   1.00
  3     0.03   0.64   1.00


++RESIDUAL ACF++
  COEF    T-VAL LAG _____0_____
  0.09    0.55   1              [              0>>>>              ]
  0.36    2.23   2              [              0>>>>>>>>>>>>>>>>]>>
  0.01    0.03   3        [                    0                 ]
 -0.02   -0.09   4        [                   <0                 ]
  0.18    0.98   5        [                    0>>>>>>>>>        ]
 -0.07   -0.36   6        [                 <<<0                 ]
  0.24    1.27   7        [                    0>>>>>>>>>>>>     ]
 -0.04   -0.21   8   [                        <<0                     ]
  0.05    0.24   9   [                         0>>                     ]
  0.09    0.46  10   [                         0>>>>>                  ]
  CHI-SQUARED* =    11.11 FOR DF =   7
```

Figure C15.9 Estimation and diagnostic-checking results for model (C15.4).

at the remaining lags. Therefore, we represent the nonseasonal part of the data with an MA(2) model. Adding this to (C15.2) using the multiplicative form gives an ARIMA$(0, 0, 2)(0, 1, 1)_4$:

$$\tilde{w}_t = \left(1 - \Theta_4 B^4\right)\left(1 - \theta_1 B - \theta_2 B^2\right)a_t \qquad (\text{C}15.3)$$

The residual autocorrelations at lags 1 and 2 in Figure C15.8 could be correlated with each other so that adding either θ_1 or θ_2, but not both, to (C15.2) could clear up the residual pattern. The principle of parsimony dictates that we should not use two coefficients where one provides equally good results. Since the residual autocorrelation at lag 1 is the larger, we consider an ARIMA$(0, 0, 1)(0, 1, 1)_4$ as an alternative to (C15.3):

$$\tilde{w}_t = \left(1 - \Theta_4 B^4\right)(1 - \theta_1 B)a_t \qquad (\text{C}15.4)$$

Further estimation and diagnostic checking. Figure C15.9 shows that (C15.4) is not adequate. Although $\hat{\theta}_1$ is significant, the residual autocorrela-

```
+ + + + + + + + +ECOSTAT UNIVARIATE B-J RESULTS+ + + + + + + + +
+ FOR DATA SERIES:   EXPORTS                                        +
+ DIFFERENCING:      4                            DF   = 34        +
+ AVAILABLE:         DATA = 38   BACKCASTS = 6    TOTAL = 44        +
+ USED TO FIND SSR:  DATA = 38   BACKCASTS = 6    TOTAL = 44        +
+ (LOST DUE TO PRESENCE OF AUTOREGRESSIVE TERMS:            0)      +
```

COEFFICIENT	ESTIMATE	STD ERROR	T-VALUE
THETA 1	-0.428	0.159	-2.69
THETA 2	-0.460	0.152	-3.02
THETA* 4	0.893	0.150	5.96
CONSTANT	80.72	6.34428	12.7233
MEAN	80.72	6.34428	12.7233

ADJUSTED RMSE = 83.7   MEAN ABS % ERR =   5.38

```
     CORRELATIONS
      1      2      3      4
 1   1.00
 2   0.28   1.00
 3  -0.31  -0.05   1.00
 4  -0.18  -0.01   0.61   1.00
```

```
++RESIDUAL ACF++
  COEF   T-VAL LAG _____0_____
 -0.02  -0.13   1         [              <0                ]
  0.03   0.16   2         [               0>               ]
  0.08   0.49   3         [               0>>>>            ]
 -0.01  -0.04   4         [               0                ]
  0.09   0.54   5         [               0>>>>            ]
 -0.07  -0.43   6         [           <<<<0                ]
  0.26   1.58   7         [               0>>>>>>>>>>>>>   ]
 -0.07  -0.43   8     [           <<<<0                    ]
 -0.02  -0.13   9     [               <0                   ]
  0.12   0.70  10     [               0>>>>>>              ]
      CHI-SQUARED* =    5.45 FOR DF =   6
```

Figure C15.10 Estimation and diagnostic-checking results for model (C15.3).

tion at lag 2 now has a larger t-value than before. This result brings us to (C15.3).

The results in Figure C15.10 show that (C15.3) performs well. All estimated coefficients have absolute t-values well in excess of 2.0; the reader should check to see that all invertibility conditions are satisfied. The estimated coefficients are not too highly correlated. The residual acf is consistent with the hypothesis that the shocks in (C15.3) are independent; all absolute t-values are less than the relevant warning levels and the chi-squared statistic is satisfactory. The residual plot in Figure C15.11 does not suggest problems for the model, though the large residual for the first quarter of 1965 calls for an investigation to see if that observation is recorded correctly.

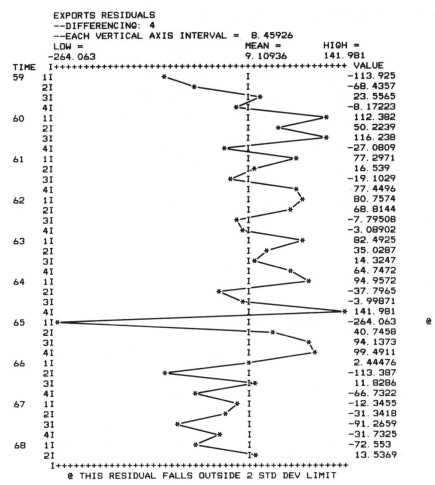

```
EXPORTS RESIDUALS
  --DIFFERENCING: 4
  --EACH VERTICAL AXIS INTERVAL =   8.45926
     LOW =                       MEAN =          HIGH =
     -264.063                    9.10936         141.981
TIME   I++++++++++++++++++++++++++++++++++++++++++++++++++  VALUE
59    1I                    *      I                        -113.925
      2I                 *         I                        -68.4357
      3I                        I-*                          23.5565
      4I                      *-I                            -8.17223
60    1I                         I                *          112.382
      2I                         I     *                      50.2239
      3I                         I                *          116.238
      4I                   *-----I                           -27.0809
61    1I                         I          *                77.2971
      2I                        I.*                           16.539
      3I                    *--I                             -19.1029
      4I                         I              *             77.4496
62    1I                         I              *            80.7574
      2I                         I       *                    68.8144
      3I                     *-I                              -7.79508
      4I                     *-I                              -3.08902
63    1I                         I            *              82.4925
      2I                         I  *                         35.0287
      3I                       I*                             14.3247
      4I                         I        *                   64.7472
64    1I                         I               *           94.9572
      2I                   *-----I                           -37.7965
      3I                    *-I                               -3.99871
      4I                         I                *          141.981
65    1I*------------------------I                          -264.063    @
      2I                       I--*                           40.7458
      3I                         I             *             94.1373
      4I                         I              *            99.4911
66    1I                         *                            2.44476
      2I         *---------------I                          -113.387
      3I                         I*                           11.8286
      4I                 *-------I                           -66.7322
67    1I                         *-I                         -12.3455
      2I                      *--I                           -31.3418
      3I            *----------I                             -91.2659
      4I                      *--I                           -31.7325
68    1I                      *--I                           -72.553
      2I                         I.*                          13.5369
       I++++++++++++++++++++++++++++++++++++++++++++++++++
       @ THIS RESIDUAL FALLS OUTSIDE 2 STD DEV LIMIT
```

Figure C15.11 Residuals from model (C15.3).

Forecasting. Forecasts from (C15.3) are shown in Table C15.1. They are calculated from the difference-equation form of (C15.3) as discussed in Chapter 10. That form is

$$\hat{z}_t = z_{t-4} + \hat{C} - \hat{\Theta}_4 \hat{a}_{t-4} - \hat{\theta}_1 \hat{a}_{t-1} - \hat{\theta}_2 \hat{a}_{t-2} + \hat{\theta}_1 \hat{\Theta}_4 \hat{a}_{t-5} + \hat{\theta}_2 \hat{\Theta}_4 \hat{a}_{t-6}$$

$$(C15.5)$$

Table C15.1 Forecasts from model (C15.3)

Time		Forecast Values	80% Confidence Limits		Future Observed Values	Percent Forecast Errors
			Lower	Upper		
68	3	1450.9571	1343.8211	1558.0931	n.a.[a]	n.a.
	4	1648.0558	1531.5322	1764.5793	n.a.	n.a.
69	1	1592.8197	1466.3237	1719.3157	n.a.	n.a.
	2	1662.6479	1536.1519	1789.1439	n.a.	n.a.
	3	1556.2652	1429.2463	1683.2840	n.a.	n.a.
	4	1723.2240	1596.1098	1850.3383	n.a.	n.a.
70	1	1673.5397	1546.3154	1800.7640	n.a.	n.a.
	2	1743.3679	1616.1436	1870.5922	n.a.	n.a.

[a]n.a. = not available.

From origin $t = 42$ the forecasts for lead times $l = 1$ and 2 are

$$\hat{z}_{42}(1) = 1299.3 + 80.72 - 0.893(-91.2659) + 0.428(13.5369)$$

$$+ 0.460(-72.553) + (-0.428)(0.893)(-31.3418)$$

$$+ (-0.460)(0.893)(-12.3455)$$

$$= 1450.9571$$

$$\hat{z}_{42}(2) = 1485.1 + 80.72 - 0.893(-31.7325) + 0.428(0)$$

$$+ 0.460(13.5369) + (-0.428)(0.893)(-91.2659)$$

$$+ (-0.460)(0.893)(-31.3418)$$

$$= 1648.0558$$

(C15.5) shows how the model captures both the seasonal pattern and the upward trend in the data. The first term (z_{t-4}) in (C15.5) starts each forecast from the same quarter one year earlier to begin at the proper

```
+ + + + + + + + + + + + AUTOCORRELATIONS + + + + + + + + + + + + +
+ FOR DATA SERIES: EXPORTS                                          +
+ DIFFERENCING:  4   1                        MEAN    =  5.78919    +
+ DATA COUNT =   37                           STD DEV =  137.163    +
  COEF   T-VAL LAG _____0_____
 -0.37  -2.24   1              <<[<<<<<<<<<<<<<<<<0              ]
  0.21   1.11   2              [              0>>>>>>>>>>         ]
 -0.09  -0.45   3              [       <<<<0                     ]
 -0.46  -2.37   4          <<<[<<<<<<<<<<<<<<<<<<<<<0             ]
  0.34   1.54   5              [              0>>>>>>>>>>>>>>>>>  ]
 -0.36  -1.54   6     [     <<<<<<<<<<<<<<<<<<0                   ]
  0.38   1.53   7     [              0>>>>>>>>>>>>>>>>>>>>        ]
 -0.14  -0.52   8              <<<<<<<0                          
  0.05   0.21   9              0>>>                              
  0.12   0.45  10              0>>>>>>                           
 -0.29  -1.07  11       <<<<<<<<<<<<<<0                          
  0.20   0.71  12              0>>>>>>>>>>                       
        CHI-SQUARED* =    43.31 FOR DF =   12

+ + + + + + + + + + PARTIAL AUTOCORRELATIONS + + + + + + + + + + +
  COEF   T-VAL LAG _____0_____
 -0.37  -2.24   1              <<[<<<<<<<<<<0             ]
  0.08   0.50   2              [         0>>>             ]
  0.02   0.09   3              [         0>               ]
 -0.58  -3.53   4         <<<<<<<<<<<[<<<<<<<<<0          ]
  0.04   0.25   5              [         0>               ]
 -0.13  -0.78   6              [     <<<<0               ]
  0.10   0.59   7              [         0>>>             ]
 -0.24  -1.44   8              [ <<<<<<<<<0              ]
  0.08   0.51   9              [         0>>>             ]
  0.01   0.07  10              [         0               ]
 -0.06  -0.34  11              [       <<0               ]
 -0.19  -1.14  12              [     <<<<<<0             ]
```

Figure C15.12 Estimated acf and pacf for the export data after seasonal and nonseasonal differencing ($D = 1$ and $d = 1$).

overall seasonal level. Then the estimated constant term raises the forecast by 80.72 units each time period to account for the upward trend. The rest of the terms account for the remaining seasonal and nonseasonal patterns, including the interaction between the two represented by the multiplicative terms.

An alternative model. Suppose we ignore the seasonal pattern at the beginning of our analysis and start from the acf and pacf in Figure C15.2. They are consistent with an AR(2): the acf decays and the pacf has two significant spikes at lags 1 and 2. Estimation results (t-values in parentheses) are

$$(1 - 0.386B - 0.616B^2)\tilde{z}_t = \hat{a}_t$$
$$(2.97) \quad (4.66)$$

```
+ + + + + + + + +ECOSTAT UNIVARIATE B-J RESULTS+ + + + + + + + + +
+ FOR DATA SERIES:   EXPORTS                                       +
+ DIFFERENCING:      4  1                          DF    = 35      +
+ AVAILABLE:          DATA = 37    BACKCASTS = 12   TOTAL = 49      +
+ USED TO FIND SSR:  DATA = 37    BACKCASTS = 11   TOTAL = 48      +
+ (LOST DUE TO PRESENCE OF AUTOREGRESSIVE TERMS:              1)   +

COEFFICIENT    ESTIMATE     STD ERROR     T-VALUE
PHI    1        -0.374        0.154        -2.44
THETA* 4         0.960        0.107         8.98

ADJUSTED RMSE =   87.5628    MEAN ABS % ERR =    5.61
      CORRELATIONS
        1      2
1     1.00
2    -0.07   1.00

++RESIDUAL ACF++
   COEF   T-VAL  LAG _____0_____
  -0.09  -0.53   1                    <<<<<<<<<O
   0.01   0.08   2                             O>
  -0.21  -1.25   3     <<<<<<<<<<<<<<<<<<<<<<<<O
  -0.11  -0.65   4                 <<<<<<<<<<<O
   0.13   0.73   5                             O>>>>>>>>>>>>>
  -0.08  -0.46   6                    <<<<<<<<O
   0.20   1.08   7                             O>>>>>>>>>>>>>>>>>>>>
  -0.08  -0.43   8                    <<<<<<<<O
   0.02   0.08   9                             O>>
   0.14   0.72  10                             O>>>>>>>>>>>>>
    CHI-SQUARED* =     6.85 FOR DF =   8
```

Figure C15.13   Estimation and diagnostic-checking results for model (C15.6).

Since $\hat{\phi}_1$ and $\hat{\phi}_2$ sum to 1.002 the model is nonstationary and first differencing is required. We have already seen from the estimated acf for the first differences in Figure C15.4 and the plot of the first differences in Figure C15.5 that seasonal differencing is warranted.

Seasonal and nonseasonal differencing together produce the estimated acf and pacf in Figure C15.12. The acf spike at lag 4 followed by a cutoff at lags 8 and 12 calls for an $MA(1)_4$ seasonal model. The decaying pacf values at the seasonal lags confirm this. The alternating spikes at lags 1, 2, and 3 in the acf and the single spike at lag 1 in the pacf suggest an $AR(1)$ with $\phi_1 < 0$. Our model is

$$(1 - \phi_1 B)(1 - B)(1 - B^4)\tilde{z}_t = (1 - \Theta_4 B^4)a_t \qquad (C15.6)$$

Estimation and diagnostic-checking results (Figure C15.13) show that (C15.6) is acceptable, though the RMSE and chi-squared are both slightly larger than those for (C15.3) in Figure C15.10.

Final comments.   We have found two adequate models, (C15.3) and (C15.6). The key difference between them is that (C15.3) accounts for the trend in the data with a deterministic drift represented by the constant term $\hat{C} = 80.72$. But (C15.6) involves first differencing as well as seasonal differencing and has no constant term. Any drifting in the forecasts derived from (C15.6) will be purely stochastic.

It is unlikely that data series in economics or other social sciences contain truly deterministic trend factors. However, a model is not necessarily the truth: a model is only an imitation of reality. If a model with a deterministic trend component provides a superior representation of the behavior of a data series, then we may use that model, keeping in mind that we have merely found a way of describing the behavior of the available data; we need not believe that the underlying process contains a truly deterministic trend element.

In the present case we can rationalize the deterministic component of (C15.3) as representing a growth factor resulting from inflation or from the long-term growth of world trade as economies continue to expand over the years. This result would be even more defensible if the model were in log form. Then a constant represents growth by a fixed percentage amount rather than growth by a fixed absolute amount per time period. In either case, a deterministic trend is more defensible in this instance than it is in Case 6. There we find a significant negative constant after first differencing, but this implies that the price of AT & T stock trends downward by a fixed amount each week; this has no sensible long-term interpretation.

If we dislike the notion that the data contain a deterministic growth factor, we can turn to model (C15.6). Its forecasts are not tied to any deterministic element, and it fits the data nearly as well as (C15.3). The reader can experiment with the idea of a deterministic trend by modeling the data in Case 9 using only seasonal differencing.

TABLES

Table A   Student's *t*-distribution

The tabled values are two-tailed values $t(\alpha; \nu)$ such that

$$\text{prob}\{|t_\nu \text{ variate}| > t(\alpha; \nu)\} = \alpha$$

The entries in the table were computed on a CDC Cyber 172 computer at the University of Minnesota using IMSL subroutine MDSTI

	α				
ν	0.200	0.100	0.050	0.010	0.001
1	3.08	6.31	12.71	63.66	636.62
2	1.89	2.92	4.30	9.92	31.60
3	1.64	2.35	3.18	5.84	12.92
4	1.53	2.13	2.78	4.60	8.61
5	1.48	2.02	2.57	4.03	6.87
6	1.44	1.94	2.45	3.71	5.96
7	1.41	1.89	2.36	3.50	5.41
8	1.40	1.86	2.31	3.36	5.04
9	1.38	1.83	2.26	3.25	4.78
10	1.37	1.81	2.23	3.17	4.59
11	1.36	1.80	2.20	3.11	4.44
12	1.36	1.78	2.18	3.05	4.32
13	1.35	1.77	2.16	3.01	4.22
14	1.35	1.76	2.14	2.98	4.14
15	1.34	1.75	2.13	2.95	4.07

	α				
ν	0.200	0.100	0.050	0.010	0.001
16	1.34	1.75	2.12	2.92	4.01
17	1.33	1.74	2.11	2.90	3.97
18	1.33	1.73	2.10	2.88	3.92
19	1.33	1.73	2.09	2.86	3.88
20	1.33	1.72	2.09	2.85	3.85
21	1.32	1.72	2.08	2.83	3.82
22	1.32	1.72	2.07	2.82	3.79
23	1.32	1.71	2.07	2.81	3.77
24	1.32	1.71	2.06	2.80	3.75
25	1.32	1.71	2.06	2.79	3.73
26	1.31	1.71	2.06	2.78	3.71
27	1.31	1.70	2.05	2.77	3.69
28	1.31	1.70	2.05	2.76	3.67
29	1.31	1.70	2.05	2.76	3.66
30	1.31	1.70	2.04	2.75	3.65
31	1.31	1.70	2.04	2.74	3.63
32	1.31	1.69	2.04	2.74	3.62
33	1.31	1.69	2.03	2.73	3.61
34	1.31	1.69	2.03	2.73	3.60
35	1.31	1.69	2.03	2.72	3.59
36	1.31	1.69	2.03	2.72	3.58
37	1.30	1.69	2.03	2.72	3.57
38	1.30	1.69	2.02	2.71	3.57
39	1.30	1.68	2.02	2.71	3.56
40	1.30	1.68	2.02	2.70	3.55
41	1.30	1.68	2.02	2.70	3.54
42	1.30	1.68	2.02	2.70	3.54
43	1.30	1.68	2.02	2.70	3.53
44	1.30	1.68	2.02	2.69	3.53
45	1.30	1.68	2.01	2.69	3.52
46	1.30	1.68	2.01	2.69	3.51
47	1.30	1.68	2.01	2.68	3.51
48	1.30	1.68	2.01	2.68	3.51
49	1.30	1.68	2.01	2.68	3.50
50	1.30	1.68	2.01	2.68	3.50
60	1.30	1.67	2.00	2.66	3.46
70	1.29	1.67	1.99	2.65	3.44
80	1.29	1.66	1.99	2.64	3.42
90	1.29	1.66	1.99	2.63	3.40
100	1.29	1.66	1.98	2.63	3.39
120	1.29	1.66	1.98	2.62	3.37
∞	1.28	1.64	1.96	2.58	3.29

Source: Sanford Weisberg, *Applied Linear Regression*, John Wiley & Sons, New York, 1980. Reprinted by permission.

Table B χ^2 critical points

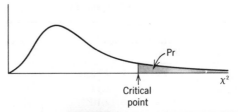

<table>
<tr><th>Pr
d.f.</th><th>.250</th><th>.100</th><th>.050</th><th>.025</th><th>.010</th><th>.005</th><th>.001</th></tr>
<tr><td>1</td><td>1.32</td><td>2.71</td><td>3.84</td><td>5.02</td><td>6.63</td><td>7.88</td><td>10.8</td></tr>
<tr><td>2</td><td>2.77</td><td>4.61</td><td>5.99</td><td>7.38</td><td>9.21</td><td>10.6</td><td>13.8</td></tr>
<tr><td>3</td><td>4.11</td><td>6.25</td><td>7.81</td><td>9.35</td><td>11.3</td><td>12.8</td><td>16.3</td></tr>
<tr><td>4</td><td>5.39</td><td>7.78</td><td>9.49</td><td>11.1</td><td>13.3</td><td>14.9</td><td>18.5</td></tr>
<tr><td>5</td><td>6.63</td><td>9.24</td><td>11.1</td><td>12.8</td><td>15.1</td><td>16.7</td><td>20.5</td></tr>
<tr><td>6</td><td>7.84</td><td>10.6</td><td>12.6</td><td>14.4</td><td>16.8</td><td>18.5</td><td>22.5</td></tr>
<tr><td>7</td><td>9.04</td><td>12.0</td><td>14.1</td><td>16.0</td><td>18.5</td><td>20.3</td><td>24.3</td></tr>
<tr><td>8</td><td>10.2</td><td>13.4</td><td>15.5</td><td>17.5</td><td>20.1</td><td>22.0</td><td>26.1</td></tr>
<tr><td>9</td><td>11.4</td><td>14.7</td><td>16.9</td><td>19.0</td><td>21.7</td><td>23.6</td><td>27.9</td></tr>
<tr><td>10</td><td>12.5</td><td>16.0</td><td>18.3</td><td>20.5</td><td>23.2</td><td>25.2</td><td>29.6</td></tr>
<tr><td>11</td><td>13.7</td><td>17.3</td><td>19.7</td><td>21.9</td><td>24.7</td><td>26.8</td><td>31.3</td></tr>
<tr><td>12</td><td>14.8</td><td>18.5</td><td>21.0</td><td>23.3</td><td>26.2</td><td>28.3</td><td>32.9</td></tr>
<tr><td>13</td><td>16.0</td><td>19.8</td><td>22.4</td><td>24.7</td><td>27.7</td><td>29.8</td><td>34.5</td></tr>
<tr><td>14</td><td>17.1</td><td>21.1</td><td>23.7</td><td>26.1</td><td>29.1</td><td>31.3</td><td>36.1</td></tr>
<tr><td>15</td><td>18.2</td><td>22.3</td><td>25.0</td><td>27.5</td><td>30.6</td><td>32.8</td><td>37.7</td></tr>
<tr><td>16</td><td>19.4</td><td>23.5</td><td>26.3</td><td>28.8</td><td>32.0</td><td>34.3</td><td>39.3</td></tr>
<tr><td>17</td><td>20.5</td><td>24.8</td><td>27.6</td><td>30.2</td><td>33.4</td><td>35.7</td><td>40.8</td></tr>
<tr><td>18</td><td>21.6</td><td>26.0</td><td>28.9</td><td>31.5</td><td>34.8</td><td>37.2</td><td>42.3</td></tr>
<tr><td>19</td><td>22.7</td><td>27.2</td><td>30.1</td><td>32.9</td><td>36.2</td><td>38.6</td><td>32.8</td></tr>
<tr><td>20</td><td>23.8</td><td>28.4</td><td>31.4</td><td>34.2</td><td>37.6</td><td>40.0</td><td>45.3</td></tr>
<tr><td>21</td><td>24.9</td><td>29.6</td><td>32.7</td><td>35.5</td><td>38.9</td><td>41.4</td><td>46.8</td></tr>
<tr><td>22</td><td>26.0</td><td>30.8</td><td>33.9</td><td>36.8</td><td>40.3</td><td>42.8</td><td>48.3</td></tr>
<tr><td>23</td><td>27.1</td><td>32.0</td><td>35.2</td><td>38.1</td><td>41.6</td><td>44.2</td><td>49.7</td></tr>
<tr><td>24</td><td>28.2</td><td>33.2</td><td>36.4</td><td>39.4</td><td>32.0</td><td>45.6</td><td>51.2</td></tr>
<tr><td>25</td><td>29.3</td><td>34.4</td><td>37.7</td><td>40.6</td><td>44.3</td><td>46.9</td><td>52.6</td></tr>
<tr><td>26</td><td>30.4</td><td>35.6</td><td>38.9</td><td>41.9</td><td>45.6</td><td>48.3</td><td>54.1</td></tr>
<tr><td>27</td><td>31.5</td><td>36.7</td><td>40.1</td><td>43.2</td><td>47.0</td><td>49.6</td><td>55.5</td></tr>
<tr><td>28</td><td>32.6</td><td>37.9</td><td>41.3</td><td>44.5</td><td>48.3</td><td>51.0</td><td>56.9</td></tr>
<tr><td>29</td><td>33.7</td><td>39.1</td><td>42.6</td><td>45.7</td><td>49.6</td><td>52.3</td><td>58.3</td></tr>
<tr><td>30</td><td>34.8</td><td>40.3</td><td>43.8</td><td>47.0</td><td>50.9</td><td>53.7</td><td>59.7</td></tr>
<tr><td>40</td><td>45.6</td><td>51.8</td><td>55.8</td><td>59.3</td><td>63.7</td><td>66.8</td><td>73.4</td></tr>
<tr><td>50</td><td>56.3</td><td>63.2</td><td>67.5</td><td>71.4</td><td>76.2</td><td>79.5</td><td>86.7</td></tr>
<tr><td>60</td><td>67.0</td><td>74.4</td><td>79.1</td><td>83.3</td><td>88.4</td><td>92.0</td><td>99.6</td></tr>
<tr><td>70</td><td>77.6</td><td>85.5</td><td>90.5</td><td>95.0</td><td>100</td><td>104</td><td>112</td></tr>
<tr><td>80</td><td>88.1</td><td>96.6</td><td>102</td><td>107</td><td>112</td><td>116</td><td>125</td></tr>
<tr><td>90</td><td>98.6</td><td>108</td><td>113</td><td>118</td><td>124</td><td>128</td><td>137</td></tr>
<tr><td>100</td><td>109</td><td>118</td><td>124</td><td>130</td><td>136</td><td>140</td><td>149</td></tr>
</table>

Source: Ronald J. Wonnacott and Thomas H. Wonnacott, *Econometrics*, 2nd ed., John Wiley & Sons, New York, 1979. Reprinted by permission.

REFERENCES

1. G. E. P. Box and G. M. Jenkins. *Time Series Analysis*: *Forecasting and Control*, 2nd ed. San Francisco: Holden-Day, 1976.
2. G. M. Jenkins and A. S. Alavi. "Some Aspects of Modelling and Forecasting Multivariate Time Series." *Journal of Time Series Analysis*, **2**, 1 (1981).
3. G. C. Tiao and G. E. P. Box. "Modeling Multiple Time Series with Applications." *Journal of the American Statistical Association*, **76**, 802 (1981).
4. R. L. Cooper. "The Predictive Performance of Quarterly Econometric Models of the United States." In B. C. Hickman, Ed., *Econometric Models of Cyclical Behavior*. New York: National Bureau of Economic Research, 1972, pp. 813–926.
5. T. H. Naylor, T. G. Seaks, and D. W. Wichern. "Box–Jenkins Methods: An Alternative to Econometric Forecasting." *International Statistical Review*, **40**, 123 (1972).
6. C. R. Nelson. "The Prediction Performance of the FRB–MIT Model of the U. S. Economy." *American Economic Review*, **62**, 902 (1972).
7. G. M. Jenkins and D. W. Watts. *Spectral Analysis and its Applications*. San Francisco: Holden-Day, 1968.
8. E. Mansfield. *Statistics for Business and Economics*. New York: Norton, 1980.
9. T. H. Wonnacott and R. J. Wonnacott. *Econometrics*, 2nd. ed. New York: Wiley, 1979.
10. J. Durbin. "The Fitting of Time Series Models." *Review of the International Institute of Statistics*, **28**, 233 (1960).

554   References

11. M. S. Bartlett. "On the Theoretical Specification of Sampling Properties of Autocorrelated Time Series." *Journal of the Royal Statistical Society*, **B8**, 27 (1946).

12. T. H. Wonnacott and R. J. Wonnacott. *Introductory Statistics*, 3rd ed. New York: Wiley, 1977.

13. R. L. Anderson. "Distribution of the Serial Correlation Coefficient." *Annals of Mathematical Statistics*, **13**, 1 (1942).

14. M. H. Quenouille. "Approximate Tests of Correlation in Time Series." *Journal of the Royal Statistical Society*, **B11**, 68 (1949).

15. G. M. Jenkins. "Tests of Hypotheses in the Linear Autoregressive Model." I. *Biometrika*, **41**, 405 (1954); II. *Biometrika*, **43**, 186 (1956).

16. H. E. Daniels. "The Approximate Distribution of Serial Correlation Coefficients." *Biometrika*, **43**, 169 (1956).

17. C. W. J. Granger and P. Newbold. *Forecasting Economic Time Series*. New York: Academic, 1977.

18. K. O. Cogger. "The Optimality of General-Order Exponential Smoothing." *Operations Research*, **22**, 858 (1974).

19. G. E. P. Box and D. R. Cox. "An Analysis of Transformations." *Journal of the Royal Statistical Society*, **B26**, 211 (1964).

20. D. W. Marquardt. "An Algorithm for Least Squares Estimation of Non-linear Parameters." *Journal of the Society for Industrial and Applied Mathematics*, **11**, 431 (1963).

21. P. Newbold and C. F. Ansley. "Small Sample Behavior of Some Procedures Used in Time Series Model Building and Forecasting." *Proceedings of the Twenty-Fourth Conference on the Design of Experiments in Army Research Development and Testing*, ARO Report No. 79-2.

22. J. Durbin. "Testing for Serial Correlation in Least-Squares Regression When Some of the Regressors are Lagged Dependent Variables." *Econometrica*, **38**, 410 (1970).

23. G. E. P. Box and D. A. Pierce. "Distribution of Residual Autocorrelations in Autoregressive-Integrated Moving Average Time Series Models." *Journal of the American Statistical Association*, **64**, 1509 (1970).

24. G. M. Ljung and G. E. P. Box. "On a Measure of Lack of Fit in Time Series Models." *Biometrika*, **65**, 297 (1978).

25. N. Davies, C. M. Triggs, and P. Newbold. "Significance Levels of the Box–Pierce Portmanteau Statistic in Finite Samples." *Biometrika*, **64**, 517 (1977).

26. I. Olkin, L. J. Gleser, and C. Derman. *Probability Models and Applications*. New York: MacMillan, 1980.

27. C. R. Nelson. *Applied Time Series Analysis for Managerial Forecasting*. San Francisco: Holden-Day, 1973.

28. C. W. J. Granger. *Forecasting in Business and Economics*. New York: Academic, 1980.

29. A. Zellner. "Folklore versus Fact in Forecasting with Econometric Methods." *Journal of Business*, **51**, 587 (1978).

30. A. Zellner. "Statistical Analysis of Econometric Models." *Journal of the American Statistical Assocation*, **74**, 628 (1979).

31. G. E. P. Box and G. C. Tiao. "Intervention Analysis with Applications to Economic and Environmental Problems." *Journal of the American Statistical Assocation*, **70**, 70 (1975).

32. R. McCleary and R. A. Hay, Jr. *Applied Time Series Analysis for the Social Sciences*. Beverly Hills, Calif.: Sage Publications, 1980.

33. J. P. Cleary and H. Levenbach. *The Professional Forecaster*. Belmont, Calif.: Lifetime Learning Publications, 1982.

34. G. M. Jenkins. *Practical Experiences with Modelling and Forecasting Time Series*. Jersey, Channel Islands: Gwilym Jenkins and Partners, 1979.

35. S. J. Deutsch and F. B. Alt. "The Effect of Massachusetts' Gun Control Law on Gun-Related Crimes in the City of Boston." *Evaluation Quarterly*, **1**, 543 (1977).

36. R. A. Hay, Jr. and R. McCleary. "Box–Tiao Time Series Models for Impact Assessment: A Comment on the Recent Work of Deutsch and Alt." *Evaluation Quarterly*, **3**, 277 (1979).

INDEX

Air-carrier freight, case study, 441
Alavi, A. S., fn 20
Alt, F. B., 472, 479
Anderson, R. L., 68
Ansley, C. F., fn 220
AR process, *see* Autoregressive process
AR(1) process, 47, 58, 95, 124, 142, 318
  as moving-average process of infinitely
    high order, 108
  theoretical autocorrelation functions, 56,
    60, 123, 124, 142, 147, 301
  theoretical partial autocorrelation func-
    tions, 56, 123, 124, 301
AR(2) process, 95
  and quasi-periodic patterns, 386
  theoretical autocorrelation functions, 123,
    126, 302
  theoretical partial autocorrelation func-
    tions, 123, 126, 302
ARIMA models, *see* UBJ-ARIMA models
ARIMA(0,1,1) model as exponentially-
    weighted moving average (EWMA),
    109, 433, 462
ARIMA(p,d,q) notation, *see* Notation,
    ARIMA(p,d,q)
ARMA process, 127, 391
ARMA(1,1) process, 95, 233
  theoretical autocorrelation functions, 123,
    129, 305
  theoretical partial autocorrelation func-
    tions, 123, 129, 305
Armed robberies (Boston), case study, 472
AT&T stock price, case study, 167, 402
Athletic shoe production data, 6
Autocorrelation coefficients:
  estimated, 35

correlation among, 64
defined and explained, 35
example of calculation, 36
formula for calculation, 35
maximum useful number, 37
standard error, 68
statistical significance, 38, 68, 82
warning t-values, 308
theoretical, 54, 140, 146
Autocorrelation function (acf):
  estimated, 37, 58
  advantages, 38
  graphical representation, 37
  and stationarity of mean, 42, 108
  formed from estimated autocorrelation
    coefficients, 37
  sampling error, effects of, 60
  theoretical, 54
    for AR(1) process, 56, 60, 123, 124, 301
    for AR(2) process, 123, 126, 302
    for ARMA(1,1) process, 123, 129, 305
    derived for AR(1) process, 142
    derived for MA(1) process, 136
    for MA(1) process, 57, 64, 123, 127, 303
    for MA(2) process, 123, 128, 304
    for pure AR process, 55, 122
    for pure MA process, 55, 122
    see also Identification stage
Autocovariances, 52, 138, 144
  defined, 52
  for a stationary process, 54
  matrix representation, 53
  standardized, 53
Autoregressive (AR) operator:
  nonseasonal, 99
  seasonal, 280

557